G000025135

BIOMASS PROCESSING, CONVERSION AND BIOREFINERY

RENEWABLE ENERGY: RESEARCH, DEVELOPMENT AND POLICIES

Additional books in this series can be found on Nova's website
under the Series tab.

Additional e-books in this series can be found on Nova's website
under the e-book tab.

ENERGY SCIENCE, ENGINEERING AND TECHNOLOGY

Additional books in this series can be found on Nova's website
under the Series tab.

Additional e-books in this series can be found on Nova's website
under the e-book tab.

RENEWABLE ENERGY: RESEARCH, DEVELOPMENT AND POLICIES

BIOMASS PROCESSING, CONVERSION AND BIOREFINERY

BO ZHANG

AND

YONG WANG

EDITORS

nova publishers

New York

Copyright © 2013 by Nova Science Publishers, Inc.

All rights reserved. No part of this book may be reproduced, stored in a retrieval system or transmitted in any form or by any means: electronic, electrostatic, magnetic, tape, mechanical photocopying, recording or otherwise without the written permission of the Publisher.

For permission to use material from this book please contact us:
Telephone 631-231-7269; Fax 631-231-8175
Web Site: http://www.novapublishers.com

NOTICE TO THE READER

The Publisher has taken reasonable care in the preparation of this book, but makes no expressed or implied warranty of any kind and assumes no responsibility for any errors or omissions. No liability is assumed for incidental or consequential damages in connection with or arising out of information contained in this book. The Publisher shall not be liable for any special, consequential, or exemplary damages resulting, in whole or in part, from the readers' use of, or reliance upon, this material. Any parts of this book based on government reports are so indicated and copyright is claimed for those parts to the extent applicable to compilations of such works.

Independent verification should be sought for any data, advice or recommendations contained in this book. In addition, no responsibility is assumed by the publisher for any injury and/or damage to persons or property arising from any methods, products, instructions, ideas or otherwise contained in this publication.

This publication is designed to provide accurate and authoritative information with regard to the subject matter covered herein. It is sold with the clear understanding that the Publisher is not engaged in rendering legal or any other professional services. If legal or any other expert assistance is required, the services of a competent person should be sought. FROM A DECLARATION OF PARTICIPANTS JOINTLY ADOPTED BY A COMMITTEE OF THE AMERICAN BAR ASSOCIATION AND A COMMITTEE OF PUBLISHERS.

Additional color graphics may be available in the e-book version of this book.

Library of Congress Cataloging-in-Publication Data

ISBN: 978-1-62618-346-9

Published by Nova Science Publishers, Inc. † New York

CONTENTS

Preface   ix

Section 1. Biomass Overview and Processing   1

Chapter 1   Biomass – An Overview on Classification,
                Composition and Characterization   3
                Sonil Nanda, Janusz A. Kozinski
                and Ajay K. Dalai

Chapter 2   Lignocellulosic Biomass Feedstock Supply
                Logistics and System Integration   37
                Zewei Miao, Yogendra Shastri,
                Tony E. Grift and K.C. Ting

Chapter 3   Lignocellulosic Biomass: Feedstock Characteristics,
                Pretreatment Methods and Pre-Processing for Biofuel and
                Bioproduct Applications, U.S. and Canadian Perspectives   61
                Kingsley L. Iroba and Lope G. Tabil

Chapter 4   Biomass Harvest and Drying   99
                Bo Zhang

Chapter 5   Biomass Size Reduction   111
                Bo Zhang

Chapter 6   Biomass Pelletization for Energy Production   125
                Hui Wang, Lijun Wang
                and Abolghasem Shahbazi

Section 2. Biomass Thermochemical and Biochemical Conversion Technologies   137

Chapter 7   Biomass Pyrolysis for Bio-Oil   139
                Hui Wang, Lijun Wang
                and Abolghasem Shahbazi

Chapter 8   Biomass to Bio-oil by Liquefaction   153
                Huamin Wang and Yong Wang

Chapter 9   Biomass Gasification: Process Overview,
              History and Development   **167**
              Daniel T. Howe

Chapter 10 Hydrogen Production from Catalytic Steam
              Co-gasification of Waste Tyre and Palm Kernel
              Shell in Pilot Scale Fluidized Bed Gasifier   **181**
              Suzana Yusup, Reza Alipour Moghadam,
              Ahmed Al Shoaibi, Murni Melati, Zakir Khan,
              Lim Mook Tzeng and Wan Azlina A. K. GH.

Chapter 11 Review of Upgrading Researches of Biomass
              Pyrolysis Oil to Improve its Fuel Properties   **193**
              Junming Xu, Jianchun Jiang and Kang Sun

Chapter 12 Bio-Oil Upgrading   **201**
              Ying Zhang and Jianhua Guo

Chapter 13 Biofuel and Bio-Oil Upgrading   **221**
              Changjun Liu and Yong Wang

Chapter 14 Biodiesel Production   **251**
              Oscar Marin-Flores, Anna Lee Tonkovich
              and Yong Wang

Chapter 15 Utilization of *Ceiba Pentandra* Seed Oil
              As Potential Feedstock for Biodiesel Production   **263**
              S. Yusup, M. M. Ahmad, Y. Uemura, S. Abu Bakar,
              R. Nik Mohamad Kamil, A. T. Quitain and S. Shari

Chapter 16 Microalgae for Biodiesel Production and
              Wastewater Treatment   **277**
              Rifat Hasan, Lijun Wang and Bo Zhang

Chapter 17 Thermochemical Conversion of Fermentation-Derived
              Oxygenates to Fuels   **289**
              Karthikeyan K. Ramasamy and Yong Wang

Chapter 18 Microbial Conversion of Bio-Based Chemicals:
              Present and Future Prospects   **301**
              Huibin Zou, Guang Zhao and Mo Xian

Chapter 19 Pretreatment Technologies for Production
              of Lignocellulosic Biofuels   **329**
              Bo Zhang

Chapter 20 Biochemical Conversion of Ethanol from Lignocellulose:
              Pretreatment, Enzymes, Co-Fermentation, and Separation   **347**
              Xian-Bao Zhang and Ming-Jun Zhu

Chapter 21   Bioprocessing: The Use of Thermophilic
and Anaerobic Bacteria   **385**
Jing-Rong Cheng and Ming-Jun Zhu

Section 3. Integrated Biorefinery Processes   **407**

Chapter 22   Economic Analysis of Waste to Power:
A Case Study of Greensboro City   **409**
Ransford R. Baidoo, Abolghasem Shahbazi,
Matthew Todd and Harith Rojanala

Chapter 23   Process Design for Biological Conversion
of Cattails to Ethanol   **423**
Bo Zhang

Chapter 24   Green Biorefining of Green Biomass   **435**
Shuangning Xiu and Abolghasem Shahbazi

Index   **447**

PREFACE

Biomass presents an attractive source for the production of fuels and chemicals, mainly due to the concerns over the depleting fossil fuel, growing awareness of environmental issues associated with fossil fuel consumption, and increasing world energy demand. Biomass resources include agricultural and forest residues, energy crops, livestock residues as well as municipal solid waste. These biomass resources are first processed into in a conversion-friendly form, followed by the transformation to a wide range of energy and/or chemical products using two primary biorefinery platforms: biochemical and thermochemical. This book covers the most recent advances in biomass processing, biochemical and thermochemical conversion technologies, and thus, serves as a useful reference to agriculture engineers, chemical engineers, biotechnology engineers and engineering students.

The contents of the book are divided into three sections: biomass overview and processing, biomass thermochemical and biochemical conversion technologies, and integrated biorefinery processes.

Section 1 provides an overview of biomass concepts, supply logistics, and processing technologies. This section begins with a chapter on different biomass sources along with their compositions and properties (Chapter 1), followed by discussions on lignocellulosic feedstock supply logistics (Chapter 2), biomass resources in Canada and U.S. (Chapter 3), the harvesting system for biomass and drying (Chapter 4), and biomass size reduction (Chapter 5). This section ends with a chapter on techniques for biomass pelletization (Chapter 6).

Section 2 focuses on biomass conversion technologies and biomass-derived fuels/products. This section starts with the overview of three primary thermochemical conversion technologies: pyrolysis (Chapter 7), liquefaction (Chapter 8), and gasification (Chapter 9). Steam gasification of biomass is used as an example for hydrogen production (Chapter 10). Three chapters (Chapters 11-13) in this section provide different aspects of pyrolysis oil/biofuel upgrading, including hydrodeoxygenation and catalytic cracking of pyrolysis oils. Recent advances in bio-diesel production from seed oil and microalgae are summarized in Chapters 14-16. Chapter 17 summarizes the current status in the thermochemical conversion of fermentation-derived oxygenates to fuels. Biochemical conversion includes microbial conversion of bio-based chemicals (Chapter 18), pretreatment technologies (Chapter 19) for the conversion of cellulosic biomass to ethanol (Chapter 20), and biodegrading lignocellulosic feedstocks using thermophilic and anaerobic bacteria (Chapter 21).

Section 3 emphases the importance of integrated biorefinery concept and applications. The issues covered in this section are economic analysis of municipal solid waste to power

(Chapter 22), process design for biological conversion of cattails to ethanol (Chapter 23), and green biorefining of green biomass (Chapter 24).

The completion of this book would not have been possible without assistance from a large number of people. Most important are the contributors, who prepared their work in a timely and professional manner. We are also very grateful to all the peer reviewers whose time and efforts in evaluating individual chapters have enhanced the quality of this book. We express our sincere appreciation to Voiland School of Chemical Engineering and Bioengineering and the Agricultural Research Center at Washington State University for the general support of this book project, to Ms. Jun Chen for the skillful assistance in the acquisition, editing, and production processes.

Bo Zhang
Yong Wang

Editors
Dr. Bo Zhang

Dr. Bo Zhang is the Chutian Scholar Distinguished Professor in Chemical Engineering at Wuhan Institute of Technology, China and the Senior Scientist of Biological Engineering Program at North Carolina Agricultural and Technical State University. He earned his Ph.D. at the prestigious Department of Chemical Engineering and Materials Science of University of Minnesota. He has published over 30 peer-reviewed research articles in internationally renowned journals, and has many patents granted in US and China. He is the Editor of *Journal of Petroleum and Environmental Biotechnology*. He is the senior member of American Institute of Chemical Engineers (AIChE), member of the Sigma Xi, The Scientific Research Society, and American Chemical Society. His main research activities include developing new and improving existing biological and thermochemical technologies for converting biomass into useful products for fuels or chemical applications, and improving fundamental understanding and developing new technologies for upstream bioprocess engineering.

Dr. Yong Wang

Dr. Yong Wang is the Voiland Distinguished Professor in Chemical Engineering at Washington State University, and a Laboratory Fellow/Associated Director of Institute for Integrated Catalysis at Pacific Northwest National Laboratory. He is best known for his leadership in the development of novel catalytic materials and reaction engineering for the conversion of fossil and biomass feedstocks. He has received three R&D 100 awards, two PNNL Inventor of the Year awards, Asian American Engineer of the Year Award, and Presidential Green Chemistry Award. Professor Wang is an AAAS Fellow and ACS (American Chemical Society) Fellow. He has authored more than 150 peer reviewed publications and holds numerous issued patents (including 85 issued US patents). Professor Wang is a co-editor of *Microreactor Technology and Process Intensification* (ACS, 2005).

SECTION 1. BIOMASS OVERVIEW AND PROCESSING

In: Biomass Processing, Conversion and Biorefinery
Editors: Bo Zhang and Yong Wang

ISBN: 978-1-62618-346-9
© 2013 Nova Science Publishers, Inc.

Chapter 1

BIOMASS – AN OVERVIEW ON CLASSIFICATION, COMPOSITION AND CHARACTERIZATION

*Sonil Nanda,[1] Janusz A. Kozinski[1] and Ajay K. Dalai[2],**

[1]Lassonde School of Engineering, York University, Ontario, Canada
[2]Department of Chemical and Biological Engineering,
University of Saskatchewan, Saskatchewan, Canada

ABSTRACT

Waste biomasses are promoted in a wide-scale as a means of reducing greenhouse gas emissions by supplying next generation biofuels. The use of biomass, particularly from lignocellulosic residues is becoming progressively essential as a renewable energy source. Concerns over the exhausting fossil fuel supplies, global warming due to excessive fossil fuel combustion and increasing world energy demand are some chief driving factors in using biomass for sustainable fuels.

Renewable energy can be recovered from a variety of biomass resources including agricultural and forest residues, energy crops, livestock residues as well as municipal organic waste.

Through a variety of pathways including physiochemical, biochemical or thermochemical, these waste biomasses can be directly used to produce heat, electricity and fuels. Much attention is being focused on identifying a suitable biomass species that can provide recurring high energy outputs to substitute conventional fossil fuels. The main purpose of this chapter is to discuss different biomass sources along with their composition and properties.

Since the biomass system is a broad entity for classification, we present a classification scheme based on biomass efficiency, its origin and source. Moreover, it is important to identify the type of biomass in order to understand its biopolymeric composition of cellulose, hemicellulose and lignin. Nevertheless, the physical and chemical characteristics of biomass determine the conversion process and any imminent process complexities that may arise.

---

* Phone (306) 966-4771. Fax (306) 966-4777. E-mail: ajay.dalai@usask.ca.

With this knowledge on the perspective of biomasses and their diversity, this chapter will discuss on their selection and characterization to understand their basic make-up for subsequent conversion and environmentally safe utilization.

INTRODUCTION

The world's present economy is vastly driven by various kinds of fossil energy sources such as oil, natural gas and coal. These resources are the primary components in the generation of fuel and electricity for domestic and industrial purpose. With the expansion of human population and prosperity in industrialization, global energy consumption has also increased gradually.

The energy consumption rate includes an individual's share of electricity and fuel used in making food, for transportation and other varying necessities. The worldwide energy consumption in 2008 was 533 EJ (exajoules), and with the increasing demand, the projection is about to rise by 653 EJ in 2020 and 812 EJ in 2030 [1]. Excessive consumption of fossil fuels indicates that they are being exhausted rapidly leading to drastic climate change. Increased use of fossil fuels has led to an unprecedented increase in the level of greenhouse gases (GHG) in the earth's atmosphere along with an alleged increase in the average global temperatures causing global warming.

Due to changes in atmospheric CO_2 levels, the global average surface temperature has increased by 0.8°C in the past century with the current warming rate of about 0.2°C per decade [2].

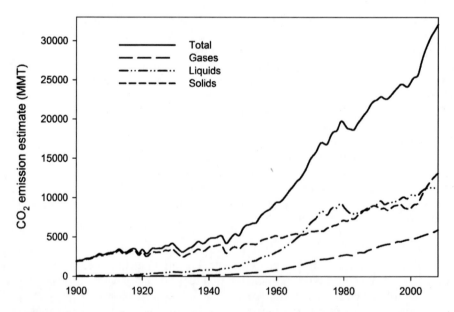

Figure 1. Global CO_2 emissions (in MMT CO_2) from fossil fuel burning, 1900 to 2008 (Source of statistical data [5]).

The consumption of fossil fuels accounts for the majority of worldwide anthropogenic GHG emissions as the CO_2 concentrations have augmented from 280 ppm in 1750 [3] to over

390 ppm or 39% above preindustrial levels in 2010 [4]. The total CO_2 emission by burning of fossil fuels in 1900 was 1,958 MMT (million metric tons), however in 1950 it was 5,977 MMT and in 2008 it drastically rose to 32,083 MMT (Fig. 1). With this unparalleled rise in CO_2 level, the carbon cycle is now being disrupted by anthropogenic activities such as burning of large quantities of petroleum and coal.

The annual global oil production is believed to decline in the near future [6] and the fossil energy consumption is expected to escalate exceptionally than the current trend. This will create possibilities for exploiting new petroleum reserves in the Alaskan frontier or Siberia and developing technologies for conversion of enormous amounts of oil shale into fuels for existing power infrastructures [7]. However, it raises matters of conservation versus consumption since these fuel reserves are located in ecologically preserved zones. In this scenario, renewable resources seem to be sustainable alternatives to meet the increasing consumption patterns of energy.

Various sources of renewable energy such as wind, water, sun, geothermal heat and biomass can supply electricity, thermal and mechanical energy as well as produce fuels to satisfy multiple energy service needs. The energy from wind, water and sun can be harnessed to meet the domestic and industrial requirements.

On the other hand, fuel production and chemical industry largely depend on biomass as an alternative source. While wind energy harnesses the kinetic energy of moving air; ocean energy derives the potential, kinetic, thermal and chemical energy of seawater to provide electricity and thermal energy. In the same way, hydropower harnesses the kinetic energy of water from higher to lower elevations to generate electricity. Solar energy is a source for thermal energy to produce electricity using photovoltaics for direct lighting needs in urban and rural areas. Furthermore, geothermal energy utilizes the thermal energy from the earth's interior to meet multiple heat and power requirements. Apart from all, biomass has a potential to replace petroleum-based fuels by supplying renewable fuels such as bioethanol, biobutanol, biodiesel, biohydrogen etc.

Bioenergy can be recovered from a variety of biomass resources including agricultural and forest residues, energy crops, livestock residues, organic component of municipal solid waste as well as other organic waste streams. Through a variety of pathways including physicochemical, biochemical or thermochemical, these feedstocks can be directly used to produce heat, electricity and fuels. The main purpose of this chapter is to discuss different biomass sources along with their composition and properties. With this knowledge on the perspective of biomasses and their diversity, we will discuss on their selection and characterization to understand their basic make-up for the subsequent conversion process and environmentally safe utilization.

1. BIOMASS CLASSIFICATION

Biomass is a term for all organic material that is available on a recurring basis including trees, crops and algae. More specifically, it refers to agricultural food and crop residues, wood residues, dedicated energy crops, animal wastes and organic municipal wastes. Biomass is a general category for non-fossil and composite biogenic solid organic product formed by natural and anthropogenic processes. The biomass arising through natural processes include components from the vegetation in the terrestrial and aquatic habitats via photosynthesis as well as the organic waste generated via animal and human food digestion. Anthropogenic processes of biomass generation refer to the products derived in processing the above natural constituents.

Biomass from green plants is produced by converting solar energy into plant material through photosynthesis. As a result of photosynthesis in chlorophyll containing organisms such as green plants, algae and some bacteria, CO_2 reacts with water in presence of sunlight to produce carbohydrates that form the building blocks of biomass. A generalized chemical equation for photosynthesis can be represented as follows.

$$2n \; CO_2 + 4n \; H_2\alpha \xrightarrow{\text{Sunlight}} 2(CH_2O)_n + 2n \; H_2O + 2n \; \alpha_2$$

In this photosynthetic reaction, $H_2\alpha$ refers to the compound that is oxidized. In plants and algae that perform oxygenic photosynthesis, $H_2\alpha$ is water (H_2O) and α_2 is oxygen (O_2), whereas in some sulfur-oxidizing bacteria, $H_2\alpha$ is hydrogen sulphide (H_2S). $(CH_2O)_n$ is a general formula for the carbohydrates.

In plants, the energy from sunlight is stored in the chemical bonds within their molecular framework. This solar energy stored in plants as chemical energy can only be recovered as bioenergy when the bonds between adjacent carbon, hydrogen and oxygen molecules are broken by combustion, digestion or decomposition. As a result of biochemical or thermochemical conversion, the energy stored within the plant's structural components can be extracted in form of biofuels which on subsequent utilization (combustion) get oxidized to produce CO_2 and water. This resulting CO_2 is made available in the atmosphere to produce new biomass. This process of CO_2 flow in the environment makes biomass a carbon-neutral source for renewable energy.

More than 40 million tons of non-edible plant material, including agricultural residues (e.g., wheat straw, corn stover, flax straw, barley straw, sugarcane bagasse etc.), energy crops (e.g., switch grass, timothy grass, rye grass etc) and woody residues including forest harvest and wood shavings from logging are produced abundantly in a global scale [8]. In 2008, biomass attributed to about 10% (50.3 EJ/yr) of the primary energy supply worldwide [4]. It is predictable that a major proportion of the future energy supply (250-500 EJ per year by 2050) will be provided by lignocellulosic biomass [9]. Lignocellulosic biomass is a generic term for the non-edible plant biomass that is found abundantly across the globe and has the potential to support the sustainable production of liquid transportation fuels.

The biomass system is a broad entity for classification. Various types of biomass can be categorized in different ways but we present here a simplified method of classification based on two modules. Module 1 classifies biomasses on the basis of their efficiency. On the other hand, module 2 classifies biomasses on the basis of their origin and source. Fig. 2 illustrates different categories of biomass classification.

In terms of efficiency, the biomass can be classified into two major categories i.e., low-efficiency traditional (LET) biomass and high-efficiency modern (HEM) biomass [4]. LET biomasses are mostly used by rural populations for basic necessities such as cooking, lighting and space heating. They comprise mostly of wood, straw, bagasse, cattle dung and other manures.

The energy from these feedstocks is mostly obtained through combustion but their use has many negative implications on health and living conditions such as production of huge amount of smoke and odor. Moreover, these biomasses are also beneficial in adding carbon to the soil due to the production of char from combustion. The total estimated LET biomass supply for energy is 37-43 EJ/yr and the energy delivered is 3.6-8.4 EJ/yr [4].

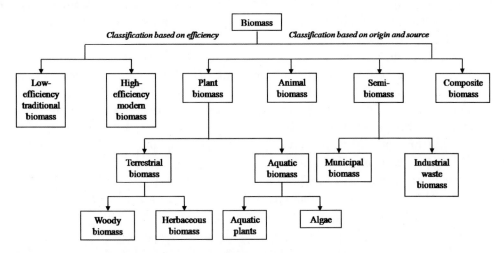

Figure 2. Classification of biomass on the basis of efficiency, origin and source.

In contrast to LET biomass, HEM biomass system aims to satisfy the energy needs relatively efficiently. Waste biomass either chopped or in form of chips, lumps, pellets, briquettes or sawdust are subjected to chemical conversions for generation of liquids and gaseous energy carriers to supply transport fuels (e.g., bioethanol, biobutanol and biodiesel), electricity and heat. Biomass derived gases, mostly methane from anaerobic digestion of plant biomass and municipal solid waste, and syngas from biomass gasification are used to generate heat, electricity or combined heat and power. The total estimated HEM biomass supply for bioenergy is 11.3 EJ/yr and the energy delivered is 6.6 EJ/yr [4].

The second module of biomass classification is based on its source and origin. This divides the biomass sector into four main groups, namely plant biomass, animal biomass, semi-biomass and composite biomass. Within this categorization, the biomass can be further subdivided into terrestrial or land-based biomass and aquatic or water-based biomass. Some authors also refer this method of division with regards to the moisture content i.e., land-based feedstocks as low-moisture containing biomass and aquatic feedstocks as high-moisture containing biomass [10]. The land-based biomass comprise of woody and herbaceous feedstocks. There are many supply sources of land-based biomass for bioenergy purposes, scattered across large and diverse geographical areas.

Wood residues from forest harvest practices or wood processing facilities are suitable feedstocks for biofuel production. Considerable amount of mill residues are generated in the forest during logging operations. The major byproducts of forest industries are used to produce fuel wood, char and black liquor (an organic waste of pulp mills), which are major fuel sources for electricity in Brazil, U.S., Canada, Finland and Sweden [11]. The forest residues range from soft or hard wood such as stems, barks, branches (twigs), leaves (foliage) and bushes (shrubs) and cones (e.g., pine cone), whereas the forest mill residues can be in the form of chips, lumps, bales, pellets, briquettes and sawdust. The mill residues can be considered as HEM biomass for having high calorific value of wood densified in small packages. The annual forest residues in the U.S. range from 60-100 MDT (million dry tons) [12], whereas in Canada they are in the range of 9.8-46 MDT [13]. In British Columbia, Canada, a new source of non-renewable forest residue is generated due to the Mountain pine beetle insect outbreak called disturbance wood. The availability of disturbance wood in Canada is between 9.3-12.3 million tons per annum, which has the potential to produce 2.8-3.6 billion liters ethanol per annum [13].

Enormous amount of herbaceous biomass sub-ordered as agricultural refuse and perennial grasses are accessed throughout the world. The agricultural residues vary from straws (e.g., wheat, corn, barley, flax, paddy, rape, others cereal and pulse crops), stalks (e.g., corn, cotton, mustard, alfalfa, bean, sesame, sunflower), fibers (e.g., flax, jute bast, coconut coir, palm), shells and husks (e.g., coconut, rice, cotton, sunflower, walnut, almond, coffee, cashew nut, hazelnut, peanut, millet, olive), pits (e.g., peach, plum, olive, apricot) and other processed refuse (e.g., coir, cobs, bagasse, de-oiled cakes, pulps, seeds). The estimate for annual agricultural residue harvest in the U.S. and Canada ranges up to 1,147 MDT [14] and 2.7-18 MDT [12], respectively. Every year about 15-40% of the agricultural residues, particularly straw refuse, is accessible to industries for biofuels after considering the factors of soil conservation, livestock feed and variation in seasonal harvest [15].

The last segment in the terrestrial biomass category is reserved to energy crops. These crops are those that are specifically grown for the purpose of converting them into energy. The energy crops mostly include perennial grasses such as switchgrass, timothy hay, reed canary grass, ryegrass, miscanthus, alfalfa and bamboo. Hybrid poplar is an example of a fast-growing wood species that can be harvested on a short rotation of round 7-20 years depending upon the geographical location [16]. What makes the grasses ideal as energy crop are their high yield (maximum production of dry matter per hectare), round the year availability, low energy input in cultivation, low cost, low nutrient requirements, potential for regeneration on degraded soils, resistance to extreme weather conditions (e.g., heat, cold and drought) and requirement of no intense farming practices. The annual availability of energy crops in the U.S. and Canada ranges up to 3383 MDT [14] and 433 MDT [17], respectively.

Marine and freshwater macroalgae (seaweed) and microalgae including various hydrophytes and other aquatic vegetation such as diatoms, duckweed, kelp and salvinia constitute potential feedstocks for biofuels, particularly biodiesel. Hydrophytes, especially water hyacinth [18], water caltrop, Chinese water chestnut, Indian lotus, water spinach and watercress are prospective resources for biofuel and biogas production [19]. The aquatic vegetations as a group are an interesting bioenergy feedstock because they are perennial, naturally grown, easily degradable and do not compete with arable crop plants for space, light and nutrients. There are various advantages of using microalgae for bioenergy, a few of which are their high growth rate and utilization of a large fraction (up to 10%) of the solar energy. Algae are capable in producing 61,000 liters per hectare of biodiesel compared to 200-450 liters of biodiesel from soya and canola [20]. Despite being single-celled organisms, algae have the tendency to metabolize CO_2, H_2 and N_2 into carbohydrates, lipids and proteins which makes it an interesting energy crop. There are many studies on algae for the production of different biofuels including bioethanol, vegetable oils, biodiesel, syngas and biohydrogen [21]. Biofuels from algae have higher yields and have a potential to produce 30-100 times more energy per hectare compared to land-based biomass.

In the recent time, there is a requirement for safe disposal of large quantities of animal manure generated at poultry, dairy, swine and other animal farms due to expansion in the livestock industries. This makes animal biomass our next category in classification whose basic ingredients are bones, bone meals, chicken litter, livestock and poultry manure, dairy wastewater, feedlot runoff, silage juices and wasted feed. Although under research, animal waste biomass is considered as efficient sources for biogas, particularly methane. Anaerobic digestion of animal waste via biological means is advantageous in producing methane and generating odor-free residues rich in nutrients as fertilizers [22].

Semi-biomass refers to both municipal solid waste as well as industrial biomass waste. In combination, they include municipal paper waste, fiberboard, plywood, wood pallets and boxes, oriented strand board, railway sleepers, paper and pulp sludge, sewage sludge, tannery waste and pharmaceutical waste. These residues are termed semi-biomass because they are not completely natural in origin and are also composed of certain additives such as lacquers, lignosulphonates, paints and preservatives [19]. Last in the list is the composite biomass which is a heterogeneous division and is a blend of plant, animal and semi-biomass. Poultry litter, although described in animal waste biomass can also serve as an example of composite biomass because of its physical nature. Poultry litter is a mixture of bedding materials (e.g., wood shavings, sawdust, peanut hull etc.), bird excreta and feather, feed spills and chemical treatment agents such as alum, sodium bisulphate [23]. The annual generation of poultry litter in the U.S. is 12 MDT but the majority of the produce is used as an organic fertilizer due to its high organic matter content [24].

In order to achieve a successful biofuel supply to meet the increasing fuel demands, it is desirable to extend the exploitation of bioenergy feedstocks beyond the plant biomasses towards animal-, semi- and composite biomasses. However, the biofuel that is expected to be most widely used in the near future is ethanol and butanol, which can be produced from plentiful supplies of biomass from all land plants and plant-derived materials. The next section describes the composition of plant biomass that makes it an exciting raw material for biofuel industries.

2. BIOMASS COMPOSITION

Lignocellulosic biomass is a complex heterogeneous mixture of organic matter, inorganic matter, containing various solid and fluid intimately associated phases. The fundamental organic components of biomass include three key structural biopolymers, which are cellulose, hemicellulose and lignin. Biomass extractives are a class of compounds not completely secluded from the organic matter in biomass although they are associated with certain inorganic phases. The inorganic matter in biomass comprises of different mineral matter and mineraloids from various mineral groups that could be indigenous or exogenous in origin to the feedstock. However, the basic composition of lignocellulosic biomass depends on a number of factors, a few of which are discussed as follows.

(i)   Type and class of biomass
(ii)   Plant species and plant part
(iii)   Age of the plant

(iv)   Plant metabolism and physiology
(v)   Environmental and growth conditions of the plant
(vi)   Geographical variations
(vii)   Farming practices
(viii)   Handling, storage and transport

It is very important to identify the type of biomass to understand its composition. For instance, plant biomass generally exhibits high organic matter content and low inorganic matter compared to semi-biomass and vice versa. Additionally, the proportions of lignocellulosic and mineral components vary within the same biomass type e.g., within species level and part of plant [25–27]. The age of plant during biomass harvest is equally vital in determining its chemical composition [28, 29]. The variations in cellulose, hemicellulose and lignin contents between two rye grass harvests can be compared (see Table 1). The age of the plant is correlated to its metabolism and physiology such as the ability to uptake metabolites from soil, water and air through roots and stomatal openings and to transport and deposit them within its tissues [30]. The uptake of organic and ionic nutrients by the plant is influenced by the environmental conditions such as sunlight [10], O_2 and CO_2, available water, soil type and pH [31, 32].

Table 1. Composition of cellulose, hemicellulose and lignin in different biomass samples (expressed in wt%)

Biomass varieties[a]	Cellulose	Hemicellulose	Lignin	Extractives	Reference
a) Straws					
Barley straw	42.0	21.9	19.4	6.8	[34]
Flax straw	28.7	26.8	22.5	19.5	[39]
Legume straw	28.1	34.1	34.0	2.0	[40]
Oat straw	39.6	22.4	18.2	10.1	[34]
Rape straw	45.0	19.0	8.0	-	[36]
Rice straw	40.0	18.0	5.5	-	[41]
Rye straw	34.0	28.5	17.5	-	[42]
Wheat straw	38.7	21.1	18.2	13.2	[34]
Average	37.0	24.0	17.9	10.3	
b) Grasses					
Bamboo	42.4	27.9	23.1	4.4	[28]
Bermuda grass	25.0	37.7	6.4	-	[41]
Elephant grass	22.0	24.0	23.9	-	[42]
Esparto grass	35.5	29.5	18.0	-	[42]
Orchard grass	32.0	40.0	4.7	-	[42]
Reed canary grass	35.2	27.3	8.2	-	[43]
Rye grass (early leaf)	21.3	15.8	2.7	-	[42]
Rye grass (seed setting)	26.7	25.7	7.3	-	[42]
Sabai grass	-	23.9	22.0	-	[42]
Sweet sorghum	35.0	17.0	17.0	23.0	[44]
Switchgrass	39.5	25.0	17.8	-	[45]
Average	31.5	26.7	13.7	13.7	

Biomass varieties[a]	Cellulose	Hemicellulose	Lignin	Extractives	Reference
c) Wood residues					
Albizzia wood	58.3	8.1	33.2	1.9	[46]
Aspen wood	60.7	19.1	14.8	-	[47]
Beech wood	44.2	33.5	21.8	2.6	[48]
Birch wood	56.5	24.8	12.2	-	[47]
Eucalyptus wood	48.0	14.0	29.0	2.0	[44]
Oak wood	54.0	29.0	9.4	-	[47]
Pine wood	52.1	15.4	27.5	-	[47]
Premna wood	63.2	10.5	21.1	3.5	[46]
Pterospermum wood	59.1	8.7	28.5	1.9	[46]
Spurce wood	43.0	29.4	27.6	1.7	[48]
Subabul wood	39.8	24.0	24.7	9.7	[49]
Syzygium wood	60.2	5.5	31.0	1.4	[46]
Average	53.3	18.5	23.4	3.1	
d) Twigs					
Albizzia twig	15.4	32.8	22.2	1.9	[46]
Premna twig	19.4	29.6	20.0	2.0	[46]
Pterospermum twig	13.6	31.5	24.2	1.0	[46]
Syzygium twig	12.2	35.3	21.6	1.6	[46]
Average	15.2	32.3	22.0	1.6	
e) Barks					
Albizzia bark	21.5	16.7	31.1	4.5	[46]
Premna bark	18.5	34.7	20.1	2.6	[46]
Pterospermum bark	20.1	20.6	25.3	3.1	[46]
Syzygium bark	21.9	18.6	30.1	3.0	[46]
Average	20.5	22.7	26.7	3.3	
f) Leaves					
Albizzia leaves	24.5	10.6	30.0	3.0	[46]
Premna leaves	28.9	8.5	18.0	4.6	[46]
Pterospermum leaves	21.3	11.7	26.3	3.7	[46]
Syzygium leaves	27.2	7.0	28.0	3.3	[46]
Average	25.5	9.5	25.6	3.7	
g) Stalks					
Corn stalk	42.7	23.6	17.5	9.8	[49]
Oreganum stalk	33.8	10.9	9.3	-	[50]
Sunflower stalk	43.1	7.4	9.7	-	[50]
Tobacco stalk	21.3	32.9	30.2	5.8	[40]
Average	35.2	18.7	16.7	7.8	
h) Shells					
Almond shell	50.7	28.9	20.4	2.5	[51]
Apricot stone	22.4	20.8	51.4	5.2	[40]
Cashewnut shell	36.2	16.4	18.3	8.4	[52]
Coconut shell	36.3	25.1	28.7	8.3	[49]
Groundnut shell	35.7	18.7	30.2	10.3	[49]
Hazelnut shell	25.2	28.2	42.1	3.1	[48]
Sunflower shell	48.4	34.6	17.0	2.7	[51]
Walnut shell	25.6	28.9	52.3	2.8	[51]
Average	35.1	25.2	32.6	5.4	

Table1. (Continued)

Biomass varieties[a]	Cellulose	Hemicellulose	Lignin	Extractives	Reference
i) Husks					
Millet husk	33.3	26.9	14.0	10.8	[49]
Olive husk	24.0	23.6	48.4	9.4	[51]
Rice husk	31.3	24.3	14.3	8.4	[49]
Average	29.5	24.9	25.6	9.5	
j) Fibers					
Bagasse	41.3	22.6	18.3	13.7	[49]
Bast fiber jute	49.0	19.5	23.5	-	[42]
Bask fiber kanaf	35.0	22.5	17.0	-	[42]
Bast fiber seed flax	47.0	25.0	23.0	-	[42]
Coconut coir	47.7	25.9	17.8	6.8	[49]
Coir pith	28.6	15.3	31.2	15.8	[49]
Corn cob	45.0	35.0	15.0	15.4	[49]
Cotton seed hair	85.0	12.5	0	-	[42]
Curaua fiber	70.2	18.3	9.3	-	[53]
Flax sieve	32.9	18.9	25.8	6.5	[34]
Average	48.2	21.6	18.1	11.6	
k) Aquatic plants					
Algae	7.1	16.3	1.5	-	[54]
Water hyacinth	18.2	48.7	3.5	13.3	[55]
Average	12.7	32.5	2.5	13.3	
Other plant wastes					
Banana waste	13.2	14.8	14.0	-	[42]
Cotton gin waste	77.8	16.0	0	1.1	[49]
Lemon peels	12.7	5.3	1.7	-	[54]
Orange peels	13.6	6.1	2.1	-	[54]
Pine cone	32.7	37.6	24.9	4.8	[56]
Potato peel	55.3	11.7	14.2	-	[57]
Tea waste	31.2	22.8	40.3	2.3	[48]
Average	33.8	16.3	13.9	2.7	
l) High efficiency modern biomass					
Pine chips	52.9	9.6	26.0	0.5	[58]
Pine chip pellets	45.0	11.4	23.2	0.6	[58]
Pine sawdust	43.8	25.2	26.4	4.3	[40]
Average	47.2	15.4	25.2	1.8	
m) Semi-biomass and composite biomass					
Dairy manure	31.4	13.9	16.0	-	[59]
Newspaper	47.5	32.5	24.0	-	[41]
Paper	92.0	0	7.5	-	[41]
Poultry litter	25.0	21.4	9.0	0.4	[58]
Sorted refuse	60.0	20.0	20.0	-	[41]
Swine waste	6.0	28.0	-	-	[41]
Waste paper pulp	65.0	15.0	7.5	-	[42]
Wastewater solids	-	11.5	26.5	-	[42]
Average	46.7	17.8	15.8	0.4	
All herbaceous biomass (average)	34.3	25.4	15.8	12.0	
All woody biomass (average)	29.7	24.5	24.0	2.7	

Biomass varieties[a]	Cellulose	Hemicellulose	Lignin	Extractives	Reference
All other plant residues (average)	34.6	19.4	22.1	6.8	
All plant biomass (average)	31.5	22.7	19.9	7.2	
All lignocellulosic biomass (average)	33.7	21.9	20.0	6.3	

[a] Some of the data presented here are the average values from a number of determinations for a given biomass variety

A few important geographical variations influencing the metabolic composition in the plant are location [33], forest edge [29], and near sea [25], climate [34] and seasonal patterns [31, 35].

The contamination of biomass with mineral matter is mostly through inorganic fertilizer [36] and pesticide application during farming [27, 38]. It is the biomass collection procedures along with transport and storage that significantly affect the biomass composition. Most of the mineral constituents in feedstocks arise as inclusions from dust and soil particles during harvesting and post-harvest processing [38].

The plant's cell wall which provides structural integrity to the plant and defense against pathogens and insects is the main source of lignocellulosic biomass [60]. The plant biomass is mainly composed of cellulose, hemicellulose and lignin along with smaller amounts extractives. Lignocellulose is a complex carbohydrate polymer consisting of cellulose $(C_6H_{10}O_5)_x$, hemicellulose $(C_5H_8O_4)_m$ and lignin $[C_9H_{10}O_3(OCH_3)_{0.9-1.7}]_n$. The composition of these three constituents vary from one plant species to another and sometimes within the single plant species with respect to the plant's age, plant part, stage of growth and other conditions as discussed earlier [61].

About 90% of dry matter in lignocellulosic biomass comprises of cellulose, hemicelluloses and lignin [63], whereas the rest consists of extractives and ash (inorganic components). A typical lignocellulosic biomass (dry matter) is determined with 30-60% cellulose, 20-40% hemicellulose and 15-25% lignin. The distribution of cellulose, hemicellulose and lignin in different types of lignocellulosic residues are given in Table 1. The distribution explains the variation of lignocellulosic contents between agricultural residues, forest-wood residues, water-based vegetation, animal biomass and semi-biomass. Additionally, it also makes a component distinction between different plant parts such as leaves, twigs, barks, shells etc.

2.1. Cellulose

Cellulose is a linear and crystalline homopolymer of repeating sugar units of glucose linked by β-1,4 glycosidic bonds. Cellulose makes up to 15-30% of the dry mass of primary cell walls and up to 40% of secondary cell walls in plants. The global plant population produces about 180 billion tons of cellulose annually which makes this polysaccharide the largest organic carbon reservoir on earth [63].

The occurrence of cellulose in plant cell walls is in form of microfibrils (Fig. 3). The microfibrils are unbranched polymers with nearly 15,000 anhydrous glucose molecules organized in a parallel alignment with β-1,4 linkages. The cellulose microfibrils consist of both crystalline and amorphous regions.

Although the amorphous or soluble region houses molecules that are exposed and less compact in nature, yet the tightly packed molecules in crystalline region make the overall cellulose structure strong [64]. The crystalline cellulose regions in the microfibrils are mostly coated with amorphous cellulose and hemicelluloses [60].

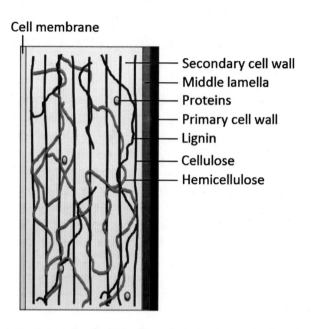

Figure 3. Arrangement of cellulose, hemicellulose and lignin in the plant cell wall.

As mentioned earlier, plants produce huge amount of cellulose through photosynthesis every year. Since there is no net accumulation of cellulose on earth, it implies their utilization by cellulose degrading microorganisms that use lignocellulose as the carbon source. This makes cellulose an exciting raw material for bioconversion to fuel alcohols as numerous deconstructing enzyme mixtures are commercially available today for degrading cellulose microfibrils to monomeric sugars (glucose) for fermentation by yeast or bacteria. The group of cellulose degrading enzymes is termed cellulases which include endoglucanase (EG, endo-1,4-D-glucanohydrolase or EC 3.2.1.4), exoglucanase or cellobiohydrolase (CBH, 1,4-β-d-glucan cellobiodehydrolase or EC 3.2.1.91) and β-glucosidase or cellobiase (EC 3.2.1.21) [41]. For an efficient cellulose hydrolysis, a series of cellulase activity is essential. In the first step, endoglucanase randomly cleaves different crystalline regions of cellulose to generate free chain ends. It may be noted that the amorphous regions do not necessitate endoglucanase activity as they naturally have loosely packed molecules with staggered chain ends. The free chain ends are attacked by exoglucanases cleaving the cellobiose units in the next step. β-glucosidase, in the final step, breaks the β-glycosidic bond between the two glucose molecules of cellobiose to produce monomers of glucose. In nature, there are six crystalline structures of cellulose, namely Cellulose I, Cellulose II, Cellulose III$_1$, Cellulose III$_2$, Cellulose IV$_1$ and Cellulose IV$_2$ [53]. Among the six polymorphs, Cellulose I is the most abundant, stable and native form of cellulose [65], which is made of repeating β-1,4-D-glucopyranose units that are the building blocks of parallel glucan chains. Lignocellulosic biomass contains two distinct polymorphs of Cellulose I, specifically Cellulose I_α and Cellulose I_β. Cellulose I_α is triclinic, metastable and predominant for lower plants, whereas Cellulose I_β is monoclinic, stable and predominant for larger plants [19]. Cellulose II is obtained from Cellulose I through mercerization (treatment with NaOH) or regeneration (solubilizing Cellulose I in a solvent followed by dilution with water for precipitation). Cellulose III_1 and III_2 are formed from Cellulose I and II, respectively in a reversible manner upon treating with liquid ammonia or some amines and evaporation of excess ammonia.

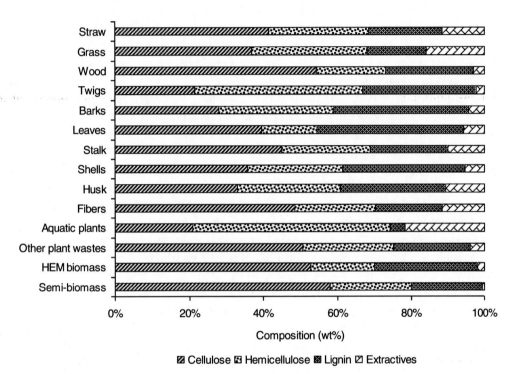

Figure 4. Component distribution of cellulose, hemicelluloses, lignin and extractives in different biomass classes.

Celluloses IV_1 and IV_2 are obtained by heating Cellulose III_1 and III_2, respectively in glycerol at 206°C [65]. Cellulose is an exciting raw material in the production of fuels (e.g., ethanol and butanol), solvents (e.g., acetone), organic acids (e.g., lactic acid), levulinic acid, 5-hydroxymethyl furfural and lubricants [34].

In the present investigation of the data obtained from various sources, it is found that the cellulose content within lignocellulosic feedstocks is highly uneven (Table 1). Among herbaceous residues, rape straw, barley straw and bamboo have high cellulosic content, whereas it is in case of Albizzia, Aspen, Premna, Pterospermum and Syzygium wood for woody biomass residues. The extremely high cellulose was found characteristic of some residues such as cotton seed hair and gin waste, curaua fiber, sorted refuse, paper and waste paper pulp. In contrast, twigs, algae and water hyacinth and some plant wastes exhibit exceptionally low cellulose content. Fig. 4 gives the trend for distribution of cellulose, hemicelluloses, lignin and extractives among different biomass groups.

2.2. Hemicellulose

Inside the plant cell wall, cellulose microfibrils are tightly and non-covalently bonded with other polysaccharides such as hemicellulose or xyloglucans (Fig. 3). Hemicellulose is a short and highly branched heteropolymer containing C_5 (β-D-xylose, α-L-arabinose) and C_6 (β-D-mannose, β-D-glucose, α-D-galactose) monosaccharide sugars along with uronic (α-D-glucuronic, α-D-4-O-methylgalacturonic and α-D-galacturonic acids) acids [66]. The five pentose (C_5) and hexose (C_6) sugars in hemicellulose are covalently linked with each other in long chains to give it a random, highly branched and amorphous structure [67]. The backbone of hemicellulose is linked by β-1,4 glycosidic bonds and occasionally by β-1,3 glycosidic bonds [68]. The homo- and heteropolymers in hemicelluloses are easily hydrolysable, which is very unlikely in case of cellulose.

Hemicelluloses, especially heteroxylans demonstrate some degree of acetylation. Xylans are the group of hemicelluloses that are most abundant in nature constituting about 20-30% of the secondary cell walls in hardwood and herbaceous biomass [66]. Unlike xylans that are predominant in hardwoods, glucomannans and galactoglucomannans are the mannan-group of hemicelluloses occurring in large quantities in the secondary cell wall of softwoods. Similar to cellulose, hemicellulose can be degraded to fermentable sugars by enzymatic hydrolysis for production of fuel alcohols. As the major constituent of hemicellulose is β-1,4-xylan, the most abundant class of hemicellulase is xylanase which has both endo- and exo-activity [69]. Hemicellulases facilitate cellulose hydrolysis by exposing the cellulose fibers making them more accessible to enzymatic degradation [70]. Since, hemicelluloses constitute 20-40% of the feedstocks it is essential to utilize the recovered hemicellulose fraction in order to make bioethanol or other alcohol-based fuel production process less expensive to compete in commercial markets. The hemicellulases studied so far are endo-β-1,4-xylanase (EC 3.2.1.8), exo-β-1,4-xylosidase (EC 3.2.1.37), α-L-arabinofuranosidase (EC 3.2.1.55), endo-α-1,5-arabinanase (EC 3.2.1.99), α-glucuronidase (EC 3.2.1.139), endo-β-1,4-mannanase (EC 3.2.1.78), exo-β-1,4-mannosidase (EC 3.2.1.25), α-galactosidase (EC 3.2.1.22), β-glucosidase (EC 3.2.1.21), endo-galactanase (EC 3.2.1.89), acetyl xylan esterase (EC 3.1.1.72), acetyl mannan esterase (EC 3.1.1.6), ferulic and ρ-cumaric acid esterases (EC 3.1.1.73) [70].

From Table 1 it can be noticed that maximum hemicelluloses are found in twigs and aquatic biomass and minimum in case of leaves. Homoxylans are homopolymers of xylose that have their only occurrence in seaweeds [66]. Lowest levels of hemicelluloses are typical in the wood and leaves of Albizzia, Premna, Pterospermum and Syzygium as well as Oreganum and Sunflower stalk, pine chips and paper refuse.

2.3. Lignin

Lignin which accounts for 10-20% of plants' secondary cell wall materials is composed of a complex of phenylpropanoids linked in a network to cellulose and hemicellulose with ester, phenyl and covalent bonds [64]. Lignin is hydrophobic in nature and tightly binds cellulose and hemicelluloses together rendering them protection from microbial attack. It is an amorphous, three-dimensional aromatic polymer of ρ, hydroxyphenylpropanoid units linked via C–C and C–O–C bonds. Lignin arises mostly through oxidative polymerization of the three monolignols, namely ρ-coumaryl, coniferyl and sinapyl alcohols [71]. The biosynthetic pathway of lignin and the mechanism of its association with cellulose and hemicelluloses are not well understood [72]. The hydrophobic nature of lignin allows plants to translocate water [73]. It delivers mechanical strength to the plants to withstand extreme environmental conditions. Lignin binds to the cellulosic microfibrils with relatively high structural rigidity. The presence of lignin makes the hydrolysis of cellulose and hemicelluloses difficult. It is one of the few organic molecules resistant to biological degradation. The most effective lignin-degrading microorganisms in nature are thought to be white rot fungi, especially *Phanerochaete chrysosporium* and *Trametes versicolour* [74]. The three main families of lignin-modifying enzymes that are produced by fungi are laccases, manganese-dependent peroxidases and lignin peroxidases [75]. High molecular weight lignin is sulfuric acid insoluble unlike low molecular weight lignin which is acid-soluble [76]. However, lignin is considered to be removed using alkaline solution [77]. Among the lignocellulosic materials presented in Table 1, lignin is generally high in barks, leaves, husks and shells. Characteristically, higher amount of lignin is found in apricot stone, hazelnut shell and walnut shell. In general, the agricultural residues including perennial grasses exhibit lower levels of lignin compared to woody biomass. Algae and water hyacinth contain the least amount of lignin.

2.4. Extractives and Inorganic Components

Extractives are the soluble non-structural components in biomass such as non-structural sugars, nitrogenous compounds, chlorophyll and waxes [68]. The biomass extractives also include certain organic acids, hydrocarbons, sterols, oils, fats, lipids, proteins, aromatics, phenols, terpenes, terpenoids, resins, alkaloids, gums, mucilages, dyes, pigments, saponins, tannins and flavonoids [19]. These components can be extracted using various polar or non-polar solvents from biomass. In most standard procedures, a series of extraction procedures involving water, ethanol, benzene, acetone, hexane and toluene are followed [78]. The recovery of extractives from the biomass largely varies with the type of biomass, extracting solvent, extraction time and extraction temperature used. In addition, the concentration of each class of compound is influenced by the plant species, type of plant tissue, growth stage and growing conditions [79]. Table 1 indicates that the amount of extractives is significantly high in agricultural and herbaceous residues than in wood and forest biomass. Flax straw, wheat straw, sweet sorghum, coir pith, corn cobs and water hyacinth show dramatically high levels of extractives. Algae generally have lipids and proteins in larger amounts. The total hexane soluble extractives in algae accounts for 58.0 wt% of the dry algal cell [21]. The inorganic components in biomass mostly comprise of minerals and mineraloids from numerous mineral groups and classes. Some commonly occurring minerals in lignocellulosic biomass are oxides (e.g., hematite, magnetite), carbonates (e.g., calcite, dolomite, magnesite), silicates (e.g., chlorite, feldspars, kaolinite, opal, quartz, zeolites), phosphates (e.g., apatite), nitrates, chlorides, sulphates (e.g., anhydrite, gypsum), sulphites and sulphides [19]. Most of the minerals are widespread soil contaminants [80]; hence their presence in feedstocks depends on geographical and environmental factors such as the distance of the plant from source of pollution such as mines, factories and highway [33,81]. Other important sources of these mineral components are inorganic fertilizer application to farmlands [27] as well as transport and handling processes [38].

3. BIOMASS CHARACTERISTICS

It is very crucial to understand the innate properties of the biomass in order to determine the conversion process and any subsequent process complexities that may arise. In the same manner, the type of biomass, plant part from which the biomass was generated and age of the plant influence their energy content. The interaction between all these characteristics enables flexibility to evaluate the energy efficiency of biomass and the energy extraction process itself. Biomass is similar to other solid fuel types in the requirement of standardized methods of analysis leading to precise and steady assessment of fuel properties [79]. The chief characteristics of interest to estimate the fuel properties of a biomass can be summarized into the following parameters.

(i)   Moisture
(ii)   Ash
(iii)   Volatile matter
(iv)   Fixed carbon
(v)   Elemental analysis
(vi)   Calorific value
(vii)   Particle size and bulk density
(viii)   Holocellulose and lignin

The biomass analysis based on moisture, ash, volatile matter and fixed carbon estimation is defined by proximate analysis.

Table 2. Proximate and ultimate composition of different biomass samples expressed in wt%

Biomass varieties[a]	Proximate analysis[b]				Ultimate analysis[b]						HHV[c] (MJ/kg)	Reference
	M	A	VM	FC	C	H	N	S	O	Cl		
a) Straws												
Alfalfa seed straw	-	7.3	72.6	20.2	46.8	5.4	-	-	40.7	-	17.1	[82]
Barley straw	6.9	9.8	78.5	4.8	41.4	6.2	0.6	0.01	51.7	-	14.7	[39]
Corn straw	7.4	7.7	73.2	19.2	44.7	5.9	0.6	0.1	40.4	0.6	17.1	[83]
Flax straw	7.9	3.0	80.3	8.8	43.1	6.2	0.7	0.1	50.0	-	15.6	[39]
Legume straw	9.8	1.6	73.7	14.8	43.3	5.6	0.6	0.1	50.4	-	14.7	[40]
Mint straw	16.8	9.0	58.1	16.2	37.6	4.6	2.1	0.2	29.8	0.4	14.6	[81]
Oat straw	8.2	5.9	80.5	13.6	47.6	5.8	0.5	-	43.5	-	17.5	[84]
Rape straw	8.7	4.7	76.5	17.8	46.2	6.1	0.5	0.1	42.5	0.03	17.6	[83]
Rice straw	-	19.8	80.2	-	36.9	5.0	0.4	-	37.9	-	13.6	[49]
Rye straw	-	2.0	83.0	15.0	46.3	5.6	-	-	45.9	-	16.4	[82]
Wheat straw	7.3	12.7	64.0	23.4	41.8	5.5	0.7	0	35.5	1.5	16.4	[51]
Average	8.1	7.6	74.6	14.0	43.2	5.6	0.6	0.1	42.6	0.2	15.9	
b) Grasses												
Arundo grass	42.0	2.0	46.5	9.5	27.3	3.4	0.4	0.1	24.9	0.1	10.1	[81]
Bamboo	13.0	0.8	71.0	15.2	51.5	5.1	0.4	0.04	42.0	0.1	18.0	[28]
Bana grass	4.5	9.4	70.1	15.9	43.0	5.2	0.8	0.1	37.0	0.8	16.1	[81]
Elephant grass	12.4	2.9	66.9	17.8	41.6	4.8	0.4	-	50.3[d]	-	13.0	[85]

Biomass varieties[a]	Proximate analysis[b]				Ultimate analysis[b]						HHV[c] (MJ/kg)	Reference
	M	A	VM	FC	C	H	N	S	O	Cl		
Miscanthus grass	12.5	2.9	70.3	14.3	41.7	5.0	0.4	0.04	37.4	0.1	15.3	[81]
Reed canary grass	6.1	4.8	43.8	15.4	44.2	6.1	0.3	-	38.4	-	17.6	[43]
Sorghastrum grass	11.3	3.7	72.4	12.6	41.9	5.3	0.3	0.04	37.4	0.04	15.8	[81]
Sudan grass	-	8.7	72.8	18.6	44.6	5.4	-	-	39.2	-	16.6	[82]
Switchgrass	-	8.9	76.7	14.4	46.7	5.9	0.8	0.2	37.4	-	18.3	[51]
Average	14.5	4.9	65.6	14.9	42.5	5.1	0.4	0.1	38.2	0.1	15.6	
c) Wood residues												
Acacia wood	-	1.7	79.3	19.0	45.4	6.1	-	-	46.4	-	16.7	[82]
Aspen wood	8.2	0.4	80.4	11.0	45.8	5.2	0.4	0.01	40.0	-	16.6	[47]
Beech wood	-	0.5	82.5	17.0	49.5	6.2	0.4	-	41.2	-	19.1	[51]
Birch wood	11.4	0.8	74.4	13.5	44.4	3.5	0.3	0	36.7	-	14.2	[47]
Casuarina wood	-	1.8	78.6	19.6	48.5	6.0	-	-	43.3	-	18.1	[82]
Eucalyptus wood	-	1.2	82.6	16.2	47.1	6.0	-	-	45.3	-	17.3	[82]
Hybrid poplar wood	6.9	2.5	79.0	11.6	46.7	5.6	0.6	0.02	37.7	0.01	17.8	[81]
Madrone wood	-	0.3	84.5	15.1	48.6	6.1	-	-	45.1	-	18.0	[82]
Mango wood	-	3.0	85.6	11.4	46.2	6.1	-	-	44.4	-	17.3	[82]
Oak wood	8.8	0.2	76.8	14.2	45.4	5.0	0.3	0.01	41.3	-	16.0	[47]
Peltophorum wood	-	1.7	82.5	15.8	46.0	6.1	-	-	45.9	-	17.0	[82]
Pine wood	12.9	0.3	71.5	15.3	41.9	4.5	0.2	0	40.2	-	14.2	[47]
Spurce wood	-	-	-	-	52.4	6.1	0.3	-	41.2	-	19.9	[51]
Subabul wood	-	0.9	85.6	-	48.2	5.9	0	-	45.1	-	17.6	[49]
Tanoak wood	-	0.2	90.6	9.2	48.7	6.0	-	-	45.0	-	17.9	[82]
Willow wood	11.4	2.0	82.6	4.0[d]	48.6	6.2	0	0.02	43.3	-	18.4	[86]
Average	9.9	1.1	76.0	13.8	47.1	5.7	0.2	0.004	42.6	0	17.3	
d) Barks												
Balsam bark	-	2.6	77.4	20.0	52.8	6.1	0.2	0	38.6	-	20.4	[38]
Beech bark	-	7.9	75.2	18.9	47.5	5.5	0.6	0	38.5	-	17.8	[38]
Birch white bark	-	1.7	80.3	18.0	57.4	6.7	0.3	0	33.8	-	23.6	[38]
Birch yellow bark	-	2.5	76.5	21.0	54.5	6.4	0.6	0	26.2	-	23.3	[38]
Black spruce bark	-	2.8	74.7	22.5	52.0	5.8	0.1	0	39.7	-	19.6	[38]
Douglas fir bark	-	1.8	65.5	32.8	53.1	6.1	-	-	40.6	-	20.2	[82]
Elm bark	-	0.1	73.1	18.8	46.9	5.3	0.6	0	39.1	-	17.2	[38]
Eucalyptus bark	12.0	4.8	78.1	17.1	47.4	5.5	0.3	-	44.1	-	16.9	[84]
Hemlock bark	-	2.5	72.0	25.5	53.6	5.8	0.2	0	37.9	-	20.4	[38]
Jack pine bark	-	2.1	74.3	23.6	53.4	5.9	0.2	0	38.9	-	20.3	[38]
Maple hard bark	-	2.5	75.1	19.9	50.4	5.9	0.5	0	39.1	-	19.3	[38]
Maple soft bark	-	3.0	78.1	18.9	50.1	5.9	0.3	0	40.7	-	18.9	[38]
Poplar bark	-	2.2	78.9	17.2	51.8	6.5	0.3	0	38.0	-	20.8	[38]

Table 2. (Continued)

Biomass varieties[a]	Proximate analysis[b]				Ultimate analysis[b]						HHV[c] (MJ/kg)	Reference
	M	A	VM	FC	C	H	N	S	O	Cl		
Red spruce bark	-	3.3	72.9	23.7	52.1	5.9	0.1	0	38.6	-	19.9	[38]
Tamarack bark	-	4.2	69.5	26.3	55.2	9.9	0.7	0	31.0	-	27.8	[38]
White spruce bark	-	3.5	72.5	24.0	52.4	6.4	0.1	0	38.4	-	20.7	[38]
Average	12.0	3.0	74.6	21.8	51.9	6.2	0.3	0	37.7	-	20.4	
e)　Stalk												
Alfalfa stalk	-	6.5	76.1	17.4	45.4	5.8	2.1	0.1	36.5	-	17.8	[51]
Corn stalk	-	6.8	80.1	-	41.9	5.3	0	-	46.0	-	14.5	[49]
Cotton stalk	-	17.2	62.9	19.9	39.5	5.1	-	-	38.1	-	14.6	[82]
Oreganum stalk	9.0	4.0	-	-	42.5	6.0	0.7	0.3	42.2	-	16.3	[50]
Sunflower stalk	4.9	11.2	-	-	36.1	5.3	1.3	0.6	39.0	-	13.6	[50]
Tobacco stalk	8.5	10.0	65.5	16.0	39.6	4.9	3.2	0.1	52.3	-	12.2	[40]
Average	7.5	9.3	71.2	17.8	40.8	5.4	1.2	0.2	42.4	-	14.8	
f)　Shells												
Almond shell	-	3.3	74.0	22.7	47.8	6.0	1.1	0.1	41.5	0.1	18.2	[51]
Apricot stone	8.5	0.2	75.1	16.2	44.4	5.7	0.4	0.1	49.5	-	15.4	[40]
Brazil nut shell	-	1.7	76.1	22.2	49.2	5.7	-	-	42.8	-	18.0	[82]
Castor seed shell	-	8.0	72.0	20.0	44.3	5.6	-	-	41.9	-	16.4	[82]
Coconut shell	-	0.7	80.2	-	50.2	5.7	0	-	43.4	-	18.2	[49]
Groundnut shell	-	5.9	83.0	-	48.3	5.7	0.8	-	39.4	-	18.2	[49]
Hazelnut shell	-	-	-	-	52.1	5.6	1.4	-	40.9	-	19.1	[48]
Palm kernel	11.0	5.1	77.3	17.6	48.3	6.2	2.6	0.3	37.4	0.2	29.2	[83]
Peach pit	-	1.0	-	19.9	53.0	5.9	0.3	0.1	39.1	-	20.1	[51]
Pistachio shell	7.5	1.3	75.5	15.7	46.4	5.8	0.6	0.2	38.1	0.01	17.9	[81]
Prune pit	33.6	0.9	53.7	11.8	32.7	4.4	0.6	0.1	27.8	0.01	12.9	[81]
Olive pit	6.1	1.6	77.0	15.3	49.6	6.3	0.4	0.1	36.0	0.04	20.0	[81]
Sunflower shell	-	4.0	76.2	19.8	47.4	5.8	1.4	0.1	41.3	0.1	17.8	[51]
Walnut shell	-	2.8	59.3	37.9	53.5	6.6	1.5	0.1	45.4	0.1	20.3	[51]
Average	13.3	2.8	73.3	19.9	47.7	5.8	0.8	0.1	40.3	0.04	18.0	
g)　Husks												
Coffee husk	9.2	2.5	68.3	18.5	45.4	4.7	1.1	0.4	48.4	-	14.4	[87]
Cotton husk	6.9	3.2	73.0	16.9	50.4	8.4	1.4	0	39.8	-	22.7	[87]
Millet husk	-	18.1	80.7	-	42.7	6.0	0.1	-	33.0	-	17.8	[49]
Mustard husk	5.6	3.9	68.6	22.0	46.1	9.2	0.4	0.2	44.7	-	21.7	[87]
Olive husk	-	4.1	77.5	18.4	49.9	6.2	1.6	0.1	42.0	0.2	19.1	[51]
Rice husk	-	23.5	81.6	-	38.9	5.1	0.6	-	32.0	-	15.4	[49]
Sunflower husk	9.1	1.9	69.1	19.9	51.4	5.0	0.6	0	43.0	-	17.7	[87]
Soya husk	6.3	5.1	69.6	19.0	45.4	6.7	0.9	0.1	46.9	-	17.5	[87]
Average	7.4	7.8	73.6	19.1	46.3	6.4	0.8	0.1	41.2	0.03	18.3	
h)　Fibers												
Coconut coir	-	0.9	82.8	-	47.6	5.7	0.2	-	45.6	-	17.0	[49]
Coir pith	-	7.1	73.3	-	44.0	4.7	0.7	-	43.4	-	14.7	[49]

Biomass varieties[a]	Proximate analysis[b]				Ultimate analysis[b]						HHV[c] (MJ/kg)	Reference
	M	A	VM	FC	C	H	N	S	O	Cl		
Corn cob	-	2.8	85.4	-	47.6	5.0	0	-	44.6	-	16.2	[49]
Palm fiber	36.4	5.3	46.3	12.0	51.5	6.6	1.5	0.3	40.1	-	20.5	[87]
Sugarcane bagasse	10.4	2.2	76.7	10.7	43.6	5.3	0.1	0.04	38.4	0.3	16.2	[81]
Average	23.4	3.7	72.9	11.4	46.9	5.5	0.5	0.1	42.4	0.1	17.0	
i) Aquatic plants												
Algae	11.5[d]	13.3	49.7	25.5	33.2	4.7	1.7	2.0	35.2	-	12.4	[88]
j) Other plant residues												
Almond hulls	6.5	5.7	69.0	18.8	47.5	6.0	1.1	0.1	36.6	0.02	18.8	[81]
Biomass mix	8.8	12.5	69.4	18.1	49.6	5.8	2.4	0.7	28.9	0.1	20.4	[83]
Cotton gin waste	-	5.4	88.0	-	42.7	6.0	0.1	-	49.5	-	15.2	[49]
Fern	-	-	-	-	49.0	5.7	1.5	1.4	40.0	-	18.4	[89]
Forest residue	49.9	2.0	43.1	7.0	25.7	2.4	0.5	0.1	20.4	0.02	8.9	[81]
Olive residue	10.6	7.2	67.4	25.5	54.2	5.4	1.3	0.2	31.7	0.2	21.0	[83]
Peanut hulls	-	5.9	73.0	21.1	45.8	5.5	-	-	39.6	-	17.1	[82]
Pepper plant	6.5	14.4	64.7	20.9	36.1	4.3	2.7	0.5	41.9	0.1	11.8	[83]
Pepper waste	9.7	7.4	58.4	24.4	45.7	3.2	3.4	0.6	47.0	-	12.6	[87]
Pine cone	9.6	0.9	77.8	11.7[d]	42.6	5.6	0.8	0.1	51.0	-	14.4	[56]
Pine needle	-	1.5	72.4	26.1	48.2	5.9	-	-	43.7	-	17.8	[82]
Rice hulls	10.9	18.1	56.6	14.4	34.6	4.2	0.5	0.1	31.7	0.1	12.7	[81]
Tea waste	-	-	-	-	49.6	5.5	0.5	-	44.4	-	17.6	[48]
Walnut hulls and blows	47.9	1.5	41.5	9.1	27.9	3.4	0.8	0.1	18.5	0.01	11.3	[81]
Average	17.8	6.9	65.1	17.9	42.8	4.9	1.3	0.3	37.5	0.04	15.5	
k) High efficiency modern biomass												
Alder-fir sawdust	52.6	2.0	36.3	9.1	24.2	2.8	0.2	0.02	18.3	0.01	9.3	[81]
Fir mill waste	63.0	0.2	30.4	6.5	19.0	2.2	0.03	0.01	0.01	43.8	9.5	[81]
Pine chips	7.6	6.0	72.4	21.7	49.7	5.7	0.5	0.1	38.1	0.1	18.9	[83]
Pine chip pellets	7.7	1.2	-	-	48.0	5.9	0.1	0.1	44.8	-	17.6	[58]
Pine sawdust	5.0	0.3	77.7	16.9	50.3	6.7	0.2	0.2	42.7	-	19.8	[40]
Red oak sawdust	11.5	0.3	76.4	11.9	44.2	5.2	0.03	0.01	38.8	0.01	16.3	[81]
Wood chips	8.1	0.4	-	-	48.0	6.2	0.5	-	43.1	-	18.3	[90]
Average	22.2	1.5	58.6	13.2	40.5	5.0	0.2	0.1	32.3	6.3	15.7	
l) Animal biomass												
Chicken broiler waste	33.4	21.5	-	-	39.6	5.1	3.4	-	34.1	-	15.3	[90]
Chicken flock waste	22.8	22.8	-	-	37.2	5.3	3.1	-	34.7	-	14.7	[90]
Meat bone meal	2.5	24.0	63.3	12.7	43.1	6.0	9.2	1.3	15.6	0.9	20.6	[83]
Average	19.6	22.8	63.3	12.7	40.0	5.5	5.2	1.3	28.1	0.9	16.9	
m) Semi-biomass												
Chicken litter	9.3	37.8	47.8	14.4	37.4	4.2	3.8	0.7	15.6	0.5	16.1	[83]
Dairy manure	-	-	-	-	41.5	-	1.23	-	-	-	-	[59]
Demolition wood	9.0	11.9	67.8	11.2	42.1	4.9	0.5	0.1	31.4	0.1	16.2	[81]
Furniture waste	12.1	3.2	73.0	11.8	43.9	5.2	0.3	0.03	35.4	0.01	16.7	[81]
Land clearing wood	49.2	8.4	35.4	7.0	21.5	2.6	0.2	0.03	18.2	0.01	8.1	[81]
Poultry litter	9.8	13.5	-	-	36.5	5.1	3.9	0.5	40.5	-	13.2	[58]

Table 2. (Continued)

Biomass varieties[a]	Proximate analysis[b]				Ultimate analysis[b]						HHV[c] (MJ/kg)	Reference
	M	A	VM	FC	C	H	N	S	O	Cl		
Refuse derived fuel	4.2	25.0	70.3	0.5	38.0	5.5	0.8	0.3	26.1	-	16.5	[81]
Sewage sludge	5.9	48.1	48.5	3.7	25.8	4.3	3.0	0.8	18.0	0.04	12.0	[91]
Yard waste wood	38.1	12.6	40.9	8.4	25.7	3.0	0.5	0.2	19.9	0.2	9.8	[81]
Waste paper	8.8	7.6	76.9	6.8	43.8	6.1	0.1	0.1	33.6	0	18.2	[81]
Average	16.3	18.7	57.6	8.0	35.6	4.5	1.4	0.3	26.5	0.1	14.1	
Bituminous coal	7.5	4.9	34.0	53.6	88.0	6.0	1.2	0.8	4.0	-	37.4	[87]
Lignite	-	5.0	-	-	56.4	4.2	1.6[e]	-	18.4	-	22.0	[10]
Peat	37.0	4.3	41.0	17.7	57.1	5.9	2.3	0.8	43.1	-	20.9	[87]

Abbreviations: Moisture (M), ash (A), volatile matter (VM), fixed carbon (FC), higher heating value (HHV). [a] Some of the data presented are the average values from a number of determinations for a given biomass variety. [b] Measured at different basis (as received and/or moisture free). [c] Calculated using Dulong's formula for HHV. [d] Calculated from the difference. [e] Combined N and S.

Alternatively, elemental analysis of biomass accessible as C, H, N, S and O together with other metal concentrations in biomass and its ash is termed as ultimate analysis. To study the physiochemical properties of biomass, it is investigated compositionally for proximate analysis and chemically for ultimate analysis. Table 2 gives the proximate and ultimate (C, H, N, S, O) analyses of some typical biomass sources along with the values for bituminous coal, lignite and peat for reference.

3.1. Moisture

The moisture in biomass is stored in spaces within the dead cells and cell walls. Upon drying the biomass the native moisture equilibrates with the ambient relative humidity, with equilibrium being usually about 20% in air dried solid fuel [33]. The moisture content in biomass is the mineralized fluid containing different cations and anions that are essential for plant growth and metabolism. Since, moisture is a mineralized aqueous solution vital in plant physiology an intensive mineral precipitation occurs upon moisture evaporation after biomass harvesting and during biomass drying [92]. During biomass drying, leaching of some major elements (e.g., Ca, Cl, K, Mg, Na, P, S) occur from the plant's dead cells, which later precipitate to form different water-soluble components such as oxides and hydroxides, carbonates, phosphates, sulphates, chlorides and nitrates.

The moisture in biomass occurs in two forms i.e., intrinsic moisture and extrinsic moisture [10]. Intrinsic moisture is the indigenous physiological solution that is not influenced by the prevailing weather conditions during harvest, whereas extrinsic moisture has a direct influence by the prevailing weather conditions during harvesting. It is important to estimate the intrinsic moisture only after eliminating the extrinsic moisture which is possible by air-drying the biomass. In most cases, the moisture content in biomass is measured at different basis, especially as received, air-dried and oven-dried basis.

The moisture content of biomass determines the conversion pathway for biofuels. High moisture containing biomass is suitable for bioconversion, whereas thermal conversion such as pyrolysis requires very low moisture containing feedstock. The presence of water in biomass has an effect on its degradation behavior during pyrolysis and affects the physical properties and quality of the pyrolysis oil [48]. Using high initial moisture containing feedstocks in thermal conversion technologies can adversely affect the overall energy balance for the conversion process. This is due to the fact that moisture in biomass decreases its heating value. The moisture content of wood residues is generally lower than herbaceous residues (Table 2). Due to their lower initial moisture content, woody and forest residues are the most efficient biomass sources for thermochemical conversion to liquid fuels. For the biochemical routes towards production of bioethanol or other alcohol-based fuels, high moisture containing herbaceous residues are more suited.

3.2. Ash

Ash is the product of chemical degradation of biomass by thermochemical means. Ash is the inorganic component of biomass that depends on the type of plant and type of soil (in terms of its mineral composition and fertilizer dose) in which the plant grows. Ash is a significant consideration for measuring the bulk inorganic matter of biomass as well as the predominant affinity of elements and compounds to inorganic or organic matter [92]. The elemental composition in biomass ash estimated both qualitatively and quantitatively gives an idea about the possible contamination of biomass.

With the degree of the ash content, available energy of the fuel reduces proportionately [10]. Biomass with high magnitudes of ash suffers from the high handling and processing costs. The first reason is due to the chemical composition of the ash which can present significant operational problems due to the formation of a viscous and sticky liquid phase at elevated temperatures that blocks the channels in the furnace and boiler plant. The second reason is due to the insoluble ionic components in the ash that act as a heat barrier dramatically affecting the heat flow in the biorefineries, thus reducing process efficiency. On the other hand, soluble ionic compounds can have a catalytic effect on the pyrolysis of the fuel and favor the formation of char [93]. Compared to the ash percentage from combustion of a biomass, its biochemical conversion would generate high amount of solid residue which is the non- or poorly-biodegradable carbon present in that biomass. This non- or poorly-biodegradable carbon is not entirely ash, rather it represents the recalcitrant carbon that cannot be degraded further biologically but could be thermochemically pyrolysed to liquid fuels or gasified to syngas. The high concentration of ash is found in case of semi-biomass and animal biomass (Table 2). Semi-biomass which comprises of waste wood and chemically treated wood and other residues such as demolition wood, furniture waste, land clearing wood, poultry litter, refuse derived fuel, sewage sludge, yard waste wood and waste paper indicate the presence of different mineral and metallic impurities due to the manufacturing process [92]. Extremely high ash concentration in chicken litter is mostly due to its physical nature. A significant fraction of chicken litter consists of bedding material for the livestocks which is the chief source of ash [90]. With the partial composting of litter, the manure content increases which subsequently raises the ash levels.

3.3. Volatile Matter and Fixed Carbon

The volatile matter in biomass commonly includes moisture, CO_2, CO, H_2, light hydrocarbons, tars, acetic acid, methanol and acetone [94]. In other words, it is the volatile fraction of biomass that is driven-off as a gas by heating to 950°C for 7 min. The fixed carbon is the material in the biomass other than the volatile matter (including moisture) and ash [10]. Fixed carbon is usually calculated from the difference of moisture, ash and volatile matter. Volatile matter and fixed carbon are significant in measuring the thermal properties of the biomass related to its ignition, gasification and combustion. In most cases, algae present a modest volatile matter proportion with a high fixed carbon (Table 2).

3.4. Elemental Analysis

Elemental analysis of biomass includes the measurement of common organic elements in biomass such as C, H, N, S along with the various other inorganic metals and metalloids. The percentage of O is mostly calculated from the difference of C, H, N, S and ash. The proportions of C, H and O are critical in determining the energy value of the fuel. This is because of the fact that in photosynthetic organisms, the interaction between hydrogen and carbon indicates occurrence of carbohydrates and in other fuel-based compounds as hydrocarbons. High levels of nitrogen in chicken broiler waste, chicken flock waste, chicken and poultry litter, meat bone meal and sewage sludge (Table 2) is an indication of N-based organic compounds such as proteins, amino acids, nucleic acid, uric acid and some chemical treatment [58]. The composition of ash is important in understanding the inorganic constituents in the biomass. The intrinsic inorganic matter is uniformly distributed in the biomass and is referred to as atomically dispersed material [79]. The release of atomically dispersed inorganic material from biomass is dependent on its volatile matter content and the reactions between the organic portions. As discussed earlier, the elemental composition, particularly alkaline elements (e.g., Na, K, Mg, Ca) in ash are important for thermochemical conversion processes. These alkaline elements along with a few other elements such as S, Cl and Fe react with silica leading to ash fouling and slagging that blocks the airways in the furnaces [38]. Although the intrinsic silica content of a feedstock may be low, the contamination with soil and dirt during harvesting, transportation and storage can considerably increase the total silica content. Various deleterious heavy metals may also be introduced into the biomass through the prevailing weather conditions (air current) at the time of harvesting or by the geographical location of the plant near mining sites, industries, mines or highways. On the other hand, certain metal constituents in the biomass tend to have catalytic properties in transforming the composition and yield of organic molecules resulting from its thermal degradation. Metals such as K and Na present in the biomass catalyze the thermochemical breakdown of cellulose and reduce the yield of levoglucosan during pyrolysis by lowering the decomposition temperature and increasing both the overall reaction rate and yield of char [95]. Similarly, Ca is known to catalyze hemicellulose pyrolysis [58]. In a study by Li et al. [96], the presence of Na and Ca in the biochars acted as catalyst in blocking its aromatic ring sites from reaction with O_2. This activity against oxidation in the char helped keep its cross-linking structure intact indicating their long-term stability in soil and relatively higher adsorption of soil nutrients.

3.5. Heating Value

The heating value (HV) or calorific value (CV) or the heat of combustion of a feedstock is an estimate of its energy or heat content released upon combustion. It is expressed as gross calorific value (i.e., higher heating value, HHV) and the net calorific value (i.e., lower heating value, LHV). HHV represents the maximum amount of energy recovered from a biomass as it is the total heat energy released when the biomass is burnt in air including the latent heat of water vapor [10]. In some contexts, the latent heat of water vapor cannot be effectively used; hence its omission from the HHV is the actual energy available for subsequent use which defines LHV. With water in the vapor phase, LHV at constant pressure measures the enthalpy change due to combustion [79], while HHV at a constant pressure estimates the enthalpy change of combustion with water condensed at initial temperature and pressure. The CV of a fuel is usually represented in MJ (megajoules)/kg of the fuel. The CV of lignocellulosic biomass and their conversion products can be determined both experimentally and theoretically. The experimental determination of CV requires an adiabatic bomb colorimeter [97], whereas the theoretical determination relies on the ultimate and/or proximate analysis. The heating value of a biomass has been correlated with its proximate, ultimate and biochemical composition by many authors [94, 98]. The combustion properties of a lignocellulosic material depend on its proximate and ultimate composition, conversion products and their overall CV. The moisture and ash content of a feedstock adversely affects its heating value. An increase in the moisture and ash content decreases the CV, hence high ash and moisture containing biomass makes it a less desirable fuel at industrial scale [79]. There is no direct correlation between HHV of the feedstock and its cellulose and hemicelluloses concentration, but the lignin content is a function of the HHV [98]. An approximate measure of CV of biomass (dry and ash-free basis) can be done in various ways as follows [94, 98].

(i)   Dulong's equation
        CV (MJ/kg) = 144.4 (C wt%) + 610.2 (H wt%) − 65.9 (O wt%) + 0.39 (O wt%)^2
(ii)   Dulong-Berthelot's equation
        CV (MJ/kg) = 81.37 + 354 [(H wt%) − (O wt% + N wt% − 1)/8] − 22.1 (S wt%)
(iii)   Modified Dulong's equation
        HHV (MJ/kg) = 33.5 (C wt%) + 142.3 (H wt%) − 15.4 (O wt%)
(iv)   LHV equation
        LHV (MJ/kg) = 33.4 (C wt%) + 139.7 (H wt%) − 15.6 (O wt%) − 2.6 (H_2O wt%)
(v)   Equation for woody biomass
        HHV (MJ/kg) = 0.0893 (Lignin wt%) + 16.9742
(vi)   Equation for non-woody biomass
        HHV (MJ/kg) = 0.0877 (Lignin wt%) + 16.4951
(vii)   Equation based on fixed carbon content
        HHV (MJ/kg) = 0.196 (Fixed carbon wt%) + 14.119
(viii) Differential HHV (dHHV) equation based on extractives content
        dHHV (MJ/kg) = 0.00639 (Extractives wt%)^2 + 0.223 (Extractives wt%) + 0.691

It may be noted that the equation for woody and non-woody biomass uses lignin content (wt%) measured on a dry, ash-free and extractive-free sample basis. In addition, the use of differential HHV equation needs precise observations as the recovery of extractives varies with the biomass depending on the type of extracting solvent, extraction time and extraction temperature.

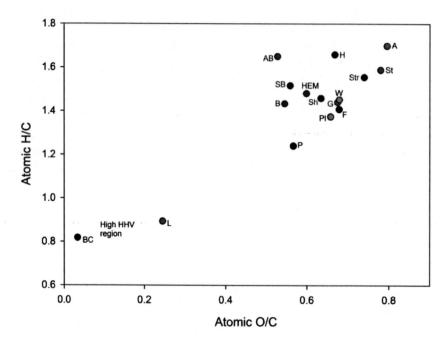

Figure 5. van Krevelen diagram for various biomass varieties along with bituminous coal (BC), lignite (L) and peat (P) as reference fuels. Abbreviatons: straws (Str), grasses (G), wood residues (W), barks (B), stalks (St), shells (S), husks (H), fibers (F), algae (A), other plant residues (Pl), HEM biomass (HEM), animal biomass (AB) and semi-biomass (SB).

The most widely used equation for theoretical calculation of HHV is the modified Dulong's equation that precisely uses the weight percent composition of C, H and O. The HHV for different lignocellulosic biomasses mentioned in Table 2 has been calculated using the modified Dulong's equation.

The correlation between the heating value and biochemical components is not very straight forward. Cellulose has a smaller heating value than lignin because of its higher degree of oxidation [51].

In contrast, presence of hydrocarbons tends to raise the heating value of fuels due to the lower degrees of oxidation. The significance of molar ratios of O and H to C on the calorific value of solid fuels can be illustrated using a van Krevelen diagram or coalification diagram (Fig. 5). The diagram makes a comparison between various biomasses with coal, lignite and peat. It illustrates that the higher concentration of O and H than C reduces the heating value due to the lower energy contained in C–O and C–H bonds than in C–C bonds [10]. The van Krevelen diagram infers the combustion aspects of the fuels along with their chemical structure [99].

3.6. Particle Size and Bulk Density

Prior to the use for any conversion, the biomass particles have to be brought into suitable size in order to make the process efficient and economically viable in recovering the process residues. Studies by Tamaki and Mazza [34] showed that with an increase in the particle size of the biomass from < 150-180 μm to > 850 μm, the recovery of cellulose, hemicelluloses and acid-insoluble lignin increased, whereas the extractives, proteins and ash yield decreased.

The bulk density or volume of biomass as-produced (on field) and as-processed (for transport in form of bales, pellets, briquettes etc.) is another important property to determine the overall cost of logistics (transport and storage) envisioned in a biorefinery. The density of the processed product affects the sizing of the materials handling system and the behavior of the material during the conversion process. The variation in different proximate, ultimate and biochemical (cellulose, hemicelluloses and lignin) composition of as-produced and as-processed biomass can be found in Tables 1 and 2 (e.g., pine wood as per as-produced; pine chips, pine chip pellets and pine sawdust as per as-processed).

3.7. Holocellulose and Lignin

The proportion of holocellulose (cellulose and hemicelluloses) and lignin fraction present in a biomass has large influence on the efficiency of its biochemical conversion. Cellulose has a higher degree of biodegradability than lignin; hence there is a better bioconversion of high cellulose-containing biomass than for biomass with a high amount of lignin [10]. This is a determining factor in selecting a variety of feedstock for biochemical processing. On the contrary, highly lignified biomass has a potential to serve as a candidate biomass for thermochemical conversion. This is because lignin is not generally fermentable [79] but it also has high energy content (about 30% higher than cellulose and hemicellulose) [73]. Holocelluloses are composed entirely of sugar units with a high level of oxidation and thus have reasonably low calorific value in the native feedstock. In an average, high lignin containing biomass have relatively high heating value. Lignin and extractives have a lower degree of oxidation and contribute to a higher heat of combustion [19].

Biomass is composed of biopolymers, particularly cellulose, hemicelluloses and lignin that embrace most of their cellular structure. However, due to this complex construction of plant cell walls, lignocellulosic biomass is more difficult to break down into sugars than starch-based biomass. The holocellulose and lignin composite is critical to many physiological functions including resistance and strength to the plant cell walls. A feedstock with high holocellulose content is required to provide a high yield of ethanol or butanol. On the other hand, a high lignin containing feedstock, although representing a potentially large energy sink, suffers from effective acid and enzymatic hydrolysis. The polymeric lignin prevents access of enzymes and chemicals to hemicellulose and cellulose, reducing the degradability of the carbohydrate material. The crystalline regions in the cellulose microfibrils are highly resistant to chemical and biological hydrolysis because of the complex arrangement of cellodextrin chains [60] and the sheath of hemicelluloses and lignin enclosing them. The magnitude of lignin as well as cellulose crystallinity and its extent of polymerization are a few factors believed to contribute to the recalcitrance of lignocellulosic biomass to saccharifying enzymes and chemicals [100].

At the molecular level of biomass, the chain conformation of the glucose units in cellulose causes the –OH groups into radial orientation and the aliphatic H-atoms into axial positions [60]. This results in a strong inter-chain H-bonding between adjacent chains in a cellulose fiber along with weaker hydrophobic interactions between them. This hydrophobic interaction between cellulose fibers leads to the formation of a dense layer of water near the hydrated cellulose surface giving resistance to the crystalline regions of cellulose towards acids and enzymes.

The effect of strong inter-chain H-bonding is not the case for the amorphous regions of cellulose and hemicelluloses which are susceptible to chemical and biological hydrolysis. The higher-order structures in plant cell walls also contribute to the biomass recalcitrance. The recalcitrance of lignocellulosic materials is the origin and nature of the feedstock itself i.e., woody or herbaceous.

Herbaceous plants that incorporate various agricultural residues and grasses are typically composed of loosely bound fibers, while woody plant species have tightly bound fibers creating mass-transport limitations for access of chemicals and biochemical catalysts for hydrolysis. A higher scale of lignification is also indicated by tightly bound fibers in some biomass varieties [10].

Nevertheless, many pretreatment methods employing various physical and chemical agents have been found helpful in reducing the recalcitrance of biomass and depolymerizing the hemicellulose. A pretreated biomass is characterized with reduced macroscopic rigidity and less physical barriers to mass transport. The main objectives of biomass pretreatment and conversion to alcohol-based fuels are removal of lignin from the biomass to liberate holocellulose, depolymerization of the carbohydrate polymers in holocellulose to produce free sugars and fermentation of liberated pentose and hexose sugars to alcohols. The pretreatments that have been widely investigated on biomass are (dilute) acid and alkaline hydrolysis, organosolv, ammonia fiber explosion, ultrasonication, ozonolysis, steam explosion and CO_2 explosion [68]. There are some new promising technologies in biomass pretreatment which include supercritical water treatment [50, 66] and ionic liquids [45, 101]. The pretreatment breaks the native hemicelluloses in biomass to monosaccharides and oligosaccharides which are then easily hydrolyzed and fermented. As the hemicelluloses are removed from over the cellulose microfibrils the crystalline regions of cellulose are exposed to cellulase enzymes during enzymatic hydrolysis of biomass prior to alcohol production by fermentation.

CONCLUSION

The use of biomass as an energy source is becoming essential to alleviate the global warming caused by burning fossil fuels. However, excessive exploitation of natural resources for biofuel production should be avoided. The potential bioenergy resources should be preferably non-edible and should not compete with food crops or arable land for cultivation. On the other hand, they should be available throughout the globe and in surplus round the year. Lignocellulosic biomass is the prospective resource that applies to all such conditions. Lignocellulosic biomass, as discussed comprehensively in this chapter, arise from non-edible agricultural residues, perennial grass systems, forest-wood residues, short-rotation energy crops (e.g., microalgae, grasses and some wood-based crops), discarded wood materials, municipal organic waste and industrial sludge. Most of these materials are regarded as waste refuse, while a few such energy crops are grown on non-arable and degraded lands and wastewater or contaminated ponds.

Due to the huge amount of lignocellulosic biomass generated from a variety of biomass crops it is often difficult to classify them in a proper genre. It is attempted to classify them based on their efficiency and origin to accommodate maximum amount of residues. In order to reach this level of understanding, an initial screening of the biomass is required to attain information about its composition and characteristics.

The existing conversion technologies worldwide prefer mostly two main types of biomass i.e., herbaceous and woody biomass. Despite the considerable amount of chemical energy stored in the semi-biomass feedstocks and animal biomass, their diversity in composition causes impediments in choosing a proper conversion technology and ends up in their application as manures.

Characterizing the biomass prior to conversion is a key parameter for a sustainable refinery. It is the interaction between various factors such as its intrinsic and extrinsic moisture content, ash and metallic composition, volatiles and fixed carbon, heating value, particle size and density and organic matter composition that helps to decide the potentiality of a feedstock for a conversion process. Most of the chemical energy obtained through photosynthesis in plants is contained in the organic components of the biomass, namely cellulose, hemicellulose and lignin, although their proportions vary according to type and/or part of plant. A number of other factors that have an impact on the organic and inorganic composition of the biomass are age, metabolism and physiology of the plant, environmental and growth conditions of the plant, geographical location, farming practices and harvesting, post-harvest handling, storage and transport.

A biofuel refinery is envisaged to cover certain main aspects such as post-harvest handling of biomass, biomass pretreatment, chemical or enzymatic hydrolysis, fermentation to alcohol-based fuels and/or pyrolysis or gasification to liquid and gaseous fuels. Existing biomass conversion schemes typically rely on a combination of chemical and enzymatic treatments.

The complex interplay between cellulose, hemicelluloses and lignin gives a recalcitrant phenomenon to the lignocellulosic biomass which is still another main technological barrier in employing them as a source of energy. Due to the biomass recalcitrant, it is difficult to breakdown the polysaccharides into fermentable sugars for a maximum fuel per unit biomass yield.

However, various pretreatments are found helpful in reducing the indigenous rigidity of biomass and depolymerizing the hemicelluloses present in them. The conversion of lignocellulosic biomass to alcohol-based fuels requires removal of lignin from the biomass to liberate holocellulose, depolymerization of the carbohydrate polymers in holocellulose to produce free sugars and fermentation of liberated pentose and hexose sugars to alcohols.

Unlike the fossil fuel resources such as coal, natural gas and petroleum that are localized geographically, biomass has a widespread occurrence and on a continual basis. A sustained diversification in the biofuel industry sector worldwide will help in continued supply of feedstocks for bioenergy. Lignocellulosic biomass sources are promising from both an economic and a GHG mitigation perspective, although the relative enormity of their mitigation potential is not well understood.

Biomass is the only renewable resource that has a potential to provide high energy density liquid and gaseous fuels. In addition to biofuels, a wide range of bio-based products can also be produced at biorefineries as byproducts of the process. These value-added bio-products have a tendency to balance the net economics related to the overall conversion process. Development of commercial bioenergy sectors across the globe and facilitation of international bioenergy trade can help achieve today's sustainable energy goals.

ACKNOWLEDGMENTS

The authors acknowledge the financial support from Natural Sciences and Engineering Research Council (NSERC) of Canada, Canada Research Chair Program, University of Saskatchewan and York University.

REFERENCES

[1] United States Energy Information Administration (USEIA). International Energy Outlook 2011. Accessed from: http://www.eia.gov/forecasts/ieo/pdf/0484(2011).pdf. 2011. Last accessed on Jan 03, 2012.

[2] Hansen J, Sato M, Ruedy R, Lo K, Lea DW, Medina-Elizade M. Global temperature change. *PNAS* 2006;103:14288-93.

[3] International Panel on Climate Change (IPCC). Climate Change 2001: The Scientific Basis. Contributions of Working Group 1 to the Third Assessment Report of the Intergovernmental Panel on Climate Change. In: Houghton JT, Ding Y, Griggs DJ, Noguer M, van der Linden PJ, Dai X, Maskell K, Johnson CA, editors. Cambridge University Press, Cambridge, UK; 2001.

[4] International Panel on Climate Change (IPCC). Renewable energy sources and climate change mitigation: Special report of the Intergovernmental Panel on Climate Change. Technical Support Unit Working Group III. In: Edenhofer O, Madruga RP, Sokona Y, Seyboth K, Matschoss P, Kadner S, Zwickel T, Eickemeier P, Hansen G, Schlomer S, Stechow CV, editors. Cambridge University Press, New York, USA; 2012.

[5] Boden TA, Marland G, Andres RJ. Global, regional, and national fossil-fuel CO2 emissions. Carbon Dioxide Information Analysis Center, Oak Ridge National Laboratory, US Department of Energy, Oak Ridge, TN, USA; 2009. Accessed from: cdiac.ornl.gov/trends/emis/overview_2007.html.

[6] Campbell CH, Laherrere JH. The end of cheap oil. *Sci. Am.* 1998;78-83.

[7] Swana J, Yang Y, Behnam M, Thompson R. An analysis of net energy production and feedstock availability for biobutanol and bioethanol. *Bioresource Technol.* 2011;102: 2112-7.

[8] Sanderson K. Lignocellulose: A chewy problem. *Nature* 2011;474:S12-S14.

[9] Berndes G, Hoogwijk M, van den Broek R. The contribution of biomass in the future global energy system: A review of 17 studies. *Biomass Bioenerg.* 2003;25:1-28.

[10] McKendry P. Energy production from biomass (part 1): overview of biomass. *Bioresource Technol.* 2002;83:37-46.

[11] Food and Agriculture Organization of the United Nations (FAO). The state of food and agriculture. Biofuels: Prospects, risks, and opportunities. Rome, Italy; 2008.

[12] Gronowska M, Joshi S, MacLean HL. A review of U.S. and Canadian biomass supply studies. *Bioresource* 2009;4:341-369.

[13] Mabee WE, Saddler JN. Bioethanol from lignocellulosics: status and perspectives in Canada. *Bioresource Technol.* 2010;101:4806-13.

[14] Jones A, O'Hare M, Farrell A. Biofuel boundries: estimating the medium-term supply potential of domestic biofuels. Research report UCB-ITS-TSRC-RR-2007-4. University of California, Berkeley; 2007.

[15] Bowyer JL, Stockmann VE. Agricultural residues: An exciting bio-based raw material for the global panels industry. *Forest Prod. J.* 2001;51:10-21.

[16] Perlack RD, Wright LL, Turhollow AF, Graham RL. Biomass as feedstock for a bioenergy and bioproducts industry: The technical feasibility of a billionton annual supply. Oak Ridge National Laboratory, Oak Ridge, Tennessee, USA; 2005.

[17] Yemshanov D, McKenney D. Fast-growing poplar plantations as a bioenergy supply source for Canada. *Biomass Bioenerg.* 2008;32:185-97.

[18] Gunnarssona CC, Petersen CM. Water hyacinths as a resource in agriculture and energy production: A literature review. *Waste Manage.* 2007;27:117-29.

[19] Vassilev SV, Baxter D, Andersen LK, Vassileva CG, Morgan TJ. An overview of the organic and inorganic phase composition of biomass. *Fuel*, 2012;94:1-33.

[20] Savage N. Alage: The scum solution. *Nature* 2011;474:S15-S16.

[21] Demirbas A. Use of algae as biofuel sources. *Energ. Convers. Manage.* 2010;51: 2738-49.

[22] Karim K, Hoffmann R, Klasson KT, Al-Dahhan MH. Anaerobic digestion of animal waste: Effect of mode of mixing. *Water Res.* 2005;39:3597-606.

[23] Song W, Guo M. Quality variations of poultry litter biochar generated at different pyrolysis temperatures. *J. Anal. Appl. Pyrol.* 2012;94:138-45.

[24] Sharpe RR, Schomberg HH, Harper LA, Endale DM, Jenkins MB, Franzluebbers AJ. Ammonia volatilization from surface-applied poultry litter under conservation tillage management practices. *J. Environ. Qual.* 2004;33:1183-8.

[25] Cuiping L, Chuangzhi W, Yanyongjie, Haitao H. Chemical elemental characteristics of biomass fuels in China. *Biomass. Bioenerg.* 2004;27:119-30.

[26] Monti A, Virgilio ND, Venturi G. Mineral composition and ash content of six major energy crops. *Biomass Bioenerg.* 2008;32:216-23.

[27] Obernberger I, Brunner T, Barnthaler G. Chemical properties of solid biofuels–significance and impact. *Biomass Bioenerg.* 2006;30:973-82.

[28] Scurlock JMO, Dayton DC, Hames B. Bamboo: An overlooked biomass resource? *Biomass Bioenerg.* 2000;19:229-44.

[29] Werkelin J, Skrifvars BJ, Hupa M. Ash-forming elements in four Scandinavian wood species. Part. 1. Summer harvest. *Biomass Bioenerg.* 2005;29:451-66.

[30] Demirbas A. Trace metal concentrations in ashes from various types of biomass species. *Energ. Source* 2003;25:743-51.

[31] Demeyer A, Voundi Nkana JC, Verloo MG. Characteristics of wood ash and influence on soil properties and nutrient uptake: an overview. *Bioresource Technol.* 2001;77:287-95.

[32] Ulery AL, Graham RC, Amrhein C. Wood-ash composition and soil pH following intense burning. *Soil Sci.* 1993;156:358-64.

[33] Demirbas A. Potential applications of renewable energy sources, biomass combustion problems in boiler power systems and combustion related environmental issues. *Prog. Energ. Combust. Sci.* 2005;31:171-92.

[34] Tamaki Y, Mazza G. Measurement of structural carbohydrates, lignins, and micro-components of straw and shives: effects of extractives, particle size and crop species. *Ind. Crop Prod.* 2010;31:534-41.

[35] Paulrud S, Nilsson C. Briquetting and combustion of spring-harvested reed canary-grass: effect of fuel composition. *Biomass Bioenerg.* 2001;20:25-35.

[36] Sander B. Properties of Danish biofuels and the requirements for power production. *Biomass Bioenerg.* 1997;12:177-83.

[37] Vamvuka D, Zografos D. Predicting the behaviour of ash from agricultural wastes during combustion. *Fuel* 2004;83:2051-7.

[38] Bryers RW. Fireside slagging, fouling and high-temperature corrosion of heat transfer surface due to impurities in steam raising fuels. *Prog. Energ. Combust. Sci.* 1996;22:29-120.

[39] Naik S, Goud VV, Rout PK, Jacobson K, Dalai AK. Characterization of Canadian biomass for alternative renewable biofuel. *Renew. Energ.* 2010;35:1624-31.

[40] Wei L, Xu S, Zhang L, Zhang H, Liu C, Zhu H, Liu H. Characteristics of fast pyrolysis of biomass in a free fall reactor. *Fuel Process. Technol.* 2006;87:863-71.

[41] Prasad S, Singh A, Joshi HC. Ethanol as an alternative fuel from agricultural, industrial and urban residues. *Resour. Conserv. Recy.* 2007;50:1-39.

[42] Abbasi T, Abbasi SA. Biomass energy and the environmental impacts associated with its production and utilization. *Renew. Sust. Energ. Rev.* 2010;14:919-37.

[43] Bridgeman TG, Darvell LI, Jones JM, Williams PT, Fahmi R, Bridgwater AV, Barraclough T, Shield I, Yates N, Thain SC, Donnison IS. Influence of particle size on the analytical and chemical properties of two energy crops. *Fuel* 2007;86:60-72.

[44] Huber GW, Iborra S, Corma A. Synthesis of transportation fuels from biomass: Chemistry, catalysts, and engineering. *Chem. Rev.* 2006;106:4044-98.

[45] Li C, Knierim B, Manisseri C, Arora R, Scheller HV, Auer M, Vogel KP, Simmons BA, Singh S. Comparison of dilute acid and ionic liquid pretreatment of switchgrass: Biomass recalcitrance, delignification and enzymatic saccharification. *Bioresource Technol.* 2010;101:4900-6.

[46] Kataki R, Konwer D. Fuelwood characteristics of some indigenous woody species of north-east India. *Biomass Bioenerg.* 2001;20:17-23.

[47] Shen DK, Gu S, Luo KH, Bridgwater AV, Fang MX. Kinetic study on thermal decomposition of woods in oxidative environment. *Fuel* 2009;88:1024-30.

[48] Demirbas A. Thermochemical conversion of biomass to liquid products in the aqueous medium. *Energ. Source,* 2005;27:1235-43.

[49] Raveendran K, Ganesh A, Khilar KC. Influence of mineral matter on biomass pyrolysis characteristics. *Fuel* 1995;74:1812-22.

[50] Yanik J, Ebale S, Kruse A, Saglam M, Yuksel M. Biomass gasification in supercritical water: Part 1. Effect of the nature of biomass. *Fuel* 2007;86:2410-5.

[51] Demirbas A. Combustion characteristics of different biomass fuels. *Prog. Energ. Combust. Sci.* 2004;30:219-30.

[52] Raveendran K, Ganesh A. Adsorption characteristics and pore-development of biomass-pyrolysis char. *Fuel* 1998;77:769-81.

[53] Correa AC, de Morais Teixeira E, Pessan LA, Mattoso LHC. Cellulose nanofibers from curaua fibers. *Cellulose* 2010;17:1183-92.

[54] Ververis C, Georghiou K, Danielidis D, Hatzinikolaou DG, Santas P, Santas R, Corleti V. Cellulose, hemicelluloses, lignin and ash content of some organic materials and their suitability for use as paper pulp supplements. *Bioresource Technol.* 2007;98:296-301.

[55] Nigam JN. Bioconversion of water-hyacinth (*Eichhornia crassipes*) hemicellulose acid hydrolysate to motor fuel ethanol by xylose–fermenting yeast. *J. Biotechnol.* 2002;97:107-16.

[56] Brebu M, Ucar S, Vasile C, Yanik J. Co-pyrolysis of pine cone with synthetic polymers. *Fuel* 2010;89:1911-8.

[57] Lenihan P, Orozco A, O'Neill E, Ahmad MNM, Rooney DW, Walker GM. Dilute acid hydrolysis of lignocellulosic biomass. *Chem. Eng. J.* 2010;156:395-403.

[58] Das KC, Garcia-perez M, Bibens B, Melear N. Slow pyrolysis of poultry litter and pine woody biomass: Impact of chars and bio-oils on microbial growth. *J. Environ. Sci. Health Part A* 2008;43:714-24.

[59] Liao W, Wen Z, Hurley S, Liu Y, Liu C, Chen S. Effects of hemicellulose and lignin on enzymatic hydrolysis of cellulose from dairy manure. *Appl. Biochem. Biotechnol.* 2005;121-124:1017-30.

[60] Himmel ME, Ding SY, Johnson DK, Adney WS, Nimlos MR, Brady JW, Foust DT. Biomass recalcitrance: Engineering plants and enzymes for biofuels production. *Science* 2007;315:804-7.

[61] Perez J, Dorado JM, Rubia TD, Martinez J. Biodegradation and biological treatment of cellulose, hemicellulose and lignin: An overview. *Int. Microbiol.* 2002;5:53-63.

[62] Balat M. Production of bioethanol from lignocellulosic materials via the biochemical pathway: a review. *Energ. Convers. Manage.* 2011;52:858-75.

[63] O'Neill MA, Ishii T, Albersheim P, Darvill AG. Rhamnogalacturonan II: Structure and function of a borate cross-linked cell wall pectic polysaccharide. *Annu. Rev. Plant Biol.* 2004;55:109-39.

[64] Carpita NC, McCann MC. The functions of cell wall polysaccharides in composition and architecture revealed through mutations. *Plant Soil* 2002;247:71-80.

[65] O'Sullivan A. Cellulose: the structure slowly unravels. *Cellulose* 1997;4:173-207.

[66] Girio FM, Fonseca C, Carvalheiro F, Duarte LC, Marques S, Bogel-Lukasik R. Hemicelluloses for fuel ethanol: A review. *Bioresource Technol.* 2010;101:4775-800.

[67] Wyman CE. Cellulosic ethanol: A unique sustainable liquid transportation fuel. *MRS Bull.* 2008;33:381-383.

[68] Kumar P, Barrett DM, Delwiche MJ, Stroeve P. Methods for pretreatment of lignocellulosic biomass for efficient hydrolysis and biofuel production. *Ind. Eng. Chem. Res.* 2009;48:3713-29.

[69] Warren RAJ. Microbial hydrolysis of polysaccharides. *Annu. Rev. Microbiol.* 1996;50:183-212.

[70] Shallom D, Shoham Y. Microbial hemicellulases. *Curr. Opinion Microbiol.* 2003;6:219-28.

[71] Chen F, Tobimatsu Y, Havkin-Frenkel D, Dixon RA, Ralph J. A polymer of caffeyl alcohol in plant seeds. *PNAS* 2012:109:1772-7.

[72] Sticklen MB. Feedstock crop genetic engineering for alcohol fuels. *Crop Sci.* 2007;47:2238-48.

[73] Novaes E, Kirst M, Chiang V, Winter-Sederoff H, Sederoff R. Lignin and biomass: A negative correlation for wood formation and lignin content in trees. *Plant Physiol.* 2010;154:555-61.

[74] D'Souza TM, Merritt CS, Reddy CA. Ligninmodifying enzymes of the white rot basidiomycete *Ganoderma lucidum. Appl. Environ. Microbiol.* 1999;65: 5307-13.

[75] Hatakka A. Lignin-modifying enzymes from selected white-rot fungi: production and role in lignin degradation. *FEMS Microbiol. Rev.* 1994;13:125-35.

[76] Sluiter JB, Ruiz RO, Scarlata CJ, Sluiter AD, Templeton DW. Compositional analysis of lignocellulosic feedstocks. 1. Review and description of methods. *J. Agric. Food Chem.* 2010;58:9043-53.

[77] Zhao X, van der Heide E, Zhang T, Liu D. Delignification of sugarcane bagasse with alkali and peracetic acid and characterization of the pulp. *Bioresource* 2010;5:1565-80.

[78] Sluiter A, Ruiz R, Scarlata C, Sluiter J, Templeton D. Determination of extractives in biomass. Technical report NREL/TP-510-42619. National Renewable Energy Laboratory (NREL), Colorado, USA; 2008.

[79] Jenkins BM, Baxter LL, Miles Jr. TR, Miles TR. Combustion properties of biomass. *Fuel Process. Technol.* 1998;54:17-46.

[80] Baxter LL. Ash deposition during biomass and coal combustion: a mechanistic approach. *Biomass Bioenerg.* 1993;4:85-102.

[81] Miles TR, Miles JTR, Baxter LL, Bryers RW, Jenkins BM, Oden LL. Alkali deposits found in biomass power plants. A preliminary investigation of their extent and nature. Report of the National Renewable Energy Laboratory (NREL/TZ-2-11226-1; TP-433-8142), Golden, CO, USA; 1995.

[82] Parikh J, Channiwala SA, Ghosal GK. A correlation for calculating elemental composition from proximate analysis of biomass materials. *Fuel* 2007;86:1710-9.

[83] Masia AAT, Buhre BJP, Gupta RP, Wall TF. Characterising ash of biomass and waste. *Fuel Process. Technol.* 2007;88:1071-81.

[84] Theis M, Skrifvars B-J, Hupa M, Tran H. Fouling tendency of ash resulting from burning mixtures of biofuels. Part 1: Deposition rates. *Fuel* 2006;85:1125-30.

[85] Strezov V, Evans TJ, Hayman C. Thermal conversion of elephant grass (*Pennisetum purpureum Schum*) to bio-gas, bio-oil and charcoal. *Bioresource Technol.* 2008;99:8394-9.

[86] Zevenhoven-Onderwater M, Blomquist JP, Skrifvars BJ, Backman R, Hupa M. The prediction of behaviour of ashes from five different solid fuels in fluidised bed combustion. *Fuel* 2000;79:1353-61.

[87] Werther J, Saenger M, Hartge EU, Ogada T, Siagi Z. Combustion of agricultural residues. *Prog. Energ. Combust. Sci.* 2000;26:1-27.

[88] Ross AB, Jones JM, Kubacki ML, Bridgeman T. Classification of macroalgae as fuel and its thermochemical behaviour. *Bioresource Technol.* 2008;99:6494-504.

[89] Carrier M, Loppinet-Serani A, Denux D, Lasnier JM, Ham-Pichavant F, Cansell F, Aymonier C. Thermogravimetric analysis as a new method to determine the lignocellulosic composition of biomass. *Biomass Bioenerg.* 2011;35:298-307.

[90] Kim SS, Agblevor FA. Pyrolysis characteristics and kinetics of chicken litter. *Waste Manage.* 2007;27:135-40.

[91] Wei X, Schnell U, Hein KRG. Behaviour of gaseous chlorine and alkali metals during biomass thermal utilisation. *Fuel* 2005;84:841-8.

[92] Vassilev SV, Baxter D, Andersen LK, Vassileva CG. An overview of the chemical composition of biomass. *Fuel* 2010;89:913-33.

[93] Demirbas A. Determination of combustion heat of fuels by using non-calorimetric experimental data. *Energy Educ. Sci. Tech.* 1998;1:7-12.

[94] Demirbas A, Giillti D, Caglar A, Akdeniz F. Estimation of calorific values of fuels from lignocellulosics. *Energ. Source* 1997;19:765-70.

[95] Szabo P, Varhegyi G, Till F, Faix O. Thermogravimetric/mass spectrometric characterization of two energy crops, Arundo donax and Miscanthus sinensis *J. Anal. Appl. Pyrol.* 1996;36:179-90.

[96] Li X, Hayashi Ji, Li CZ. Volatilisation and catalytic effects of alkali and alkaline earth metallic species during the pyrolysis and gasification of Victorian brown coal. Part VII. Raman spectroscopic study on the changes in char structure during the catalytic gasification in air. *Fuel* 2006;85:1509-17.

[97] Anon JAR, Lopez FF, Castineiras JP, Ledo JP, Regueira LN. Calorific values and flammability for forest wastes during the seasons of the year. *Bioresource Technol.* 1995;52:269-74.

[98] Demirbas A. Relationships between heating value and lignin, moisture, ash and extractive contents of biomass fuels. *Energ. Explor. Exploit.* 2002;20:105-11.

[99] Baxter LL, Miles TR, Miles Jr. TR, Jenkins BM, Milne T, Dayton D, Bryers RW, Oden LL. The behavior of inorganic material in biomass-fired power boilers: field and laboratory experiences. *Fuel Process. Technol.* 1998;54:47-8.

[100] Mansfield SD, Mooney C, Saddler JN. Substrate and enzyme characteristics that limit cellulose hydrolysis. *Biotechnol. Prog.* 1999;15:804-16.

[101] Lee SH, Doherty TV, Linhardt RJ, Dordick JS. Ionic liquid-mediated selective extraction of lignin from wood leading to enhanced enzymatic cellulose hydrolysis. *Biotechnol. Bioeng.* 2009;102:1368-76.

In: Biomass Processing, Conversion and Biorefinery   ISBN: 978-1-62618-346-9
Editors: Bo Zhang and Yong Wang   © 2013 Nova Science Publishers, Inc.

Chapter 2

LIGNOCELLULOSIC BIOMASS FEEDSTOCK SUPPLY LOGISTICS AND SYSTEM INTEGRATION

Zewei Miao[1,2,*], Yogendra Shastri[3], Tony E. Grift[1,2] and K.C. Ting[1,2]
[1]Energy Biosciences Institute, University of Illinois at Urbana-Champaign, Urbana, US
[2]Department of Agricultural and Biological Engineering,
University of Illinois at Urbana-Champaign, US
[3]Department of Chemical Engineering, Indian Institute of Technology Bombay,
Powai, India

ABSTRACT

The success of the lignocellulosic biomass-based energy sector depends significantly on an efficient and sustainable supply system of biomass feedstock. An efficient and sustainable supply system is characterized with advanced feedstock production, low economic cost, efficient energy use, reduced consumption of natural resources, lower environmental impacts, and optimum integration in terms of the entire feedstock production-supply-conversion systems. Given the fact that lignocellulosic biomass feedstock supply is relatively new and not yet commercialized, considerable scope exists to first study and learn from past research in related fields, and then adapt and innovate for the biomass supply systems. This book chapter not only overviewed lignocellulosic feedstock supply logistics, but also discussed supply issues in related sectors in an effort to draw parallels and identify lessons which lignocellulosic feedstock supply can learn from. System integrations of lignocellulosic biomass supply chains including standardization of procedures, feedstock quality assessment, regulations and configuration of feedstock supply facility, infrastructure and equipment have been discussed as well. The underlying purpose of this chapter is to set the foundation for developing a lignocellulosic biomass supply management system for liquid biofuel production.

Keywords: Mechanical pre-processing, size reduction, densification, biomass forms, liquid biofuel production, procedure and standard

---

* Corresponding author: Phone: 217-333-1106; E-mail: zmiao@illinois.edu.

1. INTRODUCTION

In the past decade, more and more attention has been paid to liquid biofuel production to alleviate energy crisis and reduce the environmental pollution caused by fossil fuel consumption. The U.S. Biomass R&D Technical Advisory Committee, a panel established by the Congress to guide the future direction of federally funded biomass R&D, envisioned a 30% replacement of the current U.S. petroleum consumption with biofuels by 2030 [1]. The European Commission Directive 2009/28/EC has set the goal of using a minimum of 10% sustainable biofuels within the transportation sector of every member state by 2020. The Chinese central government in 2007 committed US$5 billion over 10 years to ethanol development with a focus on cellulosic technologies [2]. Japan, India and Brazil have also invested a large amount of fiscal resources to facilitate lignocellulosic biofuel production technology and commercial industry.

The success of the biofuel sector depends significantly on a sustainable, reliable and cost-effective lignocellulosic feedstock production and supply system [1, 3, 4]. The total cost of herbaceous feedstock production, processing and supply accounts for 40–60% of total cost of biofuel production for a medium or large biofuel plant based on the current supply systems [4, 5, 6]. Depending on biomass densification levels, transportation costs represent between 13% and 28% of the production and supply price of biomass [1]. Even for the 1st generation sugar cane ethanol production in Brazil, transportation cost accounts 22~40% of the total sugar cane commodity prices. Therefore, it is a priority to efficiently bridge biomass production and conversion systems of liquid biofuel production.

The advanced lignocellulosic feedstock production and supply systems include a production subsystem that requires minimum tillage and resources to produce high-quality feedstock at high-yield, and a supply subsystem that is efficient economically and environmental friendly. As the largest biomass source on the earth, the cellulosic feedstock production has been commonly classified into the production systems of dedicated herbaceous energy crops, green energy crops, agricultural residues, and forest-based woody biomass with various characters. Potential herbaceous energy crops mainly include Miscanthus (*Miscanthus x giganteus*, a natural hybrid of Miscanthus sinensis and M. sacchariflorus), switchgrass (*Panicum virgatum*), reed canary grass (*Phylaris arundinacea*), prairie cord grass, big blue stem, Indian grass and alfalfa. The perennial prairie grass feedstock is characterized with broad adaption, low bulk density and low moisture at harvest season. Green energy crops consist of energy/sugar cane, energy/sweet sorghum, as well as short-rotation woody coppice. Of them, short-rotation woody coppice usually includes poplar (the genus *Populus*), willow (*Salix spp.*), Eucalyptus (*Eucalyptus* spp.*), oak and southern pines (*Pinus* L.). Green energy crops are featured with high stress tolerance, high fiber content and moisture content at harvest season [1, 8]. Sugar cane residues enclose top leaf trash and bagasse, which features that are more or less similar to green energy crop. Agricultural residues mainly include corn stover, cotton and sunflower stalks, wheat straw, oak straw, oil-seed straw, etc. Agricultural residual yield level is much lower than that of dry energy crops. Forest-based feedstock includes wood logs, branches, twigs, foliages, debris and litter. Feedstock yield level and physical properties determine the subsequent supply system of post-harvested feedstock [7, 8].

Biomass feedstock supply system provides the necessary materials input to the conversion process for liquid fuel, power, and value-added by-products. The ConSEnt (concurrent sciences, engineering and technology) platform, which integrates agricultural, biological, physical and chemical sciences with related engineering and technology, has been applied to optimize logistics and improve overall efficiency and sustainability of biomass feedstock supply chains [6]. On the engineering and technical sides, biomass feedstock supply includes feedstock collection (or harvest), mechanical processing including size reduction and densification, handling, thermal torrefaction, storage and transportation, as well as other enabling logistics [3-5, 9, 10-11]. That is, the supply system includes the feedstock transformation and delivery from farms to a local depot facility (short-distance transportation) or from farms to a centralized storage facility and then to the biorefinery (long-distance transportation). It is a costly and yet essential procedure, for example, feedstock transport over a distance of 100 km accounts for roughly 10% of the biomass energy content [1, 8, 13-16].

The logistical complexity and barriers of biomass supply systems are characterized by a large biomass demands, a wide distribution of sources (especially for agricultural residues), weather sensitive crop maturity, narrow seasonal windows for biomass collection, competitions from concurrent harvest, postharvest processing (e.g., chopping, baling, loafing or densifying) and handling (loading and unloading) operations [10]. For instance, to meet the goal by 2030, feedstock demand will go up to one billion dry tons of cellulosic feedstock annually [1, 8], more than three times the US 2011 corn production. In other words, nearly 3 million tons of lignocellulosic feedstock must be harvested, processed, stored and transported per day. With a throughput rates of 15-20 ton h^{-1} of the large horizontal or tub grinder hammer mills, more than 6250 hammer mills will be required per day to comminute the 3-million biomass feedstock through the screen with an aperture size of 2.54 cm [1, 6, 8]. To date, the definitions of lignocellulosic feedstock quality has not yet been standardized for liquid biofuel production. Feedstock quality dynamics and control practices during storage are not thoroughly investigated yet, especially for green energy biomass. This leads to an important question of whether the current transport infrastructure, storage facility, harvest equipment and supply logistics, which are mainly for grain and husbandry productions, can meet the challenges of lignocellulosic feedstock-based energy.

These issues highlight the application of systems informatics and analysis methods to biomass feedstock supply, especially for liquid biofuel production. Such a study should not only aim to optimize the existing infrastructure, facility and supply options, but also explore and propose new alternatives that will positively affect biomass feedstock supply. Since the 2nd generation bioenergy sector is in its infancy and not yet commercialized at a large scale, there is much scope to innovate and adapt strategies that have been successfully incorporated in other sectors. It is expected that ideas and lessons learned from those sectors will be valuable contributors to an optimized biomass feedstock supply system.

The chapter is arranged as follows. The next section presents a review of the existing and potential biomass supply chains. Section 3 reviews the work in the area of supply logistics from other sectors that can conceptually contribute to the biomass supply problem. This is followed by section 4 presenting systems integration containing configuration, standardization, procedures and regulation of equipment, facility and infrastructure. The chapter ends with concluding thoughts.

BIOFUEL PRODUCTION ROAD MAP AND LIGNOCELLULOSIC FEEDSTOCK SUPPLY LOGISTICS

Lignocellulosic feedstock supply logistics are determined by the road map and pathways of bioenergy production. In other words, feedstock supply systems are driven by the requirements of bioenergy plants and constrained by feedstock types, properties, local infrastructure, and geographical and climatic conditions [6].

Feedstock supply has to address two fundamental questions: 1) what are the optimum feedstock forms to be supplied to bioenergy plants? and 2) how can the feedstock be efficiently collected and delivered from farm gates to bioenergy plant gates? The first question relates to feedstock mechanical preprocessing and transformation, which were driven not only by feedstock types and properties but also by bioenergy conversion use and technological requirement. The most commonly encountered feedstock forms are mainly comprised of bale (bundle or module) and bulk solids (e.g., powder, chips, pellet, briquette cubes, torrefied pellets and charcoal powders). However, we do not yet know what the best feedstock forms are to be efficiently supplied to end-users. Unlike the 1st generation biofuel feedstock that can be used as human food or animal feed, the majority of lignocellulosic biomass feedstock crops are exclusively for bioenergy production and has relatively low market values. Therefore, advanced lignocellulosic feedstock supply has to provide the right forms to the end-users with simple, efficient and sustainable supply logistics. The second question is associated with transportation modes, handling, storage and supply procedures, which are constrained by traffic infrastructure, storage facility, bioenergy plant capacity, land-use pattern, and geographic and climatic conditions. Feedstock handling, storage and transportation all are interrelated to biomass transformation and forms. Therefore, the two questions cannot be addressed separately but systematically.

2.1. Lignocellulosic Feedstock Use and Transformation

Lignocellulosic biomass feedstock is currently used for domestic heating, combustion or co-fire for commercial electricity power generation and steam heating [1-3, 17]. Domestic heating usually requires biomass commodities in the forms of bagged-pellets, briquettes, woody chips or bundled-firewood log. Biomass chips, pellets, briquettes or bales are often burned for commercial steam heating and electricity generation through co-firing or direct combustion [17-18]. Lignocellulosic feedstock has also been transformed to charcoal through torrefaction which can be ground into coal size particles and blended well with coal. As for commercial heating or electricity with direct combustion systems, there is a burning facility available to directly combust the whole bale in addition to pellets and briquettes to generate electric power generate electric power [17-18].

For liquid transportation fuel production, lignocellulosic feedstock is converted to ethanol, methanol, butanol, biodiesel, biocrude and methane through advanced thermochemical/hydro-thermo technologies, biochemical and/or chemical technologies. Whatever the thermochemical, biochemical or chemical pathways, however, lignocellulosic feedstock must be comminuted into bulk formats, e.g., finer particles usually ≤ 25.4-mm with pretreatment and <1-mm without pretreatment [6]. There are not yet any conversion technologies which could efficiently convert a whole plant stem or the entire feedstock bale to liquid biofuel. Hence, biomass feedstock size reduction and transformation are essential for liquid biofuel production.

2.2. Lignocellulosic Biomass Feedstock Supply Logistics

Biomass feedstock supply systems include post-harvest processing and handling (weighing, dumping, loading, unloading and various conveying operations), in-field transportation, post-harvest densification, long distance transportation from farms to storage facility and from storage facility to biorefinery accompanied by loading and unloading operations as needed. These operations are schematically shown in Figure 1. The common feature of biomass feedstock delivery logistics is to collect and deliver feedstock from many locations (farms and satellite storage) to a centralized storage site and from the centralized storage to one processor (bioconversion plant). Therefore, the feedstock transportation chain is a combination of "many-to-few" and "one-to-one" (or "many-to-few") logistics [17-20] (Figure 1).

The "many-to-few" delivery logistic has been widely recognized as the most complex activity within generic product supply management [19]. When explored in detail, the logistic of feedstock delivery and supply generally incorporates harvesting and/or collection (cutting, raking), mechanical pre-processing, handling, on-farm storage, transportation from farms to centralized storage facilities and from centralized storage facilities to the biorefinery (Figure 2) [4, 9-10, 19, 20-22].

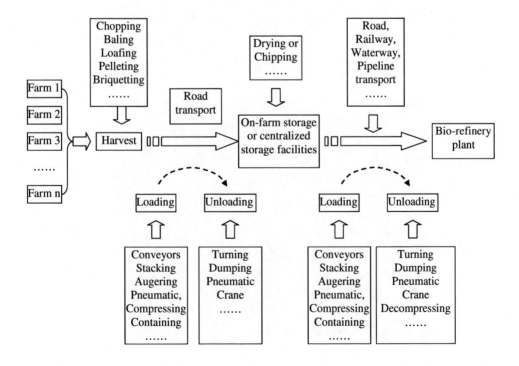

Figure 1. Schematic of lignocellulosic biomass feedstock supply chains.

Figure 2. (Continued).

(d)

(e)

Figure 2. Equipment and operation of lignocellulosic feedstock harvest (a), baling (b), in-farm transport (c), size reduction (d) and storage (e) at Energy Farm of University of Illinois at Urbana-Champaign.

Biomass feedstock mechanical pre-processing and transformation include size reduction and densification (e.g., baling, module building, cubing, pelletizing and briquetting). Feedstock handling consists of tarping (covering of bales), loading onto trailer in the field, loading, unloading, and various conveying at the storage and conversion plant [19-20]. For instance, the viable feedstock conveying methods include augers, belt conveyors, chain conveyors, pneumatic conveyors, vibrating conveyors, bucket elevator, and front end loader (e.g., crane) [25].

2.3. Lignocellulosic Biomass Feedstock Transportation Alternatives

The current transportation modes for biomass feedstock include roads, railways, waterways, pipelines and/or the combination of two or multiple modes [4, 9, 12-14]. The logistical choice of biomass delivery options depends on numerous aspects such as feedstock type and quantity, intended customer and purpose of use (e.g., liquid biofuel or combustion and co-firing), handling and processing technology, infrastructure availability, environment impact, cost, geographic feature and distance. The simplest alternative is to have direct transportation of biomass from farm to the refinery using a single mode of transport. The most likely means of biomass transportation is road transportation using tractors with trailers, trailer trucks (or truck with semi-trailers), haulage wagon or container lorry, especially for small- and medium-scale transportation requirements. This configuration has been frequently used in past work related to biomass transportation logistics [14, 15, 16, 17, 18]. Although road transportation has lower fixed costs as compared to other transportation options, it has higher variable costs, e.g., fuel consumption, driver (labor) costs, road repairmen cost, tire cost and wear cost. Based on the road load regulation of the U.S. Department of Transportation and commercial trailer dimension, the maximum bulk density of feedstock is 220-250 kg m^{-3} for road transportation. The bulk density of densified bales is able to reach the load limit but will increase the per unit baling cost.

Rail transportation usually requires a large fixed investment to build railway infrastructure between biofuel producers and farm storage units. Rail transportation, however, becomes cost-effective for longer transport distances and long-term biofuel plants involving staple and constant flows of goods due to the low variable costs, especially for logs, bales (or bundles), and industrial densified biomass type (e.g., pellet, briquette and bagged powder). Different rail wagons are available depending on the biomass type and quantity to be transported [19]. According to the rail load regulation of railway firms and rail car dimension, the maximum bulk density of feedstock ranges from 950 to 1100 kg m^{-3} for rail transportation. Therefore, rail transportation has a higher requirement for feedstock bulk density to reach maximum possible supply efficiency. In addition to the existing railway infrastructure and facility, it may be an efficient supply alternative to develop an exclusive railway between a centralized storage and processing facility (CSP) and the biorefinery plant in terms of long-term developmental strategy of a large biofuel plant.

Waterway transport is similar to rail transportation in terms of the high capital investment (e.g., to build ships or freights) but low variable costs [13]. This model of transportation is especially suitable and used for international trade of pellets and densified bales. For instance, the United States of America and Canada have exported a hay bale with a bulk density of about 400 kg m^{-3} to Japan and South Korea for animal feed.

There are numerous studies about pipeline feasibility and technology for feedstock supply. Pipeline, similar to railway, is associated with large capital investment and low per unit operating cost. In the near future, however, this supply alternative may not come into reality at commercial operational level because of its technological complexity.

Rail, water and pipeline supply alternatives actually all include road transportation for short-distance delivery of biomass feedstock from farm gates to intermediate storage and processing units. This is known as intermodal transportation which may require the development of a transport node infrastructure such as local depot or centralized distribution storage facilities of lignocellulosic feedstock.

SUPPLY CHAIN AND LOGISTICS MODELING AND OPTIMIZATION: LESSONS FROM OTHER SECTORS

The topic of supply logistics is embedded in the more general topic of material logistics and supply chain management. The concept of logistics management as a formal management discipline is often credited to Drucker [27] and has since been extensively implemented in many sectors that need to balance supply and demand, reduce transportation costs, reduce loading and hauling time and costs, and minimize the environmental impacts. This includes well known problems in the traditional field of flight and freight rail routing and process industry raw material procurement [28-31]. However, there are also some non-traditional sectors such as waste pickup and distribution [31], postal and emergency service (such as ambulance) routing [32] where these issues are very relevant. Each problem has its set of unique characteristics which call for a dedicated analysis of the problem. There are also many cross-cutting issues that are present in problems from disparate sectors. It is, therefore, prudent to explore a wide set of such problems, identify issues common with the problem at hand (biomass feedstock supply) and adopt the solution strategies that have been successful elsewhere. This section presents a brief review of a few such examples from different sectors, although certainly not an exhaustive one.

3.1. Sugar Cane

Improving sugar cane transportation and logistics from farm to a sugar mill has been an intensively researched topic in the sugar cane growing regions such as South Africa, Australia and Brazil. Apart from the seasonality issue (i.e. seasonal harvesting) typical for most agricultural products, sugar cane transportation is challenging due to its rapid post-harvest quality degradation [33-35]. Consequently, inefficiencies in transportation lead to direct financial penalties, either to the grower or the mill. The key issue identified for the sugar industry is the coordination of the harvesting, transportation and milling operations to minimize sugar cane degradation and overall cost [37-39]. Mathematical programming techniques such as linear programming and mixed integer programming have been successfully used for the analysis of sugar cane logistics. Higgins [39] formulated a mixed integer programming problem to decide the trailer pickup schedule for a set of trucks so that the queue time for the vehicles was minimized. Grunow et al. [36] used a hierarchical optimization approach where the transportation network problem was preceded by the solution of cultivation planning and harvest scheduling stages. Such a hierarchical approach had the advantage that decisions could be made using the appropriate level of modeling accuracy for different sub-systems. The transportation planning stage used optimized results from the first two stages along with constraint programming to optimize vehicle dispatch schedule. Other examples using the mathematical programming approach include [40-43]. A simulation based approach to improve the transportation logistics has been successfully used to compare scenario alternatives for performance improvement. Hansen et al. [33] developed a simulation model for minimizing harvest-to-crush delay and quantified the impacts of factors such as equipment breakdown, inclement weather and number of vehicles in the system. They used an object-oriented programming approach for the modeling system.

Stutterheim et al. [41] proposed a simplified model based on first principles that represented the complete supply chain as a single entity. They proposed the CAPCONN model for the South African sugar industry and used scenario simulations to compare different alternatives. The focus of their work was on maximizing the utilization of various equipment including harvest and transportation. Discrete event simulations have also been employed to optimize the sugar cane transportation logistics [3, 9, 43-44]. Iannoni and Morabito [42] modeled a sugar mill in Brazil and analyzed the performance in terms of three measures: truck waiting time, average unloading rates, and utilization rate of the mill. Different scenarios such as change in the truck fleet configuration were modeled and the effect on the performance measures was studied. Higgins et al. [25] argued that the optimization of the sugar cane transportation system should consider an integrated framework where all the key stakeholders (farmers, sugar mill, and transportation contractors) are incorporated. They emphasized that individual elements of this chain had been optimized in the past but that did not ensure the optimality of the complete systems. With this motivation, Thorburn et al. [26] proposed an agent based modeling approach where each stakeholder was represented as an agent. The advantage of using an agent based approach is the ability to model various interactions among the agents using simplified relations and yet reliably replicate the real interactions. Scenarios can be formulated from which to predict the consequences on the basis of modeled interactions. This review strongly points towards a focus on combining the transportation and logistics decisions with harvesting and farm management decisions, irrespective of the approach being used. Given that similar constraints in terms of harvesting horizon and biorefinery capacity will exist for the energy biomass problem, the concept of developing an integrated modeling framework similar to that for the sugar cane industry appears to be a logical approach.

3.2. Fresh Food

The logistics of the delivery chain for fresh-food products such as meat, vegetables, fruits and dairy from producers to vendors is generally complex, owing to the perishable nature of these products and the variability of the products entering the chain. It is therefore of great importance to manage the distribution chain and find transport efficiency in a practical and transparent way in order to deliver the product at the right time, while guaranteeing the desired quality level of fresh food. "Quality Controlled Logistics" (e.g., cleaning, weighing, labeling, radiation-ionizing, ozone application and packaging) has often been integrated with delivery logistics to direct the flow of goods with different quality attributes to various logistical distribution channels (with different environmental conditions) and/or different customers (with different quality demands, e.g., famers, processors, distributors, retailers, customers) [45] (Figure 3). For instance, the logistics of beef carcass delivery includes weighing, labeling, pre-chilling, refrigerating, loading, transportation, wholesale distribution, unloading and transfer to the destination cell (retailer) [46, 47]. The logistical delivery chain of the abattoir consists of activities from loading of animals at farms, transport from farm to abattoir, unloading of animals at the abattoir, and operations in the slaughter chain from lairage box to cooling room for carcasses [47]. The logistics of fruit harvesting and delivery comprises the following steps: drenching, pre-classification, waxing, classification, packaging, cold storage for overseas export transportation (air cargo) or dispatching by truck to regional or local markets [48].

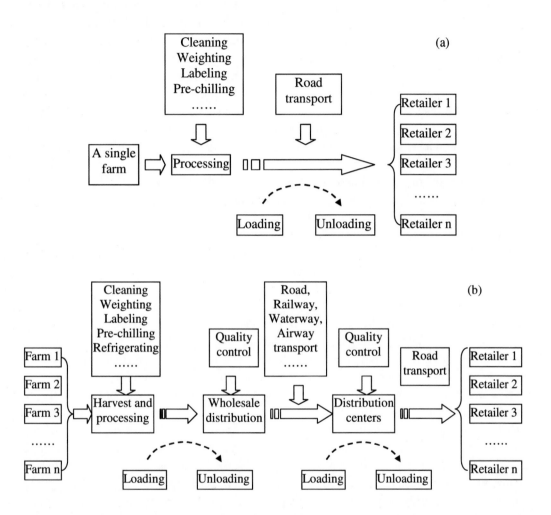

Figure 3. Flowchart of fresh food delivery chains: short distance (a) and long distance (b).

Owing to the perishable nature of fresh food, it is more important to coordinate delivery logistic chains, find the shortest path and time of vehicle travel and reduce delivery costs by using operational research and route optimizations. Road transportation is the common way to deliver fresh food in a timely manner over a short distance (e.g., restaurant food delivery services), while air cargo, rail wagon and ship have been used to deliver staple frozen dietary foods over longer distances nationally or internationally. Berruto et al. [49] developed a hybrid model with various logistic alternatives with the aim of optimizing storage and distribution chains of beef meat where discrete events and time-continuous behaviors were jointly considered. The model adopted a Markov chain to describe the meat delivery logistic movements along with the different traveling stations. At the four registration phases (pre-chilling, storage cell, transport and distribution center), a discrete time model was implemented to predict the temperature of beef carcasses and regulate the timing of the movements of beef carcasses from one cell to another [49]. Similarly, a hybrid dynamic model, which included discrete event dynamics, finite capacity constraints and time-driven dynamics, was developed to optimize the fresh-food supply chain (including pre-chilling tunnel, refrigeration cell, transportation and the destination or retailer cell) and provided solutions that were robust with respect to possible perturbation in northern Italy [44, 49]. Ljungberg et al. [47] used RouteLogiX to optimize route choice of animal and abattoir transport with respect to time, distance, road condition and traveling cost. Van der Zee and van der Vorst [44] employed an agent-based modeling approach to simulate the dynamics of the supply and transport chain of chilled salads, which included producer (stock and order processing from distribution center), distribution center (order picking from retailer, order generation to producer, food collection from producer and delivery to retailer) and retailer outlets.

In contrast to the combinatorial logistic chains of quality control and delivery management of fresh food, biomass feedstock delivery does not put biomass quality and transport time as the priorities, but emphasizes cost, environmental impacts, and transport efficiency as performance indicators. The fresh food delivery profile can span multiple supply locations to multiple demand locations (e.g., fresh vegetable to super markets), single supplier to many customers (e.g., restaurant food delivery services) or many suppliers to a centralized processing facility (e.g., fruit on-field harvest to mill). This significantly differs from the reverse logistics of biomass feedstock "many-to-few" supply. Road transportation is the typical way to deliver fresh dietary food on a timely basis over a short distance, while air cargo, rail wagons and ships may have been used to deliver staple refrigerated food over a long distance statewide or internationally. While fresh food products could be harvested frequently and supplied constantly every year, especially for greenhouse farming system, bio-energy crops (Miscanthus and switch grass) usually produce high yield but can only be harvested once within a narrow time-window per year. Therefore, on-farm satellite or centralized storage facilities are crucial in biomass delivery chains to sustain bioenergy conversion plants. Additionally, biomass feedstock supply efficiency is highly related to harvest and post-harvest processing modes (e.g., cutting, chopping, baling and loafing machinery), post-harvest densification (briquetting and pelleting), loading and unloading methods (e.g., augering, pneumatic, turning and dumping), while many of fresh food deliveries do not include compression and densification after harvest.

3.3. Cereal and Oil Crops

Cereal crop harvesting and transportation are dependent upon the intended use and are crop, season and weather sensitive [50-52]. Presently, there are two types of harvesting and transport systems for cereal (or oil) crops: the delivery of the whole crop (e.g., single pass harvesting system) or only the grain (two-pass harvesting system). The whole crop system harvests and transports the whole crop to on-farm storage (or a centralized storage) and processing plant where the crops are separated into grains, straw/stover and chaff. In comparison, the conventional crop harvesting and transportation only harvests and transports the grains and leaves the straws and chaff on the field [53-54].

For the whole crop delivery, logistic events and transport design are different from the conventional transport of only grain. Generally, the delivery chain of the whole crop system is from many supply locations to single or a few demand locations with a large supply delivery, which is similar to the biomass "many-to-few" delivery. The whole plant delivery logistics integrate the harvest and supply of stover with the well-developed grain harvest and delivery system. The stover harvest is used for heat and power, and grain is delivered to the ethanol plant or other processing facilities (e.g., food, oil, etc.) either directly from the farm store or from the depot [51-54]. The majority of logistic deliveries of corn and oilseed feedstock encompass baling, pelletizing or sealing in bins, drying if necessary, loading, transport from the field to the farm or storage facility and from storage site to the conversion facility, unloading, and ethanol/biorefinery handling process [55-58]. Delivery from the field to the farm or temporary storage facility is mainly by road transport such as tractor with wagon or truck-trailer because of the short distance and sparse distribution of supply locations. Owing to the long distance, delivery of whole crop biomass from the farm or storage to the biorefinery is influenced by many factors (e.g., postharvest handling methods, amount of biomass, distance and container availability). In Denmark, the form of the oilseeds biomass can either be in bulk or sealed in bins. The bulk seeds can be transported with container and container truck, tanker, grain or animal feed vehicle and walking floor trailer, while the sealed bins of seeds can be transported by flat bed trailers, tipper trailer or truck [47]. Sokhansanj et al. [52] created a semi-whole crop harvesting and transport model for barley, wheat and canola crops based on McLeod (McLeod Harvester Inc., Winnipeg, Manitoba, Canada) harvesting systems, where grains, chaffs and weed seeds were separated from straw when harvesting. For the whole crop supply systems, further studies on the maximum crop residues which can be harvested from farm fields are required in terms of soil production sustainability.

Conventional grain transportation delivers only grain but does not include crop stover [54, 55]. The grain delivery logistics are complex encompassing three types of delivery chains: collection and transportation from multiple farm sites to single or multiple centralized processing or storage facilities; distribution from the processing or storage location to multiple demand customers; and single-to-single staple and reliable grain delivery over a long distance. Similar to the biomass feedstock supply, storage facility is vital in the grain delivery logistics. Road transportation is the typical way to deliver grain food over a short distance, while train and ship have been used to delivery cereal goods over long distances. Apart from harvest season and market price, transport cost, environmental effects and time often are the priorities of grain delivery optimization. For example, Gebresenbet and Ljungberg [56] developed an optimization model of agricultural material and grain goods flow among farms to reduce environmental impacts. Using GPS and Route LogiX (Distribution Planning Software (1996)), the study collected, optimized and navigated 76 routes for transport of grain and grain-related commodities (e.g., fertilizer, animal fodder and seed), 15 routes for slaughter animals, 17 routes for meat distribution, 60 routes for milk collection, and 28 routes for distribution of dairy products. Farm crop yield, engine idling time, transport distance, cargo load capacity and return haulage (or back haulage), and emissions from the vehicles were some of the parameters considered in the analysis [57]. In the Netherlands, a barley-malt-beer chain delivery model included farmers, collectors, maltsters, and brewers [58]. Retailers were not included in the model. A mixed-integer linear programming model was developed to minimize the total cost including the blending, loading, transportation and inventory costs of the wheat delivery chain [59].

In comparison with cereal or oil crop transportation, biomass supply requires higher transport capacity and larger storage facilities because of high biomass yield and low energy density per unit. Thus, apart from biomass loading and unloading logistics, biomass postharvest densification and size reduction (e.g., milling and granulation) could take an important part in biomass transportation and transformation from a highly variable resource to a reliable commodity resource for bio-based industries of the future. Biomass delivery distance usually is shorter than that of cereal and oil crops because these crops can also be transported internationally.

3.4. Manufacturing/Processing

The manufacturing and process industry, driven by the globalization of supply, manufacturing and consumer sectors leading to increasingly competitive markets, represents an early practitioner of the formal supply transportation logistics concept [60-62]. The supply logistics issues for manufacturing industry are complicated owing to the presence of multiple suppliers, processing facilities, inventory locations and consumers that can be geographically dispersed [63-65]. The goal of the logistics model is to streamline the product flow while minimizing cost and maximizing returns (that accounts for profit and demand satisfaction). Given the importance of the manufacturing sector in the economy, a huge body of work researching logistics issues exists. The initial work in logistics management in this field analyzed the input side (i.e. raw material procurement, inventory etc.) and output side (i.e. product inventory, distribution and sales) independently, as transportation being a part of the analysis [66]. The subsequent effort has been to conduct an integrated analysis that accounts for the input and output sides, not only for a single production unit but also for multiple units and elements of their supply chain network [64, 65]. In recent times, there exists a trend to incorporate the traditional industry supply chain with inventory life cycle analysis networks of products to reduce industrial environmental impacts. Manzini et al. [63] classify the planning horizon for such problems into strategic (years), tactical (years to weeks) and operational (weeks and days). From the transportation perspective, the strategic level looks into issues such as fleet size determination (minimization) [64], while the tactical and operational levels analyze scheduling and routing of individual vehicle or customer [65-66]. The strategic level problems have often utilized optimization techniques (either rigorous or heuristic), while the routing and scheduling studies at the tactical and operational level have relied upon queuing theory [67], approximations of dynamic optimization [68] or variants of the famous traveling salesman problem [69]. Sarmiento and Nagi [59] present multiple possibilities of the logistics problem such as single supply and single demand location, single supply and multiple demand locations, and multiple supply and multiple demand locations. However, they do not include the possibility of multiple supply locations and single demand location, which is the scenario for biomass feedstock supply.

The above review points out a few key differences and similarities when compared with the logistics challenges for lignocellulosic feedstock supply. The manufacturing logistics problems are often solved as optimization problems, even at the operational level [70-72]. Stochastic optimization techniques are used to deal with uncertainties. This is probably because the supply and demand profiles for the manufacturing sector are not seasonal. The optimization based solutions are applicable for the complete year. On the contrary, supply profiles for agricultural products (such as energy crops) are seasonal, and therefore one sees preference towards event based simulation approaches (such as system dynamics, discrete-event simulation or queuing theory) that allow more accurate temporal representation of the system dynamics. For the manufacturing sector, supply issues are often embedded into the larger issue of the overall supply chain. An equivalent of this approach for the feedstock supply problem will be to integrate the feedstock supply logistics with the refinery operation and distribution of biofuels further downstream. It should be realized that the biofuel distribution problem will be very similar to the existing conventional fuel distribution, and hence proven solutions already exist in the literature and in practice. A significant difference for the manufacturing sector is that quality degradation of the material is often not an issue and the drive towards just-in-time delivery is more from a demand and inventory perspective. On the contrary, just-in-time delivery or high-quality storage will be more important to ensure sustained quality of feedstock supply, particularly for green energy crops. Another significant difference between feedstock supply and the manufacturing sector is that the feedstock supply delivery usually is over relatively short distances but requires a large supply at harvest seasons, while industrial products may be delivered state-wide or even globally with compact packages.

3.5. Waste Delivery System

Reuse of products and materials or recycling of garbage is known as reverse distribution, which causes goods and information to flow in the opposite direction to normal industrial logistic chain flows and activities [59, 73]. Figure 4 shows a framework for reversed distribution combining the forward flow from producer to customer, and the reverse flow from customer to producer [74, 75]. The garbage reverse distribution is the process of collection, transportation, classification, composting, recycling and disposal of used products and packages, i.e., the convergent structure of network from many sources to few demand points [76, 77]. In comparison, the traditional industrial location models typically consider a divergent network structure from few sources to many demand points [59]. Such a "many-to-few" problem is similar to the structure of the feedstock supply (biomass feedstock production) harvest-to-biorefinery chain. The waste collection and delivery are typically carried out by road transportation over short distances. The waste reverse logistic is often characterized by the higher degree of uncertainty on the supply side in terms of both quantity and quality of used products returned by the consumers, and also in terms of the more uncertain end-markets for recovered products and garbage inventory life cycle analysis [79]. These uncertainties make the garbage collection and delivery networks much more complicated because extensive transportation of low value products is more uneconomical than the centralized network structure [79]. In contrast to waste recycling systems, feedstock supply storage is pivotal to provide reliable and large quantities of biomass for biorefinery because biomass production is seasonal. In addition, the biomass production and delivery chain is dependent upon the macro-economic situation, energy markets and governmental public policy. Thus, policy and economic scenario analyses play an important role in cellulosic feedstock supply simulation.

Corresponding to the "many-to-few" network of the waste delivery chain, reverse logistic approaches have been used in garbage delivery management [79]. Some commonly-used operational algorithms including GIS-based location allocation, GIS-based routing optimization, mixed-integer linear programming, heuristic algorithms, queuing theory and neural networks were used to select optimal locations of waste disposal plants, simulate transportation routes and schedules, and calculate the amount of waste processed [73, 78-79].

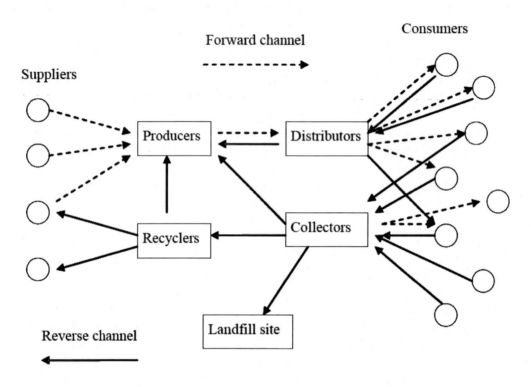

Figure 4. Framework of garbage reverse distribution (Fleischmann et al., 1997; Ghose et al. 2006).

For example, Kroon and Vrijens [71] established a mixture integer programming model to minimize the total logistic costs of container delivery in the Netherlands, which consist of distribution costs, collection costs, relocation costs and the fixed costs of the container depots. In India, Ghose et al. [73] developed a GIS-based optimal routing model to determine the minimum cost/distance efficient collection paths for transporting solid wastes to a landfill site based on information about population density, waste generation capacity, road network and the types of road, storage bins and collection vehicles. In the model, three types of vehicles were used as alternatives of transportation for calculating waste optimal path and delivery cost.

The "one-to-one" waste delivery chain is often used for delivery of large volumes of industrial waste (e.g., brewery sludge to biogas plant). For example, Rosentrater et al. [74] simulated reprocessing and transportation cost of corn masa (corn treated with either slaked lime or lye in a Mexican food process called corn nixtamalization) by-products including direct shipping to livestock feeding facilities, blending prior to shipping with a dry carrier material, extrusion processing, pellet mill processing and dehydration. This study suggested that direct shipping was by far the most inexpensive means of recycling masa processing residuals.

In summary, the waste reverse logistic delivery is carried out year-round by road transportation over short distances. Garbage classification, environmental impact analysis and sterilization are usually incorporated with the waste collection and delivery logistic chain. With seasonal harvest and low bulk density, however, biomass postharvest handling, processing, densification, and storage for quality preservation take a unique role in feedstock supply chain to provide sufficient feedstock for bioenergy conversion plant at appropriate delivery rate and efficiency.

4. SYSTEM INTEGRATION AND STANDARDIZATION OF PROCEDURES AND EQUIPMENT OF FEEDSTOCK SUPPLY

4.1. Systems Integration

In order to maximize efficiency of biomass feedstock supply, systems integration is fundamental for biofuel industry to streamline the entire chains from farm collection/harvesting through mechanical preprocessing, feedstock transformation, storage, transportation, pretreatment, conversion to liquid biofuel distribution. In the integration of feedstock supply chain, the priority is to clarify the purpose of use and requirements of biomass forms by the biorefinery [6, 8]. A variety of conventional supply management systems such as "many-to-one" systems modeling, enterprise application integration and business process management in non-bioenergy sectors can be learned to integrate the feedstock supply and conversion chains. For example, star integration (or known as Spaghetti integration) or horizontal integration seems to be more applicable to integrate lignocellulosic feedstock supply chain. However, it is prerequisite and critical for system integration to unify procedure and regulations of lignocellulosic feedstock supply and standardize harvest, processing, handling and transportation equipment and facility based on their performances [6, 8].

4.2. Standard Procedures and Regulations of Lignocellulosic Feedstock Supply

Biomass feedstock supply must follow the existing administrative regulations, legal codes and standards issued by the congress, federal, interstate agencies and professional societies for commercial transportation, commodity storage, safety and security management, fire prevention and suppression, and dust control.

However, the components infrastructure and facility management including interstate, collector and local road, railroad, barge, shield, silo (or bin) and large vertical structure should be managed and operated specifically for biomass supply and biofuel distributions. There is a need to thoroughly and quantitively investigated definitions of biomass feedstock quality, standards of feedstock assessment and classification and quality management practices during storage. Technical advances in feedstock supply system will be helpful to improve the standards and regulations of biomass feedstock supply. For example, feedstock densification can increase rail car capacity, highway trucking, rail distribution and storage efficiency by combining in-field baling with local depot processing unit or centralized storage and processing unit. Because of the features of low bulk density, low fire ignition temperature, high degradation and large-amount supply of biomass, there is a need to develop and unify standards and regulations of feedstock harvest, transport and storage security for sustainable biomass feedstock supply-conversion chains [8].

4.3. Equipment and Facility Standardization

In the "many-to-few" feedstock supply chain, equipment configuration should be standardized based on their performance to improve overall efficiency, security and safety, and sustainability of feedstock supply systems [5]. For example, round and rectangular bales are proposed to be the most common feedstock forms to be delivered for grass and short-rotation woody coppice feedstock.

However, bale size and weight varies from farm to farm as well as from crop species to species. In north America, the commonly-used feedstock or hay bale sizes include 1.14 x 0.91 x 1.14-m, 1.14 x 1.14 x 1.14-m, 1.14 x 0.76 x 1.14-m, 1.14 x 0.76 x 1.78-m, 1.32 x 0.81 x 1.93-m, 1.22 x 0.76 x 1.83-m, 1.07 x 0.76 x 1.22-m, 1.22 x 0.91 x 1.83-m and 1.37 x 0.91 x 1.52-m. For the 1.1×0.8×1.1-m, 1.2×1.2×2.4-m or 0.9×0.9×2.1-m rectangular bales, and the bale weights range from 570 to 720 kgDM and from 280 to 350 kgDM, respectively [81]. For domestic shipments, for instance, the 14.6-m trailers are commonly used, but the 15.8- and 16.2-m trailers or truck bed is frequently employed as well [81, 82].

The harvest, processing, handling equipment and transport vehicle are not yet standardized and not exclusively for biomass supply. Those inconsistences in bale size and weight attribute to the existing agricultural equipment and facility which are currently used for lignocellulosic feedstock supply. Those inconsistences will decrease the overall efficiency of mechanical processing, loading, unloading, transport and storage facilities and undermine systems integration.

CONCLUSION

The "many-to-few" lignocellulosic feedstock supply logistic comprises biomass collection, chopping, baling or pelletizing (or briquetting), unit loading, transport at short distance, unloading at storage sites as well as loading, transport at long distances and unloading at conversion plants.

The supply systems are driven by requirements of conversion technology, size of a biorefinery plant, and properties of the feedstock. The current feedstock supply systems include roads, railways, waterways, and pipelines or intermodal systems, which combine different infrastructure and facility. The supply and transportation issues have been previously addressed in many related agricultural sectors. Each sector has its set of unique characteristics, but also shares some commonalities with potential biomass feedstock sector. It is therefore, prudent to explore a wide set of such issues, identify issues common with the problem at hand (biomass feedstock supply) and adopt the solution strategies that have been successful elsewhere.

The book chapter overviewed supply logistics of sugar cane, fresh food, cereal and oil crops, manufacturing/processing and waste delivery systems. Compared to the non-bioenergy sectors, lignocellulosic feedstock supply systems are characterized by a large amount of biomass demands, narrow harvest window, a wide distribution of sources, weather sensitive crop maturity, competitions from concurrent harvest, postharvest processing (e.g., chopping, baling, loafing or densifying) and handling (loading and unloading) operations, long-duration storage.

Systems integration is fundamental for lignocellulosic feedstock supply chains to streamline the entire chain from farm collection/harvesting to mechanical preprocessing, feedstock transformation, storage, transportation, pretreatment, conversion and distribution. Except for the mathematical optimization and simulation modeling approach, it is critical for system integration to develop standard procedures of lignocellulosic feedstock supply and standardize harvest, processing, handling, storage and transportation equipment based on their performances.

Such a systems integration and standardization of procedure and equipment should lead to significant cost and energy savings, thereby contributing to the successful development of the bioenergy sector.

ACKNOWLEDGMENTS

Publication contains data produced under an Energy Biosciences Institute-funded award.

REFERENCES

[1]   USDOE. Biomass as feedstock for a bioenergy and bioproducts industry: The technical feasibility of a billion ton annual supply. Technical report, United Stated Department of Energy and United Stated Department of Agriculture 2005.

[2]   European Commission. Renewable energy-Biofuels and other renewable energy in the transport sector. Available at http://ec.europa.eu/energy/renewables/biofuels/biofuels_ en.htm (Verified on December 06, 2012).

[3]   Sokhansanj S, Kumar A. Switchgrass (Panicum vigratum, L.) delivery to a biorefinery using integrated biomass supply analysis and logistics (IBSAL) model. *Bioresource Technology,* 2007; 98: 1033–44.

[4]   Hess JR, Wright CT, Kenney KL. Cellulosic biomass feedstocks and logistics for ethanol production. Biofuels, *Bioproducts and Biorefining*, 2007; 1: 181–190.

[5]   Miao Z, Shastri YN, Grift TE, Hansen AC, Ting KC. Lignocellulosic biomass feedstock transportation alternatives, logistics, equipment configuration and modeling. Biofuels, *Bioproducts and Biorefining,* 2012; 6: 351–362.

[6]   Miao Z, Grift TE, Hansen AC, Ting KC. An overview of lignocellulosic biomass feedstock harvest, processing and supply for biofuel production (invited editorial policy). Biofuels 2013; 4: 5–8.

[7]   Sokhansanj S, Fenton J. Cost benefit of biomass supply and preprocessing. BIOCAP: A BIOCAP Research Integration Program Synthesis Paper 2006.

[8]   USDOE. U.S. Billion-Ton Update: Biomass Supply for a Bioenergy and Bioproducts Industry. USDOE Oak Ridge, TN, Oak Ridge National Laboratory: 1-227, 2011.

[9]   Sokhansanj S, Kumar A, Turhollow A. Development and implementation of integrated biomass supply analysis and logistics model (IBSAL). *Biomass and Bioenergy,* 2006; 30: 838–47.

[10] Miao Z, Grift TE, Hansen AC, Ting KC. Energy requirement for comminution of biomass in relation to particle physical properties. *Industrial Crops and Products,* 2011; 33: 504–513.

[11] Miao Z, Phillips JW, Grift TE, Mathanker SK. Measurement of pressure and energy requirements to compress Miscanthus giganteus to the bulk density of coal in railcars. *Biosystems Engineering,* 2013; 114: 21-25.

[12] Kumar A, Cameron JB, Flynn PC. Pipeline transport of biomass. *Applied Biochemistry and Biotechnology,* 2004; 113–116: 27–39.

[13] Kumar A, Cameron JB, Flynn PC. Pipeline transport and simultaneous saccharification of corn stover. *Bioresource Technology,* 2005; 96: 819–829.

[14] Khanna M, Dhungana B, Clifton-Brown J. Costs of producing Miscanthus and Switchgrass for bioenergy in Illinois. *Biomass and Bioenergy,* 2008; 32: 482-493.

[15] Venturi P, Huisman W, Molenaar J. Mechanization and costs of primary production chains for Miscanthus x Giganteus in the Netherlands. *Journal of Agricultural Engineering Research,* 1998; 69: 209–15.

[16] Huisman W, Venturi G, Molenaar J. Cost of supply chain for Miscanthus x Giganteus. *Industrial Crops and Products,* 1997; 6:353–66.

[17] Van Loo S, Koppejan J. (ed.). The handbook of biomass combustion and co-firing. Earthscan, Sterling, VA, USA. p:1–442, 2008.

[18] Georgia Institute of Technology and the Engineering Experiment Station. The industrial wood energy handbook. Van Nostrand Reinhold Company Inc. New York, NY, USA. p: 1–240, 1984.

[19] Rentizelas AA, Tolis AJ, Tatsiopoulos IP. Logistics issues of biomass: The storage problem and the multi-biomass supply chain. *Renewable and Sustainable Energy Reviews,* 2009; 13: 887–894.

[20] Ravula P, Grisso R, Cundiff J. Comparison between two policy strategies for scheduling trucks in a biomass logistic system. *Bioresource Technology,* 2008; 99: 5710–21.

[21] Pootakham T, Kumar A. A comparison of pipeline vs truck transport of bio-oil. 2007 ASABE Annual International Meeting, Minneapolis, Minnesota, 17-20 June 2007.

[22] Macharis C, Bontekoning YM. Opportunities for OR in intermodal freight transport research: A review. *European Journal of Operational Research,* 2004;153: 400–416

[23] Searcy E, Flynn P, Ghafoori E, Kumar A. Relative cost of biomass energy transport. *Applied Biochemistry and Biotechnology,* 2007;136–140: 639–652.

[24] Badger, PC, Fransham P. Use of mobile fast pyrolysis plants to densify biomass and reduce biomass handling costs—A preliminary assessment. *Biomass and Bioenergy,* 2006; 30: 321–325.

[25] Higgins A, Antony G, Sandell G, Davies I, Prestwidge D, Andrew B. A framework for integrating a complex harvesting and transport system for sugar production. *Agricultural Systems,* 2004; 82: 99–115.

[26] Thorburn P, Archer A, Hobson P, Higgins A, Sandell G, Prestwidge D, Antony G. Evaluating diversification options for sugar supply chains: Whole crop harvesting to maximize co-generation. In 7th International Conference on Management in AgriFood Chains and Networks, Ede, The Netherlands, 2006.

[27] Drucker P. The economy's dark continent. Fortune, April 1962.

[28] Matos PL, Ormerod R. The application of operational research to European air traffic flow management - understanding the context. *European Journal of Operational Research,* 2000; 123: 125–44.

[29] Boyd EA. An efficient algorithm for solving an air traffic management model of the national airspace system. *INFORMS Journal on Computing,* 1998;10: 417-426.

[30] Caron F, Marchet G. Project logistics: Integrating the procurement and construction processes. *International Journal of Project Management,* 1998;16: 311–319.

[31] Fleischmann M, Bloemhof-Ruwaard JM, Dekker R, van der Laan E, van Nunen JA, Van Wassenhove LN. Quantitative models for reverse logistics: A review. *European Journal of Operational Research,* 1997; 103: 1–17.

[32] Hoot NR, LeBlanc LJ, Jones I, Levin SR, Zhou C, Gadd CS, Aronsky D. Forecasting emergency department crowding: A discrete event simulation. *Annals of Emergency Medicine,* 2008; 52:116–125.

[33] Hansen AC, Barnes A, Lyne P. Simulation modeling of sugarcane harvest-to-mill delivery system. *Transactions of the ASAE,* 2002; 45: 531-538.

[34] Higgins A, Thorburn P, Archer A, Jakku E. Opportunities for value chain research in sugar industries. *Agricultural Systems,* 2007; 94: 611–621.

[35] Higgins AJ. Scheduling of road vehicles in sugarcane transport: A case study at an Australian sugar mill. *European Journal of Operational Research,* 2006; 170: 987–1000.

[36] Grunow M, Gunther HO, Westinner R. Supply optimization for the production of raw sugar. *International Journal of Production Economics,* 2007; 110: 224–239.

[37] Milan EL, Fernandez SM, Aragones LMP. Sugar cane transportation in Cuba, a case study. *European Journal of Operational Research,* 2006; 174: 374–386.

[38] Perry I, Wynne A. The sugar logistic improvement programme (SLIP): An initiative to improve supply chain efficiencies in the South African sugar industry. *International Sugar Journal,* 2004; 106: 559-567.

[39] Higgins AJ. Optimizing cane supply decisions within a sugar mill region. *Journal of Scheduling,* 1999;2: 229–244.

[40] Grimley S, Horton J. Cost and service improvements in harvest/transport through optimisation modelling. In Proceedings of the Australian Society of Sugar Cane Technologist, volume 19. p: 1-613, Tucson, AZ.

[41] Stutterheim P, Bezuidenhout C, Lyne P. A framework to simulate the sugarcane supply chain from harvest to raw sugar. *The International Sugar Journal,* 2008.

[42] Iannoni AP, Morabito R. A discrete simulation analysis of a logistics supply system. *Transportation Research Part E,* 2006; 42: 191–210.

[43] Arjona E, Bueno G, Salazar L. An activity simulation model for the analysis of the harvesting and transportation systems of a sugarcane plantation. *Computer and Electronics in Agriculture,* 2001;32: 247–264.

[44] van der Zee DJ, van der Vorst JGAJ. A modeling framework for supply chain simulation: Opportunities for improved decision making. *Decision Sciences,* 2005;36: 65–95.

[45] Dabbene F, Gay P, Sacco N. Optimisation of fresh-food supply chains in uncertain environments, Part I: Background and methodology. *Biosystems Engineering,* 2008;99: 348–359.

[46] Dabbene F, Gay P, Sacco N. Optimisation of fresh-food supply chains in uncertain environments, Part II: a case study. *Biosystems Engineering,* 2008;99: 360–371.

[47] Ljungberg D, Gebresenbe G, Aradom S. Logistics chain of animal transport and abattoir operations. *Biosystems Engineering,* 2007;96: 267–277.

[48] Blanco AM, Masini G, Petracci N, Bandoni JA. Operations management of a packaging plant in the fruit industry. *Journal of Food Engineering,* 2005;70: 299–307.

[49] Berruto R, Gay P, Piccarolo P, Tortia C. Hybrid models for beef meat chilling and delivering optimization. 2002 ASAE Annual International Meeting/ CIGR XVth World Congress. Chicago, IL, USA. July 28-31, 2002.

[50] Graumans CAM. Communication system barley-malt-beer chain. The 1st European Conference for Information Technology in Agriculture, Copenhagen, Denmark, June 15-18, 1997.

[51] Sokhansanj S, Mani S, Bi X. Dynamic simulation of McLeod Harvesting Systems for wheat, barley and canola crops. 2004 ASAE (American Society for Agricultural Engineering) / CSAE (the Canadian Society for Agricultural Engineering) Annual International Meeting, Ottawa, ON, Canada. Aug. 1-4, 2004.

[52] Sokhansanj S, Turhollow A. Integrating Biomass Feedstock with an Existing Grain Handling System for Biofuels. 2006 ASABE Annual International Meeting. Portland, OR, 9–12 July, 2006.

[53] Mukunda A, Ileleji KE, Wan H. Simulation of corn stover logistics from on-farm storage to an ethanol plant. 2006 ASABE Annual International Meeting. Portland, OR, 9–12 July, 2006.

[54] Petrolia DR. The economics of harvesting and transporting corn stover for conversion to fuel ethanol: A case study for Minnesota. *Biomass and Bioenergy,* 2008; 32: 603–612.

[55] Sambra A, Sørensen CG, Kristensen EF. Optimized harvest and logistics for biomass supply chain. p: 1-8. The 16th European Biomass Conference and Exhibition, Valencia, Spain, June 2-6, 2008.

[56] Gebresenbet G, Ljungberg D. Coordination and route optimization of agricultural goods transport to attenuate environmental impact. *Journal of Agricultural Engineering Research,* 2001; 80: 329–342.

[57] Busato P, Berruto R, Saunders C. Logistics and efficiency of Grain harvest and transport systems in a South Australian context. 2008 ASABE Annual International Meeting, Providence, RI. June 29 – July 2, 2008.

[58] Bilgen B, Ozkarahan I. A mixed-integer linear programming model for bulk grain blending and shipping. *International Journal of Production Economics,* 2007;107: 555–571.

[59] Sarmiento AM, Nagi R. A review of integrated analysis of production-distribution systems. IEEE Transactions 1999; 31: 1061–1074.

[60] Rudberg M, Olhager J. Manufacturing networks and supply chains: an operations strategy perspective. Omega 2003; 31: 29–39.

[61] Chen I, Paulraj A. Understanding supply chain management: critical research and a theoretical framework. *International Journal of Production Research,* 2004; 42: 131–163.

[62] Kent JJ, Flint D. Perspectives on the evolution of logistics thought. *Journal of Business Logistics,* 1997; 18: 15–29.

[63] Manzini R, Gamberi M, Gebennini E, Regattieri A. An integrated approach to the design and management of a supply chain system. *The International Journal of Advanced Manufacturing Technology,* 2008; 37:625–640.

[64] Yano C, Gerchak Y. Transportation contracts and safety stocks for Just-in-time deliveries. *Journal of Manufacturing and Operations Management,* 1989; 2: 314–330.

[65] Bell W, Dalberto L, Fisher M, Greenfield A, Jaikumar R, Kedia P, Mack R, Prutzman P. Improving the distribution of industrial gases with an on-line computerized routing and scheduling optimizer. *Interfaces,* 1983; 13: 4–23.

[66] Larson R. Transporting sludge to the 106-mile site: an inventory/routing model for fleet sizing and logistics system design. *Transportation Science,* 1988; 22: 186–198.

[67] Wu Y, Dong M. Combining multi-class queueing networks and inventory models for performance analysis of multi-product manufacturing logistics chains. *The International Journal of Advanced Manufacturing Technology,* 2008; 37: 564–575.

[68] Chandra P. A dynamic distribution model with warehouse and customer replenishment requirements. *Journal of the Operational Research Society,* 1993; 44: 681–692.

[69] Federgruen A, Zipkin P. A combined vehicle routing and inventory allocation problem. *Operations Research,* 1984; 32: 1019–1037.

[70] Kutanoglu E, Lohiya D. Integrated inventory and transportation mode selection: A service parts logistics system. Trans*portation Research Part E,* 2008;44: 665–683.

[71] Kroon L, Vrijens G. Returnable containers: an example of reverse logistics. *International Journal of Physical Distribution and Logistic Management,* 1995;25: 56–68.

[72] Fleischmann M, Bloemhof-Ruwaard JM, Dekker R, van der Laan E, van Nunen JAEE, van Wassenhove LN. Quantitative models for reverse logistics: a review. *European Journal of Operation Research,* 1997;103: 1–17.

[73] Ghose MK, Dikshit AK, Sharma SK. A GIS based transportation model for solid waste disposal–a case study on Asansol municipality. *Waste Management,* 2006; 26: 1287–1293.

[74] Rosentrater KA, Richard TL, Bern CJ, Flores RA. Economic simulation modeling of reprocessing alternatives for corn masa byproducts. Resources, *Conservations and Recycling,* 2003;39: 341–367.

[75] Dougherty M. A review of neural networks applied to transport. *Transportation Research C,* 1995;3: 247–260.

[76] Bhadeshia HKDH. Neural networks in materials science. *ISI Journal International,* 1999; 39: 966–979.

[77] Heymans BC, Oneria JP, Carriere P.E. Determining maximum traffic flow using back propagation. Proc. Int. Joint Conf. on Neural Network. Seattle, WA. 1991.

[78] Murat Celik H. Modeling freight distribution using artificial neural networks. *Journal of Transport Geography,* 2007;12: 141–148.

[79] Abdulhai B, Pringle R, Karakoulas GJ. Reinforcement learning for true adaptive traffic signal control. *Journal of Transportation Engineering,* 2003;129: 278–285.

[80] Giampietro M, Mayumi K. Complex Systems Thinking and Renewable Energy Systems. D. Pimentel (ed.), Biofuels, solar and wind as renewable energy systems. Springer Science and Business Media, B.V. Dordrecht, the Netherlands. 2008.

[81] Moore Recycling Associates Inc. Guidelines for Proper Handling, Loading, Safety and Bale Specifications. Available at http://www.plasticsmarkets.org/plastics/guidelines. html (verified on Feb 12, 2011).

[82] Miao Z, Grift TE, Hansen AC, Ting KC. Energy requirement for lignocellulosic feedstock densifications in relation to particle physical properties, pre-heating and binding agents. *Energy and Fuels,* 2013; 27: 588–595.

In: Biomass Processing, Conversion and Biorefinery   ISBN: 978-1-62618-346-9
Editors: Bo Zhang and Yong Wang   © 2013 Nova Science Publishers, Inc.

Chapter 3

Lignocellulosic Biomass: Feedstock Characteristics, Pretreatment Methods and Pre-Processing for Biofuel and Bioproduct Applications, U.S. and Canadian Perspectives

*Kingsley L. Iroba and Lope G. Tabil**

Department of Chemical and Biological Engineering, University of Saskatchewan,
Saskatoon, SK, Canada

Abstract

Renewable energy using lignocellulosic biomass is expected to become one of the key energy resources in the near future to deal successfully with global warming and depletion of conventional fossil fuel resources. This will to a large extent save our environment and human health. The challenges involved in the production of ethanol from lignocellulosic biomass must be critically addressed to enhance the digestibility and accessibility of cellulose and hemicellulose, and the subsequent conversion to simple sugars usable by fermentation yeasts. Effective conversion relies on a number of factors: the composition and structure of the feedstock, particle size, pretreatment method used, and type and loading of enzymes. The highly organized crystalline structure of cellulose poses obstacle to hydrolysis. Cellulose and hemicellulose are largely protected from enzymatic attack. This inaccessibility to attack is primarily a result of the association of these polysaccharides with lignin, which acts as a barrier, shielding the polysaccharides. Optimization of the pretreatment, hydrolysis, and fermentation processes with minimal production of inhibitors will contribute immensely to an efficient and cost-effective biorefinery industrial process. To mitigate the difficulty created by the bulky, loose, and disperse nature of lignocellulosic biomass, there is a need for pre-processing and densification for easy, economical, and efficient transportation, handling, and storage. It increases mass per unit volume and improves the convenience and accessibility of lignocellulosic biomass feedstock due to the uniform shape and size.

---

* Corresponding author: Lope G. Tabil lope.tabil@usask.ca.

Keywords: Lignocellulosic biomass; cellulose; hemicellulose; lignin; pretreatment; accessibility; biofuel; densification

1. INTRODUCTION

The world's major energy reliance is on fossil fuels. These energy sources are not renewable, because they take several years to be formed. For decades, the use of fossil fuels has been a major factor of environmental degradation and pollution. The heavy dependence on petroleum is extremely a serious energy security issue; there will be an increase in energy demand with the geometric increase in world population, and there is also the fear of depletion in the supply of petroleum [1]. The dominant energy structure reliance on coal in most parts of the world causes serious environmental problems. To reduce the dependence on fossil fuels, there is tremendous interest and emphasis towards sustainable and environmentally friendly sources of alternative fuels [2].

The present chapter aims to examine the economic and environmental benefits of biofuels and review the work that has been done in the past on biomass feedstocks. This chapter also sheds light on some of the practical approaches and methods that have been studied and adopted in the pretreatment, pre-processing, and densification of lignocellulosic biomass prior to the production of lignocellulosic biofuel. The main bottlenecks involved in lignocellulosic biomass processing, which do not encourage cost-effective and competitive production of biofuel are also discussed.

1.1. Fossil Fuel and the Environment

There is a general consensus in the scientific community that global climate change is caused by forced warming from greenhouse gases (GHGs) and due to the depletion of the ozone layer in the stratosphere, which acts as shield to the ultraviolet radiation from the sun [2]. The United Nation's Intergovernmental Panel on Climate Change [3] reported that warming of the climate system is a result of the observed increase in anthropogenic GHG concentrations. In the last century (20th century), the average global temperature increased by $0.74°C$ and sea level rise amounted to 17 cm. This is due to the thermal expansion of the ocean and melting of ice across the world [3]. According to long term forecasts, if this increase in temperature continues, the world will experience more floods, storms, heat waves, and droughts. This exposes low lying coastal countries with land surface about a meter or two above sea level to serious danger with life-threatening consequences [3]. Major GHGs include carbon dioxide, methane, and nitrous oxide. The present atmospheric CO_2 increase is primarily connected to anthropogenic emissions. The burning of fossil fuels accounts for about three-quarters of the overall GHG emissions [4]. Between 1980 and 1989, human activities released an estimated 7.1 billion tonnes of carbon into the atmosphere annually as carbon dioxide. Majority of this, 5.5 billion tonnes per year on average resulted from the burning of *fossil fuels*, with a minor contribution from cement production. The burning and clearing of tropical forests accounted for much of the remaining 1.6 billion tonnes [4]. Increasing the use of fossil fuels to meet global energy demand will significantly increase the

GHG emissions. In addition to global warming, burning of fossil fuels is a major contributor to acid rain. The high sulfur and nitrogen content in fossil fuels form sulfur dioxide and nitrous oxides when combusted and the resulting acid rain can damage fresh water sources, forests, soils, and buildings and adversely affect human health [5].

1.2. Biomass as a Potential Source of Energy

To overcome the aforementioned problems, there is a need for a more reliable, sustainable, renewable, and environmentally friendly energy source. Lignocellulosic biomass (LB) feedstock is one potential source of alternative energy. Due to the enormous quantity, availability, and renewable characteristic of LB, the efficient utilization as energy source has been receiving great attention. LB has annual production of approximately 200 billion tonnes worldwide [6], and is considered as the best option with the highest possibility to add to the energy needs of the present society and ensure fuel supply in the future for both developing and industrialized nations [7]. In other to avoid the competition between food and energy, current attention has been focused on agricultural residues (non-edible portion) as source of bioenergy and bioproducts, since they are readily available worldwide at low cost [8-10]. Therefore, it is of paramount importance to fully harness the potentials of LB. Other alternative renewable energy sources are geothermal, hydroelectric, solar, and wind. These are important components of an environmentally sustainable long-term solution to partially substitute the use of fossil fuels. Using these alternative sources of energy will enhance diversity, energy security, sustainability, and improve air quality, as well as mitigate the harmful effects of greenhouse gases [11].

Lignocellulosic biomass refers to the organic matter which comprises agricultural crop residues (such as straw/stalk, stover, and bagasse), forest residue (woody biomass such as sawdust), agricultural dedicated energy crops (short-rotation woody crops such as poplar, willow, switchgrass), municipal solid waste, animal wastes, waste from food processing, and aquatic plants and algae. These are the most important biomass energy sources which can be used for energy and in a number of different applications [12-13].

The carbon dioxide released during the combustion process or usage of biofuel is compensated by the carbon dioxide intake by plants from the atmosphere during the process of photosynthesis over their growth cycle, which converts light energy to chemical energy and stores it as carbohydrates as shown below:

$$6CO_2 + 6H_2O \rightarrow C_6H_{12}O_6 + 6O_2 \tag{1}$$

Hence, bioenergy have the potential for significant carbon sequestration to mitigate the effects of global warming. Thus, they are classed as a carbon-neutral fuel. Above all, there is no fear of depletion; rather it will encourage farmers to produce more food with the intention of making more profit from the residues, and at the same time creating more job opportunities and providing regional economic development [12, 14].

1.3. Ethanol Benefits and Considerations

Besides the benefits mentioned previously, bioethanol have some other inherent advantages over gasoline. Table 1 compares the fuel properties of bioethanol and gasoline. Flash point (the lowest *temperature* at which a material can vaporize to form an ignitable mixture in *air*) of bioethanol (12.78°C) is much higher than that of gasoline (-42.78°C) [15]. This is an indication of the safety level of bioethanol usage. Gasoline is a dangerous liquid in a vehicle crash, since it can explode or ignite. Producing, refining, transporting, and using gasoline leads to large environmental discharge of pollutants.

Table 1. Fuel property comparison for ethanol and gasoline [15]

Property	Ethanol	Gasoline
Chemical Formula	C_2H_5OH	C4 to C12
Molecular Weight	46.07	100–105
Carbon	52.2	85–88
Hydrogen	13.1	12–15
Oxygen	34.7	0
Specific gravity, 15.56°C/15.56°C	0.796	0.72–0.78
Density, kg/L @ 15.56°C	0.79	0.72-0.78
Boiling temperature, °C	77.78	26.67–225
Reid vapor pressure, kPa	15.86	55.16–103.42
Research octane no.	108	90–100
Motor octane no.	92	81–90
(R + M)/2	100	86–94
Cetane no.(1)	--	5–20
Fuel in water, volume %	100	Negligible
Water in fuel, volume %	100	Negligible
Freezing point, °C	-114	-40
Dynamic viscosity, 15.55°C	1.18×10^{-3}	$(3.7–4.4) \times 10^{-3}*$
Flash point, closed cup, °C	12.78	-42.78
Auto-ignition temperature, °C	422.78	257.22
Specific heat, kJ/kg K (C_p)	1.88	2.22

*Calculated

The *auto-ignition temperature* (which does not require an ignition source) of bioethanol (422.78°C) is also much higher than that of gasoline (257.22°C). This also shows that bioethanol is reliably safer to consumers than gasoline. The characteristic temperature at which liquids turn into solids (known as freezing point) is much lower than that of the gasoline. This shows that bioethanol can be more effectively used in the cold/artic regions of the world without impeding engine performance. The motor octane number of bioethanol (108 and 92, respectively) is higher than gasoline (90-100 and 81-90, respectively). The higher the octane number, the more compression the fuel can withstand before detonating. In principle, fuels with a higher octane rating are used in high-compression engines that generally have higher performance. The use of gasoline with low octane numbers generally results to *engine knocking* [15]. A liter of bioethanol contains less energy than a liter of gasoline. This depends on the blend. E85 (85% bioethanol/15% gasoline on a volume basis) has about 27% less energy per liter than gasoline. However, because bioethanol is a high-octane fuel, it offers increased vehicle power and performance [16]. Sheehan et al. [17] reported that for each kilometer fueled by bioethanol share of E85, the vehicle consumes 95% less petroleum as compared to a kilometer driven in the same vehicle on gasoline. Greenhouse gas emissions (fossil CO_2, N_2O, and CH_4) on a life-cycle basis are reduced by 113% using E85. Displacing gasoline with E100 reduces fossil CO_2 emissions by 267 g of CO_2/km [17].

1.4. Biomass Resources

1.4.1. A Canadian Focus

Sokhansanj et al. [18] studied the production and distribution of cereal straw (wheat, barley, oats, and flax) in the Canadian prairies over a period of 10 years (1994-2003). They reported that the provinces of Alberta, Saskatchewan, and Manitoba collectively produced approximately 37 million tonnes of wheat, barley, oat, and flax straw annually. After grain harvest, some amount of the available straw is always left on the field to maintain soil health and fertility and to avoid soil erosion (water and wind). Excess straw available can be used in a sustainable way for other economic purposes like livestock feeding and bedding, while the rest is bunched and the piles are burn on the farm land. The net average available straw from the prairies that was burnt annually after satisfying soil conservation and livestock requirements is estimated at 15 million tonnes, which can sometimes be as high as 27.6 million tonnes and as low as 2.3 million tonnes depending on the year's harvest [18]. The results of this study show that a good portion of crop straw resources is lost, with serious environmental pollution. This implies that straws (agricultural residues) existing in the waste streams from farms and commercial crop processing plants have little inherent value, which traditionally presents a disposal problem. Lignocellulosic agricultural biomass residues represent an abundant, inexpensive, and readily available source of renewable LB [8, 19]. Lafond et al. [20] studied the effects of straw removal from the field through baling and measured the long-term impact on soil quality and wheat production on the Canadian prairies. It was reported that there was no measurable impact on the amount of soil organic carbon and soil organic nitrogen. These researchers concluded that there is great potential to harvest cereal residues with a baler for ethanol production or other industrial purposes without having negative effects on the soil quality and productivity, only if acceptable and satisfactory soil and crop management practices are applied.

Canadian governments both at the national level and within specific provinces (including Alberta, British Columbia, Maritimes, Ontario, Quebec, and Saskatchewan) have shown considerable support and efforts toward developing and promoting an industry-led approach to the development of the bioproducts sector [1]. According to Saskatchewan Ministry of Agriculture [21], Saskatchewan alone produced 12.44 million tonnes of wheat, 5.63 million tonnes of canola, 4.59 million tonnes of barley, and 2.3 million tonnes of oat in 2008. The Food and Agriculture Organization [22] reported that Canada produced about 21 million tonnes of wheat, 11 million tonnes of barley, and 5 million tonnes of oat in 2007. It should be noted that straw yield depends on the specific varieties and is mostly affected by agronomic, environmental, and climatic factors. Montane et al. [23] cited that an average ratio of 1.3 kg of straw per kg of wheat grain harvested is obtained for most common varieties of wheat. The issues of biodiversity and soil conservation, however, will limit the removal of agricultural residues.

Canada plays a very significant role in the development of biorefinery sector because of its contribution as a predominant supplier or source of biomass feedstock. Canada has 402.1 million hectares of forest and other wooded land which represents 10% of the world's forest cover and 30% of the world's boreal forest [24]. There is great potential for the forestry sector to supply part of or the entire feedstock requirement for bioenergy, biofuel, and bioproduct sectors. Canadian forests predominantly supply more than 200 million m^3 of biomass annually via commercial operations. This makes Canada the second largest supplier of woody lignocellulosic biomass in the world, behind the United States [24, 25].

The relatively high availability of Canadian biomass is an indication that Canadian agriculture and forestry sectors has great potential of leading the world in the development of biorefinery sector, while meeting food and feed demands. This can generate an additional economic activity by potentially creating both direct and indirect job opportunities while making agriculture and forestry more stable and sustainable industries, and enhance tax revenue generation for provincial and federal governments. Hence, it is of paramount importance to optimize the extraction process of the necessary sugars for the production of bioethanol.

1.4.2. Biomass Resources in the U.S.

The oil crisis of the 1970s triggered an interest in the United States (U.S.) in building domestic and renewable energy resources that could decrease the country's reliance on non-renewable and foreign energy supplies. Recently, in the U.S., the use of biofuels for transportation has become subject of intense policy debate and action as a result of combination of the following factors: a) unstable political environment in oil producing countries that do not have good relationship with the U.S.; b) fluctuating energy prices; c) increasing global demand for fossil fuels resulting from the geometric increasing population; and d) issues of global warming resulting from the use of fossil fuels. More than 50% of the oil consumed in the U.S. is from foreign sources and about two-thirds of U.S. petroleum demand is in the transportation sector [15-17, 26].

In the U.S., bioethanol is produced primarily from starch in corn kernels. In 2002, 32 million hectares of corn grain were planted in the U.S. This amounts to a potential savings of roughly 740,000 barrels of crude oil/day, a savings of less than 4% of the current U.S. demand for oil [17]. However, it offers the possibility to provide oil savings which is comparable to the 800,000 barrels/day in foreign oil savings projected for the opening of the Alaskan Natural Wildlife Reserve to oil drilling and exploration [17, 27].

In 2005, majority of the 15.14 billion liters of bioethanol produced came from 13% of the U.S. corn crop (36 billion tonnes of corn grain). In 2006 and 2007, the bioethanol produced in the U.S. exceeded 18.92 billion and 41.64 billion liters, respectively, and estimated to have used over 20% of domestic corn production [28]. America uses close to 511-530 billion liters of gasoline a year [29]. The Renewable Fuels Association's [29] calculated that in 2011, the bioethanol industry replaced the gasoline produced from imported oil by more than 48.15 billion liters, which represents about 25% of domestically produced and refined motor fuel for gasoline engines. With the increase in corn use by the bioethanol industry, a serious concern has developed with respect to bioethanol's effect on corn price. Corn is a second staple food in some countries; therefore, the use of corn grains for energy creates competition between food and energy. Looking at the numbers and impacts that corn makes, it is clear that corn is not the sole answer to U.S. energy security problems. The U.S. needs to explore more of the agricultural residues (lignocellulosic biomass) to avoid such competition in future. In 2010, the U.S. produced about 316 Mt (million tonnes) of corn, 60 Mt of wheat, 4 Mt of barley, and 1.2 Mt of oats [30]. Ethanol from lignocellulosic biomass is considered to be more promising from a sustainability perspective, because of the significantly lower life cycle greenhouse gas emissions compared to grain ethanol [16, 26, 31]. Perlack et al. [32] in 2005 estimated that up to 823 Mt of cellulosic biomass could be collected from agricultural land within the next 35-40 years in the U.S. The various existing sources of cellulosic biomass (lumber industry waste and woody biomass, forage crops, animal manure, industrial waste, municipal solid waste (MSW) [trash or garbage which emanates from homes, schools, hospitals, and businesses], and crop residues) are the potential candidates for the bioenergy industry. The most abundant sources of biomass currently available for the production of bioethanol in the U.S. are crop residues. These could contribute about 446 million dry tonnes of biomass, which will amount to about 54% of the total cellulosic biomass sources in the U.S. [17, 32]. In 2010, about 250 Mt of MSW was generated in the U.S. [33]. In 2003, Graham [34] reported that the U.S. produced 521 million dry tonnes of agricultural waste (urban wood waste, mill waste, crop and forest residues, black liquor/pulp by-product, manure and bio-solids). The statistics above shows that the U.S. has a good potential for bioethanol production.

2. STRUCTURE AND COMPOSITIONS OF LIGNOCELLULOSIC BIOMASS

2.1. Challenges Associated with Lignocellulosic Biomass

Despite all the enormous benefits of using LB as bioenergy source, there are also some challenges that need to be overcome so as to make the process economically feasible and lucrative for investors, as well as to make biofuel easily affordable by the consumers. Lignocellulosic biomass is a complex formation of cellulose, hemicellulose, and lignin. The lignin acts as an external crosslink binding hemicellulose and cellulose with cellulose positioned at the inner core of the structure [35-36]. The lignocellulosic structure gives mechanical strength to plant cell walls and makes it naturally resistant to the microbial and enzymatic degradation (figure 1). This resistance is generally known as "biomass recalcitrance." It is this characteristic that is mainly responsible for the high cost of lignocellulosic conversion [35-36]. The physical, chemical, thermal, and biological processes of conversion of biomass yield a number of products. Biomass can either be used directly for combustion and co-firing for home heating, providing process heat for industrial facilities, generation of electricity, or application as liquid and gas fuels in the form of bioethanol or biogas, respectively, as well as a source of variety of bioproducts [12]. Bioproducts generated from biomass may also be utilized as means for increasing economic security. The growth of the bioenergy industry is closely tied to the availability of the technological path taken in the conversion of biomass-to cellulosic ethanol [12].

In general, lignocellulosic biomass-to-cellulosic ethanol conversion processes includes: (a) initial particle size reduction; (b) pretreatment; (c) densification (optional, depending if the feedstock need to be transported); (d) enzymatic/acid hydrolysis/saccharification; (e) fermentation; and finally, (f) distillation process.

2.2. Chemical Structure and Composition

The quantity and ratio of each chemical constituent varies depending on the type of LB. Based on mass percentage, cellulose and hemicelluloses are higher in hardwoods compared to softwoods and crop straws; the lignin content of softwoods is generally higher than that of hardwoods [35-36]. Straws have a high percentage of extractives, which will be discussed later in this section. Lignin is the binding agent that gives plants rigidity and holds it together. Hemicellulose and cellulose are complex carbohydrates bound within the cells of the plants, and these substances have the potential for bioethanol production [35].

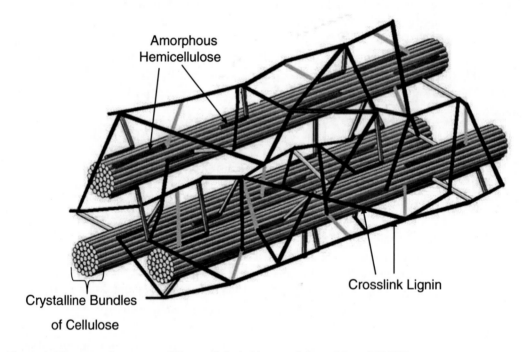

Figure 1. The general structure of lignocellulosic biomass (adapted from [37-39]).

2.3. Cellulose

Fan et al. [35] reported that cellulose content varies between species in the range of 40-50%. It is the most abundant source of carbon in biomass and on earth. Cellulose is a skeletal polysaccharide that is widely distributed in the cell wall component of wood, agricultural, and municipal cellulosic waste. It exists as an un-branched chain, a fibrous, tough, water-insoluble substance found in the cell walls of plants [35-36, 40]. Cellulose is a linear homopolymer of D-glucose residues linked by β-1,4-glycosidic bonds. This linearity results in an ordered packing of cellulose chains that interact via inter-molecular and intra-molecular hydrogen bonds involving hydroxyl groups and hydrogen atoms of neighboring glucose units (figure 2).

Consequently, cellulose exists as crystalline fibers with occasional amorphous regions. The proportion of crystalline fraction ranges between 50 to 90% [40]. The crystallinity of cellulose fibers is a major hurdle for efficient enzymatic hydrolysis. Also, the heavily lignified cell wall that surrounds the fiber reveals the cementing role of lignin and hinders the hydrolysis of cellulose [35]. Fan et al. [35, 41] reported that the hydrolytic enzyme attack on cellulose depends on its structural features, which includes: the surface area, the crystallinity of the cellulose, degree of polymerization, and the lignin seal surrounding the cellulose fibers, which leads to the structural resistance of cellulose [42]. For the enzyme activity to proceed during hydrolysis, the interaction between the enzyme molecules and the surface of cellulose particles need to be well established.

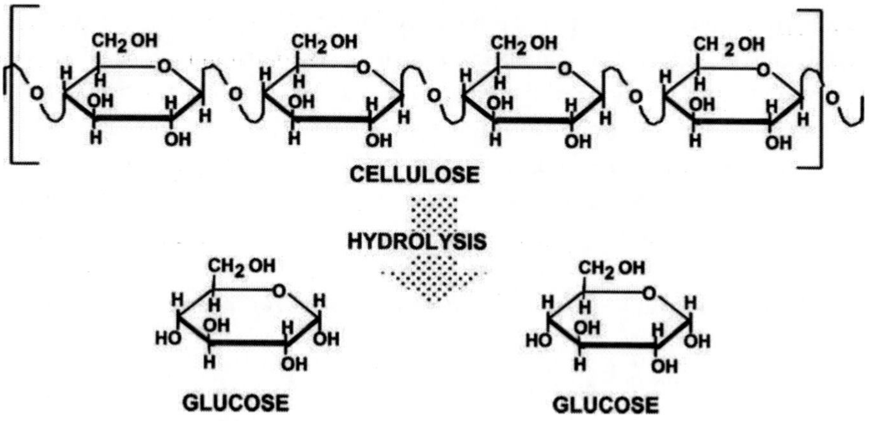

Figure 2. Structure of cellulose hydrolysis (adapted from [37-38]).

2.4. Hemicellulose

Compared to cellulose, hemicellulose is a highly branched heteropolymer, consisting primarily of five- and six-carbon sugars. It makes up 20-40% of the dry weight of LB [40]. It is composed of three hexoses: glucose, mannose, and galactose, and two pentoses: xylose and arabinose. The major hemicellulose is xylose, which consists of a xylan as the main chain (backbone) made up of β-1, 4-linked D-xylose units with the other groups mentioned above as branches [40]. The presence of the side chains minimizes hydrogen bonding. Therefore, hemicellulose has lower crystallinity and higher amorphous structure which is more easily hydrolyzed unlike cellulose. The C5 and C6 sugars are linked through glycosidic bonds, forming a loose and very hydrophilic structure. Hemicellulose is chemically bonded to lignin and it serves as an interface between the lignin and the cellulose [40-42].

2.5. Lignin and its Role in Enzymatic Hydrolysis

Lignin has very high molecular weight and it is a cross-linked complex aromatic macromolecule. It is hydrophobic and highly resistant to chemical and biological degradation [40]. It acts as cement among the plant cells, and in the layers of the cell wall. It provides the structural rigidity to plants. Forming together with hemicellulose, an amorphous matrix in which the cellulose fibrils are embedded and protected against biodegradation. Lignin content and composition vary among different plant groups [40-41]. Moreover, the lignin composition varies between the different wood tissues and cell wall layers. It contains about 50% more carbon than cellulose. It is present in cell walls of all vascular plants. The spaces between cellulose fibrils and hemicellulose and pectin are filled with lignin, therefore, it acts as binder among cell wall components [40]. Fan et al. [35, 41] reported that the mixture of lignin and the crystalline cellulose produces one of nature's most biologically resistant material. As a result, microorganisms find it very difficult to degrade the cellulose because of the covalent chemical bonds and hydrogen bonding that exists between the constituents. Lignin is a three-dimensional polyphenolic network built up of dimethoxylated (syringyl), monomethoxylated (guaiacyl) and non-methoxylated (p-hydroxyphenil) phenylpropanoid units, derived from the corresponding p-hydroxycinnamyl alcohols, which give rise to a variety of sub-units including different ether and C-C bonds. This three dimensional networks of lignin surrounding the cellulose hinder easy accessibility to enzymatic degradation. It makes up 10-25% of the dry weight of biomass [35, 40-41]. The composition of different lignocellulosic material is presented in Table 2. The separately recovered lignin after fermentation can be used for bioproducts and co-production of power and heat. Figure 3 shows a biomass-to-ethanol flowchart:

Table 2. Cellulose, hemicellulose and lignin content of different lignocellulosic biomass

Lignocellulosic materials	cellulose	hemicellulose (%)	lignin	References
Barley straw	40	20	15	a
Coastal bermudagrass	25	35.7	6.4	b
Corn cobs	45	35	15	b
Corn stalks	35	15	19	a
Cotton seed hairs	80-95	5-20	0	b
Grasses	25-40	35-50	10-30	b
Leaves	15-20	80-85	0	
Newspaper	40-55	25-40	18-30	
Nut shells	25-30	25-30	30-40	b
Oat straw	41	16	11	a
Paper	85-99	0	0-15	
Primary wastewater solids	8-15	NA	NA	
Rice hulls	36	15	20	a
Rice straw	32	24	13	a
Saw dust	55	14	21	a
Solid cattle manure	1.6-4.7	1.4-3.3	2.7-5.7	
Sorghum straw	33	18	15	a
Sorted refuse	60	20	20	
Swine waste	6.0	28	NA	
Switchgrass	45	31.4	12	b
Wheat straw	30	50	15	b
Waste papers from chemical pulps	60-70	10-20	5-10	
Hardwood:				c
Birch	38.2	19.7	22.4	
Willow	43.0	29.3	24.2	
Softwood:				c
Pine	46.4	22.9	29.4	
Spruce	43.4	18	28.1	

([40] (a); [43] (b); [44] (c)).

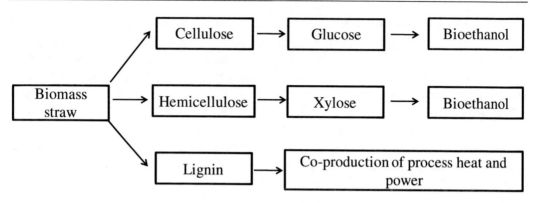

Figure 3. Lignocellulosic biomass-to-ethanol flowchart.

2.6. Extraneous Materials

In addition to the three main components, LB also contain some extraneous compounds with smaller molecular size, and available in little quantity. These include extractives and non-extractives. Agricultural residues contain little more above 8% extractives on dry weight basis. These are waxes, fats, gums, starches, alkaloids, resins, tannins, essential oils, and many other cytoplasmic constituents. Some extractives are toxic and this is an advantage of wood to resist attack by fungi and termites. The non-extractives in agricultural residues may be up to 10% of the dry weight, which includes inorganic compounds such as silica, carbonates, oxalates, etc. [40].

2. PRETREATMENT OF LIGNOCELLULOSIC BIOMASS

Pretreatment is generally aimed at breaking down the lignocellulosic matrix, loosening the highly crystalline structure of cellulose and shifting it to the amorphous state. It also increases the porosity of the biomass [45]. The cost of ethanol production from LB is relatively high based on current technologies, and the main challenges are the low yield and high cost of the hydrolysis process [45]. Considerable research efforts have been made to improve the hydrolysis of LB. Therefore, for the realization of the biofuel production potentials of LB, there are a number of challenges that need to be addressed. The enzyme-based route of hydrolysis involves biomass pretreatment, because LB is structured for strength and resistance to biological, physical, and chemical attack [46]. There is usually low yield when non-treated/native LB is subjected to enzymatic hydrolysis due to the high stability of the material to enzymatic or bacterial attacks [46]. Fan and co-workers [41] cited that most often, biodegradation of non-treated LB results in low yield of sugar from enzymatic hydrolysis, generally below 20% of the theoretical maximum.

The coexistence of lignin with cellulose in LB makes the enzymatic hydrolysis of cellulose difficult and tedious. To achieve high conversion rate, the enzymatic process then requires relatively high enzyme loadings (high cost) to produce monomeric carbohydrates that are readily fermentable by microorganisms [41, 45-46]. This low rate of conversion inhibits the establishment of an economically feasible hydrolytic process. Hence, a pretreatment process prior to enzymatic hydrolysis is necessary to significantly enhance the accessibility of the cellulosic material and adjustment of the organized structural arrangement in the LB, as shown in figure 4. Furthermore, for logistics purposes, the loose and low bulk density nature of LB requires densification. Most LB species (straws and stover) have difficulty of forming or extrusion during densification in the absence of artificial and expensive binders [47]. Past research shows that modifying the structural arrangement of cellulose-hemicellulose-lignin matrix can improve the natural binding characteristics of the lignocellulosic straw and stover biomass [47]. This implies that biomass has it own natural binder which can be harnessed through pretreatment process. Therefore, pretreatment of biomass makes it suitable for densification. Densification of biomass will be discussed in a later section.

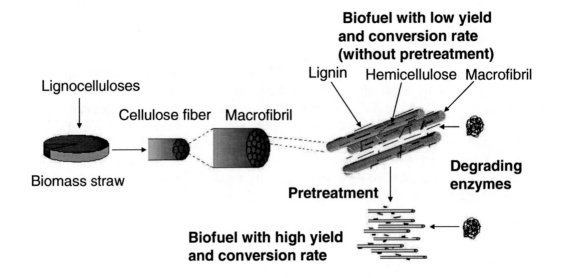

Figure 4. Effect of pretreatment on accessibility or biodegradability of lignocellulosic biomass (adapted from [48]).

A successful and cost-efficient pretreatment should satisfy the following requirements: (a) making the cellulosic fiber accessible for enzymatic activity; (b) hemicelluloses and cellulose should not be damaged; (c) prevention of the formation of substances that inhibits the subsequent activities of hydrolytic enzymes and fermenting microorganisms; (d) minimize loss of carbohydrates; (e) minimize the energy demand and the cost of pretreatment; (f) generating less residues; and (g) low operation cost such as the cost of labor, reactors, chemicals and enzymes [44, 48].

A survey of literature shows that a wide range of pretreatment methods has been investigated in recent time. Pretreatment methods are broadly classified into physical, physico-chemical, chemical, and biological based on their principal effects on the substrate [48]. Detailed descriptions of these processes have been provided by Fan et al. [41], Sun & Cheng [45], Mohammad & Karimi [48], and Mosier et al. [49]. Table 3 shows the classification of these pretreatment processes.

Table 3. Pretreatment schemes of lignocellulosic biomass (Adapted from [41, 48])

Physical	Physicochemical & Chemical	Biological
Milling: - Ball milling - Two-roll milling - Hammer milling - Colloid milling - Vibro energy milling **Irradiation:** - Gamma-ray irradiation - Electron-beam irradiation - Microwave irradiation **Others:** - Hydrothermal - High pressure steaming - Expansion - Extrusion - Pyrolysis	**Acid:** - Sulfuric acid - Hydrochloric acid - Phosphoric acid **Alkali:** - Sodium hydroxide - Ammonia - Ammonium Sulfite **Explosion:** - Steam explosion - Ammonia fiber explosion (AFEX) - CO_2 explosion - SO_2 explosion **Gas:** - Chlorine dioxide - Nitrogen dioxide - Sulfur dioxide **Oxidizing agents:** - Hydrogen peroxide - Wet oxidation - Ozone **Solvent extraction of lignin:** - Ethanol-water extraction - Benzene-water extraction - Ethylene glycol extraction - Butanol-water extraction - Swelling agents	Fungi and actinomycete

3.1. Physical Pretreatment

Physical pretreatment of lignocelluloses basically involves particle size reduction such as grinding, milling, or chopping. In general, it is classified into two categories; mechanical and non-mechanical methods [41]. For the mechanical pretreatments, physical forces are used to fractionate the LB into fine particles. Particle size reduction increases the accessible surface area of biomass particles and facilitates inter-particle bonding. It alters the crystallinity and degree of polymerization of the cellulose [41, 48]. This makes the bulk of the material to be easily degraded by enzymatic hydrolysis. The accessibility to hydrolysis is due to the increased surface-volume ratio resulting from particle size reduction [41]. The size of the material is usually 10-30 mm after chopping and 0.2-2 mm after milling or grinding [44]. The power requirement for mechanical comminution of agricultural materials depends on the final particle size and biomass characteristics [44]. In most cases, high power consumption/energy is needed to realize good accessibility to enzymatic hydrolysis. It may even be higher than the theoretical energy content that is present in the biomass [44]. Non-mechanical physical pretreatments basically expose the material to severe external forces. Such methods include irradiation, high pressure steaming, and pyrolysis [41]. Different types of physical pretreatment processes are use to improve the biodegradability of LB or enzymatic hydrolysis. Some selected types of physical pretreatment are briefly discussed in this section:

3.1.1. Ball Milling

This involves the application of shearing and compressive forces of the ball on the LB. It is an effective method of size reduction because it converts cellulose from crystalline to amorphous form, decrease the mean degree of polymerization of cellulose and hemicellulose, and increase the bulk density, thus, enhancing the enzymatic hydrolysis of LB [41]. Ball milling, as a form of physical pretreatment, has been successfully applied on cellulosic materials to enhance enzymatic hydrolysis [50]. Mais and co-workers [50] concluded that the higher the number of ball beads used in the controlled reaction volume during mechanical milling, the higher the rate of enzymatic hydrolysis of α-cellulose with a minimum enzyme loading. Fan et al. [41] cited that the rate of hydrolysis conversion is a function of the ball milling time. Ball milling effectiveness varies from material to material, with soft woods (pine, spruce) showing the least response, and hard woods (aspen, birch) showing the best response. This makes this method of physical pretreatment not a generally acceptable one.

3.1.2. Two Roll Milling

The set up consists of two cast-iron tempered surface rolls, positioned horizontally; the gap between the rolls can be adjusted by screws [41]. Recently, this method has been applied on LB to enhance the digestibility and accessibility. It reduces the crystallinity and the degree of polymerization of polysaccharides; its effects on lignin are not well understood in the scientific community [41]. Fan et al. [41] noted that the clearance between the roll mills and the duration of milling are the two parameters that control the accessibility of enzymatic attack. Fan et al. [41] reported that there was a considerable decrease in the crystallinity index and a drastic increase in the accessibility of cellulose after compression milling.

3.1.3. Extrusion and Expansion

These are similar to high pressure steaming. This pretreatment involves moist-heat expansion (extrusion) and dry-heat expansion (popping). They are used to increase feed efficiency of cereal grains in animal feeds [41]. Fan et al. [41] noted that the application of extrusion pretreatment is not effective in enhancing the digestibility of rice straw and sugarcane bagasse.

3.1.4. High Energy Radiation

The accessibility and digestibility of LB can be enhanced by the application of high energy radiation such as gamma rays, electron beam, and microwaves [48, 51]. This process results in dehydrogenation, oxidative degradation, destruction of anhydroglucose units to produce carbon dioxide, and cleavage of cellulose. Beardmore et al. [52] reported that gamma radiation of solka floc increases the surface area due to extensive depolymerization. However, it was concluded that this method is not effective in converting the crystalline to amorphous form of cellulose. Fan et al. [41] reported that the optimum accessibility and digestibility of wheat straw was realized at 2.5 x 10^8 rads using gamma radiation. Digestibility can be improved by subjecting the material to milling followed by addition of nitrate salts before irradiation, or alternatively, saturate the environment with oxygen during irradiation [53]. The high cost of this method of physical pretreatment has limited its applicability.

3.2. Physico-Chemical Pretreatment

Chandra et al. [54] reported that the combination of chemical and physical processes of pretreatment is called physico-chemical processes. The most important methods of this process will be discussed in this section:

3.2.1. Steam Explosion (Autohydrolysis)

The idea of deconstruction of LB necessitated the use of pretreatment in pulp and paper manufacturing [55]. This method has received considerable attention recently in the pretreatment of LB for both ethanol and biogas production. It may involve steaming with or without explosion (autohydrolysis) [48]. In steam explosion, the pressure is suddenly reduced, causing the LB to undergo an explosive decompression [44]. It involves fractionation of LB into its major components (hemicelluloses, cellulose, lignin, and extractives) [59]. Most of the recalcitrance of LB is removed through this process, hence, increasing the degree of accessibility and digestibility of the feedstock. Montane et al. [23] applied steam explosion as a pretreatment method on wheat straw using high pressure and high temperature between 205 and 230°C and a reaction time of 2 minutes. These researchers concluded steam explosion to be an effective pretreatment process for the fractionation of wheat straw into its constitutive polymers. During the pretreatment process, increase in temperature up to a certain level (190-210°C) can effectively release hemicellulosic sugars. However, the sugar loss steadily increases by further increasing the temperature (220-230°C), resulting in a decrease in total sugar recovery [48]. Steam explosion pretreatment process is an interesting option for LB in an ethanol production scheme. Some researchers have demonstrated the advantage of adding sulfur dioxide (SO_2) to LB prior to pretreatment. The aim is to improve the recovery of both cellulose and hemicellulose fractions. The highest overall yield of glucose and xylose is obtained at optimum conditions for one-step steam explosion pretreatment of LB (corn stover and willow wood chips) for fuel ethanol production, using SO_2 as a catalyst at temperatures between 180-210°C with residence time of 4-7 minutes [55-57].

Several other researchers such as Rai and Mudgal [58] and Bress and Glasser [59] to mention but a few, have found steam explosion as a good pretreatment process. Its energy cost is relatively moderate, and it satisfies most requirements of the pretreatment process. Sun et al. [60] pointed out that special consideration should be taken in selecting the steam explosion operating conditions so as to avoid excessive biodegradation of cellulose resulting in changes of its physical and chemical properties. They reported that in a very severe condition, lesser enzymatic digestibility of LB may also be observed after steam explosion. Like in the case where there is generation of condensation or inhibitory substances between the polymers in steam explosion of wheat straw that may lead to a more undegradable residue. Therefore, it is necessary to wash with water to remove part of these inhibitory substances; however, washing with water increases the amount of wastewater from the pretreatment process [61]

3.2.1.1. Mechanism of Steam Explosion

Steam/aqueous explosion is a thermomechanochemical process. It involves the breakdown of structural components which is aided by heat in the form of steam (thermal process), occurrence of shear forces due to the expansion of moisture (mechanical process), and finally hydrolysis of glycosidic bonds/linkages of hemicellulose and alkyl-aryl linkages of lignin to produce low molecular weight fragments (chemical process) as described by Chornet and Overend [55] and Tanahashi et al. [62].

Chornet and Overend [55] described the process of steam explosion in a reactor, under high pressure, as follows: the steam penetrates the lignocellulosic structures first via diffusion. The penetration of steam into the inner structures of the LB is permitted because vapor phase diffusion is higher than the liquid phase diffusion. At high pressure, steam condenses, and subsequently wets the "capillary-like" microporous structure of LB. Hydrolysis occurs if the temperature is high enough to thermodynamically force the dissociation of liquid water and thus build up an acidic medium, which subsequently overcome the energy barriers for hydrolysis. The condensed moisture hydrolyzes the acetyl groups of the hemicellulose fractions to form organic acids such as acetic and uronic acids. These acids act as catalysts in the depolymerization of hemicellulose, releasing xylan and limited quantity of glucan. At extreme conditions, the amorphous portion of cellulose may be hydrolyzed to a certain degree. However, at excessive conditions (at high temperatures and pressures), there is degradation of xylose to furfural and glucose to 5-hydroxymethyl furfural. Furfural is an undesirable compound in fermentation because it inhibits microbial growth and activity. After the required residence time, the "wet" biomass is "exploded" upon the sudden release of the pressure within the reactor. The "wetted"/"treated" biomass is driven at high velocities and ejected out of the reactor through a small nozzle by the induced force (sudden opening of the discharge valve, i.e. blow down). Chornet and Overend [55] further reported that as a result of the sudden decrease in pressure, the following changes occur:

- instantaneous cooling due to evaporation of the condensed moisture within the lignocellulosic structure; and
- evaporation of water and expansion of the water vapor generates and exerts shear forces on the surrounding "cylindrical tubes" structure known as fibers. If the shear force is high enough, the vapor will cause the mechanical breakdown of the lignocellulosic structure.

Chornet and Overend [55] and Wang et al. [63] reported that the retention time and temperature are the parameters required for the optimization of the process. Hydrolysis of hemicellulose is of great importance to the downstream fermentation process. The degree of hydrolysis of hemicellulose by the organic acids is determined by the length of time the biomass spends in the reactor and the operating temperature of the process. However, long retention time increases the production of inhibitory substances, which must be minimized [55, 63].

The operating temperature controls the steam pressure within the reactor. The water contained within the capillaries (before discharge) at high temperature is in equilibrium with the high pressure in the reactor. Higher temperatures translate to higher pressures, which subsequently increases the difference between the reactor pressure and atmospheric pressure. The pressure difference is proportional to the induced shear force of the evaporating moisture [55].

3.2.1.2. Severity Factor

Due to the different types of available LB, severity factor has been developed to standardize the process parameters of steam explosion so as to facilitate comparisons. It relates the net product yields and pretreatment severity [55]. Effort has been in progress to minimize product degradation resulting from pretreatment conditions. This model according to Chornet and Overend [55] defines the severity of a steam explosion pretreatment with respect to the combined effect of temperature and the residence time. The model was developed based on the assumptions that the process kinetics is first-order, which follows the Arrhenius' model [55]:

$$k = Ae^{-Ea/RT} \tag{2}$$

k = rate constant
A = Arrhenius frequency factor
Ea = activation energy (kJ/kg mol)
R = universal gas constant (8.314 kJ/kg mol K)
T = absolute temperature (K)

The reaction ordinate was developed as:

$$R_o = \int_0^t \exp\left[(T_r - T_b)/14.75\right] dt \tag{3}$$

R_o = reaction ordinate

t = residence time (min)

T_r = reaction temperature (°C)

T_b = base temperature at 100°C

(14.75 is the conventional energy of activation assuming that the overall process is hydrolytic and the overall conversion is first order)

Therefore,

$$R_o = t * \exp[(T_r - 100)/14.75] \tag{4}$$

The log value of the reaction ordinate gives the severity factor which is used to map the effects of steam explosion pretreatment on LB.

$$\text{Severity} = \log_{10}(R_o) \tag{5}$$

Heitz et al. [64] reported that it is possible to directly translate results obtained from laboratory steam explosion pretreatment process into industrial practice by employing the concept of a severity factor. This is similar to those used in pulping industry which integrates temperature and the retention time into a single measure for the purpose of controlling the process [61].

Heitz et al. [64] successfully modeled the recovery trends of lignocellulosic polymers from steam explosion of *Populus tremuloides* as a function of the severity factor, by applying high pressure steam at temperatures between the range of 180-230°C and retention time of 0.7-4 min. It was demonstrated that the higher the operating conditions (temperature and time), the higher was the severity parameter (R_o), resulting in higher recovery of cellulose and lignin, which enhances the accessibility and digestibility of cellulose. However, they stated that this recovery is at the expense of increasing the destruction and degradation of hemicellulose. It was suggested that the optimum treatment severity of each of the polymer fractions would probably involve a two-stage treatment cycle, which will require the removal of the bulk of the hemicellulose and lignin after the first stage at low severity, and second post-treatment stage to obtain high purity cellulose. However, this two-stage treatment cycle may result to low degree of polymerization [64].

Kaar et al. [65] investigated steam explosion of sugarcane bagasse as a pretreatment for conversion to ethanol by applying severity factor using a range of operating temperatures (188-243°C) and residence times (0.5-44 min) in a 10 liter batch reactor. The results indicated that the processing optimal conditions of steam explosion are highly biomass feedstock dependent, because of the varying carbohydrates composition of LB feedstocks. Feedstocks high in xylose content (hemicellulose) require milder conditions, with shorter run times, than feedstocks lower in xylose but higher in glucose (cellulose) content. These researchers observed that inhibitor compound (e.g. furfural) was produced due to the extreme treatment condition of hemicellulose. However, it was concluded that the reaction ordinate concept does not appear to be universally valid. Nonetheless, the reaction ordinate remains a useful bookkeeping method of reporting steam explosion conditions.

3.2.1.3. Physical and Chemical Effects of Steam Explosion Pretreatment on Lignocellulosic Biomass

The steam explosion process is characterized by a physical rupture of the LB structure into fine fibers and/or powder by adiabatic expansion of water in small pores in LB tissues, and autohydrolysis of the cell components [62].

Tanahashi et al. [62]) investigated the effects of steam explosion on the morphological structure and physical properties of LB using shirakanba and karamatsu wood chips. The study demonstrated that the degree of crystallinity and width of micelle increased by about 1.5 and 2.0 times, respectively, as a result of the treatment, which indicates that most of the amorphous region of cellulose are transformed to crystalline region at higher steam pressure, leading to a rise in the degree of crystallinity and micelle width of the exploded LB. At moderate severities, steam explosion promotes good delignification, which is useful for the preparation of cellulose microfibrils and lignin for the subsequent stage of the bioethanol production [62]. Tanahashi et al. [66] studied the chemical effects of steam explosion on wood. It was observed that hemicellulose fractions were found to be readily hydrolyzed to oligosaccharides by steaming at lower severities (1,962 kPa, 1 min). Higher severities further hydrolyzed the hemicelluloses to monosaccharides. However, at this severity, the concentration of furfural and 5-hydroxymethyl furfural increased.

X-ray diffractograms of steam exploded poplar feedstocks revealed that higher severity treatments resulted in increasing order of the cellulose lattice structure, thereby increasing crystallinity [67]. Excoffier et al. [68] reported that the degree of crystallinity of cellulose increased due to steam explosion treatment. These researchers attributed this observation to the crystallization of amorphous regions of cellulose during steam explosion treatment process. It was also indicated that hydrolysis and depolymerization of hemicellulose and lignin, respectively, occurred during the steam explosion process. It should be noted that the softening temperature of lignin is in the range of 130-190°C [69].

3.2.2. Ammonia Fiber Explosion (AFEX)

This process allows an explosion at a relatively low temperature (compared to steam explosion) to avoid sugar decomposition [70]. AFEX is an alkaline physico-chemical pretreatment process. The LB is placed in liquid ammonia at temperature range of 90-100°C for a time duration of about 5-30 min, followed by a decrease in pressure [70]. The required parameters in the AFEX process are ammonia and water loading, time, temperature, blow down pressure, and number of treatments [70-71]. This method produces only a pretreated solid material, unlike some other pretreatments methods (for example steam explosion) that generate slurry which can be separated into fractions of solid and liquid [49]. The lignin present in the LB can be effectively modified while the cellulose and hemicellulose fractions may remain intact. The AFEX process can substantially improve the enzymatic hydrolysis if the optimum conditions are established. The optimum conditions for AFEX depend on the LB [70]. Teymouri et al. [71] studied the effect of pretreatment using AFEX on corn stover where the highest glucan and xylan conversions and ethanol yields from were achieved. It was pointed out that enzymatic hydrolysis of corn stover treated under optimal AFEX conditions

(90°C, 60% (dry weight basis), 5 min, and ammonia-corn stover mass ratio of 1:1) yields almost 98% glucan conversion and 80% xylan conversion against 29 and 16% for untreated corn stover, respectively [71]. This method of pretreatment has one major advantage; there is no formation of some inhibitory by-products/sugar monomers, unlike some pretreatment methods (steam explosion and dilute-acid pretreatments) that produce such inhibitors, such as furans [71].

However, there are some disadvantages in using the AFEX process. It is more effective on LB that contains small amount of lignin, and it does not significantly solubilize hemicellulose as compared to other pretreatment processes (like dilute-acid pretreatment). Finally, it involves a difficult recycling of the feed ammonia as a reusable gas stream after the pretreatment so as to reduce the cost and to keep the environment safe at the same time. Since ammonia is expensive, not being able to totally reuse it is detrimental to process economy. In addition, its corrosive and toxic nature may cause problems in design and operation of the process [45, 71-72].

3.2.3. Supercritical CO_2 Explosion (SC- CO_2)

This method has been mainly used for solvent extraction. It is being considered for non-extractive purposes as a result of its several advantages [73]. CO_2 is available at low cost; it is clean and environmentally friendly, and it is not difficult to recover after use [73]. To reduce the cost observed in the ammonia explosion and to avoid the degradation of sugars such as those resulting from steam explosion due to the high temperature, the SC-CO_2 explosion seems to be an alternative, since it is operated at low temperature. Researchers have investigated the use of SC-CO_2 as a pretreatment for cellulosic hydrolysis. It was shown that this process can effectively improve the accessibility of cellulose by enzymatic hydrolysis to produce glucose [73-74]. SC-CO_2 explosion pretreatment method improves the degradation of LB upon an explosive release of carbon dioxide pressure. This process increases the accessible surface area of the cellulosic substrate to enzymatic hydrolysis [75]. It was concluded that the higher the pressure of carbon dioxide, the higher was the glucose yield; an indication that carbon dioxide molecules penetrate faster into the cellulose pores at higher pressure [75].

There are several attractive benefits of using this method, these includes: non-flammability, non-toxicity, readily available at low cost, and environmental acceptability. It is highly applicable in the biochemical and pharmaceutical industries. Simultaneous application of SC-CO_2 explosion pretreatment and enzymatic hydrolysis in one step has also been carried out to obtain 100% glucose yield [76]. However, for the application of this method, the high yield of sugar depends on the type of LB. It has a low pretreatment effectiveness for softwood. The high capital cost of the high-pressure equipment may be an obstacle to the applicability and commercialization of this pretreatment method [73].

3.2.4. Microwave-Chemical Pretreatment

Microwaves fill up a transitional region in the electromagnetic (EM) spectrum between radio-frequency and infrared radiation. This portion of EM corresponds to a frequency range of 300 MHz to 300 GHz, although a large portion of this frequency range is used for radar transmissions and telecommunication. To avoid interference, almost all domestic and industrial microwaves operate typically either at 900 MHz or 2.45 GHz [77-78]. Microwave (MW) is part of the EM spectrum that results in heating of dielectric materials by induced molecular vibration as a result of dipole rotation or ionic polarization [79]. The heating process is based on volumetric heat generation. This implies that heat transfer is from inward to outward unlike the conventional heating system. Apart from the popular domestic use of MW ovens, commercialized applications include blanching, drying, sterilization, pasteurization, tempering, rapid extraction, enhanced reaction kinetics, selective heating, disinfestations, among others [79]. To widen the scope and science behind MW radiation, it has been applied in the pretreatment of LB. The microwave-chemical pretreatment is a more effective pretreatment technique than the conventional heating chemical pretreatment due to the effective acceleration of the reactivity of cellulosic material [80]. Microwave radiation has been applied in different fields of science. It has been successfully applied in the fields of organic chemistry by Larhed et al. [81] in accelerating organic transformations at a reduced reaction time. Several researchers have applied this method on LB and coming to the conclusion that it helps in breaking down lignin, while increasing its accessibility to hydrolytic enzymes. Zhu et al. [82] investigated the effects of three different microwave-chemical pretreatment processes (microwave-alkali, microwave-acid-alkali, and microwave-acid-alkali-H_2O_2) on rice straw. The results shows that rice straw pretreated with microwave-acid-alkali-H_2O_2 had the highest rate of hydrolysis and glucose content in the hydrolysate. Donepudi et al. [83] studied the effect of microwave power level and processing time with different levels of concentrations of acid and alkaline on the yield of glucose and xylose from corn stover. It was reported that increasing the microwave processing time from 0 to 5 minutes, led to a remarkable increase in glucose and xylose yields by 58.7 and 149.6%, respectively. The effect of microwave and alkali pretreatment on rice straw and its enzymatic hydrolysis was investigated by Zhu et al. [80]. The result was compared with alkali-alone pretreated process. The results indicated that rice straw subjected to microwave-alkali pretreatment had a higher hydrolysis rate and glucose content in the hydrolysate than the method that involved alkali pretreatment alone. Pretreatment with higher microwave power with shorter processing time and lower microwave power with longer pretreatment time produced almost the same effect on the weight loss and composition at the same energy consumption [80]. However, MW is known for non-uniform heat distribution, which could cause thermal runaway during the pretreatment process [77-78, 79].

3.3. Chemical Pretreatment

The cellulose hemicellulose-lignin matrix can be broken down to smaller amorphous molecules through acid or alkaline hydrolysis as well as through wet oxidation. Alkaline or acid solutions are often used for pretreatment of biomass materials. The effect of pretreatment depends on the lignin content of biomass. Chemical pretreatments are usually effective. However, they have disadvantages which should not be ignored. These include the use of specialized corrosion resistant equipment, need for extensive washing, and disposal of chemical wastes. Different chemical methods have been investigated, which includes the use of alkalis, acids, wet oxidation, gases, oxidizing agents, cellulose solvents, extraction, and swelling agents [84].

3.3.1. Acid Hydrolysis Pretreatment

Pretreatment with acid hydrolysis can result in the improvement of enzymatic hydrolysis of LB to release the fermentable sugars. H_2SO_4, HCL, H_3PO_4, and HNO_3 have been used to hydrolyze LB. Acid pretreatment can be carried out either at low temperature (e.g. 40°C) and high acid concentration (e.g. 30-70%) or at high temperature and low acid concentration [72, 84]. Dilute-acid hydrolysis has been successfully applied in the pretreatment of LB. Studies have revealed that sulfuric acid below 4% (w/w) concentrations is usually the most interesting condition to effectively pretreat at low cost [59]. Dilute acid (H_2SO_4) was used by Saha et al. [85] for pretreatment of wheat straw at varied temperature. The enzymatic saccharification was evaluated for the conversion of cellulose and hemicellulose to monomeric sugars. Higher yield of ethanol was obtained with a corresponding decrease in the fermentation time. Optimization of dilute-acid (H_2SO_4) pretreatment of corn stover using a high-solids percolation reactor was investigated by Zhu et al. [80]. They reported that the digestibility is related to the extent of xylan removal. The effect of increasing the concentration of sulfuric acid and residence time of the pretreatment process is significantly higher for Bermuda grass than rye straw for ethanol production [86]. Concentrated acids are toxic, corrosive, dangerous, and thus require reactors that are resistant to corrosion. There is also a need to recover the used concentrated acid after hydrolysis to make the process economically feasible. These factors (high investment and maintenance costs) lower the commercial interest of this process [72, 84].

3.3.2. Wet Oxidation

This method has been used as a pretreatment for the deconstruction of lignocellulosic structure for both ethanol and biogas production. In this process, LB is treated with air or oxygen and water at temperatures between 150 and 200°C for a period of about 10 min [87]. The temperature, reaction time, and oxygen pressure are the most important factors in wet oxidation pretreatment [87, 89]. Oxygen plays a very important role in the degradation reactions and allows operation at comparatively reduced temperatures by improving generation of organic acids [90]. The main reactions are the formation of acids from hydrolytic processes, as well as oxidative reactions. However, it is an exothermic process, with fast rates of reaction and heat generation, therefore, the control of reactor temperature is necessary [87, 88]. Martin et al. [89] investigated the potential of clover and ryegrass mixtures as feedstocks for ethanol production. Wet oxidation pretreatment at three levels of temperature (175, 185, 195°C), 10 min residence time, and at two different oxygen pressures with or without the addition of sodium carbonate, was evaluated. The enzymatic hydrolysis of cellulose was significantly increased. No inhibition of cellulose enzymatic conversion by the filtrates was observed. It was revealed that no addition of nutrients was required for the fermentation of clover-ryegrass hydrolysates.

3.3.3. Alkaline Hydrolysis Pretreatment

Sodium, potassium, calcium, and ammonium hydroxides are acceptable alkaline pretreatment agents [44]. This pretreatment method involves the application of some bases/alkaline solutions to remove lignin and a portion of the hemicellulose, and successfully improve the accessibility of cellulose by enzymes [42, 48]. It enhances the saccharification process. Alkaline pretreatment can be done at low temperatures but at relatively long period of time with high concentration of the alkaline [44, 48]. It has been used as a pretreatment method in biogas production [48]. The effect of alkaline pretreatment depends on the lignin content of the materials [44, 91]. Alkaline processes cause less sugar degradation unlike the acid pretreatment. Most of the caustic salts can be recovered and/or regenerated [44]. Kumar et al. [44] stated that when biomass is treated with dilute alkaline solution, swelling occurs, which increases the internal surface area of the material. Swelling causes a decrease in the degree of polymerization, separation of structural linkages between lignin and carbohydrates, disruption of the lignin structure, and reduction of cellulose crystallinity [44-45, 49]. Among the four alkaline compounds mentioned, calcium hydroxide (slake lime) is accepted to be a suitable pretreatment agent, with the lowest cost per kilogram of hydroxide [44]. This is due to the possibility of recycling the calcium from an aqueous reaction system as insoluble calcium carbonate, by neutralizing it with low cost carbon dioxide; the calcium hydroxide thereafter can be regenerated using already discovered and well accepted lime kiln technology [44]. Alkaline pretreatment plays a very important role in making cellulose accessible to enzyme attack. The removal of lignin enhances the effectiveness of enzyme by removing non-productive adsorption sites, thereby, increasing the accessibility to cellulose and hemicellulose [44].

Kim et al. [92] investigated the pretreatment of corn stover with excess calcium hydroxide (0.5 g $Ca(OH)_2$/g raw biomass) in non-oxidative and oxidative conditions. The enzymatic degradability and digestibility of lime-treated corn stover was affected by the change of structural features such as acetylation, lignification, and crystallinity resulting from the treatment. Soybean straw soaked in ammonia liquor (10%) for 24 hours at room temperature proved that this method is very effective in reducing the lignin by 30.16% [93]. However, alkaline pretreatment indicated that it is more effective on agricultural residues than on wood materials [48, 93]. Zhang et al. [6] studied the pretreatment of rice straw using 2% NaOH to remove lignin. Their findings showed that the pretreatment increased cellulose by 54.83% and decreased lignin by 36.24%, which could improve and facilitate the process of enzymatic hydrolysis. Analysis of wheat straws indicated that hemicellulose, lignin, and silica were solubilized by NaOH pretreatment. The digestibility of different structural polysaccharides was higher for NaOH-pretreated straw than the native straw. However, at lower concentrations of alkaline solution, cellulose showed resistance to solubilization, but not at higher levels (above 7% w/w). Agricultural residues were treated at 100°C for one to two hours with lime (CaO) to increase the rate and extent of dry matter digestibility. Gandi et al. [94] concluded that lime treatment approximately doubled the digestibility making it an effective method to improve and upgrade the digestibility of agricultural residues. Alkaline pretreatment leads to increase in ash content, which increases with increasing concentrations of the alkaline [58].

3.4. Biological Pretreatment

Expensive instruments or high infrastructure cost with high energy consumption are used in most pretreatment technologies discussed previously. The physical and thermo-chemical pretreatment processes especially require excessive energy for biomass pretreatment. Biological pretreatment uses different types of lignin-degrading microorganisms (rot fungi) and it is a safe and environmentally benign method [45, 95]. The method is considered to be environmentally friendly and energy saving as it is performed at low temperature and needs no chemicals for lignin removal from the LB [45]. Microorganisms such as brown-rot, white-rot, and soft-rot fungi are used to degrade lignin and hemicellulose in waste materials [45]. Brown-rot fungi usually attack cellulose, while white- and soft-rot fungi attack both cellulose and lignin [43]. These fungi are controlled by carbon and nitrogen sources. White-rot fungi are the most effective for biological pretreatment of LB [45, 95-96]. Hatakka [96] studied the biological pretreatment of wheat straw using some white-rot fungi (*Pleurotus ostreatus, Pleurotus* sp. 535, *Pycnoporus cinnabarinus*115, *Phanerochaete sordida* 37, *Phlebia radiata* 79 and *Ischnoderma benzoinum* 108) and reported that 35% of the straw was converted to reducing sugars by *Pleurotus ostreatus* in 5 weeks. However, the rate of biological pretreatment processes is far too low for industrial use, and some materials are lost as these microorganisms consume hemicellulose and cellulose, or lignin to some extent [45, 72].

The summary of the advantages and limitations of the various pretreatment methods on lignocellulosic biomass as discussed is shown in Table 4.

Table 4. Summary of merits and demerits of various processes used for the pretreatment of lignocellulosic biomass (adapted from [Fan et al. [41] and Mohammad & Karimi [48]

Pretreatment process	Advantages	Limitations
Mechanical comminution	reduces cellulose crystallinity, increase in accessible surface area and pore size, decrease in degrees of polymerization, no chemicals are generally required for these methods	power consumption usually higher than inherent biomass energy
Steam explosion	causes hemicellulose degradation and lignin transformation; cost- effective, increase in accessible surface area, decrease in degrees of polymerization, usually rapid treatment rate, among the effective and promising processes for industrial applications	destruction of a portion of the xylan fraction; incomplete disruption of the lignin carbohydrate matrix; generation of compounds inhibitory to microorganisms, there are chemical requirements, typically need harsh conditions
Ammonia fiber explosion	increases accessible surface area, removes lignin and hemicellulose to an extent; does not produce inhibitors for downstream processes, promising processes for industrial applications	not efficient for biomass with high lignin content, there are chemical requirements
CO_2 explosion	increases accessible surface area; does not cause formation of inhibitory/toxic compounds. non-flammability, readily available at low cost, and environmental acceptability	high capital cost for high-pressure equipment, it depends on the type of LB
Acid hydrolysis	hydrolyzes hemicellulose to xylose and other sugars; alters lignin structure.	high cost; equipment corrosion; formation of toxic/inhibitory substances, there are chemical requirements
Alkaline hydrolysis	removes hemicelluloses and lignin; increases accessible surface area, utilize lower temperatures and pressures, cause less sugar degradation, many of the caustic salts can be recovered and/or regenerated.	long residence times required; there are chemical requirements, it depends on the lignin content of the LB
Biological	degrades lignin and hemicelluloses; low energy requirements. no chemical requirement, mild environmental conditions, reduction in degree of polymerization of cellulose	rate of hydrolysis is very low, very low treatment rate, not considered for commercial application

4. DENSIFICATION OF BIOMASS

In its natural form, most LB are bulky, loose, and dispersed, hence they are difficult to utilize as fuel. Biomass as an energy source and as a feedstock for biorefineries does not present easy, economical, and efficient transportation, handling, and storage characteristics due to large volume requirements [97-98]. To mitigate this difficulty, pre-processing and densification of LB is offered as a solution. Biomass densification may be defined as the compaction or compression of biomass to eliminate inter- or intra-particle pore spaces. It also reduces the moisture content of biomass during compression. The main goal of densification is to increase mass per unit volume of the biomass [99].

A conventional method of pelleting/densification entails application of pressure/pressing a material through a die. The friction between the die wall and the material creates resistance to free flow of the material through the die. This resistance causes compression of the material if the die is tapered [100]. Mani et al. [100] cited that approximately 40% of the total energy applied was used to compress the materials (straw and hay), while the remaining 60% was used to overcome friction. Therefore, a good amount of the energy required to densify biomass is used to overcome friction. The same authors reported that the frictional energy could be reduced by preheating the feed or die surfaces, so as to maintain a smooth die surface, thereby reducing the time required to extrude the biomass. Pretreatment of biomass before densification would also considerably improve the compression property of the material resulting in lower energy consumption.

As summarized by Tabil [98], upon the application of pressure, there are basically three main stages involved during the densification process: a) particle rearrangement; b) elastic and plastic deformation; and c) mechanical interlocking of particles and local melting of some constituents:

a) Particle rearrangement and breaking down the initial unstable packing arrangements is the first stage of densification, which is often called 'arches' or 'bridges'. This stage happens at low pressures resulting to a closer packing.

b) Elastic and plastic deformation happens at higher pressures and involves two sub-stages that correspond to elastic and plastic deformation. Due to the higher pressures, the particles are forced to flow into the existing void spaces, resulting in porosity decrease which increases the contact area between the particles. At this point, if the pressure is released, the particles can recover to their former state (elastic deformation and spring-back effect). Continuous application of the pressure leads to particle fracturing into fragments, followed by rearrangement of the fragments (plastic deformation), which may result to mechanical interlocking if the material is brittle.

c) Stages a) and b) progress until the compact density gets close to the specific density of the material. Local melting of the materials occurs if the melting points of the constituents (such as lignin) are reached, causing binding. If this stage is not reached, it can lead to disintegration of the pellets/briquettes.

Densification is an important strategy for the biomass market, because it improves the convenience and accessibility of biomass due to the uniform shape and size. Pelletized LB can be handled more efficiently using the handling and transportation infrastructure of commodity grains, because the handling properties of pellets are similar to grains. If the biomass were to be used for direct combustion purposes, densified biomass pellets provides clean, stable fuel, enhances its volumetric calorific value, and they can be easily adopted into the direct-combustion or co-firing with coal, gasification, pyrolysis, and biomass-based conversion reactors [99-102]. Despite the difficulties of handling, storage, and transportation, the direct combustion of loose biomass in conventional grates is associated with very low thermal efficiency. The conversion efficiencies are as low as 40% with widespread air pollution in the form of very fine particulate matter [103]. In order to develop a large-scale biomass unit, it is necessary to convert biomass into high-density and high-value solid format like pellets/briquettes [102-103]. Manufacturing good quality pellets, briquettes, and cubes is largely thought as an art rather than science by many feed mill operators [104].

Studies have demonstrated that densification process increases the bulk density and flowability of biomass products, reduces the bulkiness, decreases the porosity, essentially improving the handling characteristics of the materials for transportation and storage purposes, produces structural useful forms, enhancing appearance, and decreases spillage and wind loss [104-106]. To reduce the production costs of biofuel, it is of paramount importance to densify the biomass so as to reduce the transportation and storage costs, while increasing the heating value per unit volume.

Densification increases the bulk density of biomass from an initial bulk density (including baled density) of 40-200 kgm^{-3} to a final bulk density of 600-800 kgm^{-3} [102]. Loose plant-based biomass has low bulk density depending on the plant species, particle density, and particle size. The bulk density of biomass (dried straw) can be as low as 40 kgm^{-3} [47], 40-80 kgm^{-3} for corn stover, and 250 kgm^{-3} for some wood residues [13]. Crop straws densified into pellets increases in particle density to about 823-1011 kgm^{-3} [101]. Adapa et al. [106] reported that the mean densities of barley, canola, oat, and wheat straw compacts increased from 907 to 988 kgm^{-3}, 823 to 1003 kgm^{-3}, 849 to 1011 kgm^{-3} and 813 to 924 kgm^{-3}, respectively, upon application of pressure in the range of 31.6–138.9 MPa.

Other advantages associated with biomass fuel pellets include: a) reduced amount of emission (dust) and waste produced in consumer end-use; b) reduced storage costs, as they take up less space compared to the non-densified biomass; and c) more efficient combustion control because fuel pellets are uniform in size and their flow into combustion boilers can be better regulated [103]. Cost is a challenge in biomass densification because of the high consumption of energy. Low production costs, high quality pellets and briquettes together with safe handling of biomass will make biomass competitive with fossil fuels.

As earlier mentioned, the structure of lignocellulosic biomass consists of complex molecules of cellulose, hemicellulose, and lignin. Modifying the structural arrangement of cellulose-hemicellulose-lignin matrix during the pretreatment process can improve the binding properties of lignocellulosic straw biomass, without using any expensive artificial binder. When high molecular amorphous polysaccharides are reduced to low molecular components, the polymer becomes more cohesive in the presence of moisture [107]. Most biomass may contain natural adhesives, for example, protein in grasses and forages, lignin in crop straws and wood residues, etc [47, 104]. When such biomass is subjected to high pressure, these compounds are squeezed out of the stem and leaf walls, which are responsible for bonding of compacted particles and stabilization of pellets. These natural adhesives are released to provide sufficient binding force only within a narrow window of moisture [47,

104]. However, these natural adhesives do not form good pellets with good physical characteristics when pelletized. Sokhansanj et al. [47] and Briggs et al. [104] also reported that the hardness and durability of pelletized biomass can be enhanced by optimizing a combination of physico-chemical treatments of biomass during pretreatment and before/during densification processes.

A couple of variable parameters, namely, material properties (moisture content, particle size, and chemical constituent), and operating variables (temperature, pressure, and residence time of compression) are responsible for the physical quality (density, durability, and dimensional stability) and variability in energy requirement in the manufacture of biomass pellets and briquettes [100].

4.1. Material Properties

4.1.1. Effect of Moisture Content

An optimum window of moisture content of biomass is required for the manufacture of dimensionally stable and durable pellets and briquettes. A number of research investigations have been done on this; the results show that the optimum moisture content depends on the type of biomass with respect to the chemical constituents. Water acts as a binding agent by strengthening the bonding in the material. Water helps to develop van der Waals forces by increasing the area of contact between particles [103]. The higher the moisture content of wafers, the lower is the durability [100]. At high moisture content, water acts as a lubricant and decreases the binding characteristics of the feedstock even at high pressures; this leads to low pellet/briquette durability, stability, and density [100]. Low moisture chopped corn stover (5-10% w.b.) resulted in denser, more stable, and durable briquettes than high moisture stover (15% w.b.) [100]. The right amount of moisture develops self-bonding properties in LBs at an optimum temperature, pressure, and particle size. Colley et al. [108] also investigated the physical characteristics of switchgrass pellets. They reported that bulk density, particle density, durability, and hardness of the pellets were significantly affected by moisture content. The optimum moisture content value to obtain good and high values of the above physical characteristics of the pellet was at 8.6% (wet basis).

4.1.2. Effect of Chemical Constituents

Livestock feeds with higher amount of starch and protein composition generate denser and more stable pellets and briquettes as compared to biomass with higher amount of cellulosic material [47]. Tabil [98] cited that feeds with high natural protein can plasticize under heat generated by friction as the material passes through the die. Materials like dehydrated alfalfa, corn cob and meal, rice bran, and grain screenings with high fiber content produce good quality pellets that do not break easily, because fiber is considered to be a natural binder. However, it presents difficulty to compress and force through the die holes and at the same time, it leads to low production rates [98]. Briggs et al. [104] studied the lubricating effect of oil in the die, and reported that addition of lubricating oil increases the pellet product throughput in a pellet mill. Oil acts as a lubricant between particles and between the biomass feedstock and the die-wall, leading to a lower densification pressure.

However, the researchers demonstrated that increasing the oil content above 7.5% greatly decreased two major physical qualities of pellet, durability and hardness.

4.1.3. Effect of Particle Size

Grinding of biomass into fine particle size is one of the unit operations involved in the densification and pre-processing of biomass as a source of energy. It increases the surface area for better pretreatment, densification, and for an efficient biofuel conversion. Finely ground feedstock produces denser pellets and improves the pelleting capacity [98]. Grinders are among the largest power consuming machinery depending on the biomass material, particle size, and the grinding mechanisms [109]. The design and choice of the grinder are important for reducing the energy input in biomass pre-processing. In general, the finer the grind, the higher the energy input, the higher is the dimensional stability and durability of the biomass pellets and briquettes [102]. Fine particles usually adsorb more moisture than large particles; hence, they undergo a higher degree of conditioning. Also, large/coarse ground materials tend to produce low durability pellets because they may create natural fissure points that cause cracks and fractures in pellets [98, 102]. Mani et al. [108] investigated the grinding performance and physical properties of wheat and barley straws, corn stover, and switchgrass, and concluded that the energy consumption for grinding increases as the particle size of the ground biomass becomes finer. Kaliyan and Morey [110] studied the effects of particle size (0.56 to 0.8 mm) on the densification (briquette) characteristics of corn stover and switchgrass. They reported that decreasing the geometric mean particle size of ground corn stover from 0.80 to 0.66 mm increased the density of briquettes by 5 to 10%. It was also demonstrated that decreasing the particle size of corn stover grind from 0.80 to 0.66 mm increased the durability of briquettes by 50 to 58% at 100 MPa pressure, and by 62 to 75% using 150 MPa at a moisture content of 10%. These researchers stated that decreasing the particle size of corn stover grind from 0.80 to 0.66 mm increased the specific energy consumption from 0.8 to 1.3 MJt^{-1}

4.2. Operating Variables

4.2.1. Effect of Temperature

Temperature is one of the main operating variables that play a significant role in the stability and durability of the biomass product and energy consumption. During the densification process, heat is being added in two ways; by using the frictional heat from the die or externally by means of preheating of feed materials. Heat is also generated from friction due to compression [108]. It can be assumed that due to the presence of moisture in the biomass, there is formation of steam under high pressure condition which helps in the hydrolysis of the hemicellulose and lignin fractions of biomass into smaller sugar, lignin products, and other derivatives. These products, when subjected to heat and pressure during densification process, act as adhesive binders and provide bonding effect [103]. The addition of heat also contributes to the relaxation of the inherent fibers in biomass and apparently softens its structure, which reduces their resistance. This in turn results in an increase in the production rate during densification, reduction in wear of the contact parts, and a corresponding decrease in the specific power consumption [103]. However, densification temperature should not be more than 300°C which is the decomposition temperature of biomass [103, 111]. Under high densification pressure and temperature, lignin plays a role in the bonding of particles. When biomass is subjected to heat (temperature above 140°C), lignin tends to become soft, melts, and exhibits thermosetting binder resin properties [112]. Van Dam et al. [113] developed high strength-high density board materials from whole coconut husks, without the addition of chemical binders. The stages involved in the production are steam explosion, extrusion, milling, then followed by hot pressing. The board materials were comparable with or even superior to commercial wood-based panels.

4.2.2. Effect of Pressure

Pressure also plays a major role in determining the strength and density of pellets and briquettes. The intention of compaction is to increase the closeness between the particles, at the same time increasing the forces and strength between them, so as to provide a more densified material. Densification provides sufficient strength to withstand rough handling. If uniform pressure is not applied throughout the entire volume of the material, it causes variations in density of the pellets/briquettes [103]. Biomass density increases as the applied pressure increases depending on the moisture content of the biomass. It is important to understand that the physical properties of biomass (which includes density, moisture content, void volume, and thermal properties among others) also influence its binding mechanisms and behavior. Densification of biomass under high pressure creates mechanical interlocking and increases adhesion between the particles, generating the intermolecular bonds within the contact area [98, 114]. The application of external force (pressure) increases the contact area and the strength of the bond between the adhering partners [108]. Mani et al. [100] reported that applied pressure had the highest effect on the total energy consumption followed by moisture content. Adapa et al. [106] investigated four pressure levels on the densification of barley, canola, oat, and wheat straws. The authors observed that a pressure of 63.2 MPa for barley and wheat, and a pressure of 94.7 MPa for canola and oat produced the highest density compacts with minimal specific energy consumption values. It should be noted that the inclusion of fat/oil (animal or vegetable based) in biomass feedstock resulted in lower pellet durability [104]. The reason for this is that fat acts as a lubricant between the feedstock particles, and between the feedstock and the pellet mill die-wall. Due to the reduced friction in the die, pressure in the die is decreased. This produces pellets with lower durability. Due to the hydrophobic nature of fat, it inhibits the binding properties of the water-soluble components in the feed such as starch, protein, and fiber.

CONCLUSION

The limitation of available sites of enzymatic attack may arise from the fact that the average size of the capillaries/pores in LB is too tiny to permit the entrance of large enzyme molecules, hence, preventing effective enzymatic attack and confining it to the external surface. As a result, grinding the feedstock to a very small particle size is required before the pretreatment process. Costs and energy requirements for particle size reduction increase geometrically with decreasing particle size. The best option of pretreatment should result in reduction of highly crystalline nature of cellulose and the cross-linking features of lignin, and increasing the surface area for enzymatic reaction. Limiting the formation of inhibitory by-products is important to make the whole process cost effective and economically viable for investors. Different methods, such as physical/mechanical, chemical, physico-chemical, and biological pretreatment can be applied. The positive and negative effects of these various methods have been studied. LB are largely loose and bulky and with difficulty during handling (transportation, storage, and use). Densification is a good process to avert this problem. It is difficult to compare the performance and economics of these various approaches due to differences in the properties of different biomass feedstocks. Hence there is no generic method of pretreatment or pre-processing lignocellulosic biomass into biofuel.

REFERENCES

[1]   Mabee, W. E., Gregg, D. J., and Saddler, J. N. Assessing the Emerging Biorefinery Sector in Canada. *Applied Biochemistry and Biotechnology*, 2005;Vol. 121–124; 765-778.

[2]   Fiona, J. W., Egginton, P., Barrow, E., Desjarlais, C., Hengelveld, H., Lemmen, D. S., and Simonet. G. Chapter 2, Background Information: Concepts, Overviews and Approaches. From Impacts to Adaptation: Canada in a Changing Climate, 2007. Available from: *http://adaptation.nrcan.gc.ca/assess/2007/pdf/ch2_e.pdf*. Accessed November 11, 2009.

[3]   Intergovernmental Panel on Climate Change (IPCC) of United Nations,Chairman's speech, welcoming ceremony at COP 15/CMP5 on December 7, 2009. United Nations meeting at Bella center, Copenhagen, Denmark. Available from: *http://www.ipcc.ch/pdf /presentations/cop%2015/RKP-welc-cer-cop15.pdf*. Accessed January 14, 2010.

[4]   Environment Canada: The state of Canada's Environment Part IV, Chapter 15, Atmospheric Change: Climate change-Human activities and the greenhouse effect. 1996. Available from: *http://www.ec.gc.ca/soer-ree/English/SOER/1996report*. Accessed November 25, 2009.

[5]   Demirbas, A. *'Bioenergy, Global Warming, and Environmental Impacts', Energy Sources, Part A: Recovery, Utilization, and Environmental Effects*, 2004;26:3,225 – 236.

[6]   Zhang, Y. H. P. Reviving the carbohydrate economy via multiproduct lignocellulose biorefineries. *J. Ind. Microbiol. Biotechnology*. 2008;35, 367-375.

[7]   Demirbas, A., Fatih, M., Balat, M., Balat, H. Potential contribution of biomass to the sustainable energy development. *Energy Conversion and Management* 2009;50, 1746-1760.

[8]   Lynd, L. R., Wyman, C. E., and Gerngross, T. U. Biocommodity Engineering. *Biotechnology Prog.* 1999;15:777-793.

[9]   Lynd, L. R., Van Zyl, W. H., McBride, J. E., and Laser, M. Consolidated bioprocessing of cellulosic biomass: an update. *Current Opinion in Biotechnology*, 2005;16:577-583.

[10] Carolan, J. E., Joshi, S. V., Dale, B. E. Technical and Financial Feasibility Analysis of Distributed Bioprocessing Using Regional Biomass Pre-Processing Centers. *Journal of Agricultural & Food Industrial Organization*, 2007;volume 5, article 10.

[11] Hoekman, S. K. Biofuels in the U.S.-Challenges and Opportunities. *Renewable Energy* 2009;34, 14-22.

[12] Demirbas, A. Biomass resource facilities and biomass conversion processing for fuels and chemicals. *Energy conversion and management.* 2001;42, 1357-1378.

[13] Tabil, L. ABE 850 (Post-Harvest Technology) handout. *Department of Agriculture and Bioresource Engineering, College of Engineering,* University of Saskatchewan, Canada, 2009.

[14] Sneller, T., Durante, D., Miltenberger, M. Economic Impacts of Ethanol Production. A publication of ethanol across America. 2006; Available from: *www.ethanolacross america.net.* Accessed November 24, 2009.

[15] U.S. Department of Energy, Office of Energy Efficiency and Renewable Energy, Alternative Fuels Data Center. *http://www.afdc.energy.gov/afdc/fuels/properties.html/* accessed August 31, 2012

[16] U.S. Department of Energy/Energy Efficiency and Renewable Energy. *http://www.afdc.energy.gov/fuels/ethanol_benefits.html/* accessed September 05, 2012.

[17] Sheehan, J., Aden, A., Paustian, K., Killian, K., Brenner, J., Walsh, M., and Nelson, R. Energy and Environmental Aspects of Using Corn Stover for Fuel Ethanol. *Journal of Industrial Ecology,* 2004;Volume 7, No. 3–4, 117-146.

[18] Sokhansanj, S., Mani, S., Stumborg, M., Samson, R., and Fenton, J. , Production and distribution of cereal straw on the Canadian prairies. *Canadian Biosystems Engineering,* 2006;Volume 48, 3.39-3.46.

[19] Liu, R., Yu H., and Huang, Y. Structure and Morphology of Cellulose in Wheat Straw. *Cellulose* 2005;12: 25-34.

[20] Lafond, G. P., Stumborg, M., Lemke, R., May, W. E., Holzapfel, C. B., and Campbell, C. A. Quantifying Straw Removal through Baling and Measuring the Long-Term Impact on Soil Quality and Wheat Production. *Agronomy Journal* 2009;Volume 101, Issue 3, 529-537.

[21] Saskatchewan Ministry of Agriculture 2008;Available from: *http://www.agriculture. gov.sk.ca/Stastistics.* Accessed October 12, 2009.

[22] Food and Agriculture Organization of the United Nations, Food and Agricultural commodities production. 2007; Available from: *http://faostat.fao.org/site/339/ default.aspx.* Accessed December 11, 2009.

[23] Montane, D., Farriol, X., Salvado J., Jollez, P., and Chornet, E. Application of steam explosion to the fractionation and rapid vapor-phase alkaline pulping of wheat straw. *Biomass and Bioenergy* 1998;Vol. 14, No. 3, pp. 261-276.

[24]   NRcan Annual report The State of Canada's Forests, Natural Resources Canada, Ottawa ON, 2009;Available from: *http://canadaforests.nrcan.gc.ca/rpt#es.* Accessed December 10,2009.

[25] Food and Agriculture Organization of the United Nations, State of the World's Forests, 2003; Available from: *http://faostat.fao.org/site/339/default.aspx.* Accessed November 11, 2009.

[26] MacLean, H.L., Lave, L.B. Evaluating automobile fuel/propulsion system technologies. *Progress in Energy and Combustion Science* 2003;29, 1–69.

[27] U.S. Department of Energy. *The effects of the Alaska oil and natural gas provisions of H.R. 4 and S. 1766 on U.S. energy markets.*Washington, DC: Energy Information Administration U.S. DOE, 2002.

[28] Fortenbery, T.R. and Park H. *The Effect of Ethanol Production on the U.S. National Corn Price*. University of Wisconsin-Madison Department of Agricultural & Applied Economics Staff 2008;Paper No. 523.

[29] Renewable fuels association, accelerating industry innovation. Ethanol industry outlook, 2012; *http://ethanolrfa.3cdn.net/d4ad995ffb7ae8fbfe_1vm62ypzd.pdf/* accessed September 05, 2012.

[30] *http://faostat.fao.org/site/339/default.aspx/* accessed September 03, 2012

[31] Wu, M., M. Wang, and H. Huo. Fuel-Cycle Assessment of Selected Bioethanol Production Pathways in the United States. Energy Systems Division, Argonne National Laboratory, 2006; ANL/ESD/06-7, 2006. Available at: *http://www.transportation.anl. gov/pdfs/TA/377.pdf/* accessed July 15, 2010

[32] Perlack, R.D., Wright, L.L., Turhollow, A.F., Graham, R.L., Stokes, B.J., Erbach, D.C. *Biomass as Feedstock for a Bioenergy and Bioproducts Industry: The Technical Feasibility of a Billion-ton Annual Supply*, ORNL/TM-2005/66, U.S. Department of Agriculture and U.S. Department of Energy report, April, 2005.

[33] U.S. Environmental Protection Agency, Wastes - Non-Hazardous Waste - Municipal Solid Waste. *http://www.epa.gov/epawaste/nonhaz/municipal/index.htm/*accessed October 12, 2012.

[34] Graham R.L., Biorefinery Feedstock Information Network, *https://bioenergy.ornl. gov/main.aspx/* accessed October 12, 2012.

[35] Fan, L. T., Yong-Hyun, L., and Beardmore, H. D. Major chemical and physical features of cellulosic materials as substrates for enzymatic hydrolysis. *Advances in Biochemical Engineering/Biotechnology,* 2006;Volume 14/1980, pg 101-117.

[36] Lin, Y., and Tanaka, S. Ethanol fermentation from biomass resources: current state and prospects. *Applied Microbiology and Biotechnology* 2006;69.6: 627-642.

[37] Shleser, R. Ethanol production in Hawaii: processes, feedstocks, and current economic feasibility of fuel-grade ethanol production in Hawaii, State of Hawaii. *Department of Business, Economic Development & Tourism*, 1994.

[38] Murphy, J.D., and McCarthy, K. Ethanol production from energy crops and wastes for use as a transport fuel in Ireland, *Applied Energy* 2005;82, 148–166.

[39] Shaw, M. Feedstock and Process Variables Influencing Biomass Densification. M.Sc. Thesis, *Department of Agricultural and Bioresource Engineering*, University of Saskatchewan, 2008.

[40] Ramesh, C. K., and Singh, A. Lignocellulose Biotechnology: Current and Future Prospects, *Critical Reviews in Biotechnology*, 1993;13(2):15 1-172.

[41] Fan, L. T., Yong-Hyun, L., and Gharpuray, M. M. The Nature of Lignocellulosics and their Pretreatments for Enzymatic Hydrolysis. *Advances in Biochemical Engineering/Biotechnology,* 2006;Volume 23, 157-187.

[42] Tina, J., C., Ishizawa, I., Davis, M. F., Himmel, M. E., Adney, W. S., Johnson, D. K. Cellulase digestibility of pretreated biomass is limited by cellulose accessibility. *Biotechnology and Bioengineering*, 2007;Volume 98 Issue 1, Pages 112 – 122.

[43] Jørgensen, H., Kristensen, J. B., Felby, C. Enzymatic conversion of lignocellulose into fermentable sugars: challenges and opportunities. *Biofuels, Bioproducts and Biorefining* 2007;1:119-134.

[44] Kumar, P., Barrett, D. M., Delwiche, M. J., Stroeve, P. Methods for Pretreatment of Lignocellulosic Biomass for Efficient Hydrolysis and Biofuel Production. *Industrial and Engineering Chemistry Research*, 2009

[45] Sun, Y., and Cheng, J.J. Hydrolysis of lignocellulosic materials for ethanol production: A review. *Bioresource. Technology.* 2002;83 (1), 1-11.

[46] Söderström, J., Pilcher, L., Galbe, M., Zacchi, G. Two-step steam pretreatment of softwood by dilute H2SO4 impregnation for ethanol production , *Biomass and Bioenergy*, 2003;24 (6), 475-486.

[47] Sokhansanj, S., Mani, S., Bi, X., Zaini, P., Tabil, L. G. Binderless Pelletization of Biomass, presentation at the 2005 *ASAE* Annual International Meeting, Tampa Convention Centre, Tampa, Florida July 17-20, 2005; Paper Number: 056061.

[48] Mohammad, J. T., and Karimi, K. Pretreatment of Lignocellulosic Wastes to Improve Ethanol and Biogas Production: A Review. *International Journal Molecular Science* 2008;9, 1621-1651.

[49] Mosier, N. S., Wyman, C., Dale, B., Elander, R., Lee, Y. Y., Holtzapple, M., Ladisch, M. R. Features of promising technologies for pretreatment of lignocellulosic biomass. *Bioresource Technology*, 2005;96, (6) 673-686.

[50] Mais, U., Esteghlalian, A.R., Saddler, J.N., Mansfield, S.D. Enhancing the enzymatic hydrolysis of cellulosic materials using simultaneous ball milling. *Applied Biochemistry and Biotechnology*, 2002;98, 815-832.

[51] Millett, M. A., Baker, A. J., Feist, W. C., Mellenberger, R. W., and Satter, L. D. Modifying Wood to Increase its In Vitro Digestibility, *Journal of Animal Science.* 1970;31:781-788.

[52] Beardmorel, D. H., Fan, L. T., and Lee, Y.H. Gamma-ray irradiation as a pretreatment for the enzymatic hydrolysis of cellulose. *Biotechnology Letters* 1980;Vol. 2, No 10. 435-438.

[53] Florine, A. B., and Arthur, J. C. The Effects of Gamma Radiation on Cotton: Part I: Some of the Properties of Purified Cotton, Irradiated in Oxygen and Nitrogen Atmospheres, *Textile Research Journal*; 1958;28; 198.

[54] Chandra, R. P., Bura, R., Mabee, W. E., Berlin, A., Pan, X., Saddler, J. N., Substrate Pretreatment: The Key to Effective Enzymatic Hydrolysis of Lignocellulosics? *Advances in Biochemical Engineering/Biotechnology* 2007;108: 67-93.

[55] Chornet E., and Overend, R. P. Phenomenological Kinetics and Reaction Engineering Aspects of Steam/Aqueous Treatments. Proceedings of the International Workshop on Steam Explosion Techniques: *Fundamentals andIndustrial Applications* 1988;21-58.

[56] Öhgren, K., Galbe, M., and Zacchi, G. Optimization of Steam Pretreatment of SO2-Impregnated Corn Stover for Fuel Ethanol Production *Applied Biochemistry and Biotechnology,* 2005;volume 124 1055-1067.

[57] Sassner, P., Galbe, M., and Zacchi, G. Steam Pretreatment of Salix with and without SO2 Impregnation for Production of Bioethanol. *Applied Biochemistry and Biotechnology symposium*, 2005;1101-1117

[58] Rai, S. N., and Mudgal, V. D. Effect of Sodium Hydroxide and Steam Pressure Treatment on the Utilization of Wheat Straw by Rumen Microorganisms, *Biological Wastes* 1987;21, 203-212

[59] Brecc, K. A., and Glasser, W. G. Steam-assisted biomass fractionation. I. Process considerations and economic evaluation *Biomass and Bioenergy*, 1998;volume 14, (3), 205-218.

[60] Sun, X. F., Xu, F., Sun, R. C., Fowler, P., and Baird, M. S. Characteristics of degraded cellulose obtained from steam-exploded wheat straw. *Carbohydrate Research* 2005;340, 97–106

[61] Chundawat, S. P., Venkatesh, B., Dale, B. E. Effect of particle size based separation of milled corn stover on AFEX pretreatment and enzymatic digestibility. *Biotechnology Bioengineering*, 2006;96, issue 2, 219-231.

[62] Tanahashi, M., Takada, S., Aoki, T., Goto, T., Higuchi, T., Hanai, S. Characterization of Explosion Wood I. Structure and Physical Properties. *Wood Research* 1982;69:36-51.

[63] Wang, K., Jiang, J-X., Xu, F., Sun, R-C. Influence of steaming explosion time on the physic-chemical properties of cellulose from Lespedeza stalks (Lespedeza crytobotrya). *Bioresource Technology* 2009;100, 5288–5294.

[64] Heitz, M., Capek- Menard, E., Koeberle, P. G., Gagne J., Chornet, E., Overend, R. P., Taylor, J. D. & Yu, E. Fractionation of *Populus tremuloides* at the Pilot Plant Scale: Optimization of Steam Pretreatment Conditions using the STAKE II Technology. *Bioresource Technology* 1991;35, 23-32.

[65] Kaar, W. E., Gutierrez, C. V, and Kinoshita, C. M. Steam explosion of sugarcane bagasse as a pretreatment for conversion to ethanol. *Biomass and Bioenergy* 1998;vol. 14, no. 3, pp. 277-287.

[66] Tanahashi, M., Tamabuchi, K., Goto, T., Aoki, T., Karina, M., Higuchi, T. Characterization of Steam-Exploded Wood II Chemical Changes of Wood Components by Steam Explosion. *Wood Research* 1988;75:1-12.

[67] Atalla, R. H. Structural Transformations in Celluloses. *Proceedings of the International Workshop on Steam Explosion Techniques*: Fundamentals and Industrial Applications 1988;97-119.

[68] Excoffier, G., Peguy, A., Rinaudo, M., M. Vignon, R. Evolution of Lignocellulosic Components During Steam Explosion. Potential Applications. In B. Focher, A.Marzetti and V. Crescenzi (Eds.) *Proceedings of The International Workshop on Steam Explosion Techniques*: Fundamentals and Industrial Applications 1988;83-95.

[69] Fengel, D., Wegener, G. *Wood: Chemistry, Ultrastructure, Reactions*. Berlin: Walter de Gruyter, 1984.

[70] Alizadeh, H., Teymouri, F., Gilbert, T.I., Dale, B.E. Pretreatment of switchgrass by ammonia fiber explosion (AFEX). Applied *Biochemistry and Biotechnology*, 2005;124, 1133-41.

[71] Teymouri, F., Laureano-Pérez L., Alizadeh, H., Dale, B. E. Ammonia Fiber Explosion Treatment of Corn Stover, *Applied Biochemistry and Biotechnology*, 2004;Volume 115, 1-3, 951-963.

[72] Wyman, C.E. *Handbook on bioethanol: production and utilization*; Taylor & Francis: Washington DC, USA. 1996.

[73] Kim, K., Hong, H. J. Supercritical CO_2 pretreatment of lignocellulose enhances enzymatic cellulose hydrolysis. *Bioresource Technology*, 2001;77, 139-144.

[74] Zheng, Y., Tsao, G.T. Avicel hydrolysis by cellulase enzyme in supercritical CO_2. *Biotechnology Letters*, 1996;18, 451-454.

[75] Zheng, Y., Lin, H. M., Wen, J., Cao, N., Yu, X., Tsao, G.T. Supercritical carbon dioxide explosion as a pretreatment for cellulose hydrolysis. *Biotechnology Letters*, 1995;17, 845-850.

[76] Park, C.Y., Ryu, Y.W., Kim, C., Kinetics and rate of enzymatic hydrolysis of cellulose in supercritical carbon dioxide. *Korean Journal Chemical Engineering*, 2001;18, 475-478.

[77] Punidadas, R., D. Chantal, K. Tatiana, H. S. Ramaswamy, and G. B. Awuah. 'Radio Frequency Heating of Foods: Principles, Applications and Related Properties-A Review', *Critical Reviews in Food Science and Nutrition*, 2003;43: 6, 587-606

[78] Ryynanen S. The Electromagnetic Properties of Food Materials: A Review of the Basic Principles. *Journal of Food Engineeting* 1995;26, 409-429.

[79] Ramaswamy, H. and Tang, J. Microwave and Radio Frequency Heating. *Food Science and Technology International*; 2008;14 (5) ; 423-427.

[80] Zhu, S., Wu, Y., Yu, Z., Liao, J., Zhang, Y. Pretreatment by microwave/alkali of rice straw and its enzymic hydrolysis. *Process Biochemistry* 2005;40, 3082-3086

[81] Larhed, M., Moberg, C., and Hallberg, A. Microwave-Accelerated Homogenous Catalysis in Organic Chemistry. *Accounts of chemical research*, 2002; 35, 717-727.

[82] Zhu, S., Wu, Y., Yu, Z., Wang, C., Yu, F., Jin, S., Ding, Y., Chi, R., Liao, J., Zhang, Y. Comparison of Three Microwave/Chemical Pretreatment Processes for Enzymatic Hydrolysis of Rice Straw. *Biosystems Engineering*, 2006;93 (3), 279-283.

[83] Donepudi, A., Muthukumarappan, K. *Effect of microwave pretreatment on sugar recovery from corn stover, presentation at 2009 ASABE* Annual International Meeting, Paper Number: 097057.

[84] Sun, X. F., Xu, F., Sun, R.C., Wang, Y.X., Fowler, P., Baird, M. S. Characteristics of degraded lignins obtained from steam exploded wheat straw. *Polymer Degradation and Stability*, 2004;86, 245-256.

[85] Saha, B. C., Iten, L. B., Cotta, M. A., Wu, Y. V. Dilute acid pretreatment, enzymatic saccharification and fermentation of wheat straw to ethanol. *Process Biochemistry*, 2005; 40, 3693-3700.

[86] Sun, Y., and Cheng, J. J. Dilute acid pretreatment of rye straw and bermudagrass for ethanol production. *Bioresource Technology* 2005;96, 1599–1606

[87] Palonen, H., Thomsen, A. B., Tenkanen, M., Schmidt, A. S., Viikari, L. Evaluation of Wet Oxidation Pretreatment for Enzymatic Hydrolysis of Softwood. *Applied Biochemistry and Biotechnology*, 2004;Volume 117, Number 1 / April, 2004 1-17.

[88] Varga, E., Klinke, H. B., Reczey, K., Thomsen, A. B. High Solid Simultaneous Saccharification and Fermentation of Wet Oxidized Corn Stover to Ethanol. *Biotechnology and Bioengineering*, 2004;Volume 88 Issue 5, Pages 567 – 574.

[89] Martín, C., Mette, H. T., Hauggaard-Nielsen H., Thomsen, A. B. Wet oxidation pretreatment, enzymatic hydrolysis and simultaneous saccharification and fermentation of clover–ryegrass mixtures *Bioresource Technology*, 2008;99, 8777-8782.

[90] Taherzadeh, M.J. and Karimi, K. Enzyme-based hydrolysis processes for ethanol from lignocellulosic materials: A review. BioResources, 2007;2(4), 707-738.

[91] Kim, S., Holtzapple, M. T., Effect of structural features on enzyme digestibility of corn stover, *Bioresource Technology*, 2006;97, 583-591.

[92] Wang, Z., Keshwani, D.R., Redding, A.P., Cheng, J.J. *Alkaline Pretreatment of Coastal Bermudagrass for Bioethanol Production.* ASABE Annual International Meeting, Rhode Island, 2008;June 29 – July 2.

[93] Xu Z., Wang, Q., Jiang, Z., Yang, X., Ji, Y. Enzymatic hydrolysis of pretreated soybean straw, *Biomass and Bio*energy, 2007;31, 162-167

[94] Gandi, J., Holtzapple, M. T., Ferrer, A., Byers, F. M., Turner, N. D., Nagwani, M., Chang, S. Lime treatment of agricultural residues to improve rumen digestibility. *Animal Feed Science Technology,* 1997;68, 195-211.

[95] Okano, K., Kitagaw, M., Sasaki, Y., Watanabe, T. Conversion of Japanese red cedar (Cryptomeria japonica) into a feed for ruminants by white-rot basidiomycetes. *Animal Feed Science Technology*, 2005;120, 235–243.

[96] Hatakka, A. l. Pretreatment of Wheat Straw by White-Rot Fungi for Enzymic Saccharification of Cellulose. *European Journal of Applied Microbiology and Biotechnology*, 1983;18:350-357.

[97] Wooley, R., Ruth, M., Glassner, D., Sheehan, J. Process design and costing of bioethanol technology: a tool for determining the status and direction of research and development. *Biotechnology Progress,* 1999;15, 794–803

[98] Tabil, L. Binding and pelleting characteristics of alfalfa. Ph.D. thesis. Saskatoon, Saskatchewan: *Department of Agricultural and Bioresource Engineering*, University of Saskatchewan, 1996.

[99] Granada, E., López-González, L. M., Míguez, J. L., and Moran, J. Fuel lignocellulosic briquettes, die design and products study. *Renewable Energy,* 2002;27: 561-573.

[100] Mani, S., Tabil, L. G., Sokhansanj, S. (2006), Specific energy requirement for compacting corn stover. *Bioresource Technology* 97, 1420–1426

[101] Adapa, P., Tabil, L., Schoenau, G. Compression Characteristics of Selected Ground Agricultural Biomass. *Agricultural Engineering International*: the CIGR Ejournal. Manuscript, 2009;1347. Vol. XI.

[102] Kaliyan, N., Morey, R. V. Factors affecting strength and durability of densified biomass products. *Biomass and bioenergy,* 2009;33, 337-359.

[103] Grover, P. D., and Mishra, S. K. Biomass briquetting: technology and practices. Regional wood energy development program in Asia, GCP/RAS/154/NET, field document no. 46. Bangkok, Thailand: *Food and Agriculture Organization of the United Nations,*1996.

[104] Briggs, J. L., Maier, D. E., Watkins, B.A., and Behnke, K.C. Effects of ingredients and processing parameters on pellet quality. *Poultry Science*, 1999;78: 1464-1471.

[105] Obernberger, I, and Thek, G. Physical characterisation and chemical composition of densified biomass fuels with regard to their combustion behavior. *Biomass and Bioenergy*, 2004;27:653-69.

[106] Adapa, P., Tabil, L. Schoenau, G. Compaction characteristics of barley, canola, oat and wheat straw. *Biosystems engineering,* 2009;104, 335-344.

[107] Chen, W., Lickfield, G. C., and Yang, C. Q. Molecular modeling of cellulose in amorphous state. Part I: model building and plastic deformation study. *Polymer*, 2004;45: 1063-1071.

[108] Colley, Z., Fasina, O. O., Bransby, D., Lee, Y. Y. Moisture Effect on the Physical Characteristics of Switchgrass pellets. *American Society of Agricultural and Biological Engineer,*. 2006;Vol. 49(6): 1845–1851.

[109] Mani, S., Tabil, L. G., Sokhansanj, S. Grinding performance and physical properties of wheat and barley straws,corn stover and switchgrass. *Biomass and Bioenergy*, 2004; 27, 339 – 352

[110] Kaliyan, N., Morey, R. V. *Densification Characteristics of Corn Stover and Switchgrass. presentation at the, ASABE* Annual International Meeting, 2006; Paper Number: 066174.

[111] Ghebre-Sellassie, I. *Mechanism of pellet formation and growth. Pharmaceutical Pelletization Technology, Drugs and the pharmaceutical sciences,* 1989; volume 37,123-130. Marcel Dekke, Inc., New York and Basel.

[112] Van Dam, J. E. G., Van den Oever, M. J.A., Teunissen, W., Keijsers, E. R.P., Peralta, A. G. Process for production of high density/high performance binderless boards from whole coconut husk Part 1: Lignin as intrinsic thermosetting binder resin. *Industrial Crops and Products,* 2004; 19, 207–216.

[113] Van Dam, J. E.G., Van Den Oever, M. J.A., Keijsers, E.R.P. Production process for high density high performance binderless boards from whole coconut husk. *Industrial Crops and Products,* 2004;20, 97–101.

[114] Kaliyan, R.N., Morey, V. Natural binders and solid bridge type binding mechanisms in briquettes and pellets made from corn stover and switchgrass. *Bioresource Technology,* 2010;101, 1082–1090.

In: Biomass Processing, Conversion and Biorefinery
Editors: Bo Zhang and Yong Wang

ISBN: 978-1-62618-346-9
© 2013 Nova Science Publishers, Inc.

Chapter 4

BIOMASS HARVEST AND DRYING

Bo Zhang[*]
School of Chemical Engineering and Pharmacy,
Wuhan Institute of Technology, Hubei, China
Biological Engineering Program, Department of Natural Resources and Environmental
Design, North Carolina A and T State University, Greensboro, NC, US

ABSTRACT

Biomass encompasses a variety of biological materials with distinctive physical and chemical characteristics, such as woody or lignocellulosic materials, grasses and legumes, crop residues and animal wastes. Biomass can be converted to various forms of energy by numerous technical processes. This chapter introduces the harvesting system for woody biomass, crop, grass, and biomass drying.

The process of harvesting the trees can be broken down into five steps felling, extraction, processing, loading and trucking. When collecting woody biomass, cost effectiveness is often a challenge. The more biomass collection is integrated into the conventional system; the more cost effective it is. Today, the small-scale timber harvesting systems, which can handle woody biomass, are becoming more attractive.

Conventional multi-pass forage harvest systems are still playing the main role in the field of corn or other crops harvesting. Agricultural equipment manufacturers are developing single pass combined stream harvesting system that can be used independent of a combine. Single pass dual stream harvesting system, which could produce two streams: brain and biomass, are currently still in the research phase.

Moisture in woody biomass is normally lost by transpiration drying, through foliage or wood surfaces. Crop residues and herbaceous biomass are often field-dried. But under certain climate, artificial drying is required.

---

[*] Phone 336-334-7787. Fax 336-334-7270. E-mail: bzhang@ ncat.edu.

1. INTRODUCTION

Ever since the shortages of petroleum resources began with the global energy crisis in the 1970s, considerable attention has been focused on the development of alternative fuels. Renewable biomass sources can be converted to fuels and are a logical choice to replace oil. Unlike fossil fuel, biomass takes carbon out of the atmosphere while it is growing, and returns it as it is burned. If it is managed on a sustainable basis, biomass is harvested as part of a constantly replenished crop. This is either during woodland or agricultural management or coppicing or as part of a continuous programme of replanting with the new growth taking up CO_2 from the atmosphere at the same time as it is released by combustion of the previous harvest. This maintains a closed carbon cycle with no net increase in atmospheric CO_2 levels. Biomass encompasses a variety of biological materials with distinctive physical and chemical characteristics, such as woody or lignocellulosic materials, grasses and legumes, crop residues and animal wastes. Biomass can be converted to various forms of energy by numerous technical processes, depending on the raw material characteristics and the type of energy desired [1].

2. WOODY BIOMASS HARVESTING SYSTEM

2.1. Woody Biomass

As a feedstock for bioenergy and bio-based products, woody biomass for a logging site is commonly stored and delivered in one of three forms: unconsolidated logging residues, chips and other comminuted biomass materials, or bundled materials [2].

Unconsolidated logging residues (slash) are woody biomass in its raw form after it has been removed from the trunk of the tree. It's considered unsalable and normally left at the logging site or concentrated at the landing. While not commonly practiced, this slash can be transported to a biomass-using facility by conventional logging trucks, log trailers, or specialized containers on trailers.

Comminuted biomass materials may include chips and other comminuted materials, which are most often generated from chippers, tub grinders, or shedders. Clean chips are produced when bark, needles, leaves, and other debris are removed before or during the chipping operation. These chips tend to be uniform in size and may be required by some end-users, such as the paper production. Dirty chips, ground wood, and shredded material are all comminuted forms that require less care in processing and handling. They include some debris, for example leaves, bark, and soil. Such material is often used for hog fuel to produce energy at forest products manufacturing plants or other heat and power generating facilities.

Logging residues are often to be compressed into cylindrical bales or bundles known as "composite residue logs" (CRLs) or biomass bundles. The advantages of CRLs are compact size (a diameter of 2.0 to 2.5 ft and a length of ~10 ft), easy handling, and longer storage time without losing dry matter. However, processing logging residues into CRL form requires the use of specialized machinery, and requires additional cost too. Some brittle materials or short, large diameter pieces are not easily be bundled.

The process of harvesting the trees can be broken down into five steps: felling, extraction, processing, loading and trucking [3]. First the trees are severed from the stump and brought to the ground. The trees are pulled from the forest area to a landing or roadside, and this process is called extraction. At the handling site, the trees may be processed, including delimbing and topping the tree, and cutting the stem into logs (i.e. bucking). Then the products are sorted, stacked, loaded out onto logging trucks, and delivered to the biomass-using facility.

When collecting woody biomass, cost effectiveness is often a challenge. The capital, operational, and transportation costs of highly mechanized harvesting equipment are quite expensive. The more biomass collection is integrated into the conventional system; the more cost effective it is [4].

2.2. Conventional Harvesting Equipment

One-pass timber harvesting system is one of typical conventional woody biomass harvest systems. The system comprises of machineries, such as harvesters, skidders, and forwarders to harvest and recover woody biomass [5]. It harvests round-wood and biomass simultaneously. This extraction method is considered as the most cost effective way.

This approach is most likely to be interesting to the conventional loggers and forest landowners, because it requires few modifications to their current system, and provides additional value to traditional forest products, and reduces cost to prepare the land for reforestation.

Figure 1. Feller Buncher (http://en.wikipedia.org/wiki/File:FellerBuncher01.jpg).

In order to harvest the whole tree, the feller buncher/skidder system is the most effective method. In this process, the feller buncher fells the tree, and then skidders transport it to the log landing. The feller buncher is a self-propelled machine with a cutting head that is used strictly for cutting, holding, and placing the stems on the ground (Figure 1). The feller buncher could be wheel or track propel [6]. Tracked feller bunchers are able to be operating on wet and loose soil, and more stable on steep slopes. A skidder is any heavy vehicle used in a logging operation for pulling cut trees out of a forest in a process called "skidding", in which the logs are transported from the cutting site to a landing (Figure 2) [7]. At the landing, a chipper processes the biomass, while other merchantable products are sorted and loaded onto trucks for transport to the appropriate market. Limbs and tops from merchantable trees are also chipped.

Figure 2. Morgan SX-704 grapple skidder – a modern skidder with dual function fixed boom grapple (http://en.wikipedia.org/wiki/File:Skidderdualfunction.jpg).

Figure 3. Europe Chippers Woodchipper (http://en.wikipedia.org/wiki/File:Europe_Chippers_1.jpg).

Another one pass harvest system is the Harvester-Forwarder/Skidder system. This system utilizes a harvester that fells, delimbs, bucks trees, and piles both biomass and traditional products in separate piles (Figure 4). A forwarder or skidder carries felled logs from the stump to a roadside landing. Unlike a skidder, a forwarder carries logs clear of the ground, which can reduce soil impacts but tends to limit the size of the logs it can move. Forwarders are typically employed together with harvesters in cut-to-length logging operations (Figure 5) [8]. At the landing, traditional products are loaded with a conventional log loader and the biomass is chipped with a chipper.

Figure 4. Valmet 901-2 Harvester (http://en.wikipedia.org/wiki/File:Valmet_901_2_side.jpg).

Figure 5. A Rottne Rapid forwarder used in logging. (http://en.wikipedia.org/wiki/File:Forwarder.jpg).

The two-pass harvesting method is to harvest and recover roundwood and biomass in separate passes. This method is less popular and less cost effective than the one-pass method. During the harvest, unconsolidated biomass is either left in the woods or is piled near the landing for later processing and transportation. But this method offers the opportunity for small, specialized biomass harvesting contractors to operate, if conventional timber harvesting contractors do not wish to process the biomass [9].

2.3. Small-Scale Biomass Harvesting System

Small-scale timber harvesting systems, which can handle woody biomass, are becoming more attractive. Many timber harvest machine manufacturers are designing and manufacturing smaller scale equipment. For these small-scale systems to be feasible, some criteria have to be met. Criteria for an successful small-scale timber harvest system include: safety features, low capital cost, low transportation cost, low overhead, maneuverability, minimal access requirements, ability to optimize quickly, and ability to deal with small-diameter or irregular woody material [10].

Small-scale timber harvesting systems could be using a single machine or two machines. Using the single machine method, bunching, forwarding/skidding and loading are all done by one machine. In two-machine systems, felling and possibly processing are completed using the first machine, while skidding and loading are done by the second machine.

The small-scale system allows that independent contractors or even landowners could get into the timber business with low capital investment. However, it is rarely economically feasible to use small-scale systems to harvest lower valued material on less financially productive cuts [11]. A small-scale timber harvesting system may be more suitable to collect woody biomass, while conventional equipment harvests and collects products such as pulpwood and sawlogs.

3. CROP HARVESTING SYSTEM

Crop residues (such as corn stove) and herbaceous biomass (such as switchgrass, giant miscanthus) have been widely studied, and considered to have major potential as a biofuel feedstock once second-generation cellulosic conversion technologies are commercialized.

As an example, corn stover from existing corn production is by far the most abundant crop residue readily available today. In 2008, the United States planted corn on approximately 25 percent of its production acreage, or a total of 91 million acres. A 2005 study on biomass availability for bioenergy production jointly administered by the U.S. Department of Agriculture (USDA) and the U.S. Department of Energy (USDOE) estimated that 75 million dry tons (DT) of corn stover could be harvested sustainably from those acres. The study further estimated that with moderate-to-high yield increases that number could soar to between 170 and 256 million DT per year by 2030 [12].

To meet this stover harvesting requirement, conventional multi-pass forage harvest systems are still playing the main roll. Agricultural equipment manufacturers are developing single pass combined stream harvesting system that can be used independent of a combine.

Single pass dual stream harvesting system, which could produce two streams: brain and biomass, are currently still in the research phase.

3.1. Conventional Forage Harvest Systems

Commercial forage harvesters (also known as silage harvester, forager or chopper) are typically used in the harvest of hay and forage crops as well as for some small-scale corn stover collection to make silage [13]. Silage is chopped grass, corn or other plant that has been compacted together in the storage silo, bunker, or bags. The silage is then fermented to make livestock feed. Haylage is a similar process to silage but using dried grass [14]. Forage harvesters can be either attached to a tractor, or self-propelled machines (Figure 6). The cutting device is either a cutterhead drum or a flywheel with a number of knives, which chops and blows the silage through a chute into a wagon. Each type of silage requires different cutting equipment, so there are different types of heads available for the harvester. Grass silage is usually cut prior to harvesting to allow it to wilt, while corn and other crop silage are cut directly by the header, using reciprocating knives, disc mowers or large saw-like blades. When harvesting crops like corn and sorghum, Kernel processors that consists of two mill rolls with teeth pressed together by powerful springs, are frequently used to crack the kernels of these plant heads. Kernel processors are installed between the cutterhead and accelerator that increases material unloading speed. In most forage harvesters, the kernel processor can be quickly removed and replaced with a grass chute for chopping non-cereal crops.

Figure 6. A self-propelled John Deere 5730 Forage Harvester
(http://en.wikipedia.org/wiki/File:JD5730ForageHarvesterJuly2004.JPG).

Another type of commercial harvesters is combine harvester, which is considered as one of the most economically important labor saving inventions, and enables a small fraction of the population to be engaged in agriculture (Figure 7) [15]. The combine harvester, or simply combine, is a machine that harvests grain crops. The name derives from its combining three separate operations comprising harvesting—reaping, threshing, and winnowing—into a single process. Among the crops harvested with a combine are wheat, oats, rye, barley, corn (maize), soybeans and flax (linseed). The waste straw left behind on the field is the remaining dried stems and leaves of the crop with limited nutrients which is either chopped and spread on the field or baled for feed and bedding for livestock.

Figure 7. Combine harvester Claas Lexion 570 (http://en.wikipedia.org/wiki/File:Claas-lexion-570-1.jpg).

3.2. Single-Pass Harvest Systems

Corn stover is typically harvested as a dry product and packaged in large round or large square bales. A stove harvesting after grain harvesting involves shredding, field drying, raking into a windrow, baling, gathering bales, transporting to storage, unloading and storing [16]. Problems with the multi-pass system include slow field drying, short harvesting window, frequent weather delays, soil contamination, and low yield. The fraction of available stover harvested by conventional means ranged from 37% and 57%. The many field operations resulted in the highest costs per unit mass of the systems considered. A single-pass harvesting system that combines the harvest of corn grain and stover would further eliminate field operations and reduce costs [17].

Agricultural equipment manufacturers are developing single pass harvesting systems. AGCO Corporation of Duluth, Georgia is developing a one-pass system that marries proven combine technology and the durable, reliable Hesston large square baler to collect and package clean corn cobs, husks and leaves into 3-foot-by-4-foot square bales [18]. Hillco Technologies's single pass cob collection technology is able to capture and move cobs off the field in one simple operation [19]. Vermeer Corporation manufactures a Cob Harvester (CCX770), which is a self-contained pull-type cart simply towed by qualified combines. This system collects cobs and redistributes leaves and husks back onto the field [20].

4. DRYING

The moisture content found in woody biomass is dependent mainly on storage conditions and post-harvest handling. Moisture is normally lost by transpiration drying, through foliage or wood surfaces. The drying rate of woody biomass is a function of many factors, including ambient temperature, relative humidity, wind speed, season, rainfall pattern, tree species, and tree size [21].

Transpiration drying is also called "leaf seasoning" and "biological drying", which occurs if felled trees are left with the tops, branches, and leaves intact for several weeks [22]. During the transpirational drying, the foliage continues to pull water out of the trunk of a tree, until it wilts and falls completely off the tree. Transpiration drying is effective for most tree species. But the factors such high humidity, and low temperature will obstruct the drying efficiency [23]. In addition, drying woody biomass before processing will require extra cost.

Field-drying of crop residues and herbaceous biomass is often preferred. But the drying rate is related to the weather, relative humidity, wind speed, season, rainfall pattern, biomass species, co-harvested material, and etc. In some areas, artificial drying may be still required.

For example, the moisture content of corn stover in the standing crop in Tennessee is about 55% on September 1st and less than 15% on October 20th. When the cut stove was left on the ground at 50% moisture, field drying resulted in a linear moisture loss of 2% units/d on a grass soil and 3% units/d on a dry asphalt soil, respectively. The moisture content is below 15% after 20 days [24]. In Wisconsin, moisture content of corn stover in the standing crop is in a range of 75% (September 1st) to 55% (October 14th). However, if the cut stover was naturally field dried in Wisconsin, the moisture of stover is still over 30% after 21 days. Wet harvest of corn stover was concluded as a better option than dry harvest under Wisconsin's climate [25]. Dry conservation of corn stover would therefore require some forms of artificial drying. For this purpose, a pilot scale dryer that uses hot-air recirculation and cooling has been developed [26].

CONCLUSION

Biomass encompasses a variety of biological materials with distinctive physical and chemical characteristics, such as woody or lignocellulosic materials, grasses and legumes, crop residues and animal wastes. This chapter introduces the harvesting system for woody biomass, crop, grass, and biomass drying.

The process of harvesting the trees can be broken down into five steps felling, extraction, processing, loading and trucking. When collecting woody biomass, cost effectiveness is often a challenge. The more biomass collection is integrated into the conventional system; the more cost effective it is. Today, the small-scale timber harvesting systems, which can handle woody biomass, are becoming more attractive.

Conventional multi-pass forage harvest systems are still playing the main role in the field of corn or other crops harvesting. Agricultural equipment manufacturers are developing single pass combined stream harvesting system that can be used independent of a combine. Single pass dual stream harvesting system, which could produce two streams: brain and biomass, are currently still in the research phase.

Moisture in woody biomass is normally lost by transpiration drying, through foliage or wood surfaces. Crop residues and herbaceous biomass are often field-dried. But under certain climate, artificial drying is required.

REFERENCES

[1]   Xiu S, Zhang B, Shahbazi A. *Biorefinery Processes for Biomass Conversion to Liquid Fuel, Biofuel's Engineering Process Technology: InTech*; 2011.

[2]   Schroeder R, Jackson B, Ashton S. Biomass Transportation and Delivery Pages 145–148 In: Hubbard, W; L Biles; C Mayfield; S Ashton (Eds) *2007 Sustainable Forestry for Bioenergy and Bio-based Products: Trainers Curriculum Notebook* Athens, GA : Southern Forest Research Partnership, Inc 2007.

[3]   Visser R. Timber Harvesting (Logging) Machines and Systems, *http://web1cnrevtedu /harvestingsystems/HarvestingProcesshtm* 2007.

[4]   Ashton S, Baker S, Jackson B, Schroeder R. Conventional Biomass Harvesting Systems Pages 133–136 In: Hubbard, W; L Biles; C Mayfield; S Ashton (Eds) *2007 Sustainable Forestry for Bioenergy and Bio-based Products: Trainers Curriculum Notebook* Athens, GA : Southern Forest Research Partnership, Inc 2007.

[5]   Stokes BJ, Watson WF, Savelle IW. *Alternate Biomass Harvesting Systems Using Conventional Equipment Paper presented at Sixth Annual Southern Forest Biomass Workshop, Athens, GA* , June 5–7, 1984 1984.

[6]   Forests and Rangelands, Forest Operations Equipment Catalog *http://wwwforestsand rangelandsgov/catalog/equipment/fellerbunchershtml* 2012.

[7]   Skidder. *http://enwikipediaorg/wiki/Skidder* 2012.

[8]   Forwarder. *http://enwikipediaorg/wiki/Forwarder* 2012.

[9]   Ashton S, Jackson B. Small-scale Woody Biomass Harvesting Systems Pages 137–140 In: Hubbard, W; L Biles; C Mayfield; S Ashton (Eds) 2007 *Sustainable Forestry for Bioenergy and Bio-based Products: Trainers Curriculum Notebook* Athens, GA : Southern Forest Research Partnership, Inc 2007.

[10] Wilhoit J, Rummer B. *Application of small-scale systems: evaluation of alternatives* ASAE Paper No 99-5056 St Joseph, MI: American Society of Agricultural Engineers 1999.

[11] Visser JM, Hull RB, Ashton SF. Mechanical Vegetative Management In: Monroe, MC, LW McDonell, and LA Hermansen-Baez (Eds) *2006 Changing Roles: Wildland-Urban Interface Professional Development Program* Gainesville FL: University of Florida 2006.

[12] USDA/USDOE B*iomass as a Feedstock for a Bioenergy and Bioproducts Industry: The Technical Feasibility of a Billion-Ton Annual Supply* Report No DOE/GO-102995-2135 2005.

[13] Webster K. *Single-pass corn stover harvest system productivity and cost analysis,* Graduate Theses and Dissertations. Paper 10411. Iowa State University; 2011.

[14] Forage harvester *http://enwikipediaorg/wiki/Forage_harvester* 2012.

[15] Combine harvester *http://enwikipediaorg/wiki/Combine_harvester* 2012.

[16] Shinners KJ, Binversie BN, Muck RE, Weimer PJ. Comparison of wet and dry corn stover harvest and storage. . *Biomass and Bioenergy* 2007;31:211.

[17] Shinners KJ, Binversie BN, Savoie P. *Whole-plant corn harvesting for biomass: Comparison of single-pass and multi-pass harvest systems* ASAE Paper No 036089 St Joseph, Mich: ASAE 2003.

[18] AGCO Corporation. One pass can do it all. *http://www.agcoiron.com/file Upload /MK009573_BioBalerSpec_LR.pdf.* 2009.

[19] Hillco Technologies. COB COLLECTION SYSTEM FOR JOHN DEERE STS COMBINES. *http://www.hillcotechnologies.com/pdfs/Cob%20Collection%20Brochure .pdf.* 2010.

[20] Vermeer Corporation. CCX770 Cob Harvester. *http://www2.vermeer.com/vermeer /documents/1/207/CCX770.pdf.* 2010.

[21] Jackson B, Schroeder R, Ashton S. Pre-processing and Drying Woody Biomass. Pages 141–144. In: Hubbard, W.; L. Biles; C. Mayfield; S. Ashton (Eds.). 2007. *Sustainable Forestry for Bioenergy and Bio-based Products: Trainers Curriculum Notebook.* Athens, GA: Southern Forest Research Partnership, Inc. 2007.

[22] Stokes BJ, McDonald TP, Kelley T. *Transpirational drying and costs for transporting woody biomass — a preliminary review.* IEA/BA Task IX, Activity 6: Transport and Handling. New Brunswick, CN : IEA: 76-91. 1993.

[23] Lehtikangas P, Jirjis R. *Storage of softwood logging residues in windrows under variable conditions.* Report No 235. Sveriges Lantbruksuniversitet : Institute for Virkeslara: 45 p. 1993.

[24] Edens WC, Pordesimo LO, Sokhansanj S. Field drying characteristics and mass relationships of corn stover fractions. *ASAE Paper No. 026015. Presented at the Annual Meeting of the American Society of Agricultural Engineers in Chicago, Illinois,* July 28-31. St. Joseph, Mich.: ASAE. 2002.

[25] Shinners KJ, Binversie BN, Savoie P. Harvest and storage of wet and dry corn stover as a biomass feedstock. *ASAE Paper No. 036088. Presented at the Annual Meeting of the American Society of Agricultural Engineers in Las Vegas, Nevada,* July 27 - 30. St. Joseph, Mich.: ASAE. 2003.

[26] Savoie P, Descôteaux S. *Artificial drying of corn stover in mid-size bales.* Canadian Biosystems Engineering/Le génie des biosystèmes au Canada 2004;46:2.25.

In: Biomass Processing, Conversion and Biorefinery   ISBN: 978-1-62618-346-9
Editors: Bo Zhang and Yong Wang   © 2013 Nova Science Publishers, Inc.

Chapter 5

BIOMASS SIZE REDUCTION

Bo Zhang [*]

School of Chemical Engineering and Pharmacy,
Wuhan Institute of Technology, Hubei, China
Biological Engineering Program, Department of Natural Resources and Environmental
Design, North Carolina A and T State University, Greensboro, NC, US

ABSTRACT

Size Reduction i.e. comminution, is one of the pre-processing processes for biomass. Biomass size reduction is needed to provide predictable delivery of uniform quality biomass for conversion. Reducing energy consumption, increasing bulk handling ability, increasing biomass density, reducing transportation costs, and facilitating efficient separation are the objectives of developing an effective size reduction process. However, knowledge of mechanical biomass size reduction is limited.

This chapter introduces that the mechanism of size reduction, and the size reduction equipment including chipper, wood chunker, hammer mill, knife mill, disc mill, pin mill, ball mill, and jet mill. When selecting a mill, the properties of the material, target particle size, and the capacity of the machine are among the important factors that should be considered. The energy efficiency and particle size distribution for a mill are both a function of operating speed, mass input rate, and screen size.

1. INTRODUCTION

Size Reduction is also called comminution, which is one of the pre-processing processes for biomass. The high volume, low density characteristics of biomass including both woody biomass and agriculturally produced residues are a significant impediment to biomass utilization. Biomass size reduction is needed to provide predictable delivery of uniform quality biomass for conversion. For example, cellulosic ethanol production process needs a size of 1-6 mm of ground biomass, and the production of heating fuels such as pellets,

[*] Phone 336-334-7787. Fax 336-334-7270. E-mail: bzhang@ ncat.edu.

briquettes or cubes that is another important end use of ground biomass requires different particle size [1].

Lignocellulosic biomass can be comminuted by a combination of chipping, grinding, and/or milling to reduce size and crystallinity. The final particle size of materials is usually 10–30 mm after chipping and 0.2–2 mm after grinding or milling [2]. The energy requirement in relation to the final particle size is one of the most important parameters describing the economical point of view of this physical pretreatment. For example, Mohsenin concluded that almost all of the energy in the grinding process is wasted as heat, and only 0.06 to 1% of the input energy actually is used to disintegrate the material [3]. Reducing energy consumption, increasing bulk handling ability, increasing biomass density, reducing transportation costs, and facilitating efficient separation are the objectives of developing an effective size reduction process.

2. MECHANISM OF SIZE REDUCTION

There are four types of size reduction to comminute a material: impact, attrition, shear, and compression (Figure 1).

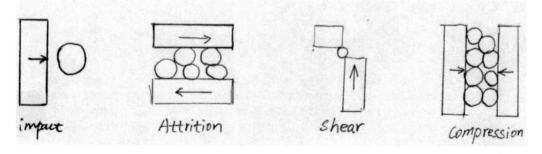

Figure 1. Types of size reduction.

Impact milling occurs when a hard object that applies a blunt force across a wide area hits a particle to fracture it. Impact-based equipment commonly uses hammers or media, and the milling action is produced by a rotating assembly that uses blunt or hammer-type blades. Another type of impact mill is a jet mill. A jet mill uses compressed gas to accelerate the particles, causing them to impact against each other in the process chamber. Impact mills can reduce both fine powders and large chunks of friable material — such as pharmaceutical powders, sugar, salt, spices, plastics, coal, limestone, chemical powders, and fertilizer — down to 50 microns with mechanical impact mills and less than 10 microns with jet mills. Mechanical impact mill types include hammer mills, pin mills, cage mills, universal mills, and turbo mills [4].

Attrition milling uses non-degradable grinding media to continuously contact the material, systematically grinding its edges down. This milling action is typically produced by a horizontal rotating vessel filled with grinding media and tends to create free-flowing, spherical particles. Attrition mills can reduce 1,000 micron (20-mesh) particles of friable materials, such as chemicals and minerals, down to less than 1 micron. One type is the media mill, which is also called a ball mill.

Shear is defined as the component of stress coplanar with a material cross section. One typical sample is a Knife milling. In knife milling, a sharp blade applies high, head-on shear force to a large particle, cutting it to a pre-determined size to create smaller particles and minimize fines. This milling action is produced by a rotating assembly that uses sharp knives or blades to cut the particles. Knife mills can reduce 2-inch or larger chunks or slabs of material, including elastic or heat-sensitive materials (such as various foods, rubber, and wax) down to 250 to 1,200 microns. Mill types include knife cutters, dicing mills, and guillotine mills.

Compression is applied via moving jaws, rolls or a gyratory cone, and force is applied by the faces of hard surfaces moving towards each other to crush the materials. The maximum discharge size is set by the clearance, which is adjustable. Compression is also called direct-pressure milling. Direct-pressure mills typically reduce 1-inch or larger chunks of friable materials — such as minerals, chemicals, and some food products — down to 800 to 1,000 microns. Types include roll mills, cracking mills, and oscillator mills.

Most mills use a combination of these principles to apply more than one type of force to a particle. This allows mill manufacturers to custom-design their equipment based on a material's characteristics and enables mills to reduce various materials.

3. SIZE-REDUCTION EQUIPMENT

There are many types of size-reduction equipment, which are often developed empirically to handle specific materials and then are applied in other situations. For example, if required size particles are larger than 1 inch, materials can be produced using drum or disk chippers and hammer hogs or tub grinders. Particles larger than 1 inch are most suitable for pulping or particle board production operations.

Woody biomass represents a significant portion of biomass that can be sustainably produced in the United States and around the world [5]. Woody biomass has many advantages as a feedstock for biofuel production. For example, hybrid poplar as a perennial, requires only 1/6 of the fossil energy needed to produce traditional feedstock, such as corn. The reduced petroleum requirement results in reduced CO_2 emissions. Poplar plantations are also a large carbon sink. High-yield, short-rotation tree farms need less fertilizer and chemical input to produce than many other cellulosic biomass crops, thus resulting in less water pollution from field run-off. In addition, poplars can be grown on land that is unsuitable for other uses. They have a relatively rapid growth cycle of 6 to 12 years, depending on climate, which is longer than annual crops but creates a more stable wildlife habitat because it is not disturbed by annual harvests. They can be harvested any time of the year and, unlike other energy crops, do not require extended storage after harvesting that exposes other biomass crops to degradation from microbial activity [6]. Thus, woody materials will be a critical part of the biomass supply mix in the future biobased economy. The most common woody biomass pre-processing operation is comminution by chipper, grinder or shredder.

3.1. Chipper

The chippers are popularly used to slice woody material, and make up most of the woody biomass size-reduction equipment in use today. The basic cutting device in chippers can be a disk to which cutting knives are attached or two horizontal high-speed drums with knives attached to them. They rely on sharp knives, but they are susceptible to knife wear from high soil content, metal contamination, rocks, and stones. Chippers are well integrated into existing timber harvesting systems. Chippers are customarily placed either off-road, at the timber harvest landing, or on-site at the mill or plant.

In disk chippers, the disk rotating speed is between 400-1000 rpm. Discs in the whole log chippers can be in a range of 140 to 160 inch in diameter, requiring 2500-5000hp. Figure 2 shows a disk chipper.

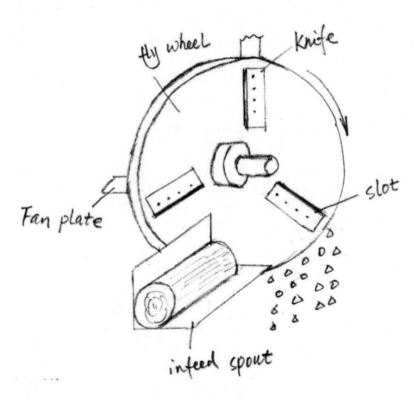

Figure 2. A disk chipper, adopted from Hakkila, P. 1989 [7].

In drum chippers, the knives are attached radially or spirally to a rotating large steel drum (Figure 3). There is typically a pocket installed in front of the knives, where the chips are stored until they can be released behind or below the drum [8].

There are two ways of feeding in these chippers: the side-feed drum chippers or end-feed drum chippers. Comparing with disk chippers, drum chippers are heavier and more expensive, but the feeding process is easier. Drum chippers can handle a wider size range of raw material. The amount of oversized chips can be controlled by employing a basket screen on the bottom of the drum chipper [9].

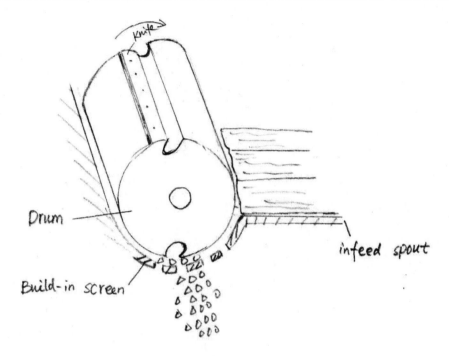

Figure 3. The principle of a cylindrical end-feed drum chipper.

3.2. Wood Chunkers

Wood chunkers are another group of size reduction equipment. Figure 4 shows three types of wood chunkers. They are A) a spiral-head wood chunker, B) a involuted disk chunker, and C) a double involuted disk chunker. When comparing to wood chippers, the particle size of ground material from a wood chunker is larger. Wood chunks are large pieces of wood, where the majority of particles have a relatively uniform length of 50-250 mm in the grain direction and a variable cross sectional area, ranging from about finger size up to the diameter of the material reduced [7].

3.3. Hammer Mill

Biomass particles less than 1 inch are most suitable for energy production. To reach such particle size, two stages of size reduction are required for woody feedstocks. First whole tree chippers are used to chip the entire tree including branches and limbs, then they are further ground using grinders.

Most grinders are derivatives of hammer mills including horizontal feed grinders. Grinders can accept a wider range of material sizes. They even accept short, non-oriented pieces including stumps, tops, brush, and large forked branches. Grinders rely on hitting a piece of wood often enough to finally break it into the desired size, usually with high speed rotating hammers. They require more energy than chippers per ton of output. Excessive soil contamination in materials will increase internal wear of a grinder.

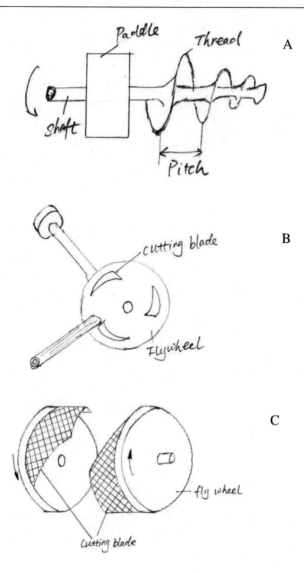

Figure 4. (A) A spiral-head wood chunker, (B) An involuted disk chunker, and (C) A double involuted disk chunker.

In a hammer mill, a rotating drum with a set of hammers or knives combines the shear, impact, attrition, and compression forces to reduce the material size, then the final particle size is controlled by built-in screens (Figure 5). The hammers might be commonly pivoted, or fixed. Forest residues are always contaminated by soil, sands and stones. These contaminates cause the sharp blades and knifes blunt, and short the life of a mill. Typically, the hammers will be reversed to provide additional life to the mill [7].

Hammer mills are relatively cheap, easy to operate, and produce a wide range of particles. The energy demand of biomass grinding depends especially on the initial and final particle sizes, moisture content, material properties, feed rate of the material, and machine variables like hammer tip speed and screen size [10].

The hammer tip speed can vary with the equipment design and size reduction requirements. Typically, tip speeds are in the range of 76 to 117m/s [11]. During the high-speed rotation, materials move around the mill in parallel to the screen surface making the openings only partially effective, whereas during a slower speed rotation materials impinge on the screen at a greater angle causing greater amounts of coarser feed to pass through [12]. As for the screen sizing, the larger hammer mill screen size requires less energy for all lignocellulosic materials. For example, the specific energy demands for grinding of 20-50mm long wheat straw using screen sizes of 0.794, 1.588, and 3.175mm were 51.55, 39.59, and 10.77kWh/t for 8.3% (wet basis) moisture content. Generally, the energy requirements increase with increasing moisture content and decreasing final particle size [11].

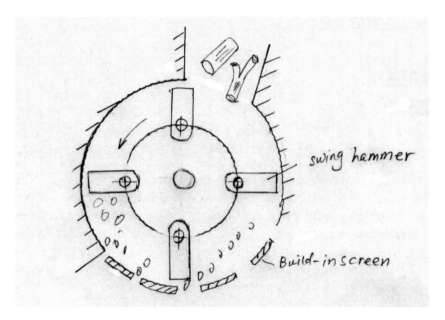

Figure 5. Schematic of hammer mill.

3.4. Knife Mill

The mechanical pretreatment of lignocelluloses using knife mills is usually applied to the treatment of dry biomass with a moisture content of up to 15% (wet basis) [13]. Knife mills are widely used for disintegration of cellulosic materials like grasses, straws, or fodder crop wastes. Lengthy straw/stalk of biomass may not be directly fed into grinders such as hammer mills and disc refiners. Hence, biomass needs to be preprocessed using coarse grinders like a knife mill to allow for efficient feeding in refiner mills without bridging and choking [14].

Typical knife mills are shown in Figure 6. A knife mill composed of rotating blades and one stationary blade. The materials are sheared by blades. The final particle size depends mainly on the feeding velocity, rotational speed of the rotor, and type of the drum screen. The energy requirement depends especially on the final particle size, rotational speed of the rotor, mounting longitudinal angle of the knife, and bevel angle of the knife [15].

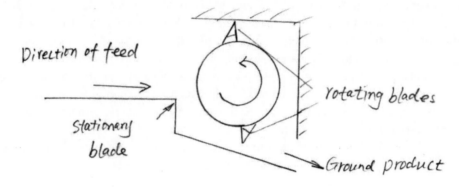

Figure 6. Schematic of knife mill.

3.5. Wiley Mill

The Wiley mill is commonly used in agriculture and soil science laboratories, and belongs to the family of knife mills (Figure 7). The term "Wiley" is the registered trademark of Eberbach Corporation, a grinding mill manufacture. Wiley mills are ideal for grinding animal and plant products (e.g. hair, leather, leaves, stalks and roots) but will also grind crystalline, fibrous, polymeric and amorphous materials (e.g. plastic). These mills prepare materials for analysis with minimal moisture loss. Well-dried samples are preferred. In the grinding mill, the material is cut into crude pieces or lumps and loaded into a hopper. From the hopper, the material drops by gravity into the path of a set of revolving hard tool steel blades driven by an electric motor. The four revolving knives work against six stationary knives and the resulting powder is forced through a steel screen. The powdered material then drops into a waiting collection vessel underneath [16].

Figure 7. Wiley mill.

3.6. Disc/Refining Mill

Disc mills are media mills that became standard particle dispersion equipment in industries. They are built in a single or double disc version with both straight and profiled blades. Material is fed into the mill through a central orifice coaxially with the rotation axis, and then due to acceleration goes through a number of rotating blades (Figure 8). Material is fragmented between discs and flows to their periphery. Other disc mills perform the final fragmentation of material between the rotating disc and a ring-like stator coaxially with the rotor. Shearing and attrition are the predominant comminuting mechanisms in these types of mills [15]. The particle size is determined by distance between discs

Millet et al. [17] studied the dependences of the final particle size on the material temperature and energy requirements during biomass disc milling. They observed that the specific energy demand decreased rapidly with increasing material temperature. For example, the energy requirements at room temperature were 750–850kWh per ton for wood chips to reach a final particle size of 0.5 mm, while to reach the same size with a temperature over 200°C a specific energy demand of only 100kWh per ton was determined.

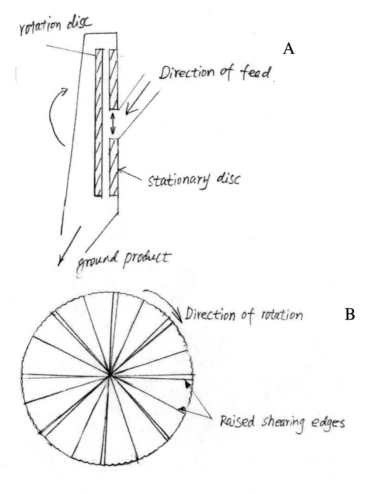

Figure 8. Schematic of disc mill (A) and Surface of rotating disc (B).

3.7. Ball Mills

A ball mill, a type of grinder, is a cylindrical device used in grinding (or mixing) materials like ores, chemicals, ceramic raw materials and paints. Ball mills rotate around a horizontal axis, partially filled with the material to be ground plus the grinding medium (Figure 9). Different materials are used as media, including ceramic balls, flint pebbles and stainless steel balls. An internal cascading effect reduces the material to a fine powder. Industrial ball mills can operate continuously, and be fed at one end and discharged at the other end. Large to medium-sized ball mills are mechanically rotated on their axis, but small ones normally consist of a cylindrical capped container that sits on two drive shafts (pulleys and belts are used to transmit rotary motion).

High-quality ball mills are potentially expensive and can grind mixture particles to as small as 5 nm, enormously increasing surface area and reaction rates. The particle size is determined by the media particles and rotating speed. The smaller the media particles, the smaller the particle size of the final product. At the same time, the grinding media particles should be substantially larger than the largest pieces of material to be ground. The grinding works on the principle of critical speed. The particle size is governed by the critical rotating speed of the media, after which grinding will not cause further particle size reduction.

Ball milling has been found to be an effective mechanical pretreatment for lignocellulose disintegration. The shearing and compressive forces cause reduction in crystallinity, decrease of the degree of cellulose polymerization, decrease of the particle size, increase of the bulk density, and increase of the external surface area. In addition, the increase of bulk density allows the treatment of highly concentrated substrates and reduces the reactor volume and investment costs. Milling at elevated temperatures raises the hydrolysis effectivity compared to milling at room temperature. However, the effectiveness of ball milling varies with the cellulose content, e.g., milling of softwood has the least efficiency [18].

Figure 9. Illustration of Ball mill (adapted from http://upload.wikimedia.org/wikipedia/commons/c/c2/Ball_mill.gif).

3.8. Pin/Impact Mills

Impact mills are screenless, high-speed beater mills for pulverising and micro-pulverising (Figure 10). The material is fed to the mill centrally via an inlet box at the top to rotating disk with revolving hammers/studs perpendicular to the surface. Impact mills use revolving hammers to strike incoming particles and to break or fling them against the machine case. The hammers also accelerate the product, moving it into the milling zone proper, at the side of the rotor. There the grinding stock fluidized in the air flow is comminuted by the grinding tools (rotor, stator). The stator is formed by a cover enclosing the rotor. The inside of this cover is provided with toothed grooves running vertical, i.e. crosswise to the sense of rotation of the rotor. The outside of the rotor is covered by numerous U-shaped sections which form a deep cassette-type structure. This geometry creates extreme air whirls in the rotor's grinding zone which induce intense secondary comminution processes due to the particles crashing into each other and due to friction and shearing forces. The final particle size can be adjusted over a wide range by changing the grinding rotor clearance, air flow and rotor.

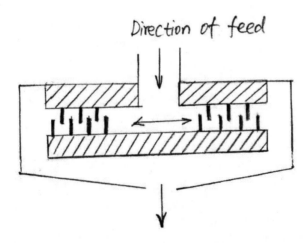

Figure 10. Schematic of pin/impact mill.

3.9. Jet Mills

In jet mills, the powder particles are fed into a flat circular milling chamber and subjected the same tangentially with a stream of pressurized air or nitrogen through a venturi. Particles collide with each other as they are transported in the stream. The material can be milled to an average particles size of 0.2 to 5 micron or coarse ground to 400 mesh depending upon characteristics of the product. Jet mills are widely used in the manufacture of drugs and pharmaceuticals, organic and inorganic pigments, optical brighteners, fluorescent pigments, food colour synthetic dyes, metal (except malleable metals) carbides & borides, tobacco, herbicides and pesticides, ceramic glasses, alumina, feldspar, frit, kaolin, mullite, electronic grade materials such as materials required for semiconductors, phosphors, photoelectronic, high temperature critical materials, and rocket solid fuels [19].

4. MILL SELECTION

When selecting a mill, some important factors to consider are properties of the material, target particle size, and the capacity of the machine.

Knowing the properties of the material to be processed is essential. Probably the most important characteristic governing size reduction is hardness, because almost all size-reduction techniques involve somehow creating new surface area, and this requires adding energy proportional to the bonds holding the feed particles together. A common way of expressing hardness is the Mohs scale, on which talcum is a 1 and diamond is a 10.

Other characteristics include particle-size distribution, bulk density, abrasiveness, moisture content, toxicity, explosiveness and temperature sensitivity. Flow properties can be major factors, too, because many size-reduction processes are continuous, but often have choke points at which bridging and flow interruption can occur. For instance, most size-reduction equipment is fed by chutes, which might constrict flow. Often, the feed flows adequately, but the crushed product will compact and flow with difficulty. Intermediate storage bins might aggravate flow issues by causing compaction and bridging.

For a given feed material, it is important to determine the desired particle-size distribution of the product. In mining, for example, very fine particles can interfere with separation processes, such as froth flotation, and might result in loss of valuable components. In other operations, the objective might be to produce very fine particles. Sometimes, as in sugar grinding, very fine particles are agglomerated to increase the share of larger particles.

Many particle-size distributions can be represented by the Gaudin-Schuhmann equation:

$$y = 100 \ (x/x_m)Â^a$$

where y is the cumulative percentage of material that is finer than size x, x_m is the theoretical maximum size, and $Â^a$ is the distribution modulus, which is related to hardness and has lower values for softer materials (0.9 for quartz and 0.3 for gypsum, for instance). The equation indicates that softer materials produce more fines [20].

Nearly all size-reduction techniques result in some degree of fines. So unless producing very fine particles is the objective, it usually is more efficient to perform size reduction in stages, with removal of the desired product after each operation.

CONCLUSION

Size Reduction i.e. comminution, is one of the pre-processing processes for biomass. Reducing energy consumption, increasing bulk handling ability, increasing biomass density, reducing transportation costs, and facilitating efficient separation are the objectives of developing an effective size reduction process. Knowledge of mechanical biomass size reduction is limited. When selecting a mill, the properties of the material, target particle size, and the capacity of the machine are among the important factors that should be considered.

Biomass size reduction is needed to provide predictable delivery of uniform quality biomass for conversion. Optimizing the energy efficiency of size reduction for a mill may be a function of operating speed, mass input rate, and screen size. For example, a knife mill with a classifying screen, total specific energy increased with operating speed, decreased with increasing mass input rate, and increased with reduced screen sizes [21]. Particle size distribution was also attributed to operating speed, mass input rate, and screen size.

ACKNOWLEDGMENTS

The authors acknowledge Ms. Jun Chen for drawing the mill pictures.

REFERENCES

[1]   Van Draanen A, Mello S. *Production of ethanol from biomass*, US Patent No. 5677154. 1995.

[2]   Sun Y, Cheng J. Hydrolysis of lignocellulosic materials for ethanol production: a review. *Bioresource Technology* 2002;83:1.

[3]   Mohsenin N. *Physical Properties of Plant and Animal Materials,* Gordon and Breach publishers Inc., Amsterdam, The Netherlands.; 1986.

[4]   Wennerstrum S, Kendrick T, Tomaka J, Cain J. Size reduction solutions for hard-to-reduce materials. *Powder and Bulk Engineering* January 2002:1.

[5]   Perlack RD, Wright LL, Turhollow A, Graham RL, Stokes B, Erbach DC. *Biomass as Feedstock for a Bioenergy and Bioproducts Industry: The Technical Feasibility of a Billion-Ton Annual Supply*. Oak Ridge National Laboratory. 2005.

[6]   USDOE. Improving Hybrid Poplars as a Renewable Source of Ethanol Fuel, *http://www1.eere.energy.gov/office_eere/pdfs/sbir_greenwood_case_study.pdf*. 2010.

[7]   Hakkila P. *Utilization of Residual Forest Biomass*. Heidelberg, Berlin: Springer-Verlag; 1989.

[8]   COFORD. Wood Harvesting Equipment *http://www.woodenergy.ie/woodharvesting equipment*. 2006.

[9]   CWC. Wood Waste Size Reduction Technology Study. Final report. Report No. CDL-97-3. Seattle, Washington. Available at *http://infohouse.p2ric.org/ref/13/12639.pdf* 1997.

[10]   Mani S, Tabil LG, Sokhansanj S. Grinding performance and physical properties of wheat and barley straws, corn stover and switchgrass. *Biomass and Bioenergy* 2004;27:339.

[11]   Yu M, Womac AR, Pordesimo LO. *ASAE Meeting*. June 2003.

[12]   Bitra VSP, Womac AR, Chevanan N. A*SABE Meeting*, Rhode Island, . July 2008.

[13]   Taherzadeh MJ, Karimi K. Pretreatment of Lignocellulosic Wastes to Improve Ethanol and Biogas Production: A Review. *Int J Mol Sci* 2008;9:1621.

[14]   Chevanan N, Womac AR, Bitra VSP, Igathinathane C, Yang YT, Miu PI, et al. Bulk density and compaction behavior of knife mill chopped switchgrass, wheat straw, and corn stover. *Bioresource Technology* 2010;101:207.

[15]   Miu PI, Womac AR, Igathinathane C. *ASABE Meeting,* Portland. 2006.

[16]   Dunn CE. *Biogeochemistry in Mineral Exploration*. Elsevier. pp. 159–160. ISBN 0-444-53074-6. 2007.

[17] Millett MA, Baker A, J,, Satter LD. Physical and chemical pretreatments for enhancing cellulose saccharification. *Biotechnol Bioeng Symp* 1976;6:125.

[18] Pandey A. *Handbook of Plant-Based Biofuels, 1st ed.*, CRC Press, New York. 2009.

[19] Promas Engineers, Jet Mills *http://www.promasjetmill.com/jet-mill.html*.

[20] Clark JP. Process engineering: Particle size reduction techniques and equipment, *http://www.chemicalprocessing.com/articles/2005/399.html*.

[21] Womac AR, Igathinathane C, Bitra P, Miu P, Yang T, Sokhansanj S, et al. *Biomass Pre-Processing Size Reduction with Instrumented Mills*, 2007 ASABE Annual International Meeting, Minneapolis, Minnesota. 2006.

In: Biomass Processing, Conversion and Biorefinery   ISBN: 978-1-62618-346-9
Editors: Bo Zhang and Yong Wang   © 2013 Nova Science Publishers, Inc.

Chapter 6

BIOMASS PELLETIZATION FOR ENERGY PRODUCTION

Hui Wang, Lijun Wang and Abolghasem Shahbazi*

Biological Engineering Program, Department of Natural Resources and Environmental
Design, North Carolina A and T State University, Greensboro, NC, US

ABSTRACT

The low density, low heating value, high volume and high transportation cost
characteristics of biomass including both woody biomass and agriculturally produced
residues are a significant impediment to biomass utilization. These difficulties can be
largely overcome through densification, also called pelletization. It is an effective
approach to use low density cellulosic feedstocks and offers an opportunity to make
biomass easier to handle and transport. This chapter focuses on the recent developments
in the biomass pelletization technology and application in energy production.

1. INTRODUCTION

1.1. Biomass Pellet

Biomass pellet is a cylindrical organic fuel made by the compression of biomass.
Biomass usually comes from agricultural residues, forest by-products and wood industries. It
is a renewable, clean-burning and cost stable home heating alternative currently used
throughout North America. There are approximately 1,000,000 homes in the U.S. using wood
pellets for heat, in freestanding stoves, fireplace inserts, furnaces and boilers. Pellet fuel for
heating can also be found in such large-scale environments as schools and prisons. North
American pellets are produced in manufacturing facilities in Canada and the United States,
and are available for purchase at fireplace dealers, nurseries, building supply stores, feed and
garden supply stores and some discount merchandisers.

---

* E-mail: hwang2@ ncat.edu.

1.2. Properties of Biomass Pellet

Biomass pellets are generally a superior fuel when compared to their raw feedstock. Not only are the pellets more energy dense, they are also easier to handle and use in automated feed systems. These advantages, when combined with the sustainable and ecologically sound properties of the fuel, make it very attractive for use. The standard shape of a fuel pellet is cylindrical, with a diameter of 6 to 8 millimeters and a length of no more than 38 millimeters. Larger pellets are also occasionally manufactured; if they are more than 25 millimeters in diameter, they are usually referred to as "briquettes." Typical properties of biomass pellet from different sources are summarized in Table 1.

Table 1. Typical properties of biomass pellet from different sources

Feedstock	Bulk density (kg/m3)	Energy content (MJ/Kg)	Ash content (%)
Sawdust	606	20.1	0.45
Bark	676	20.1	3.7
Logging leftovers	552	20.8	2.6
Switchgrass	445	19.2	4.5
Wheat straw	475	16	6.7
Barley straw	430	17.6	4.9
Corn stover	550	17.8	3.7

A high-quality pellet is dry, hard, and durable, with low amounts of ash remaining after combustion (Fig 2). "Premium" quality pellets (which are the most common pellets currently on the market) must have an ash content of less than 1 percent, whereas "standard" pellets may have as much as 2 percent ash. All pellets should have chloride levels of less than 300 parts per million and no more than 0.5 percent of fines (dust).

Figure 2. Biomass pellet.

1.3. Why Pelletizing?

The majority of North America's forest is second-growth, and requires periodic treatment in order to address forest health and fire mitigation. A tremendous amount of unusable material remains on the forest floor after such treatment. This material is rejected by high-end wood product manufacturers but is a perfect resource for commercial pellet manufacturers. Besides, engineering crops and waste such as cornstalks, straw can also be used to produce pellet. With proper sustainable forest initiatives and agricultural management, biomass is virtually limitless, and has proven to be price stable in comparison with fossil fuels.

Biomass pelletization has prompted significant interest in developed countries in recent years as a technique for utilization of residues as an energy source. Utilization of agricultural and forestry residues is often difficult due to their uneven characteristics. This drawback can be overcome by means of compaction of the residues into pellets of high density and regular shape [1]. Biomass pellets are competitive against oil fuels, natural gas and electricity not only because of the low cost but also because they are ease in usage and storage. Pellets are an excellent and inexpensive solution to replacing oil; they are renewable and inexhaustible and do not pollute the environment. Pellets can be produced from one type or mixed biomass. Mixed biomass pellets (MBP) have a great potential in enlarging the use of biomass for energy conversion. With the exhaustion of biomass residues for wood pellets production (particularly sawdust) the production of MBP is of increasing interest for project developers and biomass producers [2].

1.4. Markets for Biomass Pellets

The major wood pellet-producing countries in North America and Europe are Canada, Germany, Sweden, and United States. The major consumer countries for wood pellets are Belgium, Denmark, the Netherlands, Sweden, and the United States. The major wood pellet importing countries are Belgium, Denmark, the Netherlands, and Sweden, which have the most highly developed pellet markets. The others like Italy, Belgium, France and UK recently have been following that trend. In 2006 the production of pellets in Europe was about 4,500,000 tons, with Sweden, Austria and Germany as main producers [3].

USDA estimates that more than 80 pellet mills in North America currently produce in excess of 1 million tons of pellets annually. Since 2002 to 2006, the internal market demand of wood pellet in the United States increased by 200%. Moreover, production forecasts for 2012 were set at 6.0 million tons [4, 5]. Markets for wood pellets are well established in USA, especially in the northeastern United States. In Canada, 34 wood mills produce 2663,000 tons pellet in 2011. The North American Wood Fiber Review reported that the United States and Canada shipped a combined 1.6 million tons of pellets to the European Union in 2010. Exports are expected to increase this year as many countries look to biomass power since the EU mandated 20 percent of energy consumption must come from renewable resources by 2020.

The Asian wood pellet market is expanding at a steady pace. This can be seen in the quantity of wood pellet exports from the Port of Prince Rupert, British Columbia, to Asia, which increased 33 percent in the first half of 2010 over the same period in the previous year. The west coast is in a strong position to supply Asia with wood pellets, drawing on both timber supply and proximity to Asian markets. China's production of wood pellets was 800,000 metric tons in 2008 and 1 million metric tons in 2009. China's wood pellet consumption relies primarily on domestic production, and imports are minimal. In 2008, China imported approximately US$10.3 million dollars of wood fuel [6, 7].

2. PELLETS PRODUCTION TECHNOLOGIES

Biomass pellets are usually made from dry, untreated, industrial, agricultural and forest waste such as sawdust, shavings and wheat stover. These materials under high pressure and temperature are compressed into small pellets, cylindrical in shape. The manufacturing process is determined by the raw material but usually includes the following steps: reception of raw material, drying, grinding, pelletizing, cooling, and packaging. Fig. 1 provides the flow sheet of a typical pellet plant [8].

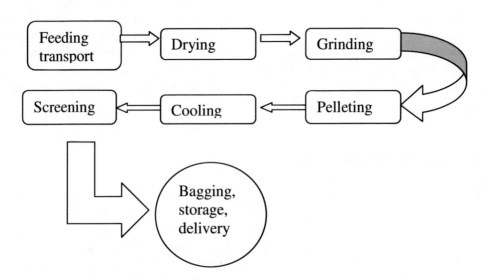

Figure 1. Flow sheet of the pelletizing process.

2.1. Feedstocks

Managing the feedstock is one of the greatest areas of concern for a pelletizing facility. Raw materials have to be sourced locally because their low bulk density makes them too costly to transport over long distances. The sources of supply vary across countries. It could be residues from forests, from arable land or wood that has already been used as products. It could also be virgin material like energy crops. The demand is mainly dependent on the cost for using it. The cost is partly a result of political decisions like environmental taxes, trade with emission quotas, etc.

2.2. Drying

Maintaining an appropriate moisture level in biomass feedstock is vital for overall quality of the final pellets. For wood, the required moisture level of the feedstock is at or near 15 percent. Moisture can be removed from the feedstock by oven-drying or by blowing hot air over or through the particles. Drying consumes a large amount of energy. This raises concerns with the net energy value of wood pellets as a fuel source. Wood fiber can be pelletized at moisture contents as high as 17%. However, the optimal level is 12% or less if a final product with a moisture content of 6–8% is to be achieved. Raw materials can also be too dry to pelletize. If the feedstock is too dry, moisture can be added by injecting steam or water into the feedstock. Drying is a focal point of research as the industry attempts to minimize costs and improve the quality of wood pellet energy [3].

2.3. Grinding

The grinding process is also known as milling. Material should be ground to a size no bigger than the diameter of the pellet (6 mm) producing a substance with a consistency similar to bread crumbs. Raw material should be filtered before grinding to remove materials like stone or metal [3]. Several types of equipment are available to carry out this task. If the biomass is quite large and dense (e.g., wood), the material is first run through a "chipper," and then run through a hammer mill or similar device to reduce the particles to the required size. Smaller and softer biomass (e.g., straw) can be fed directly into the hammer mill without first being chipped.

2.4. Pelletizing

In order to make the pellet, a roller is used to compress the biomass against a heated metal plate called a "die". The die includes several small holes drilled through it, which allow the biomass to be squeezed through under high temperature and pressure conditions. If the conditions are right, the biomass particles will fuse into a solid mass, thus turning into a pellet. A blade is typically used to slice the pellet to a predefined length as it exits the die.

Some biomass tends to fuse together better than other biomass. Sawdust is an especially suitable feedstock for pelletizing because the lignin that is naturally present in the wood acts as a glue to hold the pellet together. Grasses tend to not fuse nearly as well, and the resulting pellets are less dense and more easily broken. The proper combination of input material properties and pelletizing equipment operation may minimize or eliminate this problem. It is also possible to add a "binder" material to the biomass to help it stick together, or to mix a fraction of sawdust, with similar results. Distillers Dry Grains (a product of the corn ethanol industry) are reported to improve the binding properties of some biomass.

Pelletizing machines, also known as extruders, are available in a range of sizes. Generally, every 100 horsepower provides a capacity of approximately one ton of pellets per hour. Pelletizing machines come in two common forms [3]:

a) Flat: where raw material is pressed through the top of a horizontally mounted die.
b) Rotary: where two (or more) rotary presses push raw material from inside a ring die to the outside where it can be cut into the desired length.

Both systems create a pellet by using a great deal of pressure to force the raw material through holes in the die. As pressure and friction increase so does the temperature of the wood. This allows lignin to soften and the fiber to be reshaped into pellet form [3].

2.5. Cooling

Pellets, as they leave the die, are quite hot (~150°C) and fairly soft. Therefore, they must be cooled and dried before they are ready for use. This is usually achieved by blowing air through the pellets as they sit in a metal bin, which allows the lignin to solidify and strengthen the pellets. The final moisture content of the pellets should be no higher than 8 percent. The cooling process is critical to the pellets strength and durability.

2.6. Screening

Once pellets have cooled, they are passed over a vibrating screen to remove any fine material. These "fines" are augured back into the pelletizing process to ensure that no raw material is wasted. Screening ensures the fuel source is clean and as near to dust free as possible. Once screened, pellets are ready to be packaged for the desired end use.

2.7. Packaging

Pellets are typically sold in 18-kilogram bags, which can be easily filled using an overhead hopper and conveyor belt arrangement. The bags should be clearly labeled with the type of pellet, their grade (i.e., premium or standard), and their heat content.

3. MODERN TECHNOLOGIES FOR IMPROVED PELLET'S PRODUCTION

For the improvement of pellet production new methods have been developed recently.

3.1. Torrefaction

Torrefaction is a new technology to upgrade biomass for combustion and gasification applications. It is a thermal pre-treatment technology carried out at atmospheric pressure in the absence of oxygen. It occurs at temperatures between 200 and 300 °C where a solid uniform product is produced. This product has very low moisture content and a high calorific value compared to fresh biomass [9].

In the initial heating stage, biomass moisture content evaporation is very slow nonetheless, the biomass temperature increases. In the pre-drying stage, moisture content decreases dramatically while the biomass temperature stays constant. Following this stage, post-drying and intermediate heating occurs. The temperature increases up to 200 °C and the physically bounded water is released. Above 200 °C torrefaction reaction occurs. Devolatilization takes part in this stage. And finally, solid product is cooled to below 200 °C.

3.2. Improved Drying Methods

Renergi LTK Dryer system was developed to dry biomass pellets. The system consists mainly of a condenser and a low temperature dryer. The condenser makes it possible to transform waste energy and utilize that energy to preheat fresh air, which constitutes the drying medium. The air of the drying process has a temperature of 70 °C, meaning that it is a well-suited for drying of chips, sawdust, etc., since no hydrocarbons or other chemical contaminants are released. The gas can be emitted without advanced cleaning, assuming that any requirements on particle content are satisfied. With the Renergi system, either a fuel reduction on the order of 40-50% can be achieved, or productivity can be increased by up to 100% with retained fuel amounts [10].

3.3. Cold Pelletizing Process

A cold biomass pelletizing technology has been developed through the co-operation between Beijing Hui-Zhong-Shi Bio-Energy Company and Tsinghua University [11]. This technology simplifies existing hot pelletizing technologies, and does not need drying of raw materials that are naturally dried and extrusion. The technology has the following advantages: a) Lower specific energy consumption and operation cost, b) Less investment due to short production equipment close to production places of raw material, leading to lower transportation cost of raw materials and final products.

4. FACTORS AFFECTING PROPERTIES OF BIOMASS PELLETS

Effectiveness of a pelletizing process can be determined by testing the strength (compressive resistance, impact resistance, and water resistance), and durability (abrasion resistance) of the pellets. These tests can indicate the maximum force/stress that pellet can withstand, and the amount of fines produced during handling, transportation, and storage.

4.1. Feedstock

Factors related to the feedstock (starch, protein, fiber, fat, lignin and extractives, moisture content, and particle size and its distribution), pre-conditioning processes (steam conditioning/ preheating, and addition of binders), and pelletizing equipment (forming pressure, and pellet mill and roll press variables) would affect the strength and durability of the pellets. Due to the interactions of the above densification factors, the optimum densification variables may need to be determined using an optimization procedure. Also, post-production conditions such as cooling/drying and high humidity storage conditions would influence the strength and durability of the pellets [11].

4.2. Operation Conditions

Operation conditions related to the feed material, such as hardness, moisture, density and compacting pressure should be optimized for manufacturing high quality pellets.

Samson et al. and Jannasch et al. found that the hardness of the pellet increased moderately as the particle size decreased. The size of the mesh varied between 3.2 and 2.8 mm. Olsson, and MacMahon indicated the importance of avoiding the use of large particles, as this causes the pellet to fracture more easily. The fineness of the particle size improves the binding of the particles, up to a value of 0.5 mm. From this point on, the process becomes more difficult.

Moisture is another factor which affects the quality of the pellets. Mani and Ortiz analyzed its influence, and determined that at a moisture content of over 50% (wet basis) it became impossible to form a bound aggregate from forestry or woody residues. Water is also an important factor in the compacting process, as it can intervene chemically in the particle binding process, facilitating binding between the hydrogen bridges in the cell wall. Moreover, excessively dry lignocellulose material acts as thermal insulation which impedes heat transmission, a key element in the compacting process.

Gustafson and Kjelgaard studied the consolidation of biomass material for a wide range of moistures (28–44% wt.), and found that the density of the product diminished with the volume of moisture. Rehkugler and Buchele reported that the density of the pellet decreased with the percentage of moisture in the material when it varied between 6% and 25% (wt.). The compacting pressure used to obtain the pellets varies between 110 and 250 N/mm^2, depending on the characteristics of the species of wood. The density of the pellet increases with the quantity of material compressed and the pressures generated during the process. The retraction of the material, and therefore the density of the pellet, is lower as the length of the material obtained increases. The specific energy required is not affected by the quantity compressed [12].

4.3. Pellet Standard

There is a need to develop standards on the minimum acceptance levels for the strength (i.e., compressive resistance, impact resistance, and water resistance), and durability (i.e., abrasive resistance) of the pellets made from biomass feedstock. Many European countries have developed standards for the quality, storage, transport, and combustion of pellet. These include parameters and guidelines pertaining to the following: particle and bulk density, moisture content, crushing resistance or hardness, particle number, particle size (length and diameter), chemical composition, ash content, and heating value. As trade between countries becomes more widespread, it is also necessary to create international standards to facilitate the purchase and sale of pellet. Accordingly, the European Committee for Standardization, CEN, received a mandate from the European Commission to develop standards for pellets within the Technical Committee (TC) 335 Solid Biofuels. Nevertheless, current national standards regarding pellet quality are very heterogeneous. This is reflected in the differences detected in their quality control parameters and guidelines [13].

5. APPLICATION AND ECONOMIC ANALYSIS OF PELLET

5.1. Application of Biomass Pellet

5.1.1. Solid Fuel for Energy Production

Biomass pellet fuel can be an alternative source of heat for homes, businesses and commercial steam generation systems. Depending on the way how the pellets are fed into the furnace, three basic principles of pellet combustion systems can be distinguished: underfed burners, horizontally fed burners and overfed burners. In underfed (also called "underfeed stoker" or "underfeed retort furnace") and horizontally fed burners a so-called stoker screw feeds the fuel into the combustion chamber from below. Depending on the design, the flame burns horizontally (as for the horizontally fed burner) or upwards. In overfed burners the pellets conveyed from the storage tank by the conveying screw fall through a shaft into the glow zone at the grate. An advantage of this system is that a very accurate dosing of pellets according to the current power demand is achieved. A disadvantage is the impact of the falling pellets on the glowing bed of embers, leading to an increased entrainment of dust and unburned particles as well as to unsteady combustion behavior on the grate. [14].

5.1.2. Pyrolysis for Liquid Biofuel Production

Pyrolysis is a thermochemical method to use pellets as fuel. The products of pellet pyrolysis are liquid, gas and char, and can be used as fuel or into a source of higher value products; the first one if treated properly for the production of biodiesel and the second for energy production. Char may be useful as a fuel, either directly (briquettes) or as char-oil and char-water slurries and also in activated carbon production. Virtually, charcoal is the carbon derived from photosynthesis, which is why its use as a soil amendment sequesters carbon and reduces the greenhouse gas. Except of wood pellets, mixed biomass pellets are also challenging.

Mixed biomass pellets (MBP) have a great potential in enlarging the use of biomass for energy conversion, particularly in central and south European countries. With the exhaustion of biomass residues for wood pellets production (particularly sawdust), the production of MBP is of increasing interest for project developers and biomass producers [15].

5.2. Economic Analysis of Pelletization

The cost of setting up a pellet plant is not cheap; as a rule of thumb, expect to pay $70,000 to $250,000 per ton-per-hour capacity. The wide variation in costs is a function of the size, quality, and availability of the equipment. Larger capacity equipment is often more expensive on a per-ton basis because of the greater durability of the equipment and (usually) higher quality of the resulting pellets. Another important factor to consider is the availability of spare parts and repair professionals. In general, about half of the purchase cost of equipment will be for the pellet machine, and half for the other devices. Operating costs will include the cost of feedstock, energy, labor, and maintenance of the equipment. Typically, pellet dies will need to be replaced after every 1,000 to 1,500 hours of operation.

Comprehensive investigations and calculations of the production costs of wood pellets under consideration of all relevant parameters and for different framework conditions have been performed within the EU-ALTENER project "An Integrated European Market for Densified Biomass Fuels (INDEBIF)". The calculations are based on data from planned and already realized pellet production plants. Furthermore, data obtained from a questionnaire survey of producers of wood pellets in Austria, South Tyrol and Sweden has been considered for the calculation of the production costs [8].

The following parameters must be considered in a detailed calculation of the pellet production costs:

- The investment costs of all units of the pellet production process as well as of construction, offices and data processing, market introduction and planning as well as the utilization period and maintenance costs of all units and facilities.
- The raw material costs as well as the water content and the bulk density of the raw material used.
- The price for electricity, the electrical power required for all electrical installations and a simultaneity factor, which considers the fact that not all electrical installations operate on full load at the same time.
- The interest rate.
- The equipment availability, which considers both scheduled and unscheduled shutdowns.
- In case a preceding dryer is installed at the start of the process line, the specific heat costs and the heat demand for drying, furthermore the recoverable heat and the profit from heat sales in case that heat recovery takes place.
- The costs and the demand on bio-additives that may be used and the corresponding dosing system.
- If a conditioning unit working with steam is used, the demand and the costs of the steam.
- The storage costs, depending on the storage capacity and the kind of storage system used (storehouse and/or silo storage).
- The kind of shift work operated (plant utilization).
- The personnel cost both in production, marketing and administration.
- The annual pellet production rate as well as the water content and the bulk density of the pellets produced.
- Other costs.

CONCLUSION

Pellets are friendly to the environment, proposal of alternative form of fuels, and with important finances competitive advantages. They are transported and stored easily because of the form and their size. They constitute a very promising form of bio-fuel, which is responding in the needs of domestic market, mainly with regard to domestic users but also in wider scale consumers, as industries, having crowd of applications. The pellet markets in many counties are not well developed, but it has potential. If the proper actions are taken, the expansion of the market is sure. The biggest issue to be solved is the legislative aspect of pellets.

REFERENCES

[1]   Bhattacharya S C. *Biomass energy use and densification in developing countries. Pellets 2002: The First World Conference on Pellets; 2002* Sep 2–4; Stockholm, Sweden.

[2]   Sikkema R, Steiner M, Junginger M, and Hiegl W. Development and promotion of a transparent European Pellets Market Creation of a European real-time Pellets Atlas. *Final report on producers, traders and consumers of wood pellets*; 2009.

[3]   Peksa-Blanchard M, Dolzan P, Grassi A, Heinimö J, Junginger M, and Ranta T. Global wood pellets markets and industry: policy drivers, market status and raw material, *Potential IEA Bioenergy Task 40*; 2007.

[4]   Mani,s. Simulation of biomass pelleting operation. *Bioenergy Conference and Exhibition.* 2006.

[5]   Spelter H, and Toth D. *North America's wood pellet sector.* Research paper, United states Department of Agriculture. 2009.

[6]   Swaan J. *European wood pellet import market.* Pellet Fuels Institute annual conference, 2008, Arlington, VA.

[7]   Joseph A R and Allen M T*he Asian Wood Pellet Markets.* 2012 USDA Forest Service

[8]   Karkania V, Fanara E, and Zabaniotou A. Review of sustainable biomass pellets production – A study for agricultural residues pellets' market in Greece. *Renewable and Sustainable Energy Reviews* 2012, 16: 1426– 1436

[9]   Uslu A, Faaij A P C, and Bergman P C A. Techno-economic evaluation of torrefaction, fast pyrolysis and pelletisation. *Energy* 2008, 33(8):1206–1223.

[10]   Granstrand L. Increased production capacity with new drying system. *Proceedings of 2nd World conference on pellets.* 2006.

[11]   Dingkai Z, and Peihua W. Development and characteristics of a cold biomass pelletizing technology in China. *Proceedings of 2nd World conference on pellets.* 2006.

[12]   Relova I, Vignote S, . Leo´n M A, Ambrosio Y. Optimisation of the manufacturing variables of sawdust pellets from the bark of Pinus caribaea Morelet: Particle size, moisture and pressure. *biomass and bioenergy* 2009, 33:1351-1357

[13]   García-Maraver A, Popov V, Zamorano M. A review of European standards for pellet quality. *Renewable Energy* 2011, 36:3537-3540

[14] Obernberger I, and Thek G. Recent developments concerning pellet combustion technologies - a review of Austrian developments. *Preceedings of 2nd World conference on pellets*. 2006.

[15] Zabaniotou A, and Ioannidou O. Thermochemical conversion of biomass pellets. *Proceedings of Colloquium on Production and uses of Biofuels*, 2006

SECTION 2. BIOMASS THERMOCHEMICAL AND BIOCHEMICAL CONVERSION TECHNOLOGIES

In: Biomass Processing, Conversion and Biorefinery ISBN: 978-1-62618-346-9
Editors: Bo Zhang and Yong Wang © 2013 Nova Science Publishers, Inc.

Chapter 7

BIOMASS PYROLYSIS FOR BIO-OIL

Hui Wang, Lijun Wang and Abolghasem Shahbazi*

Biological Engineering Program, Department of Natural Resources and Environmental
Design, North Carolina A and T State University, Greensboro, NC, US

ABSTRACT

Fast pyrolysis utilizes biomass to produce a product that is used both as an energy
source and a feedstock for chemical production. Considerable efforts have been made to
convert wood biomass to liquid fuels and chemicals since the oil crisis in mid-1970s.
This chapter focuses on the recent developments in the biomass pyrolysis, factors
influencing the pyrolysis and application of resulting bio-oils.

1. INTRODUCTION

1.1. What is Pyrolysis

Pyrolysis is thermal decomposition of materials in the absence of oxygen or when
significantly less oxygen is present than required for complete combustion. Pyrolysis is
difficult to precisely define, especially when applied to biomass. The older literature generally
equates pyrolysis to carbonization, in which the principal product is a solid char. Today, the
term pyrolysis often describes processes in which oils are preferred products [1]. The general
features that occur during pyrolysis are enumerated below:

[1] Heat transfer from a heat source, to increase the temperature inside the fuel;
[2] The initiation of primary pyrolysis reactions at this higher temperature releases
volatiles and forms char;
[3] The flow of hot volatiles toward cooler solids results in heat transfer between hot
volatiles and cooler unpyrolyzed fuel;

---

* E-mail: hwang2@ ncat.edu.

[4] Condensation of some of the volatiles in the cooler parts of the fuel, followed by secondary reactions, can produce tar;

[5] Autocatalytic secondary pyrolysis reactions proceed while primary pyrolytic reactions (item 2, above) simultaneously occur in competition;

[6] Further thermal decomposition, reforming, water gas shift reactions, radicals recombination, and dehydrations can also occur, which are a function of the process's residence time/temperature/pressure profile.

1.2. Type of Pyrolysis

1.2.1. Conventional Pyrolysis

Conventional slow pyrolysis has been applied for thousands of years and has been mainly used for the production of charcoal. In slow wood pyrolysis, biomass is heated to 500°C. The vapor residence time varies from 5 min to 30 min [2, 3]. Vapors do not escape as rapidly as they do in fast pyrolysis. Thus, components in the vapor phase continue to react with each other, as the solid char and any liquid are being formed. The heating rate in conventional pyrolysis is typically much slower than that used in fast pyrolysis.

1.2.2. Fast Pyrolysis

Fast pyrolysis is a high-temperature process in which biomass is rapidly heated in the absence of oxygen [4, 5]. Biomass decomposes to generate vapors, aerosols, and some charcoal-like char. After cooling and condensation of the vapors and aerosols, a dark brown mobile liquid is formed that has a heating value that is about half that of conventional fuel oil. Fast pyrolysis processes produce 60-75 wt % of liquid bio-oil, 15-25 wt % of solid char, and 10-20 wt % of non-condensable gases, depending on the feedstock used. No waste is generated, because the bio-oil and solid char can each be used as a fuel and the gas can be recycled back into the process.

Fast pyrolysis uses much faster heating rates than traditional pyrolysis. Advanced processes are carefully controlled to give high liquid yields. There are four essential features of a fast pyrolysis process [6]. First, very high heating and heat transfer rates are used, which usually requires a finely ground biomass feed. Second, a carefully controlled pyrolysis reaction temperature is used, often in the 425-500 °C range. Third, short vapor residence times are used (typically <2 s). Fourth, pyrolysis vapors and aerosols are rapidly cooled to give bio-oil.

Heating rates of 1000 °C/s, or even 10000 °C/s, at temperatures below 650 °C have been claimed. At higher fast pyrolysis temperatures, the major product is gas. The main pyrolysis variants are listed in Table 1.

Over the last two decades, fundamental research on fast or flash pyrolysis has shown that high yields of primary, nonequilibrium liquids and gases, including valuable chemicals, chemical intermediates, petrochemicals, and fuels, could be obtained from carbonaceous feedstocks. Thus, the lower value solid char from traditional slow pyrolysis can be replaced by higher-value fuel gas, fuel oil, or chemicals from fast pyrolysis [2].

Table 1. Pyrolysis Methods and Their Variants

pyrolysis technology	residence time	heating rate	temperature (°C)	products
carbonization	days	very low	400	charcoal
conventional	5-30 min	low	600	oil, gas, char
fast	0.5-5 s	very high	650	bio-oil
flash-liquid	<1s	high	<650	bio-oil
flash-gas	<1s	high	<650	chemicals, gas
ultra	<0.5s	very high	1000	chemicals, gas
vacuum	2-30 s	medium	400	bio-oil
hydro-pyrolysis	<10 s	high	<500	bio-oil
methanol-pyrolysis	<10 s	high	>700	chemicals

1.3. Bio-Oil

Table 2. Typical Properties and Characteristics of Wood Derived Crude Bio-oil [11]

property	characteristics
appearance	From almost black or dark red-brown to dark green, depending on the initial feedstock and the mode of fast pyrolysis. Varying quantities of water exist, ranging from 15 wt % to an upper limit of 30-50 wt % water, depending on production and collection.
miscibility	Pyrolysis liquids can tolerate the addition of some water before phase separation occurs. Bio-oil cannot be dissolved in water. Miscible with polar solvents such as methanol, acetone, etc., but totally immiscible with petroleum-derived fuels.
density	bio-oil density is 1.2 kg/L, compared to 0.85 kg/L for light fuel oil.
viscosity	viscosity (of as-produced bio-oil) varies from as low as 25 cSt to as high as 1000 cSt (measured at 40 °C) depending on the feedstock, the water content of the oil, the amount of light ends that have collected, the pyrolysis process used, and the extent to which the oil has been aged.
distillation	It cannot be completely vaporized after initial condensation from the vapor phase at 100 °C or more, it rapidly reacts and eventually produces a solid residue from 50 wt % of the original liquid.
ageing of pyrolysis liquid	it is chemically unstable, and the instability increases with heating It is always preferable to store the liquid at or below room temperature; changes do occur at room temperature, but much more slowly and they can be accommodated in a commercial application; causes unusual time-dependent behavior properties such as viscosity increases, volatility decreases, phase separation, and deposition of gums change with time.

Bio-oil is a complex mixture of oxygenated hydrocarbons that may be burned in a furnace, or with some modification, combusted in industrial turbines for power generation [7]. It is obtained in yields of up to 80 wt% in total (wet basis) on dry feed, together with by-product char and gas which are, or can be, used within the process so there are no waste streams [8]. Further upgrading bio-oil in a "bio-refinery" to produce valuable chemicals for purposes other than combustion is an ongoing research concern [9] and may have strong economic potential. Raw bio-oil has a number of unique characteristics, including high density (specific gravity of 1.2), low pH (2.5), a moderate heating value (18 MJ/kg), and a high water content (15-30%). For reference, heavy fuel oil, the nearest market analogue, has a lower density (specific gravity of 0.94), very low water content (0.1%), and substantially higher heating value (40 MJ/kg). A further drawback relating to its high water content is that bio-oil will not form a stable emulsion with other hydrocarbon fuels without the addition of an emulsifier [10]. Thus, it cannot be easily blended with other petroleum fuels.

Bio-oil has many special features and characteristics. These require consideration before any application, storage, transport, upgrading or utilization is attempted. These features are summarized in Table 2.

2. TYPE AND COMPOSITION OF BIOMASS FOR PYROLYSIS

2.1. Type of Biomass for Pyrolysis

Biomass is mainly composed of cellulose, hemicellulose, lignin, mineral and ash. The detailed composition depends on the variety, producing area and planting conditions. Table 3 and Table 4 show the chemical composition, element composition and mineral composition of several types of biomass. Cellulose content in wood and corn straw is almost half of the organic content, while it is a little bit less in wheat straw and rice straw. Hemicellulose and lignin content are about 30% and 20% respectively.

Mineral content varies with the variety and planting condition of biomass. Mineral in the biomass includes K, Ca, Na, Mg, Si and other elements, of which content of Si, K and Ca is the highest. Basically, ash content of straws is much higher than woods. Rice straw has the highest ash content among the straws. By comparing corn straw and woods, it is found that the content of cellulose, hemicellulose and lignin is similar. On the other hand, Mg, Al and Na content in corn straw is much higher than woods. As a result, the yield and composition of pyrolysis products are different. As shown in Table 5, the yield of fast pyrolysis bio-oil is 60-70% for woods, higher than 50-60% for straws. For example, yield of fast pyrolysis bio-oil for poplar wood is 10% higher than corn straw. Besides, content of acids and low molecule products is much less than corn straw. The composition of cellulose, hemicellulose and lignin is similar in poplar wood and corn straw (Table 3). So, the difference in yield and composition of bio-oil is probably due to the diversity in structure and ash content.

Table 3. Chemical composition and ultimate analysis of several biomass samples (wt%, dry basis) [12, 13]

Biomass	Chemical composition analysis				Ultimate element analysis			
	cellulose	hemicellulose	Lignin	ash	C	H	O	N
Wheat straw	30.5	28.9	16.4	11.2	47.5	5.4	35.8	0.1
Rice straw	37.0	22.7	13.6	19.8	36.9	5.0	37.9	0.4
Corn straw	42.7	23.6	17.5	6.8	41.9	5.3	46.0	0
Poplar wood	42.3	31.0	16.2	0.4	49.6	6.1	43.7	0.6

Table 4. Composition of mineral in biomass (wt%, dry basis)

Biomass	Ash	Composition of mineral						
		Al_2O_3	CaO	Fe_2O_3	MgO	Na_2O	K_2O	SiO_2
Wheat straw	10-15	1.7	10	1.9	2.5	1.2	25	46
Corn Straw	5-10	6	14	1.4	18	20	0.6	40
Poplar wood	2-3	1	47	0.5	4.4	0.2	20	2.6
Deal wood	2-3	4.5	49	3.5	0.5	0.4	2.6	32.5
Maple wood	3-5	4	56	2	20	–	6	10

Table 5. Comparison of the composition of bio-oil of hardwood and corn straw [14]

Pyrolysis	Poplar wood (500-510°C)	Corn straw (480-500°C)
Gas	9-13	23-24
Char	12-15	18-21
Water	9-12	
Bio-oil	63-68	42-52
Acetic acid	5.4-6.3	13
Acids	8.4-11.4	15.8
Methanol and aldehydes	1.5-4.0	11.7
Hydroxyacealdedhyde	6.5-10.0	–
Pyrolysis lignin	16-25	–

2.2. Composition of Biomass for Pyrolysis

2.2.1. Cellulose

Cellulose, as the main component of plant cell wall, is the most widely distributed polysaccharose. Cellulose is a high molecular-weight (106 or more) linear polymer of ß-(1→4)-D-glucopyranose. The fully equatorial conformation of ß-linked glucopyranose residues stabilizes the chair structure, minimizing flexibility. Glucoseanhydride, which is formed via the removal of water from each glucose, is polymerized into long cellulose chains that contain 5000-10000 glucose units. The basic repeating unit of the cellulose polymer consists of two glucose anhydride units, called a cellobiose unit [15].

Cellulose degradation occurs at 240-350°C to produce anhydrocellulose and levoglucosan. When cellulose is pyrolyzed at a heating rate of 12 °C/min under helium in DTA (Differential Thermal Analysis) experiments, an endothermal reaction is observed at 335°C (temperature of maximum weight loss). The reaction is complete at 360°C [16]. Levoglucosan is produced when the glucosan radical is generated without the bridging oxygen from the preceding monomer unit [17]. Pakhomov also proposed that intermediate radicals form and degradation occurs through one of two biradicals. Specifically, a hydroxy group in radical I is transformed from C-6 to C-4, which occurs during the C-1 to C-6 oxygen bridge formation [18].

The content of cellulose has a great influence on the biomass pyrolysis. In the biomass feedstock with higher content of cellulose, more volatile component and higher bio-oil are produced. The composition of pyrolysis bio-oil is mainly levoglucosan and other components, like glycollic aldehyde, acetol, methanol, acetic acid [8].

2.2.2. Hemicellulose

Hemicellulose is the second major chemical constituent in biomass, which is also known as polyose. Hemicellulose is a mixture of various polymerized monosaccharides such as glucose, mannose, galactose, xylose, arabinose, 4-O-methyl glucuronic acid and galacturonic acid residues. Hemicelluloses exhibit lower molecular weights than cellulose. The number of repeating saccharide monomers is only 150, compared to the number in cellulose (5000-10000). Cellulose has only glucose in its structure, whereas hemicellulose has a heteropolysaccharide makeup and some contains short side-chain "branches" pendent along the main polymeric chain.

Hemicellulose decomposes at temperatures of 200-260°C, giving rise to more volatiles, less tars, and less chars than cellulose [19]. Most hemicelluloses do not yield significant amounts of levoglucosan. Much of the acetic acid liberated from biomass during pyrolysis is attributed to deacetylation of the hemicellulose. Hardwood hemicelluloses are rich in xylan and contain small amounts of glucomannan. Softwood hemicelluloses contain a small amount of xylan, and they are rich in galactoglucomannan.

The onset of hemicellulose thermal decomposition occurs at lower temperatures than crystalline cellulose. The loss of hemicellulose occurs in slow pyrolysis of wood in the temperature range of 130-194°C, with most of this loss occurring above 180 °C [20]. However, the relevance of this more rapid decomposition of hemicellulose versus cellulose is not known during fast pyrolysis, which is completed in few seconds at a rapid heating rate.

2.2.3. Lignin

The third major component of biomass is lignin, which is an amorphous cross-linked resin with no exact structure. It is the main binder for the agglomeration of fibrous cellulosic components while also providing a shield against the rapid microbial or fungal destruction of the cellulosic fibers. Lignin is a three-dimensional, highly branched, polyphenolic substance that consists of an irregular array of variously bonded "hydroxy-" and "methoxy-" substituted phenylpropane units [21]. These three general monomeric phenylpropane units exhibit the p-coumaryl, coniferyl, and sinapyl structures. In lignin biosynthesis, these units undergo radical dimerization and further oligomerization, and they eventually polymerize and cross-link. The resonance hybrids of the radical formed on oxidation of coniferyl alcohol illustrates the positions where radicals dimerizations occur during lignin formation.

The physical and chemical properties of lignin differ, depending on the extraction or isolation technology used to isolate them. Because lignin is inevitably modified and partially degraded during isolation, thermal decomposition studies on separated lignin will not necessarily match the pyrolysis behavior of this component when it is present in the original biomass. Lignin decomposes when heated at 280-500°C [19]. Lignin pyrolysis yields phenols via the cleavage of ether and carbon-carbon linkages. Lignin is more difficult to dehydrate than cellulose or hemicelluloses. Lignin pyrolysis produces more residual char than does the pyrolysis of cellulose. In differential thermal analysis (DTA) studies at slow heating rates, a broad exothermic plateau extending from 290°C to 389°C is observed, followed by a second exothermic, peaking at 420°C and tailing out to beyond 500 °C [22]. Lignin decomposition in wood was proposed to begin at 280°C and continues to 450-500°C, with a maximum rate being observed at 350-450°C [23].

2.2.4. Inorganic Minerals

Biomass also contains a small mineral content that ends up in the pyrolysis ash. Table 6 shows some typical values of the mineral components in wood chips, expressed as a percentage of the dry matter in the wood. Inorganic minerals have a great influence on the pyrolysis, although their content in the biomass is very little. The effects are expressed in three different ways:

Table. 6. Typical Mineral Components of Plant Biomass

element	percentage of dry matter
potassium, K	0.1
sodium, Na	0.015
phosphorus, P	0.02
calcium, Ca	0.2
magnesium, Mg	0.04

a)   Pyrolysis feature:

Raveendran investigated effect of inorganic minerals on pyrolysis of 13 types of biomass. He found that the inorganic minerals decreased the activation energy and reaction temperature [12].

b)   Distribution of pyrolysis products

Ash and added inorganic minerals in biomass decrease the volatile components of pyrolysis products, thus decreasing the yield of bio-oil and increasing the yield of char and gas. Scott investigated the relationship between ash content and bio-oil yield based on 8 types of biomass. He found the maximum bio-oil yield significantly decreased with the increase of ash content. The maximum bio-oil yield went up to 80 wt% when the ash was removed [24].

c)   Composition of bio-oil

Inorganic minerals increase the low molecule components in the bio-oil, like glycollic aldehyde and decrease the levoglucosan content. Gray reduced the ash content by pretreatment. It was found that the low molecule components in bio-oil, such as formic acid, acetic acid, methanol and acetone, were decreased and bio-oil yield were increased [25].

3. FACTORS INFLUENCING PYROLYSIS

3.1. Temperature

Pyrolysis temperature is an important factor influencing the composition of pyrolysis products. Basically, low temperature, slow heating rate and long residence time facilitates the production of char. Fast pyrolysis with high temperature reduces the char and increase the condensable gas. There exists a maximum value of bio-oil yield when the pyrolysis temperature is increased. Optimum temperature for the maximum bio-oil yields of woods and non-wood feedstocks is very close, 450~550°C. However, the maximum bio-oil yield differs from 50~60% for straws to 60~70% for woods. So, the optimum temperature is the primary factor to optimize the pyrolysis process.

3.2. Particle Size

The particle size is one of the most important variables affecting the pyrolysis. Several investigators studied the effect of the particle size on the oil yield and weight loss. The size of the particles has a direct impact on the nature of the reactions of biomass decomposition. In the literature, significant particle size effects were observed in the pyrolysis of coal and oil shales [26].In contrast to coal and oil shale, the pyrolysis yield are fairly independent of particle size for biomass sample. These results suggest that mass transfer restrictions to the volatile evolution and escape of the evolved volatiles from the inside of particle are much less important for rapeseed compared to coals and oil shales [27]. In Şensöz's study, the pyrolysis product yields were fairly independent of particle size for rapeseed [28]. Islam also found the similar result for palm shell [29]. They obtained the maximum oil yield with a relatively large particle size. This suggested that there was no significant effect of the particle size on the pyrolysis oil yield for biomass feed stock. It might because mass transfer restrictions to the volatile evolution and escape of the evolved volatiles from the inside of particle were much less important.

3.3. Residence Time

Solid particles break into smaller particles and decompose into volatile material when the feedstocks are heated. Volatile materials will occur secondary cracking and producing the incondensable gas, thus reducing the bio-oil yield. Residence time of volatile components is very important for the yield and composition of the bio-oil. Very short residence times result in incomplete depolymerisation of the lignin due to random bond cleavage and inter-reaction of the lignin macromolecule resulting in a less homogenous liquid product, while longer residence times can cause secondary cracking of the primary products, reducing yield and adversely affecting bio-oil properties.

3.4. Heating Rate

Low heating rate facilitates the production of char and reduce the bio-oil. High heating rate inhibits the secondary cracking of the gas and can improve the bio-oil yield to more than 70%. Heating rates of 1000°C/s, or even 10000°C/s, at temperatures below 650 °C have been used.

Rapid heating and rapid cooling produced the intermediate pyrolysis liquid products, which condense before further reactions break down higher-molecular-weight species into gaseous products. High reaction rates minimize char formation. Under some conditions, no char is formed [30]. At higher fast pyrolysis temperatures, the major product is gas.

3.5. Pressure

Pressure can influence the residence time of volatile materials, thus affecting the extent of the secondary cracking.

As a result, the molecule distribution will also be affected. Researchers found that yields of bio-oil and char of cellulose are 19.1% and 34.2% at 300°C, nitrogen and 1 atm, while yields of 55.8% and 17.8% at 200Pa. Low pressure can facilitate removal of volatile gas from the surface of the feedstock and reduce the opportunity of secondary reaction, improving the bio-oil yield.

3.6. Pretreatment

Feedstock pretreatment has become an important tool to increase cellulosic ethanol production yields. Pretreatment processes break down lignin and disrupt the crystalline structure of cellulose, rendering it more accessible to enzymes to allow more ready fermentation to sugars [31].

Numerous pretreatment methods for cellulosic ethanol production have been investigated, including physical [32], chemical [33], and a combination of both [34]. For example, steam explosion [35], hot compressed water treatments [36], and applications of dilute acids [37] or bases [38] have been utilized. These pretreatment processes could have potential to be used prior to fast pyrolysis to improve bio-oil characteristics, produce useful selected chemicals, and increase the yield of sugars [39, 40].

Hassan investigated six chemical pretreatments: dilute phosphoric acid, dilute sulfuric acid, sodium hydroxide, calcium hydroxide, ammonium hydroxide, and hydrogen peroxide. Bio-oils were produced from untreated and pretreated 10-year old pine wood feed stocks in an auger reactor at 450°C.

Results showed that the physical and chemical characteristics of the bio-oils produced from pretreated pine wood feed stocks were influenced by the biomass pretreatments applied. Acid pretreated pine wood produced bio-oils with lower pH, higher acid values, and viscosity compared to the bio-oils of untreated and alkaline pretreated pine wood.

The HHV for all bio-oils ranged from 19.7 to 24.4 MJ/kg except that, for the bio-oil of hydrogen peroxide pretreatment, the HHV was 7.28 MJ/kg due to very high water content (24%).

The average molecular weights for the bio-oils of acid pretreated pine wood were higher than for untreated and alkaline pretreated pine wood. GC/MS chemical characterization showed also that the chemical concentrations for most bio-oil components that arise from the degradation of hemicelluloses, cellulose, and lignin were lower than that in the bio-oil of untreated pine wood.

Calcium hydroxide pretreatment increased the concentration of laevoglucose and other anhydrosugars in the produced bio-oil. FTIR spectra indicated that most bio-oils had similar chemical compositions with the possibility of increased hydrocarbon compounds in the bio-oil of calcium hydroxide, sodium hydroxide, and hydrogen peroxide pretreated pine wood [41].

4. APPLICATION OF PYROLYSIS BIO-OIL

4.1. Heat Production

Liquid products are easier to handle and transport than solids and gases and this is important in combustion applications. Existing oil fired burners cannot be fuelled directly with solid biomass without major reconstruction of the unit. However bio-oils are likely to require only relatively minor modifications of the equipment or even none in some cases. Bio-oil is similar to light fuel oil in its combustion characteristics, although significant differences in ignition, viscosity, energy content, stability, pH, and emission levels are observed. Fractional condensation of biomass pyrolysis vapors can be used to remove a portion of the water, thereby increasing the heating value and reducing the weight of oil per unit of heat generated upon combustion.

4.2. Electricity Production

Fast pyrolysis for bio-oil production can be de-coupled with electricity power generation. The use of a bio-oil requires engine modifications, including fuel pump, the linings, and the injection system. Slight modifications of both the bio-oil and the diesel engine can render bio-oils an acceptable substitute for diesel fuel in stationary engines. Bio-oil blends with standard diesel fuels or bio-diesel fuel is also possible. Experience with bio-oil combustion in gas turbines has also been reported. Emulsions were also developed from biomass pyrolysis oil.

The economics of power generation suggest that a niche of up to 10 MW is available for exploitation. Bio-oil has been successfully fired in a diesel test engine, where it behaves very similar to diesel in terms of engine parameters and emissions. Over 400 h operations have been achieved [42]. A diesel pilot fuel is needed, typically 5% in larger engines, and no significant problems are foreseen in power generation up to 15 MW per engine. A 2.5 MW gas turbine has also been successfully tested although for not many hours operation to date [43].

4.3. Synthesis Gas Production

Gasification of bio-oil with pure oxygen and further processing of the crude synthesis gas in Fischer-Tropsch processes is technically and economically feasible. The production of synthetic gas/high-Btu gaseous fuel from bio-oil was studied. Pyrolysis of bio-oil at various temperatures in a small tubular reactor was performed at atmospheric pressure. Bio-oil was fed at a flow rate of 4.5-5.5 g/h, along with nitrogen (18-54 mL/min) as a carrier gas. Bio-oil conversion to gas was 83 wt %, whereas gas production was 45 L/100 g of bio-oil at 800 °C and a constant nitrogen flow rate of 30 mL/min. The gas consisted of H_2, CH_4, CO, CO_2, C2, C3, and C4-Cn hydrocarbons. The compositions of product gases were in the following ranges: syngas, 16-36 mol %; CH_4, 19-27 mol %; and C_2H_4, 21-31 mol %. Heating values were in the range of 1300-1700 Btu/SCF. Clearly, a strong potential exists for making syngas, methane, ethylene, and high-heating-value Btu gas from the pyrolysis of biomass-derived oil.

4.4. Chemical Production from Bio-Oil

Several hundred chemical constituents have been identified to date, and increasing attention is being paid to recovery of individual compounds or families of chemicals. Chemicals that have been reported as recovered include polyphenols for resins with formaldehyde, calcium and/or magnesium acetate for biodegradable de-icers, fertilizers, levoglucosan, hydroxyacetaldehyde, and a range of flavourings and essences for the food industry. The potentially much higher value of specialty chemicals compared to fuels could make recovery of even small concentrations viable. An integrated approach to chemicals and fuels production offers interesting possibilities for shorter term economic implementation.

CONCLUSION

The fast pyrolysis of biomass has the potential to contribute to the world's need for liquid fuels and, ultimately, for chemicals production. There are some interesting challenges to be faced in developing and modifying fast pyrolysis technology, in upgrading the liquids and adapting applications to accept the unusual behavior and characteristics of the liquid product. Higher added value products than fuels offer the most challenging and interesting opportunities including bulk chemicals such as fertilizers. Pyrolysis also offers potential for waste treatment by fixing contaminants in the char while also producing a clean liquid fuel.

REFERENCES

[1]   Dinesh M, Charles U, Pittman J, and Philip H. Steele, Pyrolysis of Wood/Biomass for Bio-oil: A Critical Review, *Energy & Fuels* 2006; 20: 848-889.

[2]   Bridgwater A V, Czernik S, and Piskorz J. An overview of fast pyrolysis. In: Bridgwater A V, editors. In *Progress in Thermochemical Biomass Conversion, Blackwell Science*, London, 2001, pp 977-997.

[3]   Bridgwater A V. Catalysis in thermal biomass conversion. *Applied Catalysis A.* 1994; 116: 5-47.

[4]   Boucher M E, Chaala A, Pakdel H, and Roy C. Bio-oils obtained by vacuum pyrolysis of softwood bark as a liquid fuel for gas turbines. Part II: Stability and ageing of bio-oil and its blends with methanol and a pyrolytic aqueous phase. *Biomass Bioenergy* . 2000; 19: 351-361.

[5]   Bridgwater A V and Kuester J L. Research in Thermochemical Biomass Conversion. In: Bridgwater, A. V. *Catalyst Today.* Elsevier Science Publishers, London, 1996, 29: 285-295.

[6]   Bridgwater A V. Renewable fuels and chemicals by thermal processing of biomass. *Chemical Engineering Journal* 2003, 91: 87-102.

[7]   Czernik S, and Bridgwater AV. Overview of Applications of Fast Pyrolysis Bio-oil. *Energy and Fuels* 2004, 18: 590–598.

[8]   Bridgwater A V, Meier A D, and Radlein B D. An overview of fast pyrolysis of biomass. *Organic Geochemistry* 1999, 30: 1479-1493

[9]   Magrini-Blair K, Czernik S, French R, Parent Y, Ritland M, ad Chornet E. Fluidizable
      Catalysts for Producing Hydrogen by Steam Reforming Biomass Pyrolysis Liquids,
      Proceedings of the 2002 U.S. DOE Hydrogen Program Review, 2002

[10] Bridgwater A V, Czernik S, and Piskorz J. Status of Biomass Fast Pyrolysis, In:
      Bridgwater. A.V editors. *Fast Pyrolysis of Biomass: A Handbook*, CPL Press, 2002,
      Vol 2.

[11] Bridgwater A V. Pyrolysis of Wood/Biomass for Bio-oil: A Critical Review. *Journal of
      Analytical and Applied Pyrolysis* 1999, 51: 3-22.

[12] Raveendran k, Ganesh A, and Khilart K C. Influence of Mineral Matter on Biomass
      Pyrolysis Characteristics. *Fuel* 1995, 74(12): 1812-1822

[13] Scott D S, and Piskorz J. The continuous flash pyrolysis of biomass. *The Canada
      Journal of Chemical Engineering* 1984, 62 (6): 404-412

[14] Piskorz J, Majerski P, Radlein D. Composition of oils obtained by fast pyrolysis of
      different woods, In: Soltes E J, and Miline T A, editors, in *ACS Symposium Series 376:
      Pyrolysis Oils from Biomass, Producing, Analyzing, and Upgrading*, 1988: 167-178.

[15] Rowell R. M. *The Chemistry of Solid Wood; American Chemical Society*, Washington,
      DC, 1984

[16] Zugenmaier, P. Conformation and packing of various crystalline cellulose fibers.
      Progress in Polymer Science. 2001, 26: 1341-1417

[17] Modorsky S L, Hart V E, and Stravs S. *Journal of research of the National Bureau of
      Standards*. 1956, 54: 343-354.

[18] Pakhomov A M, Akad-Nauk I, Otdel S. S. S. B. *Khim Nauk* 1957, 1457-1449

[19] Soltes E J, Elder T J, *Pyrolysis in organic chemicals from Biomass*, In: Goldstein I S,
      Editors, CRC Press, Boca Raton, FL, 1981; 63-95

[20] Runkel R O H, Wilke K D. Zur Kenntnis des thermoplastischen Verhaltens von Holz.
      Holzals Rohund Werkstoff 1951, 9:260-270

[21] McCarthy J, and Islam A. Lignin chemistry, technology, and utilization: a brief history.
      In: Glasser W G, Northey R A, and Schultz T P. Editor. *Lignin: Historical, Biological
      and Materials PerspectiVes, ACS Symposium Series 742, American Chemical Society*,
      Washington, DC, 2000;, 2-100

[22] Berkowitz, N. On the differential thermal analysis of coal, *Fuel* 1957, 36: 355-373.

[23] Kudo, K, Yoshida, E. On the Decomposition Process of Wood Constituents in the
      Course of Carbonization. I. The Decomposition of Carbohydrate and Lignin in
      MIZUNARA (Quercus crispla BLUME). *Journal of Japanese Wood Research*
      1957,3(4): 125-127.

[24] Scott D S, Piskorz J, Radlein D. Liquid products from the continuous flash pyrolysis of
      biomass. *Industrial & Engineering Chemistry Process Design and Development*, 1985,
      24: 581-588.

[25] Gray M R, Corcoran W H, Gavalas G R. Pyrolysis of a wood-drived material: effect of
      moisture and ash content. *Industrial & Engineering Chemistry Process Design and
      Development*, 1985, 24: 646-651

[26] Gavalas G R, In: Anderson L L, editor. *Coal science and technology 4: Coal pyrolysis*.
      Amsterdam, Elsevier Science, 1982, 92-93

[27] Ekinci E, Putun A E, Ctroglu M, Love G D, Laerty C J, Snape C E. *International
      Conference on Coal Science*, Newcastle upon Tyne, 1991, 520-523

[28] Şensöz S, Angin D, and Yorgun S. Influence of particle size on the pyrolysis of rapeseed (Brassica napus L.): fuel properties of bio-oil. *Biomass &Bioenergy*. 2000, 19, 271-279

[29] Islam M N, Zailani R, Ani FN. Pyrolytic oil from fluidised bed pyrolysis of oil palm shall and its characterization, *Renewable Energy* 1999, 17: 73-84.

[30] Demirbas, A. Pyrolysis of ground beech wood in irregular heating rate conditions. *Journal of Analytical and Applied Pyrolysis* 2005, 73: 39-43.

[31] Mosier N, Wyman C, Dale B, Elander R, Lee YY, Holtzapple M, Ladisch M. Features of promising technologies for pretreatment of lignocellulosic biomass. *Bioresource Technology* 2005, 96: 673-686

[32] Mandels M, Hontz L, Nystrom J. Enzymatic hydrolysis of waste cellulose. *Biotechnology and Bioengineering* 1974, 16: 1471–1493.

[33] Yang Y, Sharma-Shivappa R, Burns J C, Cheng J. Dilute Acid Pretreatment of Oven-dried Switchgrass Germplasms for Bioethanol Production. *Energy & Fuels* 2009, 23: 3759–3766

[34] Sun Y, Cheng J, Hydrolysis of lignocellulosic materials for ethanol production: a review. *Bioresource Technology* 2002, 83: 1–11.

[35] Kaar E, Gutierrez CV, Kinoshita C M. Steam explosion of sugarcane bagasse as a pretreatment for conversion to ethanol. *Biomass and Bioenergy* 1998, 14: 277–287.

[36] Mok W S L, and Antal J. Uncatalyzed solvolysis of whole biomass hemicellulose by hot compressed liquid water. *Industrial & Engineering Chemistry Research* 1992, 31: 1157–1161.

[37] Nguyen Q A, Tucker M P, Keller F A, Eddy F P. Two-stage dilute-acid pretreatment of softwoods. *Applied Biochemistry and Biotechnology* 2000, 84-86: 561–576.

[38] Kim T H, Kim J S, Sunwoo C, Lee Y Y. Pretreatment of corn stover by aqueous ammonia. *Bioresource Technology* 2003, 90: 39–47.

[39] Piskorz J, Radlein D, Scott D S, Czernic S. Pretreatment of wood and cellulose for production of sugars by fast pyrolysis. *Journal of Analytical and Applied Pyrolysis* 1989, 16: 127–142.

[40] Dobele G, Dizhbite T, Rossinskaja G, Telysheva G, Meie D, Radtke S, Faix O J. Pre-treatment of biomass with phosphoric acid prior to fast pyrolysis: A promising method for obtaining 1, 6-anhydrosaccharides in high yields. *Journal of Analytical and Applied Pyrolysis* 2003, 68: 197–211.

[41] Hassan M E, Steele P, and Ingram L. Characterization of Fast Pyrolysis Bio-oils Produced from Pretreated Pine Wood. *Appl Biochem Biotechnol* 2009, 154: 182–192.

[42] Leech J. Running a dual fuel engine on crude pyrolysis oil. In: Kaltschmitt M, and Bridgwater A V. Editors, *Biomass Gasification and Pyrolysis*, 199, CPL press, 495-497.

[43] Andrews R, Patnaik P C, Liu Q. Firing fast pyrolysis oils in turbines. In: Milne, T. Editor, *Proceedings of the Biomass Pyrolysis Oil Properties and Combustion Meeting*, NREL/CP-430-7215 , 1994, 383-391

In: Biomass Processing, Conversion and Biorefinery   ISBN: 978-1-62618-346-9
Editors: Bo Zhang and Yong Wang   © 2013 Nova Science Publishers, Inc.

Chapter 8

BIOMASS TO BIO-OIL BY LIQUEFACTION

Huamin Wang[1,*] *and Yong Wang*[1,2]
[1]Pacific Northwest National Labortatory, Richland, WA, US
[2]Voiland School of Chemical Engineering and Bioengineering,
Washington State University, Pullman, WA, US

ABSTRACT

Significant efforts have been devoted to develop processes for the conversion of biomass, an abundant and sustainable source of energy, to liquid fuels and chemicals in order to replace diminishing fossil fuels and mitigate global warming. Thermochemical and biochemical methods have attracted the most attention. Among the thermochemical processes, pyrolysis and liquefaction are the two major technologies for the `direct conversion of biomass to produce a liquid product, often called bio-oil. This chapter focuses on liquefaction, a medium-temperature and high-pressure thermochemical process for the conversion of biomass to bio-oil. Water has been most commonly used as a solvent and the process is known as hydrothermal liquefaction (HTL). Fundamentals of the HTL process, key factors determining HTL behavior, the role of catalysts in HTL, properties of the produced bio-oil, and the current status of the technology are summarized. The liquefaction of biomass using organic solvents, a process called solvolysis, is also discussed. A wide range of biomass feedstocks have been tested for liquefaction including wood, crop residues, algae, food processing waste, and animal manure.

1. INTRODUCTION

Biomass, one of the most abundant carriers of renewable energy on the planet, is a sustainable source of carbon-based materials that can be converted to liquid hydrocarbon fuels for transportation. Significant research is being conducted worldwide to develop technologies for the conversion of biomass into liquid fuels in order to replace fossil fuels in

* Phone 509-371-6705. Fax 509-372-4732. E-mail: huamin.wang@pnnl.gov.

the current energy production and consumption system. Thermochemical methods such as pyrolysis, liquefaction, and gasification [1, 2] and biochemical methods [3] have attracted the most attention currently. Among these routes, pyrolysis and liquefaction are the two major technologies that produce a liquid, which is called bio-oil, from biomass. Pyrolysis occurs at atmosphere pressure and high temperatures (400-550 °C) in the absence of oxygen. Liquefaction occurs at high pressures (5.0-25.0 MPa) and medium temperatures (< 400 °C) using a solvent, primarily water, and perhaps some catalysts and reducing gases. During liquefaction, a number of complex reactions including depolymerization of biomass, decomposition of biomass fragments, and rearrangement of reactive fragments occur [4, 5], leading to the liquefaction of the biomass to bio-oil, a complex mixture of more than 300 oxygenated compounds.

As early as the 1930s, Berl suggested the production of petroleum-like products from various biomass feedstocks in hot water using alkali as catalysts [6, 7]. Since then biomass liquefaction has attracted significant attention from both fundamental and practical points of view and has been conducted primarily at lab- or bench-scale with a few pilot/demonstration scale processes [4]. A review by Moffatt et al. [8] in 1985 focused on the biomass liquefaction research from 1920 to 1980. A review by Elliott [9] in 1991 provided a summary of research efforts during 1983-1990. Demirbas [5] summarized the mechanisms of biomass liquefaction. A recent review by Tooret al. [4] emphasized the current status of technology of hydrothermal liquefaction (HTL) of biomass. The most recent review by Akhtar et al. [10] focused on the process conditions for optimum bio-oil yield in the hydrothermal liquefaction (HTL) of biomass. Other more general reviews [2, 11, 12] epitomized biomass liquefaction as one of their important sections.

Biomass feedstocks for liquefaction include wood, crop residues, algae, food processing waste, and animal manure. Wood and crop residue primarily contains lignocellulose (carbohydrate polymer and lignin); food processing waste contains triglyceride, protein, and small amounts of lignocellulose; animal manure contains a moderate triglyceride and high carbohydrate content; and algae contains triglyceride, protein, and carbohydrate. Lignocellulose, the structural component of biomass, is composed of oxygen-containing organic polymers: cellulose (28-55%), hemicellulose (17-35%), and lignin (17-35%). Cellulose is a high-molecular-weight linear crystalline polymer of β-D-anhydroglucopyranose units (AGUs, ~ 10 000 AGUs for cellulose chain in wood) [13]. Hemicelluloseis derived from several sugars in addition to glucose, such as xylose, mannose, galactose, and arabinose [2]. Lignin is an amorphous highly-branched and substituted polymer consisting of an irregular array of "hydroxyl-" and "methoxyl-" substituted phenyl-propane units [2]. Triglyceride, an ester derived from glycerol and three fatty acids, is the main constituent of vegetable oil and animal fats. Proteins are macromolecules formed from amino acids, which have an amino group bonded to carbon atom next to the carboxyl group. In the liquefaction of biomass, especially hydrothermal liquefaction, water is an important reactant and therefore the above biomass can be directly treated without an energy consuming drying process.

There are two major liquefaction processes: hydrothermal processing (HTL, water or aqueous solvent; section 2) and reactive solvent solvolysis (using organic reactive liquid solvent; section 3). Hydrothermal liquefaction uses low-cost water or aqueous solvent and therefore does not require a drying process which is ideal for treating wet biomass. Hydrothermal liquefaction is generally conducted at 280-370 °C and 10-25 MPa. At these conditions water is in subcritical state and has a range of exotic properties [4, 14]. Reactive solvent solvolysis, however, uses a number of different solvents including alcohols [15, 16], acetone [15], creosote oil [17], ethylene glycol [18], glycerol [19], tetralin [20], and recycled bio-oil [8]. Various catalysts have been used for liquefaction, including alkali (from the alkaline ash in wood, alkaline oxides, and carbonates), metals (nickel, copper, and zinc salts), and heterogeneous catalysts (metal oxides and supported metals) [2, 4]. Reducing gases, such as H_2 and CO, have also been used during liquefaction [10]. The two liquefaction processes and the role of solvents, catalysts, and reducing gases will be summarized in detail in the following sections.

Table 1. Typical compositions and physical properties of wood liquefaction bio-oil, wood pyrolysis bio-oil, and conventional fuel oil

	Wood liquefaction bio-oil [21, 23]	Wood fast pyrolysis bio-oil [23, 24]	Conventional fuel oil [25, 26]
Elementary analysis, wt.%			
Carbon	72	40-50	85
Hydrogen	7-9	6.0-7.6	11-13
Oxygen	16-20	36-52	0.1-1.0
Sulfur	0.1-0.2	0.00-0.02	1.0-1.8
Nitrogen	<1.0	0.00-0.15	0.1
Water, wt.%	5.1	17-30	0.02-0.1
Viscosity, cP	15,000 (61°C)	13-30 (50 °C)	180 (50°C)
HHV, MJ/kg	36	16-20	40
Density, kg/m^3	1.15	1.2-1.3	0.9-1.0

Bio-oil, or bio-crude, the liquid product of biomass liquefaction, is usually a dark brown, free-flowing liquid [21]. The exact composition of the bio-oil depends on the feedstock and the process of liquefaction. For instance, hydrothermal liquefaction of lignocellulose produces bio-oils containing carbohydrates, acetic acid, alcohols, ketones, aldehydes, pyran derivatives, phenol and its derivatives, long-chain carboxylic acids/esters, benzene derivatives, and long-chain alkanes [21, 22], which are derived primarily from the depolymerization and fragmentation of cellulose, hemicellulose, and lignin. Table 1 compares the typical elemental compositions and physical properties of wood liquefaction bio-oil, wood pyrolysis bio-oil, and conventional fuel oil. Compared to conventional fuel oil, bio-oils from wood liquefaction usually have high oxygen content (~20 wt.%) and reactive oxygen-containing chemicals with carbonyl, carboxyl, and hydroxyl functional groups, which result in significant problems such as low heating value, high water content, poor stability, poor volatility, high viscosity, and corrosiveness (Table 1) [2]. These problems limit the utilization of bio-oils as transportation fuels in standard equipment for using the petroleum-derived fuels. Therefore, extensive oxygen removal is required for upgrading of bio-oils to liquid transportation fuels with similar properties as petroleum fuels, such as high energy density, high stability, and high volatility [1, 2]. Other products formed from biomass liquefaction include char and tar, aqueous organics, and gases such as CO_2, CO, and H_2.

Biomass liquefaction process has been scaled-up to pilot scale and demonstrated at several places, such as Pittsburgh Energy Research Center (PERC process; pilot scale),

Lawrence Berkeley Laboratory (LBL process; demonstration scale), Shell Research Laboratory (HTU process; pilot scale), SCF Technologies (CatLiq process; pilot-scale), EPA's Water Engineering Research Laboratory (STORS process; demonstration scale), and Changing World Technologies (TDP process; pilot scale) [4, 8, 9]. However, most of the biomass liquefaction process development projects have been terminated and it is questionable if the liquefaction process could be commercialized because of a lack of scientific understanding and the requirement of high-cost upgrading of bio-oil. A technoeconomic assessment of direct biomass liquefaction to transportation fuels, including liquefaction in pressurized solvent (LIPS) or atmospheric flash pyrolysis (AFP) to bio-oils and then bio-oil hydrotreating and refining, was conducted by the working group of the International Energy Agency direct biomass liquefaction activity [27]. The capital cost for bio-oil production with LIPS was much higher than for AFP, whereas the capital cost for catalytic upgrading of the oil from LIPS was lower than for AFP because of the lower oxygen content in LIPS bio-oils than AFP bio-oils. However, upgrading of the LIPS bio-oils did not show significant advantage [1]. Therefore, the LIPS process was less economical than the AFP process for conversion of biomass to transportation fuels. However, liquefaction-derived bio-oils may prove to be more beneficial in the long term since they have properties more similar to conventional fuels compared to fast pyrolysis bio-oil (Table 1) [2].

2. HYDROTHERMAL LIQUEFACTION (HTL)

Hydrothermal liquefaction (HTL) is a high-pressure (10-25 MPa) and medium temperature (280-370°C) thermochemical process using subcritical water and perhaps some catalysts to convert biomass to bio-oil by breaking down and reforming the chemical building blocks of biomass. HTL mimics the natural geological processes which produced fossil fuels and allows for the conversion of a wide range of biomass feedstocks. Because water is the reaction medium and can self-separate from the produced bio-oil, HTL does not require drying of the feedstock as in the case of gasification or pyrolysis and therefore is best suited for wet biomass.

2.1. Fundamentals of HTL

Water plays a critical role in HTL. At the temperatures and pressures in typical HTL, water is in subcritical state and becomes a highly reactive medium promoting the breakdown and cleavage of chemical bonds. Subcritical water behaves very differently from water at room temperature, especially regarding to the following properties: permittivity (dielectric constant), dissociation, and density [4]. For instance, the dielectric constant of water decreases from 78.5 F m^{-1} at 25 °C and 0.1 MPa to 14.07 F m^{-1} at 350 °C and 20 MPa (subcritical water), the ionic product of water (K_W) increases from 10^{-14} to 10^{-12}, and the density decreases from 1 to 0.6 g cm^{-3} [4]. The decrease of the permittivity results in fairly nonpolar water molecules and therefore increasesthe solubility of hydrophobic organic compounds. The increase of the ionic product in subcritical water gives high levels of H$^+$ and OH$^-$, which accelerates many acid- and base-catalyzed reactions, such as biomass hydrolysis

[28, 29]. The high dissociation constant, together with the relatively high density of subcritical water, favors ionic reactions, such as the dehydration of carbohydrates and alcohols and aldol splitting [4].

Because of the dramatic changes in the physical and chemical properties in the subcritical condition, water becomes a good media for biomass liquefaction. The basic reaction mechanisms of biomass HTL can be described as [4, 5]: depolymerization of the biomass to micellar-like fragments by hydrolysis; degradation of the broken down fragments by cleavage, dehydration, dehydrogenation, decarboxylation, deoxygenation, and deamination; and rearrangement of the produced reactive compounds by condensation, cyclization and polymerization.

The specific reaction pathway and final product differ as the composition of feedstock varies. Cellulose and hemicellulose, the polymer of carbohydrates, undergo rapid decomposition by hydrolysis to form glucose, other saccharides, and oligomers under HTL conditions [4, 30]. Kinetic studies on cellulose decomposition in subcritical water indicated that the main pathways were dehydration of the reducing-end glucose via pyrolytic cleavage of the glycosidic bond in cellulose in subcritical water and hydrolysis of the glycosidic bond via swelling and dissolution of cellulose in near-critical and supercritical water [30, 31]. The reactions of microcrystalline cellulose, however, require an extra step to break down the crystalline. The formed glucose can be easily decomposed to form stable products, such as 5-hydroxylmethyl furfural (5-HMF) in subcritical water, by the sequential reactions of glucose isomerization, dehydration, and decomposition to intermediates and further dehydration of the intermediates to produce 5-HMF [30, 32]. Dehydration (C–O bond splitting) and retro-aldol condensation (namely, C–C bond breaking) were found to be the key reactions [30, 32]. Hemicellulose has a much lower degree of crystallinity and more abundant site groups than cellulose and therefore can be easily solubilized and hydrolyzed by both acid- and base-catalyzed reactions in water at HTL conditions [4]. The saccharides released during hemicellulose hydrolysis are also degraded to furfural and other products in a way similar to the degradation of glucose [33]. Lignin, which is relatively stable, can be converted to phenols and methoxy phenols by the hydrolysis of ether-bonds and the secondary hydrolysis of methoxyl group with the formation of significant amounts of solid residues [4, 34]. Triglycerides are readily hydrolyzed in subcritical water with the production of relative stable free fatty acids and glycerol [35]. Proteins can be depolymerized by hydrolyzing the peptid bond (C-N) to form amino acids with relative low yields [11]. Further degradation of amino acids by decarboxylation and deamination occur to form acids and amines [36].

2.2. The Role of the Catalysts

Catalysts have been frequently used in biomass liquefaction processes. Homogenous catalysts in the form of alkali salts have been extensively utilized and found to be useful in promoting bio-oil formation. The heterogeneous catalysts such as supported metals and some metal oxides have been less frequently used in HTL process.

Alkali salts, such as the carbonates and hydroxides of sodium, potassium, calcium, and barium, are proven to have a positive effect on HTL by increasing liquid yields and suppressing char and tar formation as well as improving oil quality. For instance, the study on the effects of homogeneous catalysts (H_2SO_4 and NaOH) on glucose in hot compressed water

(200 °C) showed that the basic catalyst promoted isomerization of glucose to fructose while the acid catalyst promoted dehydration [37]. Na_2CO_3 was found effective for enhancing the formation of oil products and also gas products while suppressing the formation of solid residue in the liquefaction of paulownia at 280-360 °C [38]. Addition of 1.0 wt% Na_2CO_3 during the liquefaction of corn stalks increased the yield of bio-crude (from 33.4 to 47.2%) as well as improved the quality of the liquid product [39]. The use of alkali catalyst (0.25 M to 1 M K_2CO_3) suppressed char formation from 31-44 to 1-5% and increased liquid yields from 8-16 to 43-50% for the HTL of wood biomass at 280 °C [40].

The mechanism of the effect of alkali salts is not well understood. It was believed alkali salts promote gasification and the water-gas shift reaction and inhibit dehydration reaction. The water-gas shift reaction favors H_2 formation at the expense of CO and the produced hydrogen gas acts as a reducing agent increasing the heat value and quality of the oil product. Alkali is thought to combine with carbon monoxide and form active reducing agents, which are suggested to be nascent hydrogen or formate (For instance: $K_2CO_3 + CO + H_2O \rightarrow KHCO_3 + HCOOK$ [4]), that can in turn hydrogenate groups such as carbonyls. In addition, the inhibited dehydration of biomass monomers suppresses the formation unsaturated compounds which easily polymerize to char and tar [4].

Heterogeneous catalysts, which were proven to have a significant positive effect on gasification processes, have also been studied in HTL processes. The effect of heterogeneous catalysts on HTL, however, is controversial. The research on the HTL of wood at 350 °C indicated that a Ru/TiO_2 catalyst was able to convert char to gas, while leaving the oil product practically unaltered with respect to composition and yield [41]. The HTL of microalga *Dunaliellasalina* at 200 °C showed a significant increase in bio-oil yield over a Ni/REHY catalyst [42]. Recent research conducted the HTL of microalga *Nannochloropsis sp.* at 350 °C in the presence of six different heterogeneous catalysts (Pd/C, Pt/C, Ru/C, $Ni/SiO_2-Al_2O_3$, sulfided $CoMo/\gamma-Al_2O_3$, and zeolite) under inert (helium) and high-pressure reducing (hydrogen) conditions. In the absence of H_2, all of the catalysts tested produced higher yields of bio-oil with similar elemental compositions and heating values than that without a catalyst, whereas in the presence of high-pressure H_2 the bio-oil yield and heating value as well as the gas yield were insensitive to the presence or identity of the catalysts [43].

2.3. Effect of Process Parameters on Bio-oil Production

The composition of biomass feedstocks strongly influences the composition and yield of produced bio-oil. For instance, in lignocellulosic biomass, cellulose and hemicelluloses are favorable for the bio-oil yield, whereas lignin tends to form char residue. Besides the compositional variations of feedstock, the processing conditions of HTL, including temperature, residence time, biomass heating rate, pressure, biomass particle size, and type of gas, are also important to the yield and composition of produced bio-oil [10].

Temperature is the most important parameter for the HTL process. The effect of temperature on bio-oil yield depends on the type of the biomass feedstock. A volcano type dependence of bio-oil yield against temperature was reported for HTL of lignocellulosic biomass by various researches [10]. Initial increase of the temperature triggers conversion of biomass to bio-oil and then leads to maximum bio-oil yields. Further increase in temperature suppresses the formation of bio-oil, because of the gas formation by secondary decomposition and char formation by repolymerization of active species in the bio-oil. It was assumed that 300-330°C would be a suitable temperature range for the efficient production of bio-oils from lignocellulosic biomass by HTL process [10]. For biomass other than softwood and grass, a different optimum temperature was expected. A higher temperature of around 360°C was reported as the optimum temperature for HTL of microalgae *Dunalielatertiolecta* [44].

Residence time is another critical parameter for the HTL process. Similar to the temperature dependence, the effect of residence time also follows volcano-plot type behavior [14, 45]. Short residence times are beneficial for effective degradation of biomass at near critical conditions because of the fast hydrolysis and decomposition reactions. Long residence times would lead to the secondary and tertiary reactions and the subsequent conversion of liquids to gas and solids. The optimum residence time depends on the reaction temperature. Longer residence times are more suitable in low temperature liquefaction. The composition of bio-oil also depends on the residence time, due to the different extent of the secondary and tertiary reactions.

Table 2. Summary of some recent results for HTL of various biomass feedstocks

Biomass		HTL conditions				Bio-oil product			Ref.
		T (°C)	Residence time (min)	Catalyst	Gas	Yield (%)	Composition (C, H, O, N; wt%)	Heating value (MJ/kg)	
Wood and crop residue	Jack pine sawdust	300	30	None	N_2	30	79.3, 6.7, 17.1, 0	32.3	[48]
				Ba(OH)$_2$		45	73.4, 7.0, 19.6. 0	31.3	
				Ca(OH)$_2$		65	77.3, 7.5, 15.2, 0	34.1	
	Sawdust	350	10	None,	N_2	61	$C_1H_{0.95}O_{0.42}$	25.8	[41]
				Ru/TiO$_2$		63	$C_1H_{1.08}O_{0.39}$		
	Wood flour	350	30	Formic acid	-	45.4	65.2, 6.0, 28.8, -	-	[45]
	Oil palm fruit press fiber	278	40	NaOH	N_2	76.2	-	-	[49]
	Silver birch	380	-	Na$_2$CO$_3$	N_2	53.3	76.2, 5.4, 18.3, -	30.1	[50]
	Paulownia	340	10	Iron catalyst	-	36.3	-	-	[22]
Algae	ScenedesmusSpirulina	300	30	None	N_2	31-45	72, 9, 10, 6.5-8	35	[47]
	Desmodesmus sp.	300	60	None	He	47	75, 8.8, 10.2, 6.0	34.9	[46, 51]
		375	5			49	74.5,8.6,10.,6.3	35.4	
	Dunaliellatertiolecta	360	50	Na$_2$CO$_3$	-	25.8	63.6,7.7,25.1,3.7	30.7	[44]
	Nannochloropsis sp.	350	60	None	He	35	75.3, 10.2, 9.0, 4.2	38.5	[43]
					H$_2$	46	75.5, 10.5, 9.2, 4.1	39.0	
				Pd/C	He	56	73.4, 10.8, 9.0, 3.9	38.6	
					H$_2$	55	74.9, 10.6, 9.0, 4.2	38.9	
Food processing waste	Cornelian cherry stones	250-300	0-30	None	N_2	15-30	60-67, 6.7-8, 25-31, 0.4-0.7	25-28	[52]
	Waste	330-350	10	K$_2$CO$_3$, ZrO$_2$	-	0-13	-	-	[53]
Animal manure	Cattle manure	310	15	NaOH	N_2, CO, H$_2$	38-48	73.7, 8.1, 16.8, 1.4	35.5	[54]

Pressure is important during HTL for maintaining a condensed phase to avoid the large enthalpy requirement during evaporation and for promoting oil production by increasing solvent density and thus enhancing decomposition and extraction reactions [10, 14]. Heating rates and the size of biomass particles are relatively less influential parameters. A moderate heating rate may be sufficient because the subcritical water helps the dissolution and stabilization of fragmented species while overcoming the heat transfer limitation [10]. Particle sizes in the 4-10 mm range are considered to have reasonable grinding cost and to be suitable to overcome heat and mass transfer limitation in water [10]. Reducing gases are expected to stabilize the free radicals and active fragmented products of HTL and therefore reduce char formation. However, their usage requires specialized reactors to eliminate the gas channeling and maldistribution [10].

2.4. HTL of Various Biomass Feedstocks

A significant number of studies on biomass HTL have been conducted over the past few decades. Only a few of the most recent studies are summarized in Table 2, involving HTL of various feedstocks at different conditions in the absence or presence of some catalysts and reducing gases. Biomass feedstocks include wood and crop residue, algae, food processing waste, and animal manure. Algae biomass is particularly the focus of the most recent research for HTL [46], probably because algae are generally collected as wet matrix (5-20% of algae in water) and therefore HTL is more favorable than other processes such as pyrolysis which requires water volatilization [47]. Catalysts studied for biomass HTL include homogenous catalysts in form of alkali salts and heterogeneous catalysts such as Ru/TiO_2 and ZrO_2. Gases investigated include inert gases such as N_2 and He and reducing gases such as H_2 and CO. Bio-oil yield and properties vary widely between different feedstocks and processes. In general, bio-oil yield varies from 13 to 76%, carbon content of HTL bio-oil varies from 60 to 80%, hydrogen content varies from 5 to 11 %, and oxygen content varies from 9 to 30%. The bio-oils derived from algae normally have high nitrogen level of 4 to 8%. The energy density is in the range of 30-40 MJ/kg, a value considerably higher than the biomass feedstock. HTL bio-oil also contains water (4-8%). The high content of oxygen, especially as the active oxygen-containing chemicals, and water in HTL bio-oils result in multiple significant problems, limiting the direct utilization of bio-oil and requiring the subsequent bio-oil upgrading.

3. SOLVOLYSIS

Solvolysis is a liquefaction process similar to HTL but using an organic solvent. Alcohols, including methanol [55-57], ethanol [57-63], butanol [64], andoctanol [65-67], polyols, including glycerol [19] and ethylene glycol [67-70], and other solvents, including phenol [70-72], acetone [57, 64], 1,4-dioxane [55, 57, 73], tetralin [20, 74, 75], and creosote oil [17] are used for the solvolysis of various biomass to produce bio-oils. Preferred solvents are those that provide good performance during liquefaction, can be readily recovered for reuse, and are produced by biomass liquefaction.

Solvents greatly influence the yield and composition of bio-oil from biomass liquefaction. A comparison of solvents, including water, acetone, and ethanol, in the liquefaction of pinewood at 250-450 °C showed that the highest conversion rate was obtained when acetone was used as a solvent. The maximum oil yield was found to be 26.5% at 200 °C using ethanol, and the solvent efficiency can be sequenced as: ethanol > acetone > water in terms of the oil yield [63]. This study also showed that solvents strongly influenced the distribution and relative abundance of produced compounds. Liquefaction of oil palm fruit press fiber (FPF) with sub/supercritical methanol, ethanol, acetone, and 1,4-dioxane at 210 to 330 °C indicated that the sequence of the highest conversion followed methanol (81.5%) >1,4-dioxane (80.0%) >ethanol (77.8%) > acetone (67.9%) while the sequence of the highest bio-oil yield followed 1,4-dioxane (50.9%) > acetone (38.5%) ≈ methanol (38.0%) > ethanol (36.9%), indicating that subcritical 1,4-dioxane treatment was the most effective solvent in the degradation of FPF [57]. However, a study on the effect of different organic solvents, such as methanol, ethanol and 1,4-dioxane, on liquefaction of *Spirulina* (a kind of high-protein microalgae) indicated that methanol and ethanol (polar protic solvent) resulted in a higher conversion rate and bio-oil yield than that obtained using 1,4-dioxane (dipolar aprotic solvent) [55].

The compositions of bio-oil products were greatly affected by the type of solvent used and the bio-oil generated in methanol contained higher carbon and hydrogen concentrations but a lower oxygen content, resulting in a higher heating value [55]. A study on the liquefaction ofpoplar at 140-190 °C showed that *n*-octanol had the highest liquefaction efficiency among the three alcoholic solvents: *n*-octanol, dihydric ethylene glycol, and trihydric glycerol [67].

A biomass composed of a mixture of wastes (straw, wood and grass) was liquefied using Nickel Raney as a catalyst and tetralin as a solvent at relatively low hydrogen pressure and the products were primarily straight long chain alkanes (C13–C26), which are much different compared to conventional liquefaction bio-oils [74]. Some articles have also reported that the presence of organic solvents could lower the viscosity of bio-oil derived from biomass liquefaction [5].

The effect of organic solvents still requires further understanding. It was proposed that the difference in density and the polarity of solvent might be the main reasons for the different behavior during liquefaction. Increasing the density of the solvent would lead to stronger dissolving ability and better dilution of the reaction products which consequently improved bio-oil formation [55].

Most of the reactions in biomass liquefaction are ionic reactions and free radical reactions, which are related to the polarity of the reaction system. The polar solvents (such as methanol and ethanol) contain hydrogen bonds and thus could release hydrogen free radical (H·) for the breakdown of long-chain polymers by hydrocracking and for stabilizing small biomass-derived intermediates by hydrogenation [55]. These reactions prevented the formation of residue and improved the bio-oil yield [60]. Some solvents, such as tetralin, are hydrogen donor solvents, which can not only donate hydrogen but also act as a hydrogen transport vehicle [20].

Therefore, hydrogen donor solvents may stabilize the free radicals produced during liquefaction by hydrogenation, which prevents their further condensation into heavy molecules. Some solvents, such as *n*-octanol, can be combined by esterification with acidic fragments from biomass cracking, which forms volatile and simple components in the produced bio-oil [65, 67].

CONCLUSION

Among the thermochemical processes for biomass conversion to liquid fuels including gasification, pyrolysis, and liquefaction, liquefaction is attractive because of its ability to treat wet biomass feedstock and the produced bio-oil has properties more similar to conventional fuels. Liquefaction occurs at high pressures (5.0-25.0 MPa) and low temperatures ($<400^{\circ}C$) using a solvent and perhaps some catalysts and reducing gases. During liquefaction, a number of complex reactions including depolymerization of biomass, decomposition of biomass monomers, and the rearrangement of reactive fragments occurs, leading to the formation of liquefaction bio-oil, a complex mixture of oxygenates and hydrocarbons. Water has been most commonly used as a solvent (known as hydrothermal liquefaction), but the use of organic solvents (called solvolysis) has also been studied. The HTL process uses subcritical water, which is a highly reactive medium promoting the breakdown and cleavage of chemical bonds. The reaction pathway and the yield and composition of the final product differ as the composition of feedstock varies. The processing conditions of HTL, including temperature, resident time, biomass heating rate, pressure, biomass particle size, and type of gas also influence the yield and composition of the produced bio-oil. Catalysts have been frequently used in biomass liquefaction process. Homogenous catalysts in the form of alkali salts have been extensively utilized and found to be useful in promoting bio-oil formation. The heterogeneous catalysts such as metals and some metal oxides have been less frequently used in the HTL process. Solvolysis of biomass uses organic solvent, including alcohols, polyols, acetone, 1,4-dioxane, phenol, tetralin, and creosote oil, which influences the yield and composition of produced bio-oil because of their different densities and polarities. Biomass liquefaction processes have been studied for the conversion of various biomass feedstocks including wood and crop residue, algae, food processing waste, and animal manure. The produced bio-oil requires upgrading prior its utilization as liquid transportation fuels.

REFERENCES

[1]   Elliott D. C. Historical developments in hydroprocessing bio-oils. *Energy & Fuels.* 2007; 21:1792.
[2]   Huber G. W., Iborra S., Corma A. Synthesis of transportation fuels from biomass: Chemistry, catalysts, and engineering. *Chemical Review* 2006;106:4044.
[3]   Saxena R. C., Adhikari D. K., Goyal H. B. Biomass-based energy fuel through biochemical routes: A review. Renewable and Sustainable *Energy Reviews 2009*; 13:167.
[4]   Toor S. S., Rosendahl L., Rudolf A. Hydrothermal liquefaction of biomass: A review of subcritical water technologies. *Energy* 2011;36:2328.
[5]   Demirbas A. Mechanisms of liquefaction and pyrolysis reactions of biomass. *Energy Conversion and Management* 2000;41:633.
[6]   Berl E. Production of oil from plant material. *Science* 1944;99:309.
[7]   Berl E. Origin of asphals, oil, natural gas and bituminous coa. *Science* 1934;80:227.
[8]   Moffatt J. M., Overend R. P. Direct liquefaction of wood through solvolysis and catalytic hydrodeoxygenation - an engineering assessment. *Biomass* 1985;7:99.

[9]   Elliott D. C., Beckman D., Bridgwater A. V., Diebold J. P., Gevert S. B., Solantausta Y. Development in direct thermochemical liquefaction of bipmass - 1983-1990. *Energy & Fuels* 1991;5:399.

[10]   Akhtar J., Amin N. A. S. A review on process conditions for optimum bio-oil yield in hydrothermal liquefaction of biomass. *Renewable & Sustainable Energy Reviews* 2011; 15:1615.

[11]   Peterson A. A., Vogel F., Lachance R. P., Froeling M., Antal M. J., Jr., Tester J. W. Thermochemical biofuel production in hydrothermal media: A review of sub- and supercritical water technologies. *Energy & Environment Science* 2008;1:32.

[12]   Demirbas A. Biorefineries: Current activities and future developments. *Energy Conversion and Management* 2009;50:2782.

[13]   van de Vyver S., Geboers J., Jacobs P. A., Sels B. F. Recent Advances in the Catalytic Conversion of Cellulose. *ChemCatChem* 2011;3:82.

[14]   Behrendt F., Neubauer Y., Oevermann M., Wilmes B., Zobel N. Direct liquefaction of biomass. *Chemical Engineering & Technology* 2008;31:667.

[15]   Erzengin M., Kucuk M. M. Liquefaction of sunflower stalk by using supercritical extraction. *Energy Conversion and Management* 1998;39:1203.

[16]   Mun S. P., Hassan E. B. M. Liquefaction of lignocellulosic biomass with mixtures of ethanol and small amounts of phenol in the presence of methanesulfonic acid catalyst. *Journal of Industrial and Engineering Chemistry* 2004;10:722.

[17]   Karaca F. Molecular mass distribution and structural characterization of liquefaction products of a biomass waste material. *Energy & Fuels* 2006;20:383.

[18]   Rezzoug S. A., Capart R. Liquefaction of wood in two successive steps: solvolysis in ethylene-glycol and catalytic hydrotreatment. *Applied Energy* 2002;72:631.

[19]   Demirbas A. Liquefaction of biomass using glycerol. *Energy Sources Part a-Recovery Utilization and Environmental Effects* 2008;30:1120.

[20]   Wang G., Li W., Chen H., Li B. The direct liquefaction of sawdust in tetralin. *Energy Sources Part a-Recovery Utilization and Environmental Effects* 2007;29:1221.

[21]   Elliott Douglas C., Sealock L. J., Butner R. S. Product Analysis from Direct Liquefaction of Several High-Moisture Biomass Feedstocks. In: Soltes EJ, Milne TA, editors. *Pyrolysis Oils from Biomass:* ACS Symp. Seri. 376; 1988, p. 179.

[22]   Sun P., Heng M., Sun S.-H., Chen J. Analysis of liquid and solid products from liquefaction of paulownia in hot-compressed water. *Energy Conversion and Management* 2011;52:924.

[23]   Schiefelbein D. C. E. G. F. Liquid hydrocarbon fuels from biomass. *Preprint Papers - American Chemical Society, Division of Fuel Chemistry* 1989;34:1160.

[24]   Hassan E. B. M., Steele P. H., Ingram L. Characterization of Fast Pyrolysis Bio-oils Produced from Pretreated Pine Wood. *Applied Biochemistry and Biotechnology* 2009; 154:182.

[25]   Czernik S., Bridgwater A. V. Overview of applications of biomass fast pyrolysis oil. *Energy & Fuels* 2004;18:590.

[26]   Furimsky E. Catalytic hydrodeoxygenation. *Applied Catalysis a-General* 2000;199:147.

[27]   Elliott D. C., Baker E. G., Beckman D., Solantausta Y., Tolenhiemo V., Gevert S. B. et al. Technoeconomic accessment of direct biomass liquefaction to transportation fuels. *Biomass* 1990;22:251.

[28] Hunter S. E., Savage P. E. Recent advances in acid- and base-catalyzed organic synthesis in high-temperature liquid water. *Chemical Engineering Science* 2004; 59: 4903.

[29] Akiya N., Savage P. E. Roles of water for chemical reactions in high-temperature water. *Chemical Review* 2002;102:2725.

[30] Yu Y., Lou X., Wu H. Some Recent Advances in Hydrolysis of Biomass in Hot-Compressed Water and Its Comparisons with Other Hydrolysis Methods. *Energy & Fuels* 2007;22:46.

[31] Matsumura Y., Sasaki M., Okuda K., Takami S., Ohara S., Umetsu M. et al. Supercritical water treatment of biomass for energy and material recovery. *Combustion Science and Technology* 2006;178:509.

[32] Kabyemela B. M., Adschiri T., Malaluan R. M., Arai K. Kinetics of Glucose Epimerization and Decomposition in Subcritical and Supercritical Water. *Industrial & Engineering Chemistry Research* 1997;36:1552.

[33] Sasaki M., Hayakawa T. KATA. Measurement of the rate of retro-aldol condensation of d-xylose in subcritical and supercritical water. *The Proceeding of the 7th International Symposium on Hydrothermal Reactions;* 2003.

[34] Wahyudiono, Kanetake T,, Sasaki M,, Goto M. Decomposition of a Lignin Model Compound under Hydrothermal Conditions. *Chemical Engineering & Technology* 2007;30:1113.

[35] Holliday R. L., King J. W., List G. R. Hydrolysis of vegetable oils in sub- and supercritical water. *Industrial & Engineering Chemistry Research* 1997;36:932.

[36] Klingler D., Berg J., Vogel H. Hydrothermal reactions of alanine and glycine in sub- and supercritical water. *The Journal of Supercritical Fluids* 2007;43:112.

[37] Watanabe M., Aizawa Y., Iida T., Aida T. M., Levy C., Sue K. et al. Glucose reactions with acid and base catalysts in hot compressed water at 473 K. *Carbohydrate Research* 2005;340:1925.

[38] Sun P., Heng M., Sun S., Chen J. Direct liquefaction of paulownia in hot compressed water: Influence of catalysts. *Energy* 2010;35:5421.

[39] Song C., Hu H., Zhu S., Wang G., Chen G. Nonisothermal Catalytic Liquefaction of Corn Stalk in Subcritical and Supercritical Water. *Energy & Fuels* 2003;18:90.

[40] Bhaskar T., Sera A., Muto A., Sakata Y. Hydrothermal upgrading of wood biomass: Influence of the addition of K2CO3 and cellulose/lignin ratio. *Fuel* 2008;87:2236.

[41] Knezevic D., van Swaaij W., Kersten S. Hydrothermal Conversion Of Biomass. II. Conversion Of Wood, Pyrolysis Oil, And Glucose In Hot Compressed Water. *Industrial & Engineering Chemistry Research* 2010;49:104.

[42] Yang C., Jia L., Chen C., Liu G., Fang W. Bio-oil from hydro-liquefaction of Dunaliella sauna over Ni/REHY catalyst. *Bioresource Technology* 2011;102:4580.

[43] Duan P., Savage P. E. Hydrothermal Liquefaction of a Microalga with Heterogeneous Catalysts. *Industrial & Engineering Chemistry Research* 2011;50:52.

[44] Zou S., Wu Y., Yang M., Kaleem I., Chun L., Tong J. Production and characterization of bio-oil from hydrothermal liquefaction of microalgae Dunaliella tertiolecta cake. *Energy* 2010;35:5406.

[45] Yilgin M., Pehlivan D. Poplar wood–water slurry liquefaction in the presence of formic acid catalyst. *Energy Conversion and Management* 2004;45:2687.

[46] Alba L. G., Torri C., Samori C., van der Spek J., Fabbri D., Kersten S. R. A. et al. Hydrothermal Treatment (HIT) of Microalgae: Evaluation of the Process As Conversion Method in an Algae Biorefinery Concept. *Energy & Fuels* 2012;26:642.

[47] Vardon D. R., Sharma B. K., Blazina G. V., Rajagopalan K., Strathmann T. J. Thermochemical conversion of raw and defatted algal biomass via hydrothermal liquefaction and slow pyrolysis. *Bioresource Technology* 2012;109:178.

[48] Xu C., Lad N. Production of heavy oils with high caloric values by direct liquefaction of woody biomass in sub/near-critical water. *Energy & Fuels* 2008;22:635.

[49] Mazaheri H., Lee K. T., Bhatia S., Mohamed A. R. Subcritical water liquefaction of oil palm fruit press fiber in the presence of sodium hydroxide: An optimisation study using response surface methodology. *Bioresource Technology* 2010;101:9335.

[50] Qian Y., Zuo C., Tan H., He J. Structural analysis of bio-oils from sub-and supercritical water liquefaction of woody biomass. *Energy* 2007;32:196.

[51] Torri C., Alba L. G., Samori C., Fabbri D., Brilman D. W. F. Hydrothermal Treatment (HTT) of Microalgae: Detailed Molecular Characterization of HTT Oil in View of HTT Mechanism Elucidation. *Energy & Fuels* 2012;26:658.

[52] Akalin M. K., Tekin K., Karagoz S. Hydrothermal liquefaction of cornelian cherry stones for bio-oil production. *Bioresource Technology* 2012;110:682.

[53] Hammerschmidt A., Boukis N., Hauer E., Galla U., Dinjus E., Hitzmann B. et al. Catalytic conversion of waste biomass by hydrothermal treatment. *Fuel* 2011;90:555.

[54] Xiu S., Shahbazi A., Shirley V., Cheng D. Hydrothermal pyrolysis of swine manure to bio-oil: Effects of operating parameters on products yield and characterization of bio-oil. *Journal of Analytical and Applied Pyrolysis* 2010;88:73.

[55] Yuan X., Wang J., Zeng G., Huang H., Pei X., Liu Z. et al. Comparative studies of thermochemical liquefaction characteristics of microalgae using different organic solvents. *Energy* 2011;36:6406.

[56] Yang Y., Gilbert A., Xu C. Production of Bio-Crude from Forestry Waste by Hydro-Liquefaction in Sub-/Super-Critical Methanol. *AIChE Journal* 2009;55:807.

[57] Mazaheri H., Lee K. T., Bhatia S., Mohamed A. R. Sub/supercritical liquefaction of oil palm fruit press fiber for the production of bio-oil: Effect of solvents. *Bioresource Technology* 2010;101:7641.

[58] Pei X., Yuan X., Zeng G., Huang H., Wang J., Li H. et al. Co-liquefaction of microalgae and synthetic polymer mixture in sub- and supercritical ethanol. *Fuel Processing Technology* 2012;93:35.

[59] Liu H.-M., Feng B., Sun R.-C. Enhanced Bio-oil Yield from Liquefaction of Cornstalk in Sub- and Supercritical Ethanol by Acid–Chlorite Pretreatment. *Industrial & Engineering Chemistry Research* 2011;50:10928.

[60] Huang H., Yuan X., Zeng G., Wang J., Li H., Zhou C. et al. Thermochemical liquefaction characteristics of microalgae in sub- and supercritical ethanol. *Fuel Processing Technology* 2011;92:147.

[61] Chumpoo J., Prasassarakich P. Bio-Oil from Hydro-Liquefaction of Bagasse in Supercritical Ethanol. *Energy & Fuels* 2010;24:2071.

[62] Xu C., Etcheverry T. Hydro-liquefaction of woody biomass in sub- and super-critical ethanol with iron-based catalysts. *Fuel* 2008;87:335.

[63] Liu Z., Zhang F.-S. Effects of various solvents on the liquefaction of biomass to produce fuels and chemical feedstocks. *Energy Conversion and Management* 2008;49: 3498.

[64] Aysu T. Supercritical fluid extraction of reed canary grass (Phalaris arundinacea). *Biomass & Bioenergy* 2012;41:139.

[65] Zou X., Qin T., Wang Y., Huang L. Mechanisms and Product Specialties of the Alcoholysis Processes of Poplar Components. *Energy & Fuels* 2011;25:3786.

[66] Zou X.-W., Yang Z., Qin T.-F. FTIR Analysis of Products Derived from Wood Liquefaction with 1-Octanol. *Spectroscopy and Spectral Analysis* 2009;29:1545.

[67] Zou X., Qin T., Huang L., Zhang X., Yang Z., Wang Y. Mechanisms and Main Regularities of Biomass Liquefaction with Alcoholic Solvents. *Energy & Fuels* 2009;23: 5213.

[68] Zhang H., Pang H., Shi J., Fu T., Liao B. Investigation of Liquefied Wood Residues Based on Cellulose, Hemicellulose, and Lignin. *Journal of Applied Polymer Science* 2012;123:850.

[69] Zou S., Wu Y., Yang M., Li C., Tong J. Thermochemical Catalytic Liquefaction of the Marine Microalgae Dunaliella tertiolecta and Characterization of Bio-oils. *Energy & Fuels* 2009;23:3753.

[70] Yip J., Chen M., Szeto Y. S., Yan S. Comparative study of liquefaction process and liquefied products from bamboo using different organic solvents. *Bioresource Technology* 2009;100:6674.

[71] Chen H., Zhang Y., Xie S. Selective Liquefaction of Wheat Straw in Phenol and Its Fractionation. *Applied Biochemistry and Biotechnology* 2012;167:250.

[72] Mishra G., Saka S. Kinetic behavior of liquefaction of Japanese beech in subcritical phenol. *Bioresource Technology* 2011;102:10946.

[73] Mun S. P., Hassan E. B., Hassan M. Liquefaction of lignocellulosic biomass with dioxane/polar solvent mixtures in the presence of an acid catalyst. *Journal of Industrial and Engineering Chemistry* 2004;10:473.

[74] Beauchet R., Pinard L., Kpogbemabou D., Laduranty J., Lemee L., Lemberton J. L. et al. Hydroliquefaction of green wastes to produce fuels. *Bioresource Technology* 2011; 102:6200.

[75] Vasilakos N. P., Austgen D. M. Hydrogen-donor solvents in biomass liquefaction. *Industrial & Engineering Chemistry Process Design and Development* 1985;24:304.

In: Biomass Processing, Conversion and Biorefinery
Editors: Bo Zhang and Yong Wang

ISBN: 978-1-62618-346-9
© 2013 Nova Science Publishers, Inc.

Chapter 9

BIOMASS GASIFICATION: PROCESS OVERVIEW, HISTORY AND DEVELOPMENT

Daniel T. Howe[*]

Pacific Northwest National Laboratory,
Operated by Battelle for the U. S. Department
of Energy Richland, Washington, US

ABSTRACT

The current unsustainable use trends of petrochemical fuels has increased the demand for renewable energy alternatives, and thermochemical processes such as gasification will be required to meet these demands. Gasification is a process that involves the high-temperature partial oxidation of carbon-containing fuels such as lignocellulosic biomass to produce a product gas consisting of hydrogen, carbon monoxide, carbon dioxide, methane, and condensable aromatic liquids known as tars. The non-condensable product gas can be either purified into synthesis gas, a mixture of hydrogen and carbon monoxide used in Fischer-Tropsch synthesis of gasoline or diesel fuel, or sent to gas engines for the generation of heat and electricity. Gasification has been in use for over 180 years, with coal being the primary feedstock. Over the course of its development gasification technology has resulted in three general types of reactors: moving bed gasifiers, fluidized bed gasifiers, and entrained bed gasifiers. Moving bed gasifiers employ a fixed bed of fuel that moves downward under gravity while contacting an oxidant in either counter-current or co-current flow. Fluidized bed gasifiers use a bubbling fluidized bed of inert particles as a heat transfer medium while the oxidant serves as the source of fluidization.

Entrained flow gasifiers use a dense suspension of fine solid particles in the oxidant to create a high-temperature, high-throughput system. The choice of a specific type of gasifier is determined by the end use of the gas as well as engineering constraints. While the use of biomass in gasifiers is a relatively new phenomenon, a number of success stories do exist.

---

[*] Phone 509-372-4355. Fax 509-372-4252. E-mail: Daniel.howe@pnnl.gov.

1. INTRODUCTION

As the consumption of fossil fuels increases worldwide, the unsustainable rate of petrochemical use trends is receiving greater attention. The concerns being identified cover a wide range of topics, whether they are environmental [1], political [2], or economic [3]. Proposed solutions to these problems can take a wide variety of forms, each with its own set of technical and economic advantages and disadvantages. Although wind, solar, and geothermal technologies have seen a resurgence of development in both the public and private sectors [4], they cannot provide the carbon-based fuels required for the current transportation industry, which consumes 71% of all domestic and imported oil used in the United States [5]. The widespread use of coal for generating electricity also has led to numerous environmental and health concerns. Biomass, on the other hand, is cheap, abundant, and can be transformed directly into drop-in liquid transportation fuels [6], and it can be used in combined heat and power (CHP) generation units. Bioconversion processes have been developed to convert biomass-derived starches and sugars into alcohol fuels, and active process-development programs are underway on processes to biologically convert the cellulosic and hemicellulosic fractions of woody and herbaceous feedstocks into alcohols [7]. However, these advanced processes have a number of short comings, including their inability to convert the lignin in the biomass into fuels [8]. Fermentation of biomass-derived sugars will leave a lignin-rich residue that is not amenable to bioconversion processing. Thermochemical processes such as gasification, however, offer a way to convert either whole biomass or lignin-rich residues into liquid fuels, to complement the product output from an integrated biorefinery [9]. In addition, gasification can be used to generate heat or electricity from a wide variety of biomass feedstocks. The use of thermochemical conversion processes such as gasification is considered by many to be the most feasible route for the production of second generation biofuels [10], and the growing number of biomass gasification success stories illustrates its importance in the field of renewable energy.

2. GASIFICATION CHEMISTRY

2.1. Gasification Process Chemistry

Gasification is a high-temperature (> 700°C) partial oxidation of carbon-containing feedstocks into a gaseous form that consists of permanent (non-condensable) gases and condensable vapors known as tars. Partial oxidation refers to the amount of oxygen present for reaction, which is less than that stoichiometrically needed for complete combustion. Oxygen is usually fed to a gasifier either as air or steam. When the biomass is gasified, the non-condensable gas produced is a mixture of hydrogen, carbon monoxide, carbon dioxide, and methane, with trace quantities of higher hydrocarbons such as ethane, ethylene, and propane. This gas is often referred to as "synthesis gas" or "syngas," although pure syngas is a mixture of hydrogen and carbon monoxide without any other compounds. Many reactions taking take place in a gasifier, however the major reactions are shown below.

$$C + \tfrac{1}{2} O_2 \rightarrow CO \qquad (\Delta H = \text{-}111 \text{ MJ/kmol}) \tag{1}$$

$$C + O_2 \rightarrow CO_2 \qquad (\Delta H = \text{-}394 \text{ MJ/kmol}) \tag{2}$$

$$CO + \tfrac{1}{2} O_2 \rightarrow CO_2 \qquad (\Delta H = \text{-}283 \text{ MJ/kmol}) \tag{3}$$

$$H_2 + \tfrac{1}{2} O_2 \rightarrow H_2O \qquad (\Delta H = \text{-}242 \text{ MJ/kmol}) \tag{4}$$

Reactions (1) through (4) are often referred to as "combustion reactions", and generally proceed to near completion under gasification conditions. Other important reactions existing at equilibrium include

$$C + H_2O \leftrightarrow CO + H_2 \qquad (\Delta H = +131 \text{ MJ/kmol}) \quad \text{Water-Gas Reaction} \tag{5}$$

$$C + CO_2 \leftrightarrow 2CO \qquad (\Delta H = +172 \text{ MJ/kmol}) \quad \text{Boudouard Reaction} \tag{6}$$

$$C + 2H_2 \leftrightarrow CH_4 \qquad (\Delta H = \text{-}75 \text{ MJ/kmol}) \quad \text{Methanation Reaction} \tag{7}$$

When the carbon conversion is high, as is normal under gasification conditions, equilibrium reactions (5) through (7) can be combined into the following two reactions [11].

$$CO + H_2O \leftrightarrow CO_2 + H_2 \qquad (\Delta H = \text{-}41 \text{ MJ/kmol}) \qquad \text{Water-Gas Shift Reaction} \tag{8}$$

$$CH_4 + H_2O \leftrightarrow CO_2 + 3H_2 \quad (\Delta H = +131 \text{ Mj/kmol}) \qquad \text{Steam-Methane Reforming Reaction} \tag{9}$$

2.2. Biomass Chemistry

The term biomass, when used in the renewable energy sense, refers to biological material from living or recently living organisms. It has a major environmental advantage over fossil fuels such as coal or petroleum when gasified because the gasification process is carbon neutral. It releases the same amount of carbon dioxide into the atmosphere as is absorbed by the growing plants. Hence, there is no net increase in the amount of carbon dioxide in the atmosphere. While the history of gasification has focused primarily on coal, environmental concerns have resulted in lignocellulosic biomass being considered the best new potential feedstock for gasification. The term "lignocellulosic" refers to the three main constituents of biomass: lignin, cellulose, and hemicellulose.

The principal structural units in most plants are fibers formed from cellulose. Cellulose is a semicrystalline polysaccharide made up of D-glucopyranose units linked together by β-(1-4)-glucosidic bonds. Because of the large number of hydroxyl groups in its polymer chains, cellulose is highly hydrophilic. The structure of cellulose can be seen in Figure 1. In addition to cellulose, significant amounts of hemicelluloses also are present. Hemicelluloses are highly amorphous polymers of hexoses, pentoses, and uronic acid residues, with a low degree of polymerization. Hemicelluloses can include glucose, xylose, galactose, arabinose, and mannose monomers. Hemicelluloses contain a large number of hydroxyl and acetyl groups, and are partly soluble in water. The hemicelluloses serve as a bridge to the lignin in the plant. Lignin is an amorphous polymer that consists of repeated phenylpropane units held together by a large variety of different bonds. Lignin monomers are called p-hydroxyphenyl, guiacyl, and syringyl, respectively, and their structures are shown in Figure 2 [12]. Lignin has a high molecular weight, and because of its aromatic nature, it is highly hydrophobic. Lignin serves as the glue that holds fibers together, and it cements individual fibers into fiber bundles [13].

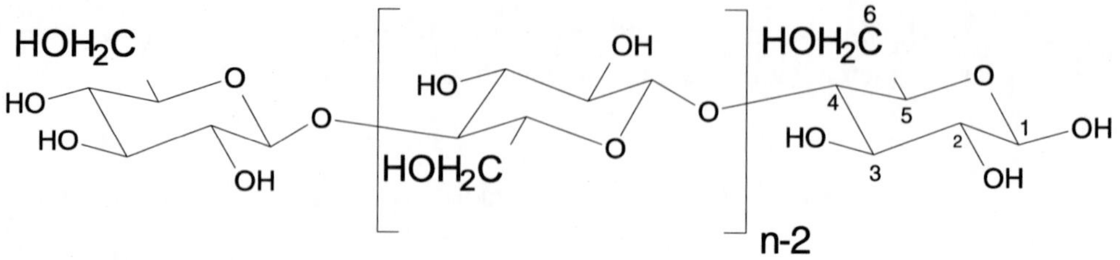

Figure 1. Cellulose structure with carbons labeled 1-6. Terminal hydroxyl group at the 1 position is referred to as the "reducing end," terminal hydroxyl groups at the 3 and 4 positions are referred to as the "non-reducing end."

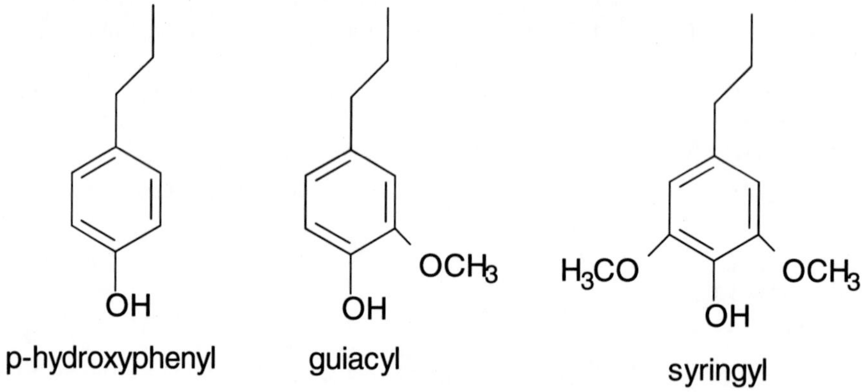

Figure 2. Structures of the lignin monomers.

Table 1. Typical fractions of cellulose, hemicellulose, and lignin found in some biomass feedstocks [14]

Biomass	Cellulose (%)	Hemicellulose (%)	Lignin (%)	Other – Protein, Minerals, Starch, Fat (%)
Wood	39.8	23.3	24.8	12.1
Corn Stover	35.3	25.3	10.8	28.6
Switchgrass	36.9	30.6	9.6	22.9
Wheat Straw	37.8	26.5	17.6	18.1
Corn Kernel	3.0	6.7	0.2	90.1(72% starch)

While lignocellulosic biomass is constructed of the three main constituents discussed above, the fractions of these compounds can differ widely depending on the type of plant. Softwoods and hardwoods generally have a larger fraction of both cellulose and lignin than grasses and agricultural residues. The average compositions for a number of different feedstocks are provided in Table 1.

3. THE HISTORY OF GASIFICATION

Although Flemish scientist Jan Baptist van Helmont was the first to observe a flammable "wild spirit" produced by heating coal in 1609, Scottish engineer William Murdoch is generally considered the inventor of coal gasification. In 1792, he began using a process in which he heated coal in the absence of air to produce a gas that he used to provide lighting for his home. With his employers Matthew Boulton and James Watt, he began providing gas for industrial lighting. In 1812, the London Gas, Light, and Coke Company was formed to produce town gas, a combustible gas that could be used for lighting, cooking, and pre-heating the combustion air used in blast furnaces. This early gasification process used coal and peat as the feedstock, and produced a gas of relatively low heating value (3.5 to 6.0 MJ/m^3) [15]. At the same time in the United States, the Baltimore Gas Company opened in 1816 and became the first gasification company outside of Europe. Use of the technology spread quickly, and by the 1880s, thousands of gasifiers were operating in the United States. By the early 1900s, electricity was replacing gas as a source of light, and the use of gasification to produce town gas became limited to rural areas without access to natural gas or electricity [11]. In 1920, Carl Von Linde commercialized the cryogenic separation of air, which allowed for the development of syngas. In 1926, Franz Fischer and Hans Tropsch developed the Fischer-Tropsch process in Germany. This process allowed syngas to be used in the production of liquid fuels such as gasoline and diesel. Given that the main product of gasification is syngas, interest in gasification technology was renewed. In 1926, the Winkler Fluid Bed process was developed, followed by the Lurgi Moving-Bed Pressurized Gasification Process in 1931 and the Koppers-Totzek Entrained-Flow Process in 1940. These technologies played a critical role in Germany's ability to manufacture fuel during World War II, and on a broader basis in the worldwide development of the ammonia industry [16].

From 1940 to 1970, little technical progress was made in gasification technology because of the increased production of natural gas and naptha. In 1950, Texaco and Shell developed the oil gasification process, which used gaseous and liquid feedstocks such as residual oils from refineries; however, the technology was not commercialized to a significant degree. Also in 1950, the South African Coal, Oil and Gas Corporation Limited constructed a plant that used coal gasification combined with Fischer-Tropsch synthesis to produce diesel, gasoline, and liquefied petroleum gas. While not considered economically viable elsewhere, this facility was necessary because of the political and geographic isolation of South Africa at that time. With the advent of the first major worldwide oil crisis, interest in coal hydrogenation (also known as hydrogasification) to produce methane and liquid fuels increased dramatically. Although this technology did not result in widespread commercial success because of its need for high-pressure processing conditions, it did lead to resurgence in gasification research. In addition to the use of coal as a feedstock, research using lignocellulosic feedstocks such as wood and grasses as a feedstock was initiated. In 1972, Shell constructed a 6 ton per day (tpd) pilot plant in Amsterdam using an integrated gasification combined cycle (IGCC) process. Also in 1972, Steag opened an IGCC plant in

Lunen, Germany. By 1978, Shell had constructed a 15 tpd plant in Harburg, Germany, and they joined forces with U.S. company Koppers to produce a pressurized version of the Koppers-Totzek Gasifier [11].

In the early 1980s, gasification research continued to thrive, with a number of techno-logical improvements being developed and larger-scale production plants being constructed. In 1981, Rheinbraun developed the high-temperature Winkler fluidized bed reactor, which improved carbon conversion and syngas quality. Eastman Chemicals began using a new coal gasification technology in 1983 to produce specialty chemicals. In 1984, Lurgi and British Gas developed a slagging version of its existing gasification technology, while the first IGCC demonstration plant in the United States was opened at the Cool Water Plant near Barstow, California. Texaco extended its oil gasification process to accept a coal slurry feed in 1984, and in 1987, a 250 tpd IGCC plant was opened at Deer Park in Houston, Texas,. With the collapse of the oil prices in the late 1980s, however, attention to research on gasification decreased.

After the end of the first Gulf War, oil prices increased. This increase, coupled with increasing political volatility in the Middle East and concerns over the environmental impacts of fossil fuel power generation, led to renewed interest in gasification as a plausible fuel option. The Polk Power Station near Mulberry, Florida, began operation in 1996. Plants also were opened in the Netherlands and at the Wabash River Coal Gasification Repowering Project near Terre Haute, Indiana. Because gasification using plant material is considered a carbon neutral process, research into the use of lignocellulosic feedstocks such as pulp and paper mill waste, agricultural waste like corn stover, and dedicated biofuel crops such as switchgrass gained increasing attention [11, 17].

In the new millennium, gasification technology has become widely used, with large-scale potential, especially in countries like China and India that have large rural populations. As of 2010, there were 144 operating gasification plants in 27 countries, with a total of 427 gasifiers being operated. Technical and economic hurdles still exist, but research is being heavily funded in both the private and public sectors.

4. TYPES OF GASIFIERS

In the development of gasification, a wide variety of different reactor configurations have been, and continue to be, used; however, the reactor types can be grouped into one of three major types: 1) moving bed gasifiers (occasionally known as fixed bed gasifiers), 2) fluidized bed gasifiers, and 3) entrained flow gasifiers. Each gasifier type is discussed in greater detail below.

4.1. Moving Bed Gasifiers

One of the oldest types of gasifiers, moving bed gasifiers consist of a bed in which the feedstock slowly moves downward under the force of gravity while being gasified by a contact agent, usually steam, oxygen, or air. The direction in which the contact agent flows defines the two different types of moving bed gasifiers: counter-current (also known as an "updraft gasifier") or co-current (also known as a "downdraft" gasifier). In an updraft gasifier, the steam and oxygen or air is introduced at the bottom of the reactor and flows in the direction that is counter to the feedstock flow. A simplified schematic of an updraft gasifier is shown in Figure 3. In a downdraft gasifier, the gases are introduced at the top of the reactor. Both configurations have advantages and disadvantages, but both are characterized by four distinct zones in which the feedstock reacts. At the top of the reactor, the feedstock will initially enter the "drying zone" where the feedstock is dried and heated. As the bed moves downward, the feedstock enters the "carbonization zone" where it is further heated and begins to devolatilize. Next, the feedstock enters the "gasification zone" where the devolatilized feedstock is gasified by reactions with steam and carbon dioxide. Finally, the remaining material enters the combustion zone. This is the highest temperature zone of the gasifier where any remaining char reacts with oxygen. For all moving bed gasifiers, the amount of oxidant required is relatively low. Reactor throughputs are low also, but thermal efficiency is high.

Updraft gasifiers are by far the more common of the moving bed gasifier configurations, and can be operated in two different modes. These operational modes are defined by how the inorganic material present in the feedstock is removed from the reactor, either in a dry state or as slag. Dry ash operation, also known as Lurgi Dry Ash Gasification, requires the reactor temperature to be maintained lower than the ash-slagging temperature of the char. Typically, excess steam is added to achieve this condition. Below the combustion zone, the ash will be cooled by the entering steam and oxygen or air, and then removed via an ash lock. The outlet gas temperature for a dry ash gasifier is relatively low, meaning that high levels of tar and methane will be present in the gas. This can be somewhat offset by operating the gasifier in slagging mode. In slagging-mode operation, the steam-to-fuel ratio is considerably lower, resulting in much higher temperatures in the combustion zone. At these higher temperatures the ash is melted, and must be removed from the reactor as solid slag. The higher temperatures result in lower methane and tar content in the product gas.

Downdraft gasifiers, while not as common as the updraft configuration, do possess some advantages over the updraft configuration. The product gas is removed at a very high temperature after passing through the hot bed of char. This means that tar and methane levels are significantly lower than those in counter-current gasifiers. The heat from the product gas often is used to heat the incoming steam and oxygen or air, resulting in a thermal efficiency that is only slightly lower than that in updraft gasifiers. The main disadvantage to downdraft gasifiers is that the drying zone is not hot enough to function without adding heat either by combusting a small amount of the feedstock or by using external heaters [18].

4.2. Fluidized Bed Gasifiers

A fluidized bed gasifier functions through the use of a bed of inert particles that are fluidized by the incoming steam. A simplified schematic is shown in Figure 4. The steam must be introduced into the bottom of the reactor at a high enough velocity to fully suspend the reactor bed. This condition also is known as fluidization. The suspended bed particles, often alumina or silica sand, result in extremely good mixing within the bed. This promotes excellent heat and mass transfer, and allows for a constant temperature across the bed that can be precisely controlled. The biomass is fed into the side of the reactor above the bottom of the fluidized bed. Feedstock particle size is critical because large particles can disrupt fluidization of the bed. Particle sizes less than 6 mm are most commonly employed, thus the feedstock must be ground prior to processing. Because fluidization must be maintained, fluidized bed gasifiers are operated below the slagging temperature of the ash. One of the most significant disadvantages of a fluidized bed gasifier is low carbon conversion. Approximately 5 to 10% of the inlet carbon remains in the reactor as char. Because of the high gas velocities, some of the char formed will be swept out of the reactor along with entrained bed material. This char must be captured in filters or cyclone separators and recycled to the reactor. Numerous methods have been developed to overcome this limitation including recirculating the bed material through a combustion chamber to burn off the char, a system known as a circulating fluidized bed reactor. Despite these disadvantages, fluidized bed gasifiers can process a variety of feedstocks, making them among the most versatile and commonly used processes for biomass gasification [19].

4.3. Entrained Flow Gasifiers

In an entrained flow gasifier, the feedstock is introduced into the top of the reactor along with steam and/or oxygen. A simplified schematic is shown in Figure 5. Air is rarely used as the oxidant source; most commercial entrained flow gasifiers use pure oxygen as the oxidant. The feedstock must be in the form of very fine solids, atomized liquids, or fuel slurry. When fed slurry, the system must be operated under pressure, and the need to evaporate the water results in a lower thermal efficiency. Because the feed must be entrained in the oxidant flow, entrained flow gasifiers require the highest usage of steam and/or oxygen. As the particles are entrained, they react with the oxidant at very high temperatures that are well above the ash-slagging temperature. The high temperatures combined with the small particle size of the feedstock means that both reaction rates and carbon conversion are very high. Residence times are short, on the order of a few seconds, so process throughput is high. Carbon conversion in an entrained flow gasifier is usually between 98 to 99.5%. The high operating temperatures also result in a product gas that is very low in methane and tar, although cooling the gas results in a low thermal efficiency [20].

Table 2. Qualitative comparison of system performance levels for the major types of gasifiers

Type of Gasifier	Product Gas Temperature	Thermal Efficiency	Carbon Conversion	Throughput	Oxidant Requirement	Product Gas Methane and Tar Levels
Updraft	Low	High	Low	Low	Low	High
Downdraft	High	High	Medium	Low	Low	Low
Fluidized Bed	Medium	Medium	Medium	Medium	Medium	Medium
Entrained Flow	High	Low	High	High	High	Low

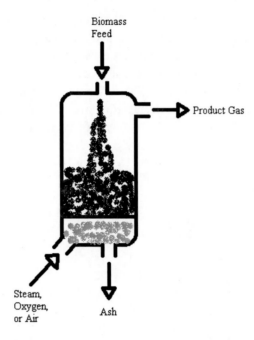

Figure 3. Simplified schematic of a moving bed (updraft) gasifier operating in a dry ash configuration.

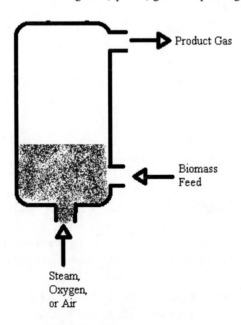

Figure 4. Simplified schematic of a fluidized bed gasifier.

4.4. Summary

The final end use of the gas produced is extremely important when building a gasification system. A gas that will be combusted in a turbine will have significantly different quality requirements than one sent to a Fischer-Tropsch synthesis reactor. As such, the different types of gasifiers exist because the economic viability of a system depends on a number of factors that affect the final gas quality. The amount of tar and methane present in the gas, the thermal efficiency of the system, the outlet gas temperature, system throughput, carbon conversion, and oxidant requirement are all engineering considerations that must be accounted for when designing a system for a specific purpose. The different design considerations are given in Table 2 for the major gasifier categories discussed above. This Table is qualitative in nature, and presents a summary of the major factors that must be taken into account.

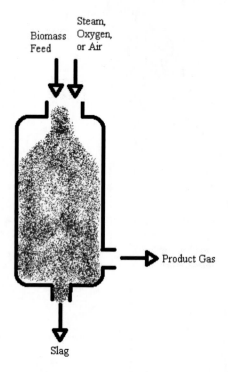

Figure 5. Simplified schematic of entrained flow gasifier.

5. CURRENT STATE OF TECHNOLOGY

Although coal gasification has been successful in the past, the use of biomass as a fuel source has met with mixed results. Numerous technical hurdles still exist, and the economic viability of the process in the absence of public subsidies has been questioned. Nevertheless, in this section, several different "success stories" detailing promising advances in the state of the technology are discussed in this section. These examples cover full-scale demonstration plants with more than 2000 hours per year of operation.

Construction of the Güssing Plant, located in Güssing, Austria, began in September of 2000 and was completed a year later in 2001. The plant operates as a dual fluidized bed reactor using steam as the oxidant, and the synthesis gas is cooled, cleaned, and used in a gas engine to produce electricity and heat. At full operation it can provide 4.5 MW of thermal energy or 2.0 MW of electricity. Since beginning operation, it has produced over 105 GWh of electricity. Its primary fuel source consists of wood chips provided by the local forestry industry. One novel aspect of the Güssing Plant is its scrubbing and recycling of tars. The outlet gas passes through a heat exchanger that cools it from 900°C to 150°C. It passes through a fabric filter to remove entrained particles and some char. The filtered gas then is sent to a scrubber where the tars are fully removed from the gas stream and the gas is cooled to 40°C. The scrubber liquid, which is saturated with tars, is recycled to a combustion zone located next to the gasifier. This provides process heat to the system, and eliminates most of waste products. In addition to the CHP cycle, the Güssing Plant also has produced liquid transportation fuels using Fischer-Tropsch reactors, natural gas, mixed alcohols, and hydrogen, and the syngas has been used to fuel a solid oxide fuel cell [21].

The Viking Gasifier, built in 2002 and operated by the Technical University of Denmark, is a CHP plant that uses wood chips as fuel in a two-stage fixed bed system. Their technology uses two separate reactors, the first being a pyrolysis reactor and the second a gasification reactor. The pyrolysis reactor, which is operated at 600°C, produces a tar-laden gas that is gasified with air in a second reactor at 1100°C. The high temperature decomposes the tars and reduces methane the content. By passing this clean syngas through a number of heat exchangers, almost all of the process heat can be recycled, and solid contaminants then are removed using a filter. The final gas is combusted in a gas engine to produce heat and electricity. The Viking Gasifier demonstrated the ability to produce 39 kW of heat or 17.5 kW of electricity, with over 37 MWh being generated during its first year of operation. In addition to the extremely low levels of tar and ash that are generated, the Viking Gasifier is novel in that it is fully automated (i.e., it is automated to the degree that in can run unattended). A scaleup of the technology is ongoing, with a 500 kW plant under construction in Hillerød, Denmark [22].

Also located in Denmark, the Harboore Plant feeds wood chips to an updraft gasifier that sends its product gas to a gas engine. The plant can produce 1.9 MW of heat, 1 MW of electricity, or 0.4 MW of bio-oil power. Since the addition of two gas engines to the system in 2000, the plant has produced over 48,000 MW of electricity. The wood fuel is provided locally but generally has very high moisture content (i.e., between 35 and 55%). A novel process design allows the moisture in the fuel to be used to generate steam as an oxidant [23].

The Enamora Gasification Plant at Campo de Criptana, Spain, is very novel because it uses olive pulp, olive pits, and grape pomace as fuel sources. The use of these feedstocks completely eliminates the liquid waste stream from a nearby alcohol distillery operated by Movialsa. Constructed in 2008, the plant has been running continuously since May 2011. The Enamora plant uses a fluidized bed gasifier, and the product gas is sent to gas engines for CHP. The electrical output from the plant is 5922 kW. When generating electricity the efficiency of the plant is higher than originally expected, and it can generate 1.3 kW/kg of biomass. This performance level is believed to be a direct consequence of the high quality of the product gas, which enables the engines to operate 226 kW in excess of the manufacturer's nominal output rating [24].

The final example (although this list is far from comprehensive) of a biomass gasification success story is the Kymiarvi Power Station located in Lahti, Finland. This circulating fluidized bed gasifier is extraordinarily versatile, and has been operated using wood, particle board, paper, plastics, and refuse derived fuels. Because the fuel source will directly impact the amount of energy produced, the plant produces a range between 40 and 70 MW of heat and around 20 MW of electricity. During 2003, it produced 125 GWh of electricity. The Lahti plant has been so successful that a new waste-to-energy CHP plant based on gasification was commissioned in the spring of 2012. Located on the same site, the new plant will use solid recovered fuels as its only feed source. Solid recovered fuels consist of industrial, commercial, or household waste. The use of this technology will allow 250,000 tons of solid recovered fuels to be consumed annually, providing 90 MW of heat and 50 MW of electricity [25].

CONCLUSION

Although gasifiers have been successfully operated for over 180 years using coal as a feedstock, the use of biomass as a fuel source is a relatively new phenomenon. Hence, a large amount of research still must be done to optimize the technology. Engineers need not "reinvent the wheel," however, as much of the technology developed for coal gasification can be built upon to use biomass. A number of CHP generation plants have been demonstrated at the commercial scale using biomass as a feedstock. Their versatility has demonstrated that, in addition to being a viable method for generating heat, power, and fuel, gasification also can make these high-value products from materials that otherwise would be considered waste. The ability to turn waste into sustainable energy ensures that gasification research will continue to be considered a worthwhile investment, and gasification will continue to be an integral factor in the renewable energy equation.

REFERENCES

[1]   IPCC. Climate Change 2007: Synthesis Report section 2.4, Attribution of Climate Change. *Intergovernmental Panel on Climate Change* (IPCC); 2007.

[2]   Watkins E. Watching the World: Iran Woos Latin America. *Oil & Gas Journal* 1/16/2012;110.

[3]   Jimenez-Rodriguez R., Sanchez M. Oil Price Shocks and Real GDP Growth: Empirical Evidence for Some OECD Countries. *Applied Economics* 2005;37:201.

[4]   ACORE. The Outlook on Renewable Energy in America. In: Energy ACOR, editor; 2007.

[5]   EIA. February 2012 - Monthly Energy Review. Washington DC: Department of Energy - U. S. *Energy Information Administration;* 2012.

[6]   Logan B. E. Extracting Hydrogen and Electricity from Renewable Resources: a roadmap for establishing sustainable processes. *Environmental Science & Technology* 2004;38:160A.

[7]   Galbe M., Zacchi G. A review of the production of ethanol from softwood. *Applied microbiology and biotechnology* 2002;59:618.

[8]   Wright J. D. *Ethanol from lignocellulose-An overview.* American Institute of Chemical Engineers, New York, NY; 1987.

[9]   Do E. Biofuels, Biopower, and Bioproducts: *Integrated Biorefineries.* 2010.

[10]   Mourant D., Wang Z., He M., Wang X. S., Garcia-Perez M., Ling K. et al. Mallee wood fast pyrolysis: Effects of alkali and alkaline earth metallic species on the yield and composition of bio-oil. *Fuel* 2011;90:2915.

[11] Higman C., van der Burgt M. Chapter 1 - Introduction. *Gasification* (Second Edition). Burlington: Gulf Professional Publishing; 2008, p. 1.

[12] Sjostrom E. *Wood Chemistry:* Fundamentals and Applications. 2nd ed. San Diego: Academic Press; 1993.

[13] Rowell R. M., Young R. A., Rowell J. K. *Paper and Composites from Agro-based Resources:* CRC Press; 1997.

[14] ECN. *Phyllis: Database for biomass and waste.* Version 4.13 ed: Energy research Centre of the Netherlands; 2012.

[15] Miller B. G. Chapter 2 - Past, Present, and Future Role of Coal. *Coal Energy Systems.* Burlington: Academic Press; 2005, p. 29.

[16] Lesch J. E. The German chemical industry in the twentieth century: Springer; 2000.

[17] Smeenk J., Brown R. C. Experience with atmospheric fluidized bed gasification of switchgrass. *Bio Energy;* 1998, p. 4.

[18] Phillips J. *Different types of gasifiers and their integration with gas turbines.* Electric Power Research Institute, Charlotte, available at www netl doe gov/technologies/ coalpower/turbines/refshelf/handbook/12 2006;1.

[19] McKendry P. Energy production from biomass (part 3): gasification technologies. *Bioresource Technology* 2002;83:55.

[20] Higman C., van der Burgt M. Chapter 5 - *Gasification Processes.* Gasification (Second Edition). Burlington: Gulf Professional Publishing; 2008, p. 91.

[21] Hofbauer H., Rauch R., Bosch K., Koch R., Aichernig C, Tremmel H, et al. *Biomass CHP-plant Güssing: a success story.* TUV, Paper presented at Pyrolysis and gasification of biomass and waste expert meeting, Strasbourg; 2002.

[22] Gassner M., Maréchal F. Thermodynamic comparison of the FICFB and Viking gasification concepts. *Energy* 2009;34:1744.

[23] Ahrenfeldt J., Thomsen T. P., Henriksen U., Clausen L. R. Biomass gasification cogeneration − A review of state of the art technology and near future perspectives. *Applied Thermal Engineering* 2013;50:1407.

[24] Knoef H. A. M., editor. *Handbook of Biomass Gasification.* 2nd ed. Enschede, The Netherlands: BTG Biomass Technology Group BV; 2012.

[25] Basu P., Butler J., Leon M. A. Biomass co-firing options on the emission reduction and electricity generation costs in coal-fired power plants. *Renewable energy* 2011;36:282.

In: Biomass Processing, Conversion and Biorefinery
Editors: Bo Zhang and Yong Wang

ISBN: 978-1-62618-346-9
© 2013 Nova Science Publishers, Inc.

Chapter 10

HYDROGEN PRODUCTION FROM CATALYTIC STEAM CO-GASIFICATION OF WASTE TYRE AND PALM KERNEL SHELL IN PILOT SCALE FLUIDIZED BED GASIFIER

Suzana Yusup[1]*, Reza Alipour Moghadam[1], Ahmed Al Shoaibi[2], Murni Melati[1], Zakir Khan[1], Lim Mook Tzeng[1] and Wan Azlina A. K. GH.[3]

[1]Department of Chemical Engineering, Biomass Processing Laboratory, Centre of Biofuel and Biochemical Research, Universiti Teknologi Petronas, Perak, Malaysia
[2]Department of Chemical Engineering, Petroleum Institute, Abu Dhabi, United Arab Emirates
[3]Department of Chemical and Environmental Engineering, Universiti Putra Malaysia, Selangor, Malaysia

ABSTRACT

In order to recover energy and recycle waste materials, an experimental study was performed on utilization of waste materials in catalytic steam gasification system. In current research, an alternative method of hydrogen production is achieved by blending of palm kernel shell (PKS) with waste tyre as a promising energy resource. Experiments were carried out at 600-800°C with steam/feedstock ratio (S/F) of 2-4 (kg/kg) and waste tyre/PKS ratio of 0-0.3 (kg/kg). This paper reports the results obtained from series of experiments that have been performed, on a pilot plant to improve hydrogen production efficiency. The highest H_2 and total syngas content of 66.15 vol% and 83.8 vol% was achieved respectively under condition of 800°C and 30 wt% of waste tyre blended with PKS and S/F ratio of 4 (kg/kg). The results obtained confirmed that mixtures of PKS and waste tyre in catalytic steam fluidized bed system produced a fuel gas with a calorific value of 14.76 MJ/Nm^3 which has the potential to be utilised in engines.

* drsuzana_yusuf@petronas.com.my Tel: +605-368 8217 Fax: +605-368 8205.

Keywords: Hydrogen, Pilot plant, Gasification, fluidized bed, Waste Tyre

1. INTRODUCTION

Hydrogen-rich gas production from waste has become an interesting alternative in energy generation due to energy policies and greater understanding on the importance of green energy. Hydrogen is one of the emerging alternative fuels. In comparison to fossil fuels, 9.5 kg of hydrogen produces an energy equivalent to that produced by 25 kg of gasoline [1]. There is also a concern about the environmental pollution caused by the use of fossil fuels. According to a recent study, the world CO_2 emission which is the main cause of global warming increased by 3% in 2011, reaching an all-time high of 34 billion tonnes [2]. Apart from CO_2, other pollutants contaminants such as CO, NOx, and SOx are released during the combustion of fossil fuels. According to the Global Climate Change Initiatives (GCCI), the greenhouse gas intensity should be decreased by 18% by the year 2012 to mitigate global warming problems [3].

In Malaysia, the agricultural sector has been growing rapidly over the years. Malaysia is the world's biggest producer and exporter of palm oil, accounting for 51% of the world's palm oil and fats production and 62% of the world's exports. The palm oil industry generates huge quantity of by product such as, empty fruit bunch (EFB), palm kernel shell and fibres during post-processing with an annual generation of 9.66, 5.20 and 17.08 million tons for fiber, shell and EFB respectively [4].

Nowadays one of serious environment problem is the amount of waste tyre that is growing year by year. Currently, only small percentage of waste tyre is recycled and disposed mainly by landfill or incineration and both techniques are associated with environmental problems. The production of energy from waste tyre by thermal conversion has been proven to be an effective solution. Beside the gasification of biomass, gasification of waste tyre presents an attractive alternative for hydrogen production [5-10]. Several studies have been performed to investigate the effects of various parameters on gasification of waste tyres. Portofino et al. [9] employed a bench scale gasifier in order to study the influence of the process temperature on gas compositions and product yields in similar process. They recommended that higher temperature will result in higher syngas production. They achieved 86 wt% and 65 vol% of syngas and hydrogen fraction respectively at 1000°C. Xiao et al. [10] performed a gasification of waste tyre in fluidized bed gasifier at low temperature. They suggested suitable condition for waste tyre gasification to be around 650-700°C and 0.2-0.4 for temperature and equivalence ratio respectively and found that LHV of the product gas increased with increasing temperature. Karatas et al. [8] investigated gasification of waste tyre with air&CO_2, air & steam and steam in laboratory scale bubbling fluidized bed gasifier. They found the LHV of the product gas was 9.59, 7.34 and 15.21 MJ/Nm^3 for air-CO_2, air-steam and steam respectively. Song and Kim [11] studied on co-gasification of tire scrap and sewage sludge in circulating fluidized bed. They reported the calorific value of the product gas decreased from 5 to 15 MJ/ Nm^3 when waste tyre was co-gasified with sewage sludge. Leung and Wang [12] carried out gasification of waste tyre in fluidized bed gasifier. They achieved produced gas with a calorific value of 6 MJ/Nm^3 at optimum operation conditions. As stated above it has been proven that by increasing the temperature in presence of catalyst and steam it is advantages for the gasification process due to steam methane reforming and water gas reaction that occur and caused an increased in hydrogen fraction and a decreased of gaseous hydrocarbons fraction and tar content.

The purpose of this research was to optimize the catalytic steam gasification process in pilot scale gasifiers and utilize the mixture of waste tyre and PKS as the energy source for enriched hydrogen gas production.

2. EXPERIMENTAL SECTION

2.1. Feedstock

The biomass feedstock PKS was obtained from a palm oil factory while waste tyres were obtained from a local supplier. Waste tyres were granulated and PKS was pulverized and sieved into a specific particle size range of 1-2 mm. The proximate and ultimate analyses of the feed stock are reported in Table 1.

Table 1. Proximate and Ultimate analysis

Proximate Analysis (wt%)	Palm Kernel shell	Waste Tyre
Moisture content	12.00	0.86
Volatile matter	30.53	66.58
Fixed carbon	48.50	27.73
Ash	8.97	4.83
Ultimate analysis (wt.%)		
C	49.26	82.43
H	5.88	7.1
O	44.41	3.64
N	0.4	0.32
S	0.05	1.6
Density (kg/m3)	733	810
HHV (MJ/kg)	24.97	37.63
Holocellulose	54.30	-
Alpha-cellulose	29.60	-
Lignin	59.30	-

2.2. Experimental Procedure

Figure 1 shows the process flow diagram of pilot scale catalytic steam gasification unit. The pilot plant consists of cylindrical reactor made of (Inconel 625). The fluidized bed gasifier has a height of 2500 mm and internal diameter of 150 mm and 200 mm in gasification and free board zone respectively. The reactor has four individual electrical heaters and the temperature is controlled by temperature controllers. Pressure differential indicator is installed in the reactor and mounted on top of the reactor. Eight thermocouples are installed across the gasifier, two in the dense bed, four in the gasification zone and two in the free board zone.

In this study, the feedstock is impregnated with commercial Ni catalyst and applied in the gasification process. Series of experiments were performed in order to investigate the influence of varying different parameters on hydrogen production. The feedstock is fed to the fluidized bed gasifier via a variable speed screw feeder and two swing lock hopper which was pressurized with nitrogen. The feeding system was cooled by water to avoid any clogging due to the pyrolysis of waste tyres in the feedstock. Steam is supplied by the boiler and heated to 270°C through a super-heater. The flow rate of super-heated steam is controlled by mass flow controller then steam was applied as the gasifying agent. The produced gas passes through the cyclone to remove the fly ash. The produced gas was injected to the fixed bed reactor to enhance the gas quality and to further crack the tar. The gas is passed through the scrubber to remove any tar residual. A multi stage condenser removed all condensable components. An online gas analyzers (Teledyne 7500, 7600 and 4060) are used to determine the amount of H_2, CO, CO_2, CH_4, N_2, O_2, H_2S and NO_2 in the product gas. The main process parameters such as gas composition, temperatures, gas flow rate and pressure were recorded by data acquisition system.

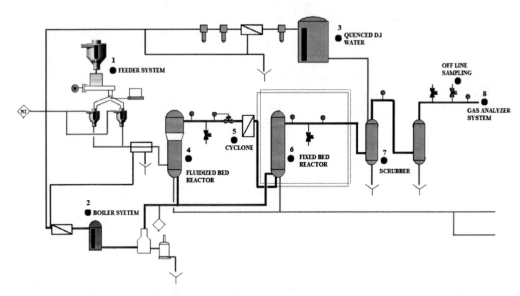

Figure 1. Process flow diagram of pilot plant catalytic steam gasification system.
1-Feeding system 2-Boiler system 3-Quenched D.I. water system 4-Fluidized bed gasifier 5-Cyclone 6-Fixed bed gasifier 7-Scrubber system 8- Gas Analyzer system.

3. RESULTS AND DISCUSSION

3.1. Effect of Temperature

Temperature profile is one of the crucial parameters in biomass gasification process and has major influence on final product composition. The reactions involve in the gasification process are endothermic and exothermic but the endothermic reactions drive the hydrogen production via gasification process.

The main steps of biomass conversion to hydrogen in pyrolysis and gasification processes are listed as indicated by the reactions below [13]:

Biomass $\longrightarrow$ Gas+ Tars + Char

Tars $\longrightarrow$ Light and Heavy hydrocarbons + CO + CO_2 +H_2

Heavy hydrocarbons $\longrightarrow$ Light hydrocarbons + H_2

Char $\longrightarrow$ CO + CO_2 + H_2 + Solid residual

The principal reactions that take place in gasification process lead to formation and consumption of H_2, CO, CO_2 and CH_4 through pyrolysis and gasification via the given reactions in (Eqs. 1-8).

$C + \frac{1}{2}O_2 \rightarrow CO$	-111 MJ/Kmol		(1)
$CO + \frac{1}{2}O_2 \rightarrow CO_2$	-283 MJ/Kmol	The Combustion reaction	(2)
$C + CO_2 \leftrightarrows 2CO$	+172 MJ/Kmol	The Boudouard reaction	(3)
$C + H_2O \leftrightarrows CO + H_2$	+131 MJ/Kmol	The Water gas reaction	(4)
$C + 2H_2 \leftrightarrows CH_4$	-75 MJ/Kmol	The Methanation reaction	(5)
$CO + H_2O \rightleftarrows CO_2 + H_2$	-41 MJ/Kmol	The Water gas shift Reaction	(6)
$CH_4 + H_2O \rightleftarrows CO +3H_2$	+206 MJ/Kmol	The steam methane reforming reaction	(7)
$CH_4 + CO_2 \rightleftarrows 2CO + 2H_2$	+260 MJ/Kmol	The carbon dioxide methane reforming reaction	(8)

According to Xiao et al. [14] biomass gasification occurs following three main steps. The first step is initial pyrolysis or devolatilisation, which occurs at low temperature and produces volatile matter and char. Second step is tar-cracking process that favours high temperature reactions and produces gases such as H_2, CO, CO_2, CH_4 and light hydrocarbons. Third step is char gasification that is enhanced by the boudouard reaction. For tyre waste, devolatilisation step is absence and increasing temperature caused decomposition of rubber to smaller molecular radicals and atom that may participate in gasification reactions. With presence of Ni catalyst, dissociated hydrocarbons and oxygen molecules is adsorbed on active surface of catalyst. CO is produced from reaction of C_1 and oxygen fractions on active site of catalyst and H_2 is dissociated from the active site [15].

H_2 production is depended on temperature variation and more H_2 is produced at higher temperatures [13, 16-18]. Referring to Figure 2 hydrogen production is favoured by high temperatures. Increasing temperature in gasifier and utilizing Ni as reforming catalyst caused reforming reactions to occurs as in reaction (9) and (10);

$$C_nH_m + nH_2O \leftrightarrows nCO + (n+m/2)H_2 \qquad (9)$$

and

$$C_nH_m + nCO_2 \leftrightarrows 2nCO + (m/2)H_2 \qquad (10)$$

thus increased the hydrogen yield. In addition by increasing the temperature, the LHV of produced gas decreased. At temperature above 700°C the hydrogen content increased to 59.65 vol% while the hydrocarbons and CO_2 content decreased to 14.7 vol% and 8.89 vol% respectively. These observations are in accordance with the results obtained by Turn at. al [18], Lucas et al. [19] and McKendry [20]. CO_2 consumption occurs due to Boudouard endothermic reaction (3) and carbon dioxide methane reforming reaction (8) that favoured high temperature thus increase formation of H_2 and CO component [21, 22].

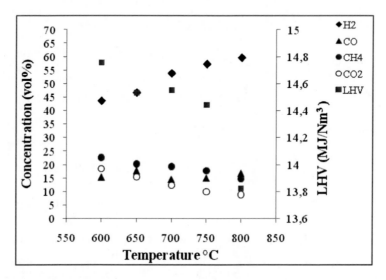

Figure 2. Effect of temperature on gas composition and LHV.

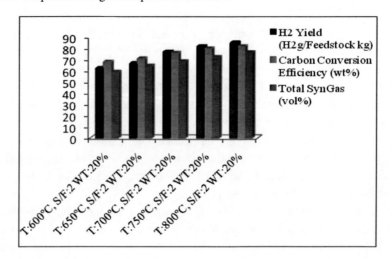

Figure 3. Effect of temperature on H_2 yield, carbon conversion efficiency and total syngas.

Tavasoli et al. [13] and Lucas et al. [19] reported with increasing temperature, thermal cracking reactions becoming dominant and caused decreasing in the concentration of CH_4 and heavy hydrocarbons. Cracking of heavy hydrocarbons is an endothermic reaction that causes an increase in concentration of H_2 in the product gas. The gasification efficiency is reflected by the carbon conversion that increased to 85.21% with increasing temperature.

Figure 3 represents the influence of temperature on product yield. H_2 yield and total syngas composition increased from 62.32 to 85.21 $H_2(g)$/feedstock(kg) and 58.95 vol% to 76.41 vol% respectively, by raising the gasifier temperature from 600°C to 800°C which is about 23% and 17.46% increased respectively.

3.2. Effect of Blending Waste Tyre with Palm Kernel Shell

Blending of waste tyre with PKS was found to significantly effect the gas composition. The hydrogen yield achieved is in the range of 86.04 - 94.5 $H_2(g)$/feedstock(kg) at 800°C by varying the amount of waste tyre from 0 – 30 wt% as shown in Figure 5.

As illustrated in Figure 4 blending of waste tyre with biomass feedstock caused an increase in hydrocarbons composition due to thermal cracking and decomposition of rubber which result in CO, CH_4, C_2H_4 and heavy hydrocarbons formation. At zero percent of waste tyre, the fraction of H_2 is 60.23 vol% while CO and CO_2 content are 13.23 vol% and 18.42 vol% respectively.

With 20% (w/w) of waste tyre blended with PKS, H_2 and CO fraction increased to 61.77 vol% and 15.68 vol% respectively. At 30 wt% of waste tyre in the feedstock caused H_2 and CO fraction to increase but reduction in CO_2 concentration is due to dominant of steam methane ($CH_4 + H_2O \rightleftarrows CO + 3H_2$) and carbon dioxide methane ($CH_4 + CO_2 \rightleftarrows 2CO + 2H_2$) reforming reactions. The LHV, total syngas and carbon conversion efficiency (CCE) increased by rising the amount of waste tyre to 30 wt% in feedstock at 12.3 MJ/Nm3, 83.63 vol% and 88 wt%, respectively. The obtained results are in accordance with Pinto et al. [23].

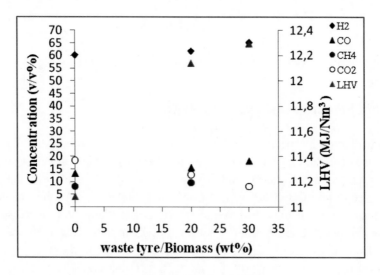

Figure 4. Effect of waste tyre on gas composition and LHV at 800°C.

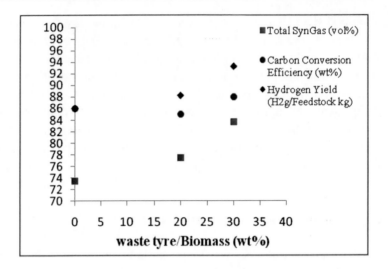

Figure 5. Effect of waste tyre on H_2 yield, carbon conversion efficiency and total syngas.

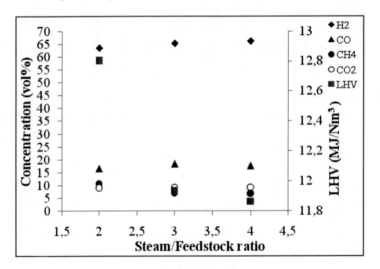

Figure 6. Effect of steam/feedstock on gas composition and LHV at 800°C, waste tyre 30 wt%.

3.3. Effect of Steam/Feedstock (S/F) Ratio

Effect of S/F ratio has been studied over the range of 1-1.5 wt%. Figure 6 and 7 show the major influence of steam on gasification reactions that enhanced hydrogen production.

At lower temperature of 600°C and S/F ratio of 2, with an increase in the amount of PE waste in the feedstock, CH_4 fraction is elevated to 22.64 vol%. As the amount of injected steam/feedstock ratio is increased from 2 to 4 and gasifier temperature increased from 600°C to 800°C. Steam methane reforming (7) became dominant and leads to increase in hydrogen and CO production while CO_2 fraction decreased due to carbon dioxide methane reforming reaction (8). At temperature range of 750-800°C, reforming reactions, water gas reaction and boudouard reaction are dominant. These reactions was enhanced as S/F ratio and temperature increased and lead to increasing H_2 fraction to 66.15 vol% and CH_4 and CO_2 fraction decreased. The findings are in agreement with previous study [16, 19, 23-26].

Maximum H_2/CO ratio of 3.94 is achieved at 800°C, S/F of 3 and WT/B of 20%. In contrast, CO/CO_2 ratio showed an increasing trend from 1.8 to 2. Pinto et al. [23] stated that injection of excess steam caused an increase in the CO_2 fraction due to occurrence of water gas shift reaction. Gao et al. [6] concluded high S/B ratio does not always favour the hydrogen production and will not be cost effective. Thus optimum S/B ratio is important. In present study, a ratio of 3 for S/B at 800°C is identified to be the optimal point.

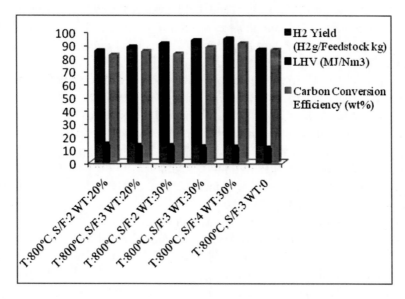

Figure 7. Effect of variation of waste tyre wt% and S/F on H_2 yield, carbon conversion efficiency and LHV.

CONCLUSION

The obtained results confirmed that temperature and steam/biomass (S/F) ratio play crucial roles in controlling the catalytic steam gasification process. Increasing the temperature enhanced the hydrogen production and reduced tar formation. By blending the waste tyre, it enhanced hydrogen production. Mixing of 30 wt% of waste tyre with palm kernel shell increased the H_2 and syngas content to 66.15 vol% and 83.8 vol%, respectively. LHV decreased by increasing the temperature of gasifier due to conversion of hydrocarbons and highest LHV is obtained at 30 wt% waste tyre in the feedstock. The reforming reactions became dominant at temperatures higher than 750°C and caused to increase in H_2 fraction and decreased hydrocarbons and CO_2 concentration. This study optimized the catalytic steam gasification process in pilot scale gasifier with the aim for lower operation cost and higher efficiency. The composition of produced gas influenced the suitable application of end-use.

ACKNOWLEDGMENTS

This work is financially supported by Universiti Technologi PETRONAS and Petroleum Institute, Abu Dhabi, for which the authors wish to record their gratitude.

REFERENCES

[1]   Mirza, U.K., N. Ahmad, K. Harijan, and T. Majeed, A vision for hydrogen economy in Pakistan. *Renewable and Sustainable Energy Reviews*.2009; 13(5): p. 1111-1115.

[2]   Olivier Jos G.J., J.-M.G., Peters Jeroen A.H.W, *Trends in global CO2 emission*, in *PBL Netherlands Environmental Assessment Agency*2012.

[3]   Granite, E.J. and T. O'Brien, Review of novel methods for carbon dioxide separation from flue and fuel gases. *Fuel Processing Technology*.2005; 86(14–15): p. 1423-1434.

[4]   Nasrin, A.B., Ma, A. N., Choo, Y. M., Mohamad, S., Rohaya, M. H., Azali, Oil palm biomass as potential substitution raw materials for commercial biomass Briquettes production. *American Journal of Applied Sciences*. 2008; 5(3): p. 5.

[5]   Elbaba, I.F., C. Wu, and P.T. Williams, Hydrogen production from the pyrolysis–gasification of waste tyres with a nickel/cerium catalyst. *International Journal of Hydrogen Energy*. 2011; 36(11): p. 6628-6637.

[6]   Gao, N., A. Li, and C. Quan, A novel reforming method for hydrogen production from biomass steam gasification. *Bioresource Technology*. 2009; 100(18): p. 4271-4277.

[7]   He, M., Z. Hu, B. Xiao, J. Li, X. Guo, S. Luo, F. Yang, Y. Feng, G. Yang, and S. Liu, Hydrogen-rich gas from catalytic steam gasification of municipal solid waste (MSW): Influence of catalyst and temperature on yield and product composition. International *Journal of Hydrogen Energy*. 2009; 34(1): p. 195-203.

[8]   Karatas, H., H. Olgun, and F. Akgun, Experimental results of gasification of waste tire with air&CO2, air&steam and steam in a bubbling fluidized bed gasifier. *Fuel Processing Technology*. 2012; 102(0): p. 166-174.

[9]   Portofino, S., A. Donatelli, P. Iovane, C. Innella, R. Civita, M. Martino, D.A. Matera, A. Russo, G. Cornacchia, and S. Galvagno, Steam gasification of waste tyre: Influence of process temperature on yield and product composition. Waste Management.2012.

[10] Xiao, G., M.-J. Ni, Y. Chi, and K.-F. Cen, Low-temperature gasification of waste tire in a fluidized bed. *Energy Conversion and Management*. 2008; 49(8): p. 2078-2082.

[11] Song, B.H. and S.D. Kim, *Gasification of tire scrap and sewage sludge in a circulating fluidized bed with a draft tube*, in *Studies in Surface Science and Catalysis*, I.-S.N. Hyun-Ku Rhee and P. Jong Moon, Editors. 2006, Elsevier. p. 565-568.

[12] Leung, D.Y.C. and C.L. Wang, Fluidized-bed gasification of waste tire powders. *Fuel Processing Technology*. 2003; 84(1–3): p. 175-196.

[13] Tavasoli, A., M.G. Ahangari, C. Soni, and A.K. Dalai, Production of hydrogen and syngas via gasification of the corn and wheat dry distiller grains (DDGS) in a fixed-bed micro reactor. *Fuel Processing Technology*. 2009; 90(4): p. 472-482.

[14] Xiao, R., B. Jin, H. Zhou, Z. Zhong, and M. Zhang, Air gasification of polypropylene plastic waste in fluidized bed gasifier. *Energy Conversion and Management*. 2007; 48(3): p. 778-786.

[15] Weil, S., S. Hamel, and W. Krumm, Hydrogen energy from coupled waste gasification and cement production—a thermochemical concept study. *International Journal of Hydrogen Energy.* 2006; 31(12): p. 1674-1689.

[16] González, J.F., S. Román, D. Bragado, and M. Calderón, Investigation on the reactions influencing biomass air and air/steam gasification for hydrogen production. *Fuel Processing Technology.* 2008; 89(8): p. 764-772.

[17] Rapagnà, S., N. Jand, and P.U. Foscolo, Catalytic gasification of biomass to produce hydrogen rich gas. *International Journal of Hydrogen Energy.* 1998; 23(7): p. 551-557.

[18] Turn, S., C. Kinoshita, Z. Zhang, D. Ishimura, and J. Zhou, An experimental investigation of hydrogen production from biomass gasification. *International Journal of Hydrogen Energy.* 1998; 23(8): p. 641-648.

[19] Lucas, C., D. Szewczyk, W. Blasiak, and S. Mochida, High-temperature air and steam gasification of densified biofuels. *Biomass and Bioenergy.* 2004; 27(6): p. 563-575.

[20] McKendry, P., Energy production from biomass (part 2): conversion technologies. *Bioresource Technology.* 2002; 83(1): p. 47-54.

[21] Encinar, J.M., J.F. González, J.J. Rodríguez, and M.a.J. Ramiro, Catalysed and uncatalysed steam gasification of eucalyptus char: influence of variables and kinetic study. *Fuel.* 2001; 80(14): p. 2025-2036.

[22] Mathieu, P. and R. Dubuisson, Performance analysis of a biomass gasifier. *Energy Conversion and Management.* 2002; 43(9–12): p. 1291-1299.

[23] Pinto, F., C. Franco, R.N. André, M. Miranda, I. Gulyurtlu, and I. Cabrita, Co-gasification study of biomass mixed with plastic wastes. *Fuel.* 2002; 81(3): p. 291-297.

[24] André, R.N., F. Pinto, C. Franco, M. Dias, I. Gulyurtlu, M.A.A. Matos, and I. Cabrita, Fluidised bed co-gasification of coal and olive oil industry wastes. *Fuel.* 2005; 84(12–13): p. 1635-1644.

[25] Li, J., J. Liu, S. Liao, and R. Yan, Hydrogen-rich gas production by air–steam gasification of rice husk using supported nano-NiO/γ-Al2O3 catalyst. *International Journal of Hydrogen Energy.* 2010; 35(14): p. 7399-7404.

[26] Pinto, F., C. Franco, R.N. André, C. Tavares, M. Dias, I. Gulyurtlu, and I. Cabrita, Effect of experimental conditions on co-gasification of coal, biomass and plastics wastes with air/steam mixtures in a fluidized bed system. *Fuel.* 2003; 82(15–17): p. 1967-1976.

In: Biomass Processing, Conversion and Biorefinery   ISBN: 978-1-62618-346-9
Editors: Bo Zhang and Yong Wang   © 2013 Nova Science Publishers, Inc.

Chapter 11

REVIEW OF UPGRADING RESEARCHES OF BIOMASS PYROLYSIS OIL TO IMPROVE ITS FUEL PROPERTIES

Junming Xu, Jianchun Jiang and Kang Sun*

Institute of Chemical Industry of Forest Products, CAF, Nanjing, China

ABSTRACT

Biomass fast pyrolysis has aroused great attention and interests in recent years. However, its direct use as liquid fuel raises many problems because of its acidity, high oxygen content and high water content. Due to its poor properties, several methods of upgrading the original bio-oil were developed in recent years. This paper reviews the main methods of upgrading biomass pyrolysis oil such as hydrodeoxygenation, catalytic cracking of pyrolysis vapors, emulsification and esterification. The advantages and shortcomings of upgrading methods were discussed and compared. Furthermore, the problems and focuses were summarized with some suggestions presented on upgrading and applications of bio-oil.

Keywords: Biomass; Bio-oil; Pyrolysis; Upgrading; Review

1. INTRODUCTION

With the depletion of fossil fuels and concerns over CO_2, NOx, SOx emissions, renewable energy sources are now being considered as an attractive solution to the energy problem of both industrialized and developing countries. Among the utilization of renewable energy sources, pyrolysis processes are thought to have great promise as a means for efficiently and economically converting biomass into higher value fuels. It is attractive because solid biomass and wastes, which are difficult and costly to manage, can be readily converted to liquid products. These liquids have advantages in transport, storage, combustion,

---

* Corresponding author: XU Junming lang811023@163.com.

retrofitting and flexibility in production and marketing. Pyrolysis also gives gas and solid (char) products, the relative proportions of which depend very much on the pyrolysis method and process conditions.

However, bio-oil is not a product of thermodynamic equilibrium during pyrolysis but is produced with short reactor times and rapid cooling or quenching from the pyrolysis temperatures. Rapid quenching then "freezes in" the intermediate products of the fast degradation of hemicellulose, cellulose, and lignin. Chemically, bio-oil is a complex mixture of water, guaiacols, catecols, syringols, vanillins, furancarboxaldehydes, isoeugenol, pyrones, acetic acid, formic acid, and other carboxylic acids [1, 2]. These products are also not at thermodynamic equilibrium at storage temperatures. The chemical compositions of the bio-oil tend to further react (degrade, cleave, or condensate with other molecules) toward thermodynamic equilibrium during storage [3].

Due to its special compositions, many problems arise in their handling and utilization. For possible future use as replacements for hydrocarbon chemical feedstocks and fuels, the oils will require considerable upgrading to improve its characteristics. The recent upgrading techniques are hydrodeoxygenation, catalytic cracking of pyrolysis vapors, emulsification, esterification, etc. This paper reviews the main methods of upgrading biomass pyrolysis oil and the advantages and shortcomings of upgrading methods were discussed and compared.

2. UPGRADING OF BIO-OIL

Biomass pyrolysis oils are highly oxygenated, viscous, corrosive, relatively unstable and chemically very complex. The direct substitution of the biomass derived pyrolytic oils for conventional petroleum fuels may therefore be limited. Consequently, researches have been conducted in upgrading the oils to produce a derived fuel product similar in quality to a refined petroleum fuel. The recent upgrading techniques are described as follows.

2.1. Catalytic Cracking of Pyrolysis Vapors

Several kinds of catalysts have been used to improve the characteristics of pyrolysis products. In particular, oxygenates in original bio-oil can be reduced using ZnO [4], ZSM-5 and mesoporous materials. Earlier publications usually concerned about zeolite type catalyst such as ZSM-5 because it is widely used in petroleum industry. The zeolite ZSM-5 catalysts have strong acidity, high activities and shape selectivity which converts the oxygenated oil to a hydrocarbon mixture in the C1 to C10 range [5]. The resultant oil is highly aromatic with a dominance of single ring aromatic compounds similar in composition to gasoline [6, 7].

The catalytic upgrading of pyrolytic oils is obtained over HZSM-5 zeolites, by which the upgraded oil may be easily separated from the aqueous phase [8]. The higher yields of oil (45~50%) were obtained at 450 °C with HZSM-5/50. The catalytic upgrading produced highly deoxygenated oil having a quite elevated heating value and a good combustibility. The formation of coke on the catalyst was not negligible and hence, may be the cause of a rapid deactivation.

Hyun Ju Park [9] provided catalytic upgrading of bio-oil using Ga modified ZSM-5 for the pyrolysis of sawdust in a bubbling fluidized bed reactor. The maximum yield of oil products was found to be about 60%. The yield of gas was increased as catalyst added. HZSM-5 showed the larger gas yield than Ga/HZSM-5. When bio-oil was upgraded with HZSM-5 or Ga/HZSM-5, the amount of aromatics in product increased. Product yields over Ga/HZSM-5 shows higher amount of aromatic components such as benzene, toluene, xylene than HZSM-5.

By using strong acidity catalyst such as HZSM-5, it was found that the catalyst used in cracking process deactivate quickly. By the analysis conducted on the catalyst [10], it was possible to assess that the loss of activity is mainly connected to the disappearance of a significant amount of acidic sites, mainly the stronger ones, due to the thermal cycling to which the catalyst was submitted. Even if the regeneration was conducted at 500°C, localized raisings of temperature above 500°C due to the combustion of coke may have caused dehydroxylation of the Brønsted acid sites that predominate in zeolites activated at 500°C with formation of Lewis acid sites. Thus, the active acid sites in the upgrading reactions are presumed to be preferentially Brønsted acid sites, which were gradually deactivated by the repeated regeneration treatments.

Moreover, the zeolite ZSM-5 catalyst has a specific pore size structure with an elliptical pore size of 5.4 to 5.6 Angstroms diameter which is shape selective. The particular pore size structure allows compounds of the approximate molecular size of a C10 molecule to enter and leave the structure. Therefore, the high molecular weight compounds in bio-oil could either be unconverted pyrolysis vapors or catalysis products formed on the outer surface of the catalyst.

In order to solve these problems, mesoporous materials with large pore width such as MCM-41、MSU、SBA-15 molecular sieves were used in cracking process in order to avoid condensation of the pyrolysis products and to ensure that the formed volatiles pass through the catalyst for upgrading of the bio-oil before undergoing secondary reactions.

Judit Adam [11] using four Al-MCM-41 type catalysts with a Si/Al ratio of 20 in catalytic cracking and barkfree spruce wood as biomass material. It was found out that levoglucosan disappeared and the yields of heavier components decreased significantly under effect of the catalysts. The overall yield of pyrolysis vapors decreased as well. The effects of MCM-41 type catalyst seem to be related to the size of the pores of the catalyst. Pore size enlargement and transition metal incorporation reduces the yield of acetic acid and water among pyrolysis products.

E.F. Iliopoulou [12] studied two mesoporous aluminosilicate Al-MCM-41 materials (Si/Al = 30 or 50) for the in situ upgrading of biomass pyrolysis vapors in comparison to a siliceous MCM-41 sample and to non-catalytic biomass pyrolysis. Due to the combination of the large surface area and tubular mesoporous (pore diameter, 2–3 nm), MCM-41 materials provided mild acidity that leads to the desired environment for controlled conversion of the high molecular weight lignocellulosic molecules. The major improvement in the quality of bio-oil was the increase of phenols concentration and the reduction of corrosive acids. Higher Si/Al ratios of the Al-MCM-41 samples enhanced the production of the organic phase of the bio-oil, while lower Si/Al ratios favored the conversion of the hydrocarbons of the organic phase towards gases and coke.

Kostas S [13] illustrated in situ upgrading of biomass pyrolysis vapors with mesoporous MSU materials. The MSU-S catalytic materials were very selective towards polycyclic aromatic hydrocarbons and heavy fractions, while they produced negligible amounts of acids, alcohols and carbonyls, and very few phenols. The MSU-S materials appeared to possess stronger acid sites than Al-MCM-41, resulting in enhanced yields of aromatics, polycyclic aromatic hydrocarbons and coke, as well as propylene in the pyrolysis gases.

Judit Adam [14] using Al-MCM-41 and SBA-15 in catalytic process compared to non-catalytic experiments. In the catalytic experiments, the hydrocarbon and acid yields increased while the carbonyl and the acid yields decreased. It was pointed out that Al-MCM-41 is the best performing catalysts.

Catalytic cracking was widely used to upgrade bio-oil in recent years. It was regarded as a cheaper route to convert oxygenated feedstock to lighter fractions. However, the yield of upgraded oils is generally low because of the high yields of coke (8~25 wt %) [3]. These undesired products, which are deposited on the catalyst and cause its deactivation, make a periodical or continual regeneration necessary.

2.2. Hydrodeoxygenation

Hydrodeoxygenation was an effective upgrading method to remove oxygen in original bio-oil [15]. It was usually carried out in hydrogen providing solvents under hydrogenation catalysts at high pressure. In earlier report, R. V. Pindoria [16] tried to combined hydropyrolysis of biomass and catalytic cracking of pyrolysis vapors. The method involved hydropyrolysis of biomass and immediate catalytic hydrocracking of pyrolytic oils in the second stage of the reactor. However, extended use of the catalyst did not result in increased yields of liquid products but in increased production of light volatiles or gas. The H-ZSM5 catalyst appeared to act as a more active cracking catalyst rather than to promote hydrogenation or deoxygenation of the liquids produced in the hydropyrolysis stage. Characterizations of the liquids by size exclusion chromatography (SEC) and UV fluoresence indicated that structural changes were relatively minor despite the significant changes in yields of liquids with process conditions.

In recent years, efficient hydrogenation catalyst was paid more and more attention. Most recently catalysts used in hydrodeoxygenation upgrading process were Co-Mo, Ni-Mo and their oxides loaded on Al_2O_3.

Zhang [17] separated the bio-oil into water and oil phases. Under the catalyst of sulphided Co-Mo-P/Al_2O_3, the reaction was operated in an autoclave filled with tetralin (as a hydrogen donor solvent) under the optimum conditions of 360 ℃ and 2 MPa of cold hydrogen pressure. The oxygen content was reduced from 41.8 % of the crude oil to 3 % of the upgraded one, and besides, the hydrophilic and hydrophobic property of product was changed. It was found that crude oil was methanol soluble while the upgraded one was oil soluble because deoxygenation.

Senol [18] eliminated oxygen in carboxylic groups with model compounds of methyl heptanoate and methyl hexanoate on sulphided NiMo/r-Al_2O_3 and CoMo/r-Al_2O_3 in a flow reactor to discern the reaction schemes. Aliphatic methyl esters produced hydrocarbons via three main pathways: the first pathway gave alcohols followed by dehydration to hydrocarbons. The de-esterification yielded an alcohol and a carboxylic acid in the second pathway. Carboxylic acid was further converted to hydrocarbons either directly or with an alcohol intermediate.

Although oxygen content could be dramatically decreased, it was clear that the hydro-treating process needs complicated equipments, superior techniques and excess cost and usually is halted by catalyst deactivation and reactor clogging.

2.3. Emulsification

The development and use of bio-oil/diesel oil emulsions represent a relatively short-term approach to the exploitation of the significant biomass resources potentially available by pyrolysis. The upgrading of the fuel itself by means of the production of stable emulsions allows the use of low-cost large-scale mass production diesel engines with only minor modifications, thus reducing significantly the investment cost in comparison to dual fuel engines.

Michio Ikura [19] provided a method to emulsify pyrolytic bio-oil with No.2 diesel fuel. The formation of stable emulsions required surfactant concentration ranging from 0.8 to 1.5 wt% of total, depending on bio-oil concentration and power input. The costs of producing stable emulsions using CANMET surfactant was 2.6 cents L^{-1} for 10% emulsion, 3.4 cents L^{-1} for 20% emulsions and 4.1 cents L^{-1} for 30% emulsions, respectively. The heating value of centrifuged bio-oil was about one third of that of No.2 diesel and cetane number of pyrolytic bio-oil was 5.6. The corrosivity of emulsion fuels was about half of the bio-oil alone.

D. Chiaramonti [20] described the production of emulsions from biomass fast pyrolysis liquid and diesel fuel for utilization in diesel engines. It was found out that the emulsions were more stable than the original bio-oil. The optimal range to have an acceptable viscosity is between 0.5 % and 2 % and the stability versus time at high temperature (approximately 70℃) of the produced emulsions was about 3 days. Thereafter, the emulsions were used in diesel engines [21]. It was recommended that the injector as well as the fuel pump should be made in stainless steel or similar corrosive resistant material.

Emulsification does not demand redundant chemical transformations and it was a relatively short-term approach to utilization of bio-oil. However, several key points for consideration in case of future activities in the following areas [21]:

- Design, production and testing of stainless steel (or other materials, as ceramic materials) injectors and fuel pumps.
- Long term testing, to verify the suitability of modified injectors and pumps to long term use.
- Detailed assessment of exhaust emissions from diesel engines fed with bioemulsions.
- Design and construction of continuous in line emulsifiers. In this regard, it has to be remarked that the process to emulsify bio-oil and diesel fuels is different on the basis of the target stability, and therefore the equipment will vary accordingly.

2.4. Esterification

This method provided another pathway to improve fuel characters. From a chemical point of view, it is anticipated that reactive molecules like organic acids and aldehydes are converted by the reactions with alcohols into esters and acetals because the oxygen content was decreased by reaction. For example, the elemental oxygen content of acetic acid is 53.3% while for its corresponding ester, ethyl acetate assuming ethanol was used for esterification, the oxygen content is 36.4%. Therefore, it was possible to obtain compounds with high heat value and hydrophobic property. Qi Zhang [22] obtained upgrading bio-oil catalyzed by solid acid and solid base. The density of upgraded bio-oil was reduced from 1.24 to 0.96 kg m^{-3}, and the gross calorific value increased by 50.7 and 51.8 %, respectively. However, water content in the upgraded bio-oil was difficult to separate even use 3A molecular sieve as dehydration method.

In order to decrease the water content in bio-oil, Xu Junming [23] provided a method to obtain upgraded bio-oil using ethanol and bio-oil as raw materials. It was an efficient method to convert low-molecular-weight acids with alcohol into esters and separate them from original bio-oil by reactive rectification. Two kinds of upgraded oil were obtained by this method. Water content decreased from 33.0 % to 0.52 % and 5.03 %, respectively. The dynamic viscosity of upgraded bio-oils was lowered from 10.5 to 0.46 and 3.65mm^2 s^{-1}. The pH value of light oil was alleviated from 2.82 to 7.06, while the pH value of heavy oil rose to 5.93. The gross calorific value of two kinds of upgraded oil increased from 14.3 to 21.5 and 24.5 MJ kg^{-1}, respectively.

F H Mahfud [24] described a method consists of treating pyrolysis oil with a high boiling alcohol like n-butanol in the presence of a (solid) acid catalyst at 323-353 K under reduced pressure (10 kPa). Water content decreased from 31.5 % to 4.9~8.7 %. The dynamic viscosity of upgraded bio-oils was lowered from 17.0 to 7.0~7.6 mm^2 s^{-1}. The gross calorific value increased from 20.6 to 27.7~28.7 MJ kg^{-1}(calculated) [25].

In this new concept, carboxyl acid and aldehyde was converted into corresponding products through esterification and aldolization. By converting the negative compounds that can react in manner which adversely affects the stability of the oil, the objective of esterification is achieved to produce an upgraded bio-oil with improved properties like a higher heating value, lower water content, lower viscosity and lower free acid content.

CONCLUSION

Bio-oils generated from biomass pyrolysis have received great attention from international energy organizations around the world. It can be used as fuels in combustors, engines or gas turbines. However, problems raised in its industrialization due to its deleterious properties of high viscosity, thermal instability, corrosiveness, and chemical complexity. Several key points for consideration in case of future activities in the following areas:

- Since bio-oils are complex and chemically unstable mixtures, more work is needed on the stabilization and upgrading of bio-oils with some modifications to equipment configuration before applying them in generating heat or power.
- Hydrodeoxygenation, catalytic cracking of bio-oils are so complicated techniques that steady, dependable, fully developed reactors are urgently desirable.
- Because the properties of the bio-oil related to its molecular structure., according to the main components and reasonable reactions, it is meaningful to convert carboxyl acid and aldehyde into esters and acetals by esterification, acetalization reactions.
- Emulsification provided a short term approach to use bio-oils, but it takes high cost of surfactant and energy consumption input and the corrosiveness to the engine is also serious. According to the molecular structure of bio-oil, it is potential to convert hydrophilic radicals (carboxyl acid and aldehyde) into hydrophobic radicals (esters and acetals) and blend with diesel in order to decrease the usage of surfactant.

ACKNOWLEDGMENTS

The authors would like to thank the "12 th national 5-year R&D plan" (2011BAD22B05) and National Natural Science Foundation of China, China (31100521) for the financial support during this investigation.

REFERENCES

[1]   Dinesh Mohan, Charles U. Pittman, Jr., Philip H. Steele. Pyrolysis of Wood/Biomass for Bio-oil: A Critical Review [J]. *Energy & Fuels* 2006, 20, 848-889.

[2]   Serdar Yaman. Pyrolysis of biomass to produce fuels and chemical feedstocks[J]. *Energy Conversion and Management.* 2004, 45, 651-671.

[3]   Zhang Qi, Chang Jie, Wang Tiejun, Xu Ying. Review of biomass pyrolysis oil properties and upgrading research. *Energy Conversion and Management* 2007, 48, 87-92.

[4]   Nokkosmaki MI, Kuoppala ET, Leppamaki EA, et al. Catalytic conversion of biomass pyrolysis vapours with zinc oxide. *J Anal Appl Pyrol* 2000, 55, 119-31.

[5]   Costa E, Aguado J, Ovejero G, Canizares P. Conversion of n-butanol-acetone mixtures to C1 to C10 hydrocarbons on HZSM-5 type catalysts. *Ind Eng Chem Res* 1992, 31, 1021-1025.

[6]   Williams PT, Horne PA. The influence of catalyst type on the composition of upgraded biomass pyrolysis oils. *J Anal Appl Pyrol* 1995, 31, 39–61.

[7]   Williams PT, Horne PA. The influence of catalyst regeneration on the composition of zeolite-upgraded biomass pyrolysis oils. *Fuel* 1995, 74, 1839-1851.

[8]   S. Vitolo, M. Seggiani, P. Frediani, G. Ambrosini, L. Politi. Catalytic upgrading of pyrolytic oils to fuel over different zeolites. *Fuel* 1999, 78, 1147-1159.

[9]   Hyun Ju Park, Young-Kwon Park, Joo-Sik Kim. Bio-oil upgrading over Ga modified zeolites in a bubbling fluidized bed reactor. *Studies in Surface Science and Catalysis* 2006, 159, 553-556.

[10] S. Vitolo, B. Bresci, M. Seggiani, M. G. Gallo. Catalytic upgrading of pyrolytic oils over HZSM-5 zeolite: behaviour of the catalyst when used in repeated upgrading–regenerating cycles. *Fuel* 2001, 80(1), 17-26.

[11] Judit Adam, Marianne Blazso´, Erika Me´sza´ros. Pyrolysis of biomass in the presence of Al-MCM-41 type catalysts. *Fuel* 2005, 84, 1494-1502.

[12] E.F. Iliopoulou, E.V. Antonakou, S.A. Karakoulia, I.A. Vasalos, A.A. Lappas, K.S. Triantafyllidis. Catalytic conversion of biomass pyrolysis products by mesoporous materials: Effect of steam stability and acidity of Al-MCM-41 catalysts. *Chemical Engineering Journal* 2007, 134(1-3), 51-57.

[13] Kostas S. Triantafyllidis, Eleni F. Iliopoulou, Eleni V. Antonakou, Angelos A. Lappas, Hui Wang, Thomas J. Pinnavaia. Hydrothermally stable mesoporous aluminosilicates (MSU-S) assembled from zeolite seeds as catalysts for biomass pyrolysis. *Microporous and Mesoporous* Materials 2007, 99(1-2), 132-139.

[14] Judit Adam, Eleni Antonakou, Angelos Lappas, Michael Stöcker, Merete H. Nilsen, Aud Bouzga, Johan E. Hustad, Gisle Øye. In situ catalytic upgrading of biomass derived fast pyrolysis vapours in a fixed bed reactor using mesoporous materials. *Microporous and Mesoporous Materials* 2006, 96(1-3), 93-101.

[15] Douglas C. Elliott. Historical Developments in Hydroprocessing Bio-oils[J]. *Energy & Fuels* 2007, 21, 1792-1815.

[16] R. V. Pindoria, A. Megaritis, A. A. Herod and R. Kandiyoti. A two-stage fixed-bed reactor for direct hydrotreatment of volatiles from the hydropyrolysis of biomass: effect of catalyst temperature, pressure and catalyst ageing time on product characteristics. *Fuel* 1998, 77(15), 1715-1726.

[17] Zhang SP, Yan Yongjie, Li T, et al. Upgrading of liquid fuel from the pyrolysis of biomass. *Bioresour Technol* 2005, 96, 545-550.

[18] Senol OI, Viljava TR, Krause AOI. Hydrodeoxygenation of methyl esters on sulphided NiMo/c-Al2O3 and CoMo/c-Al2O3 catalysts. *Catal Today* 2005, 100(3–4), 331-335..

[19] Michio Ikura, Maria Stanciulescu, Ed Hogan. Emulsifcation of pyrolysis derived bio-oil in diesel fuel[J]. *Biomass and Bioenergy* 2003,24, 221~232.

[20] D. Chiaramontia, M. Boninia, E. Fratinia et a1. Development of emulsions from biomass pyrolysis liquid and diesel and their use in engines-Part 1: emulsion production[J]. *Biomass and Bioenergy* 2003, 25, 85~99.

[21] D. Chiaramontia, M. Boninia, E. Fratinia et a1.Development of emulsions from biomass pyrolysis liquid and diesel and their use in engines—Part 2: tests in diesel engines[J]. *Biomass and Bioenergy* 2003, 25, 101~111.

[22] Qi Zhang, Jie Chang, TieJun Wang, and Ying Xu. Upgrading Bio-oil over Different Solid Catalysts. *Energy & Fuels* 2006, 20, 2717-2720.

[23] Xu Junming, Jiang Jianchun, Sun Yunjuan, Lu Yanju. Bio-oil upgrading by means of ethyl ester production in reactive distillation to removewater and to improve storage and fuel characteristics. *Biomass and Bioenergy* 2008, 32, 1056 – 1061.

[24] F H Mahfud, Melian Cabrera, R Manurung, H J Heeres. Upgrading of flash pyrolysis oil by reactive distillation using a high boiling alcohol and acid catalysts[J]. *Process Safety and Environmental Protection* 2007, 85, 466-472.

[25] Cho, K.W., Park, H.S., Kim, K.H., Lee, Y.K. and Lee, K.-H. Estimation of the heating value of oily mill sludges from steelplant. *Fuel* 1995, 74(12), 1918-1921.

In: Biomass Processing, Conversion and Biorefinery
Editors: Bo Zhang and Yong Wang

ISBN: 978-1-62618-346-9
© 2013 Nova Science Publishers, Inc.

Chapter 12

BIO-OIL UPGRADING

Ying Zhang and Jianhua Guo*

Anhui Province Key Laboratory of Biomass Clean Energy,
Department of Chemistry,
University of Science and Technology of China, Hefei, China

ABSTRACT

With the sharp depletion of fossil fuels, the utilization of renewable energy resources has attracted a lot of attention. One promising method to produce advanced alternative fuels is converting lignocellulosic biomass into bio-oil and then upgrading bio-oil into advanced biofuels. Fast pyrolysis is a relatively mature process to produce bio-oil. The bio-oil has high oxygen content, high corrosivity, low stability and low heating value, and thus isn't able to be used as transportation fuel directly. The effort to produce chemicals from bio-oil is also limited due to its complex chemical composition and resulting high processing cost. This chapter aims to focus on the main upgrading processes of bio-oil in recent years. It also points out challenges to achieve improvements of bio-oil upgrading in the future.

1. INTRODUCTION

With the rapid economic development and sharp growth of energy consumption, the world is facing the problem of petroleum shortages. The research for new sources of liquid fuels has attracted more and more attention in recent years. Fuels derived from plant biomass could be the prospective fuel in the future. Biomass is a kind of renewable, clean, CO_2-neutral, and abundant resource, which can be used to produce heat, power, fuels, and chemicals. It is also the only current sustainable source that can be used to produce liquid fuels [1-3].

Substantial technical processes have been applied to the conversion of biomass to fuels and chemicals. Among these technologies, fast pyrolysis is an effective and relatively mature

---

* Phone 86-551-6360-3463. Fax 86-551-6360-6689. E-mail: zhzhying@ustc.edu.cn.

technology to produce liquid fuel (known as bio-oil) [4]. The feedstock used in fast pyrolysis is lignocellulosic biomass, the cheapest and most abundant form of biomass. Fast pyrolysis process can give high yield of bio-oil up to 70-80 wt % and has been applied commercially [5].

Bio-oil has been recognized as a reasonable and prospective candidate to substitute fossil fuels. Compared to fossil fuels, bio-oil is a clean fuel with lower NO_x emissions and insignificant SO_x emissions. It can even be CO_2-neutral if efficient preparation methods are developed [6]. Bio-oil can be combusted directly to produce heat and power. However, it can not substitute for petroleum to produce high-quality fuels and chemicals directly for its high oxygen content, high corrosivity, low heating value and low thermal stability. Consequently, proper upgrading processes are required to improve the quality of bio-oil [7-8]. Early studies focus on producing more stable bio-oils through simple physical treatment processes. To produce fossil-like fuels, catalytic upgrading processes are introduced to improve the quality of bio-oil.

This chapter will introduce the properties of bio-oil, and the specialty of processes including both physical upgrading and catalytic upgrading.

2. PROPERTIES OF BIO-OIL

Bio-oil produced by pyrolysis is a dark brown liquid with a strong smoky odor. It has a very complex chemical composition, more than 400 organic compounds, mainly including acids, alcohols, aldehydes, ketones, esters, ethers, sugars, furans, phenols, and lignin-derived oligomers, have been identified [9-10]. The yield and specific composition of bio-oil are dependent on the feed stocks and process conditions. In fast pyrolysis process, biomass is treated at moderate temperature (400–600 °C) in absence of oxygen with a short hot vapor residence time (1–2 s). The thermal treatment mainly leads to depolymerization and fragmentation of biomass, therefore, the elemental composition of bio-oil is similar to biomass.

As shown in Table 1, the chemical composition and properties of bio-oil are quite different from petroleum derived fuels [11-13]. One main difference between bio-oil and petroleum derived fuel is oxygen content (35-40 wt % in bio-oil), which mostly affects the polarity, acidity, viscosity, and heating value of bio-oil. The high oxygen and water content lead to the high polarity of bio-oil, therefore, bio-oil is immiscible with fossil oil. The high water content, 15-30 wt % of bio-oil, contributes to the low heating value.

The pH of bio-oil is between 2 and 3, which is caused by the organic acid in bio-oil (mainly acetic acid and formic acid), leading to the corrosion of equipments used for storage, transport, and processing. The ash in bio-oil can cause erosion and blockage.

The specific gravity of bio-oil is higher than light fuel oil and heavy fuel oil. The dynamic viscosity of bio-oil is higher than light fuel oil and much lower than heavy fuel oil.

The instability of bio-oil is mainly due to the presence of reactive compounds. The reactive compounds, such as aldehydes, ketones, acids, and phenols, tend to form large molecular weight compounds during storage by interaction of these reactive functional groups. Lignin-derived oligomers, the major thermally unstable component in bio-oil, will easily cause carbon deposition and equipment blocking during application.

Table 1. Properties of bio-oil, light fuel oil and heavy fuel oil [11-13]

Property	Pyrolysis bio-oil	Light fuel oil	Heavy fuel oil
Elemental Composition (wt %)			
C	54-58	86	85
H	5.5-7.0	13	11
O	35-40	0.1	1.0
N	0-0.2	0	0.3
S	0-0.02	0-0.01	0.03
Moisture content (wt %)	15-30	0.02	0.1
pH	2.5	-	-
Specific gravity	1.15-1.25	0.70-0.85	0.82-0.95
Higher heating value (MJkg^{-1})	16-19	37	40
Viscosity (centistokes)			
at 50 °C	7	4	50
at 80 °C	4	2	41
Solids (wt %)	0.2-1	-	1
Distillation residue (wt %)	up to 50	-	1
Ash	0-0.2	0.01	0.1
Pour point	-33	-15	-18
Turbine emissions (g/MJ)			
NO$_x$	<0.7	1.4	N/A
SO$_x$	0	0.28	N/A

Overall, the bio-oil cannot substitute for fossil fuel directly for its poor quality. The upgrading processes should focus on the decrease of oxygen content and instable organic compounds, and increase of the heating value and stability of bio-oil.

3. PHYSICAL UPGRADING OF BIO-OIL

3.1. Solvent Addition

The addition of polar solvents, such as methanol, ethanol, and furfural, can improve the homogeneity and reduce the viscosity of bio-oil. Diebold investigated kinds of additives and found methanol was a good one. The stability of bio-oil increased significantly when adding 10 wt. % of methanol [14]. Most studies directly added solvents after pyrolysis, which worked well to decrease the viscosity and increase stability and heating value [4].

Solvent addition is a simple method to improve the stability of bio-oil; however, other properties such as the oxygen content and heating value haven't been improved.

3.2. Emulsification

To make bio-oil applied to internal combustion engine directly, the emulsification process has been widely investigated in past years. Bio-oil is not miscible with petroleum fuels for its high oxygen content and thus the high polarity, the addition of surfactant can make bio-oil be emulsified in petroleum fuels. A lot of researchers have demonstrated that the bio-oil would be less viscous and corrosive, and more stable after emulsification.

The selection of surfactant and parameters of emulsification bio-oil and petroleum fuel have been the focus of research. Chiaramonti et al. prepared emulsions from bio-oil and diesel fuel with different ratios, and studied their properties. Their research indicated that the emulsions are more stable than the pure bio-oil. The stability versus time at 70 °C of the emulsions was about 3 days. [15]. They also tested the fuel properties of emulsions in engines and found that the testing would be successful when using stainless steel made injector and fuel pump [16]. Ikura et al. studied the relationship of process conditions, emulsion stability and processing cost of emulsions prepared from bio-oil and diesel fuel. They demonstrated that 0.8-1.5 wt % surfactant concentration was required to the formation of stable emulsions. The costs of producing stable emulsions using CANMET surfactant were 2.6 cents/L for 10 % emulsion, 3.4 cents/L for 20 % emulsion and 4.1 cents/L for 30 % emulsions, respectively [17].

In recent years, researches are not restricted to the emulsification of bio-oil and petroleum fuel, the emulsification of bio-oil and bio-diesel is also proposed and investigated. Jiang et al. produced a stable emulsion from bio-oil and bio-diesel (an initial ratio of 4:6 by volume) in the presence of 4 % octanol at 30 °C. Great improvement has been achieved in various indexes, including acid number, viscosity, and water content compared to the original bio-oil. The emulsions could remain in a single phase throughout the aging conditions of 60 or 80 °C up to 180 h. And changes in certain properties and the chemical composition of mixtures were minimal during storage [18-19]. Recently, they further improved their emulsification process with a preprocessing of separating pyrolytic lignin to obtain a more stable emulsion [20].

Emulsification is a handy method to use bio-oil as transportation fuel directly, however, the emulsified liquid is also not stable during long-term storage, and is not suitable for a long-term operation in common engines. The high cost of the surfactants and high energy requirement for the emulsification process also hindered its development.

3.3. Chemicals Extracted from the Bio-Oil

There are many valuable compounds, such as phenols, lignin-derived oligomers, volatile organic acids, levoglucosan, and hydroxyacetaldehyde, which can be extracted from bio-oil. Phenols and lignin-derived oligomers can substitute for phenol to produce resin in industry. The phenols can be obtained by organic extraction [21] or distillation [22]. Lignin-derived oligomers can easily be obtained by water washing and separation [23]. Volatile organic acids, mainly including formic acid and acetic acid, can be collected by distillation [24]. However, the cost is obstacle to the commercialization of extracted special chemicals from bio-oil [13].

4. CATALYTIC UPGRADING OF BIO-OIL

4.1. Hydrotreating

Hydrotreating is a process which has widely been applied in the petrochemical engineering to remove sulfur and nitrogen from petroleum. In bio-oil, the content of oxygen is significantly higher than that of sulfur and nitrogen, therefore, the hydrodeoxygenation is the main method taken by researchers to reduce the oxygen content in bio-oil [25]. The hydrotreating of bio-oil always occurs at temperature between 200- 600 °C in the presence of hydrogen (100-200 bar) or hydrogen-donor solvents (methanol, formic acid, *etc.*) and heterogeneous catalysts [26-29]. After hydrotreating, the oxygen in the bio-oil can be removed in the form of H_2O, and the stability and heating value of bio-oil can be improved. However, some unwanted reactions, such as polymerization and polycondensation, may occur simultaneously at high temperature. Therefore, a wide range of investigations have been carried out on the improvement of catalysts and processes.

4.1.1. Conventional Catalysts

Sulfided Ni-Mo and Co-Mo supported on Al_2O_3 or SiO_2-Al_2O_3 are widely studied on the hydrodeoxygenation of bio-oil for their good performance in petroleum refinery [30-34]. In most cases, these catalysts were tested with model compounds, such as esters and phenols, to elucidate the main reaction pathways, the influence of the important reaction parameters, and the catalyst deactivation [35-38].

The thermally unstable compounds in bio-oil can easily cause carbon deposition on catalysts at high temperature. Elliott and co-workers proposed a two step hydrotreating process to solve this problem [29-32]. The first step (stabilization step) was taken at a low temperature (under 280 °C) in the presence of hydrogenation catalysts (noble metal catalysts or conventional hydrotreating catalysts) to reduce the active functional groups, such as aldehydes, ketones, and C=C bonds. The second step was taken at a high temperature (400 °C) to remove oxygen from bio-oil in the presence of sulfided $CoMo/Al_2O_3$ or $NiMo/Al_2O_3$. The oxygen content of refined oil was less than 1 wt %.

The Al_2O_3 support was instable in the bio-oil with high content of water and its acid sites could promote carbon deposition. To improve the activity of catalysts, the effects of supports were also investigated. Carbon, ZrO_2, TiO_2 supports were examined and showed better ability in water-resistance and anti-coking than Al_2O_3 [39-42]. The basic metallic oxide, MgO, was also tested. The dispersion of MoO_3 precursor was improved and the Co-Mo-P/MgO showed a good activity and resistance to coking [43].

However, these conventional hydrodeoxygenation catalysts may encounter problems such as possible contamination of products by incorporation of sulfur, rapid deactivation by coke formation, and potential poisoning by water [44].

4.1.2. Noble Metal Catalysts

More recently, the sulfur-free noble metal catalysts are investigated at low temperature in hydrotreating process. Ru, Rh, Pd and Pt based catalysts are the ones people most investigated [45-51].

Wildsehut and co-workers tested Ru/C, Pt/C, Pd/C, sulfided CoMo/Al$_2$O$_3$ and NiMo/Al$_2$O$_3$ in hydrotreating process and found that Ru/C and Pd/C have higher activity and stability than conventional sulfided catalysts [52-53].

Elliott *et al.* used Ru/C and Pd/C as catalysts in model compounds (acetic acid, furfural and guaiacol) hydrotreating at different temperatures and found that ruthenium catalyst and palladium catalyst had different reactivity patterns. Compared to palladium catalyst, hydrogenation could proceed at 50-100°C lower temperatures using ruthenium catalyst due to its high catalytic activity for hydrogenation. To avoid ruthenium catalyzed aqueous-phase reforming and methanation reactions, the hydrogenation temperature should be limited to less than 250 °C. Unlike ruthenium catalyst, hydrogenation could proceed at higher temperatures with palladium catalyst to improve its catalytic activity. Acetic acid can not be effectively hydrogenated to ethanol with ruthenium catalyst but can do with moderate yields using palladium catalyst at 300 °C. Furfural can be hydrogenated to tetrahydrofuran and methanol with both ruthenium and palladium catalysts at low temperatures. 1,4-pentanediol and methyl-tetrahydrofuran were the main products at 250 °C and above. The hydrogenation pathway of guaiacol with ruthenium passed through methoxycyclohexanol to cyclohexanediols at low temperatures and continued on cyclohexanol at higher temperatures. With palladium as catalysts, guaiacol was first converted into methoxycyclohexanone at 150 °C and then methoxycyclohexanol at 200 °C with some cyclohexanediol. Then the products converted to cyclohexanol and cyclohexane at 250 °C, and cyclohexane was the main products at 300 °C [32].

Recently, a new and efficient catalytic system has been developed by Lercher and Kou. In combination of Pd/C and H$_3$PO$_4$, phenolic compounds could be completely converted into C$_6$-C$_9$ cycloalkanes and methanol at 250 °C in water [44]. They also demonstrated that the phenolic compounds can be converted to alkanes in Brønsted acidic ionic liquid combined with noble metal (Ru, Rh, and Pt) nanoparticle catalysts [54].

Although the high price may limit the application of these noble metals (Ru, Rh, Pd and Pt), the good performance in hydrotreating, such as high reactivity, reproducibility, mild reaction condition requirement, and so on, could compensate this drawback and make it possible for large-scale bio-oil upgrading.

4.1.3. Other Metal Catalysts

Ni is a suitable metal in hydrotreating for its high activity and low cost. Lercher and Kou used fully heterogeneous catalyst combination, Raney® Ni and Nafion/SiO$_2$, to replace noble metal and mineral acid. And phenolic compounds can be completely converted into C$_6$-C$_9$ cycloalkanes and methanol at 300 °C [55]. Xiong *et al.* studied hydrotreating of bio-oil with Raney Ni and zeolites-supported Ru as catalysts and formic acid as hydrogen resource. A high yield of 80-90 wt % of upgraded bio-oil was obtained at 150-230 °C for 5-7 h [56]. Wang *et al.* exploited the H-transfer reactions with Raney® Ni and propan-2-ol in bio-oil under unusual, low-severity conditions and converted bio-oil to cyclic alcohols [57].

Zero valent metals can be used in bio-oil as hydrogenation catalysts. Liu *et al.*, employed Zn as catalyst and upgraded bio-oil at ambient temperature and pressure. The C=O bonds reduced from 9.8 to 3.1 mol % after upgrading [58].

The hydrodeoxygenation of model compounds has been widely investigated, however, many problems still need to be solved during hydrotreating of bio-oil. The key problems are the catalyst deactivation, reactor clogging and poor quality of upgrading oils.

4.2. Catalytic Cracking

The oxygen in bio-oil has seriously influenced the quality of bio-oil. Besides hydrodeoxygenation, catalytic cracking is also an efficient process to remove oxygen. Through catalytic cracking, the oxygen of bio-oil can be removed in the form of CO_2, CO and H_2O, accompanying with the cracking of large molecular weight compounds to low molecular weight compounds. In contrast to hydrotreating, the catalytic cracking occurs under atmospheric pressure without hydrogen at 300-600 °C, and the requirement for equipment is low [59-62].

The catalytic cracking can be performed either on liquid bio-oil or on pyrolysis vapors [63]. The reaction temperature and properties of catalysts are the main factors that influence the distribution of products. High temperature will increase the gas yield and decrease the oil yield. In the bio-oil cracking, the acidity of catalysts contributes to removing oxygen by dehydration, decarboxylation and decarboxylation [64-66]. The pore size of catalysts controls the distribution of products [67-68]. Therefore, the selection and preparation of catalysts are the main concerns of recent researches.

Zeolites are widely used as catalysts in converting petroleum crude oils to gasoline, olefinic gases and other products for their high acidity and high thermal stability in past decades. HZSM-5 is the most commonly used catalyst in early studies [69-72]. Adjaye *et al.* investigated the effect of HZSM-5, H-mordenite, H-Y, silica-alumina, and silicalite on bio-oil cracking and found that the activity of zeolites mainly depended on their acidity. HZSM-5, which is rich in Lewis and Brønsted acid sites, performed well in bio-oil cracking [67, 70, 71]. Through adjusting the Si/Al ratio or adding transition metals to zeolites, the activity of zeolites can be improved by increasing the acid sites of catalysts [65, 72]. Corma *et al.* investigated the catalytic cracking reaction pathways of biomass-derived oxygenates at 500–700 °C, and analyzed the effect of catalysts [61]. The order of conversion activity was USY~FCC (a fresh fluid catalytic cracking catalyst)> Al_2O_3 > ZSM5 (a ZSM5-based FCC additive) > ECat (an equilibrium FCC catalyst with metal impurities)>>SiC, which was in accordance with the order of acidity. However, the coke yield was also high when employing USY as catalyst. Among these catalysts, ZSM5 had the lowest yield of coke and highest yield of olefins and aromatics.

Mesoporous catalysts, which have a larger pore diameter between 2-50 nm, have a large surface area, better thermal stability and thicker wall than zeolites. The large pore size can insure the conversion of bulky molecules to hydrocarbons instead of coke in bio-oil. MCM-41 and SBA-15 are the most studied mesoporous catalysts, however, the low acidity results in quite low yield of hydrocarbons. The introduction of acid sites can improve the activity of MCM-41 and SBA-15. Al/MCM-41, Cu/MCM-41 and Al/SBA-15 can convert large molecular weight compounds to low molecular weight compounds [73-77].

The deactivation of catalyst and high coke yield are the key problems in catalytic cracking. The acid sites not only influence the catalyst activity but also contribute to the formation of coke [64]. The catalysts will be deactivated if the coke forms in micropores and blocks the acid sites. The catalysts can be regenerated by removing coke from pores at high temperature about 500-600 °C, but the number of acid sites decreases significantly [65, 78]. Some researchers found that the cracking system is more stable in hydrogen with the effect of metal modified zeolite catalysts [79-81].

Although catalytic cracking is regarded as a cheaper route to decrease the oxygen of bio-oil, the high yield of coke and deactivation of catalyst are still urgent to be resolved before application.

4.3. Catalytic Esterification

The high content of carboxylic acids (about 20 wt %), such as formic, acetic and propionic acid, in crude bio-oil is an important reason for its corrosivity and instability. Through catalytic esterification process, which can convert most of the acids into stable esters by reacting with low-molecular weight alcohols in the presence of an acid or base catalyst, the corrosiveness of bio-oil will be lowered and the heating value of bio-oil will be increased with the aid of alcohols [82-85].

The traditional catalysts used in esterification process are low-cost liquid acids (sulfuric acid, acetic acid, *etc.*), which require high quality equipment and cause environmental pollution. In recent studies, researches focus on the application of heterogeneous catalysts in esterification. Solid acid, solid base, metal based catalysts and ionic liquid catalysts have been applied to esterification process.

4.3.1. Solid Acid Catalysts

Solid acid catalysts, mainly including metallic oxide, zeolites, cation-exchange resins, heteropolyacid, and inorganic acid salt, have been mostly investigated. Among these solid acid catalysts, metallic oxide catalysts have high catalytic activity and stability. Since Hino first prepared SO_4^{2-}/M_xO_y solid acid in 1979 [86], series of SO_4^{2-}/M_xO_y (M=Fe, Ti, Sn, Zr, Hf *etc.*) catalysts have been prepared and studied. The organic acids of bio-oil decreased and the stability was enhanced obviously after esterification by SO_4^{2-}/M_xO_y solid acid and the properties can sustain after aging for several months [87].

Acidic ion-exchange resins are also effective catalysts for esterification. Researchers used strong acidic ion-exchange resins, such as pre-treated 732, NKC-9 type ion-exchange resins or Abemlyst 70, in bio-oil esterification and achieved good results [88-90]. Nafion can also be used as solid acid in esterification. Mahfud treated bio-oil with a high boiling alcohol (n-butanol) in the presence of H_2SO_4 or Nafion® SAC13 at 323-353 K under reduced pressure (<10 kPa).

In this way, the water content of the bio-oil was reduced significantly. Although the activity of Nafion is lower than H_2SO_4, the pH of the product is significantly higher (pH=3.2 versus 0.5 for H_2SO_4) [91].

Esterification can be combined with other processes, such as oxidation. Xu *et al.* oxidated the bio-oil in H_2O_2 or ozone and esterificated it subsequently with SO_4^{2-}/ZrO_2 -MCM-41 catalyst. Compared with crude bio-oil, a higher yield of upgraded bio-oil was obtained with higher thermal stability using oxidized bio-oil as feedstock by esterification [92-93]. Supercritical alcohol conditions and microwave heating also played good role in bio-oil upgrading. [94-96]. Peng *et al.* studied the esterification of bio-oil in supercritical ethanol over aluminum silicate and HZSM-5 and found that supercritical condition is good for esterification.

4.3.2. Solid Base Catalysts

Solid base catalysts, which mainly consist of inorganic acid salt, alkali metal oxide, and anion-exchange resins, are mostly used for the preparation of bio-diesel in recent studies. Alkali metal hydroxide (KOH, NaOH) and alkali metal carbonate ($MgCO_3$, K_2CO_3, Na_2CO_3, and $CaCO_3$) supported on metallic oxide shows a good performance in bio-oil esterification [97].

4.3.3. Ionic Liquid Catalysts

Ionic liquids are widely studied in organic synthesis for its non-volatility, thermostabilization, easy recycling and environmental friendliness. Xiong *et al.* first applied acidic ionic liquid to bio-oil esterification. They found that organic acid in bio-oil could be converted into esters in dual-cation $C_6(mim)_2$-HSO_4 under mild conditions, and the properties of bio-oil were improved with a decrease in water, and increases in heating value and pH. The use of acidic ionic liquid can solve the problem of acid loss of solid acid, however, further improvement should be made to enhance the esterification rate and bio-oil stability [98].

4.3.4. Metal-acid/Base Bifunctional Catalysts

In recent years, esterification of bio-oil in metal-acid/base bifunctional catalysts was investigated. Hydrogenation and cracking occur simultaneously with esterification in the presence of noble metals, such as Ru, Pd, Pt, *etc.* Compared to the bio-oil obtained from the individual process, that obtained from co-processes exhibits better properties. Tang *et al.* upgraded bio-oil with the combination of hydrotreatment, esterification, and cracking in supercritical ethanol conditions under hydrogenation atmosphere by using $Pd/SO_4^{2-}/ZrO_2/SBA$-15 catalyst and obtained the upgraded bio-oil with good quality [99]. Some researches focused on esterification and cracking of bio-oil in supercritical monoalcohols over Pt-based catalysts. PtNi/MgO and $Pt/SO_4^{2-}/ZrO_2/SBA$-15 both displayed good upgrading performance [100-102]. Xu *et al.* combined hydrotreatment and esterification over the CoRu/γ-Al_2O_3 catalyst to upgrade bio-oil. The content of esters in bio-oil increased by 2-fold after upgrading [103].

Catalytic esterification requires less energy compare to hydrogenation and cracking. However, to increase the conversion of acid in bio-oil, 2-3 times of alcohols are consumed in the process, which increase the cost. And the acid loss of catalysts is still an intractable problem to be solved.

4.4. Catalytic Reforming

Catalytic reforming is a mature technology which has been used in industry to produce products with higher octane number using fossil fuels as reactants. Now this process has been widely investigated by researchers to produce hydrogen and synthesis gas with bio-oil. Hydrogen is a promising clean energy and an important industrial chemical. Synthesis gas, which consists of H_2 and CO, can be converted into methanol or liquid fuels by Fischer-Tropsch synthesis. At present, steam reforming (SR), electrochemical catalytic reforming (ECR) and aqueous phase reforming (APR), are the main processes applied in bio-oil reforming [104-107].

4.4.1. Steam Reforming

SR is a kind of process which can put all kinds of compounds in bio-oil to effective utilization. The reactions are usually carried out by reacting steam with bio-oil in fluidized or fixed bed reactors at temperatures of 600-800 °C [106, 108]. Substantial researches on model compounds, such as alcohols, acetic acid and phenols *etc.*, and bio-oil have been carried out in the past decades. The oxygenated compounds in bio-oil can be described as $C_nH_mO_k$, and SR of bio-oil can be represented as reaction (1) [106, 109].

$$C_nH_mO_k+(n-k)H_2O \rightleftharpoons nCO+(n+m/2-k)H_2 \tag{1}$$

The SR is generally accompanied by water gas shift (WGS) reaction (2). The overall process can be present as reaction (3).

$$CO+H_2O \rightleftharpoons CO_2+H_2 \tag{2}$$

$$C_nH_mO_k+2(n-k)H_2O \rightleftharpoons nCO_2+(2n+m/2-k)H_2 \tag{3}$$

Methanation reaction is also existed and causes the consumption of hydrogen:

$$CO+3H_2 \rightleftharpoons CH_4+H_2O \tag{4}$$

Many compounds in bio-oil are unstable at high temperature. The main problem of SR is carbon deposition, which can cause the deactivation of catalysts after long term operation. High temperature is required to make sure the full conversion of feedstock but also leads to more coke deposition. Boudouard reaction (5), CO reduction (6) and thermal decomposition (7) are the main carbon formation reactions [106].

$$2CO \rightleftharpoons C+CO_2 \tag{5}$$

$$CO+H_2 \rightleftharpoons C+H_2O \tag{6}$$

$$C_nH_mO_k \longrightarrow C_nH_mO_k+gases+coke \tag{7}$$

The temperature, pressure, steam to carbon ratio (S/C) and catalyst to feed ratio (C/F) are important parameters that influence the reactions. By selecting the catalysts and controlling parameters, hydrogen or synthesis gas can be obtained selectively. The application of shift catalysts downstream the former at low temperature can increase H_2 yield by WGS reaction (2). The higher temperature will increase the CO/H_2 ratio. At higher S/C ratio, the CO/H_2 ratio decreases. The high S/C is necessary to avoid carbon formation [110].

A lot of research focuses on the SR of model compounds and bio-oil in recent years. The improvement of catalyst system is the research priority. VIII metals, such as Ni, Co, Fe, Ru, Rh, Pt and Pd, supported on metallic oxides are the catalysts mostly investigated in bio-oil SR. Although noble metals have high activity and stability, the high cost inhibits their application. Ni based catalysts are widely used for its low price and good activity. Ni/Al$_2$O$_3$ performs well in SR of bio-oil, but it is easily deactivated for coke deposition. When adding basic oxide, such as MgO, CeO$_2$ and La$_2$O$_3$, to Al$_2$O$_3$, the stability and activity can be improved [110-112]. Adding Cu, Co, Cr or Ru metals to a Ni catalyst can increase the yield of H$_2$ and decrease the cock deposition [112-116].

4.4.2. Electrochemical Catalytic Reforming

In recent years, a novel efficient ECR method for production of hydrogen from bio-oil was developed by Li *et al*. They found that the current applied to the catalyst significantly promoted the catalyst reduction and bio-oil reforming. A high carbon conversion and yield of hydrogen could be obtained via the ECR approach [117-121].

The ECR reaction was carried out in a fixed-bed reactor. SR catalysts (such as NiCuZn-Al$_2$O$_3$, CoZnAl, and Ni/Al$_2$O$_3$) were uniformly embedded around an electrified Ni-Cr wire, which was used for heating and synchronously providing the thermal electrons to catalysts during reforming. The promotion of the current can be explained as follows: 1) For the non-uniform temperature distribution in the catalytic bed and higher temperature gradients presented in ECR than conventional SR, the local close to the electrified wire was under higher temperature environment. It might partly result in the higher activity of catalysts close to electrified Ni-Cr wire. 2) The thermal electrons emitted from the surface of electrified Ni-Cr wire caused the reduction of catalyst, which result in the enhancement of catalysts activity. 3) The thermal electrons can cause the dissociation of the organic intermediates on the catalyst surface and form small unstable fragments and some active radicals, which may be useful in promoting reforming.

The ECR of different model compounds and bio-oil were investigated. Ethanol and acetic acid can be converted into hydrogen via ECR with a high yield and conversion at 400 °C. The ECR of light fraction of bio-oil and anisole require higher temperature [117-119]. Recently, the crude bio-oil and biomass char have been also investigated. When applying ECR to a dual fixed-bed system with Ni-Al$_2$O$_3$ as SR catalyst and CuZn-Al$_2$O$_3$ as WGS catalyst, an absolute hydrogen yield of 110.9 g H$_2$/kg dry biomass can be obtained [121].

4.4.3. Aqueous Phase Reforming

APR was firstly proposed by Dumesic for conversion of biomass to H$_2$ and alkanes in an aqueous solution at low temperatures (150-265 °C) [122-127]. It has been carried out with menthol, ethanol, glycol, glucose, ethylene glycol, sorbose, and light fraction of bio-oil [128-134].

Compared to the SR, there are several advantages: 1) the low temperature reduces the coke deposition; 2) the vaporization of feed and water is not necessary, which reduces the energy consumption; 3) the feed are more stable than bio-oil; 4) CO$_2$ can be extracted for recycling; 5) the process is simpler than SR.

The catalysts used in APR are mainly VIII metals supported on metallic oxides. The WGS and C-C bond breaking are the key reactions to produce H$_2$ in APR. And the C-O bond breaking, which causes the hydrogen consumption, is the reaction should be avoided. From

Dumesic's studies, Pt is the best mono metallic catalyst in terms of activity and selectivity for APR [122-125, 127]. The effect of supporter was also investigated and Pt supported on TiO_2, carbon and Al_2O_3 showed good activities [122, 127]. Furthermore, the addition of Ni, Fe or Co in Pt supported catalyst can improve its activity [127]. Non-noble metal catalyst, Raney-Ni, can be used to achieve good activity for production of H_2 by APR of biomass-derived oxygenated hydrocarbons [129]. When adding Sn to Raney-Ni, the selectivity and stability of catalysts will both increase [126].

The catalytic reforming of bio-oil are environmental processes to produce H_2 or synthesis gas, however, the short lifetime of the catalysts is still a big hindrance.

CONCLUSION

Bio-oil produced by pyrolysis has been recognized as a promising fuel to substitute petroleum fuel in the future. It can be directly combusted to produce heat and power, or can be upgraded to produce high-quality fuels and valuable chemicals. Many processes are developed for bio-oil upgrading. Nevertheless, upgrading technologies are currently confined to laboratory and pilot scale and it need some time before their application in industry. It is mainly due to the complexity of bio-oil and the high cost of upgrading process. In catalytic upgrading process, coking and deactivation of catalysts are the intractable problems urgent to be solved. The upgraded bio-oils obtained from the current technologies are less competitive than petroleum fuels in properties and price. To make the bio-oil more competitive, some recommendations are described as follows:

- Novel upgrading progresses under mild conditions are needed to be investigated to reduce energy consumption and reduce the possibility of coking and catalyst deactivation.
- Investigation on catalysts is necessary for searching stable and cheap catalysts with high activity.
- Catalytic fast pyrolysis should be developed to regulate the pyrolysis pathway and selectively obtained the target bio-oil with limited types of components which may be easier for the following upgrading process.
- Novel integrated refinery processes are required to systematically upgrade bio-oils into transportation fuels that have desirable qualities, while producing other value-added co-products to make the economics work.

REFERENCES

[1]   Bertero M, Puente G, Ulises Sedran. Fuels from bio-oils: Bio-oil production from different residual sources, characterization and thermal conditioning. *Fuel* 2012;95:263.
[2]   Mortensen PM, Grunwaldt JD, Jensen PA, Knudsen KG, Jensen AD. A review of catalytic upgrading of bio-oil to engine fuels. *Appl. Catal. A* 2011;407:1.
[3]   Naik SN, Goud VV, Rout PK, Dalai AK. Production of first and second generation biofuels: A comprehensive review. *Renew Sust. Energ. Rev.* 2010;14:578.

[4]   Xiu SN, Shahbazi A. Bio-oil production and upgrading research: A review. *Renew Sust. Energ. Rev.* 2012;16:4406.

[5]   Zhang Q, Chang J, Wang TJ, Xu Y. Review of biomass pyrolysis oil properties and upgrading research. *Energ. Convers. Manag.* 2007;48:87.

[6]   Oasmaa A, Czernik S. Fuel Oil Quality of Biomass Pyrolysis Oils State of the art for the end-users. *Energy Fuels* 1999;13:914.

[7]   Butler E, Devlin G, Meier D, McDonnell K. A review of recent laboratory research and commercial developments in fast pyrolysis and upgrading. *Renew Sust. Energ. Rev.* 2011;15:4171.

[8]   Bridgwater AV. Review of fast pyrolysis of biomass and product upgrading. *Biomass Bioenergy* 2012;38:68.

[9]   Czernik S, Bridgwater AV. Overview of Applications of Biomass Fast Pyrolysis Oil. *Energy Fuels* 2004;18:590.

[10]   Bridgewater AV. Biomass fast pyrolysis. *Therm. Sci.* 2004;8:21.

[11]   Huber GW, Corma a. Synergies between Bio- and Oil Refineries for the Production of Fuels from Biomass. *Angew. Chem. Int. Ed.* 2007;46:7184.

[12]   Czernik S, Bridgwater AV. Overview of Applications of Biomass Fast Pyrolysis Oil. *Energy Fuels* 2004;18:590.

[13]   Mohan D, Pittman CU Jr, Steele PH. Pyrolysis of Wood/Biomass for Bio-oil: A Critical Review. *Energy Fuels* 2006;20:848.

[14]   Diebold JP, Czernik S. Additives to lower and stabilize the viscosity of pyrolysis oils during storage. *Energy Fuels* 1997;11:1081.

[15]   Chiaramonti D, Bonini M, Fratini E, Tondi G, Gartner K, Bridgwater AV, Grimm HP, Soldaini I, Webster A, Baglioni P. Development of emulsions from biomass pyrolysis liquid and diesel and their use in engines-Part 1: emulsion production. *Biomass Bioenergy* 2003;25(1):85.

[16]   Chiaramonti D, Bonini M, Fratini E, Tondi G, Gartner K, Bridgwater AV, Grimm HP, Soldaini I, Webster A, Baglioni P. Development of emulsions from biomass pyrolysis liquid and diesel and their use in engines-Part 2: Tests in diesel engines. *Biomass Bioenergy* 2003;25(1):101.

[17]   Ikura M, Stanciulescu M, Hogan E. Emulsification of pyrolysis derived bio-oil in Diesel fuel. *Biomass Bioenergy* 2003;24:221.

[18]   Jiang XX, Ellis N. Upgrading bio-oil through emulsification with biodiesel: Thermal stability. *Energy Fuels* 2010;24:2699.

[19]   Jiang XX, Ellis N. Upgrading Bio-oil through Emulsification with Biodiesel: Mixture Production. *Energy Fuels* 2010;24:1358.

[20]   Jiang XX, Zhong ZP, Ellis N. Characterisation of the mixture product of ether-soluble fraction of bio-oil (ES) and bio-diesel. *Can. J. Chem. Eng.* 2012;90:472.

[21]   Chum HL, Black SK. U.S. Patent 4,942,269, 1990.

[22]   Murwanashyaka JN, Pakdel H, Roy C. Seperation of Syringol from Birch Wood-Derived Vacuum Pyrolysis Oil. *Sep. Purif. Technol.* 2001;24:155.

[23]   Sipila K, Keoppala E, Fagernas L, Oasmaa A. Characterization of biomass-based flash pyrolysis oils. *Biomass Bioenergy* 1998;14:103.

[24]   Deng L, Zhao Y, Fu Y, Guo QX. Green Solvent for Flash Pyrolysis Oil Separation. *Energy Fuels* 2009;23:3337.

[25] Huber GW, Iborra S, Corma A. Synthesis of transportation fuels from biomass: chemistry, catalysts, and engineering. *Chem. Rev.* 2006;106:4044.

[26] Choudhary TV, Phillips CB. Renewable fuels via catalytic hydrodeoxygenation. *Appl. Catal. A* 2011;397:1.

[27] Huber GW, Corma A. Synergies between Bio- and Oil Refineries for the Production of Fuels from Biomass. *Angew. Chem. Int. Ed.* 2007;46:7184.

[28] Furimsky E. Catalytic hydrodeoxygenation. *Appl. Catal. A* 2000;199:147.

[29] Elliott DC, Hart TR, Neuenschwander GG, Rotness LJ, Olarte MV, Zacher AH, Solantausta Y. Catalytic Hydroprocessing of Fast Pyrolysis Bio-oil from Pine Sawdust. *Energy Fuels* 2012;26:3891.

[30] Elliott DC, Hart TR, Neuenschwander GG, Rotness LJ, Olarte MV, Zacher AH. Catalytic hydroprocessing of biomass fast pyrolysis bio-oil to produce hydrocarbon products. *Environ. Prog. Sustainable Energy* 2009;28 (3):441.

[31] Elliott DC. Historical Developments in Hydroprocessing Bio-oils. *Energy Fuels* 2007;21:1792.

[32] Elliott DC, Hart TR. Catalytic Hydroprocessing of Chemical Models for Bio-oil. *Energy Fuels* 2009;23:631.

[33] Kwon KC, Mayfield H, Marolla T, Nichols B, Mashburn M. Catalytic deoxygenation of liquid biomass for hydrocarbon fuels. *Renew Energy* 2011;36:907.

[34] Zhang SP, Yan YJ, Ren ZW, Li TC. Study of Hydrodeoxygenation of Bio-Oil from the Fast Pyrolysis of Biomass. *Energy Source* 2003;25:57.

[35] Centeno A, Laurent E, Delmon B. Influence of the support of CoMo sulfide catalysts and of the addition of potassium and platinum on the catalytic performances for the hydrodeoxygenation of carbonyl, carboxyl, and guaiacol-type Molecules. *J. Catal.* 1995;154:288.

[36] Laurent E, Delmon B. Study of the hydrodeoxygenation of carbonyl,catalylic and guaiacyl groups over sulfided CoMo/γ-Al$_2$O$_3$ and NiMo/γ-Al$_2$O$_3$ catalysts: I. Catalytic reaction schemes. *Appl. Catal. A* 1994;109:77.

[37] Laurent E, Delmon B. Study of the hydrodeoxygenation of carbonyl,catalylic and guaiacyl groups over sulfided CoMo/γ-Al$_2$O$_3$ and NiMo/γ-Al$_2$O$_3$ catalysts: II influence of water,ammonia and hudrogen-sulfide. *Appl. Catal. A* 1994;109:97.

[38] Laurent E, Delmon B. Influence of water in the deactivation of a sulfifed nimo gamma-Al$_2$O$_3$ catalyst during hydrodeoxygenation. *J. Catal.* 1994;146:281.

[39] Ferrari M, Bosmans S, Maggi R, Delmon B, Grange P. CoMo/carbon hydrodeoxygenation catalysts: influence of the hydrogen sulfide partial pressure and of the sulfidation temperature. *Catal. Today* 2001;65:257.

[40] Ferrari M, Maggi R, Delmon B, Grange P. Influences of the hydrogen sulfide partial pressure and of a nitrogen compound on the hydrodeoxygenation activity of a CoMo/carbon catalyst. *J. Catal.* 2001;198:47.

[41] Bui V N, Laurenti D, Delichère P, Geantet C. Hydrodeoxygenation of guaiacol with CoMo catalysts. Part I: Promoting effect of cobalt on HDO selectivity and activity *Appl. Catal. B* 2011;101:239.

[42] Bui V N, Laurenti D, Delichère P, Geantet C. Hydrodeoxygenation of guaiacol: Part II: Support effect for CoMoS catalysts on HDO activity and selectivity *Appl. Catal. B* 2011; 101:246.

[43] Yang Y, Gilbert A, Xu C. Hydrodeoxygenation of bio-crude in supercritical hexane with sulfided CoMo and CoMoP catalysts supported on MgO: A model compound study using phenol. *Appl. Catal. A* 2009;360:242.

[44] Zhao C, Kou Y, Lemonidou AA, Li X, Lercher JA. Highly Selective Catalytic Conversion of Phenolic Bio-Oil to Alkanes. *Angew. Chem. Int. Ed* 2009;48:3987.

[45] Wan HJ, Chaudhari RV, Subramaniam B. Catalytic Hydroprocessing of *p*-Cresol: Metal, Solvent and Mass-Transfer Effects. *Top Catal.* 2012;55:129.

[46] Wang YX, He T, Liu KT, Wu JH, Fang YM. From biomass to advanced bio-fuel by catalytic pyrolysis/hydro-processing: Hydrodeoxygenation of bio-oil derived from biomass catalytic pyrolysis. *Bioresour Technol.* 2012;108:280.

[47] Guo JH, Ruan RX, Zhang Y. Hydrotreating of phenolic compounds separated from bio-oil to alcohols. *Ind. Eng. Chem. Res.* 2012;51:6599.

[48] Ardiyanti AR, Gutierrez A, Honkela ML, Krause AOI, Heeres HJ. Hydrotreatment of wood-based pyrolysis oil using zirconia-supported mono-and bimetallic (Pt, Pd, Rh) catalysts. *Appl. Catal. A* 2011;407:56.

[49] Wildschut J, Iqbal M, Mahfud FH, Cabrera IM, Venderbosch RH, Heeres HJ. Insights in the hydrotreatment of fast pyrolysis oil using a ruthenium on carbon catalyst. *Energy Environ. Sci.* 2010;3:962.

[50] Fisk CA, Morgan T, Ji YY, Crocker M, Crofcheck C, Lewis SA. Bio-oil upgrading over platinum catalysts using in situ generated hydrogen. *Appl. Catal. A* 2009;358:150.

[51] Gutierrez A, Kaila RK, Honkela ML, Slioor R, Krause AOI. Hydrodeoxygenation of guaiacol on noble metal catalysts. *Catal. Today* 2009;147:239.

[52] Wildschut J, Mahfud FH, Venderbosch RH, Heeres HJ. Hydrotreatment of Fast Pyrolysis Oil Using Heterogeneous Noble-Metal Catalysts. *Ind. Eng. Chem. Res.* 2009;48: 10324.

[53] Wildschut J, Arentz J, Rasrendra CB, Venderbosch RH, Heeres HJ. Catalytic hydrotreatment of fast pyrolysis oil: Model studies on reaction pathways for the carbohydrate fraction. *Environ. Prog.* 2009;28:450.

[54] Yan N, Yuan Y, Dykeman R, Kou Y, Dyson PJ. Hydrodeoxygenation of Lignin-Derived Phenols into Alkanes by Using Nanoparticle Catalysts Combined with Brønsted Acidic Ionic Liquids. *Angew. Chem. Int. Ed.* 2010;49:5549.

[55] Zhao C, Kou Y, Lemonidou AA, Li XB, Lercher JA. Hydrodeoxygenation of bio-derived phenols to hydrocarbons using RANEY® Ni and Nafion/SiO$_2$ catalysts. *Chem. Commun.* 2010;46:412.

[56] Xiong WM, Fu Y, Zeng FX, Guo QX. An in situ reduction approach for bio-oil hydroprocessing. *Fuel Process Technol.* 2011;92:1599.

[57] Wang XY, Rinaldi R. Exploiting H-transfer reactions with RANEY Ni for upgrade of phenolic and aromatic biorefinery feeds under unusual, low-severity conditions. *Energy Environ. Sci.* 2012;5:8244.

[58] Liu WJ, Zhang XS, Qv YC, Jiang H, Yu HQ. Bio-oil upgrading at ambient pressure and temp erature using zero valent metals. *Green Chem.* 2012;14:2226.

[59] Mortensen PM, Grunwaldt JD, Jensen PA, Knudsen KG, Jensen AD. A review of catalytic upgrading of bio-oil to engine fuels. *App. Catal. A* 2011;407:1.

[60] Adam J, Antonakou E, Lappas A, Stocker M, Nilsen MH, Bouzga A, Hustad JE, Oye G. In situ catalytic upgrading of biomass derived fast pyrolysis vapours in a fixed bed reactor using mesoporous materials. *Microporous Mesoporous Mater.* 2006;96:93.

[61] Corma A, Huber GW, Sauvanaud L, O'Connor P. Processing biomass-derived oxygenates in the oil refinery: Catalytic cracking (FCC) reaction pathways and role of catalyst. *J. Catal.* 2007;247:307.

[62] Graça I, Fernandes A, Lopes JM, Ribeiro MF, Laforge S, Magnoux P, Ramôa Ribeiro F. Bio-oils and FCC feedstocks co-processing: Impact of phenolic molecules on FCC hydrocarbons transformation over MFI. *Fuel* 2011;90:467.

[63] Bridgwater A V. Production of high grade fuels and chemicals from catalytic pyrolysis of biomass. *Catal. Today* 1996;29:285.

[64] Huang J, Long W, Agrawal KP, Jones CW. Effects of Acidity on the Conversion of the Model Bio-oil Ketone Cyclopentanone on H –Y Zeolites. *J. Phys. Chem. C* 2009;113: 16702.

[65] Vitolo S, Bresci B, Seggiani M, Gallo MG. Catalytic upgrading of pyrolytic oils over HZSM-5 zeolite: behaviour of the catalyst when used in repeated upgrading-regenerating cycles. *Fuel* 2001;80:17.

[66] Chiang H, Bhan A. Catalytic consequences of hydroxyl group location on the rate and mechanism of parallel dehydration reactions of ethanol over acidic zeolites. *J. Catal.* 2010;271:251.

[67] Adjaye JD, Bakhshi NN. Upgrading of a wood-derived oil over various catalysts. *Biomass Bioenergy* 1994;7:201.

[68] Sharma RK, Bakhshi NN. Catalytic upgrading of biomass derived oil to transport fuels and chemicals. *Can. J. Chem. Eng.* 1991;69:1071.

[69] Adjaye JD, Bakhshi NN. Production of hydrocarbons by catalytic upgrading of a fast pyrolysis bio-oil. part I: conversion over various catalysts. *Fuel Process Technol.* 1995;45(3):161.

[70] Adjaye JD, Bakhshi NN. Production of hydrocarbons by catalytic upgrading of a fast pyrolysis bio-oil. Part II: Comparative catalyst performance and reaction pathways. *Fuel Process Technol.* 1995;45(3):185.

[71] Williams PT, Horne PA, Characterisation of oils from the fluidised bed pyrolysis of biomass with zeolite catalyst upgrading. *Biomass Bioenergy* 1994;7:223.

[72] Gong FY, Yang Z, Hong CG, Huang WW, Ning S, Zhang ZX, Xu Y, Li QX. Selective conversion of bio-oil to light olefins: Controlling catalytic cracking for maximum olefins. *Bioresour Technol.* 2011;102:9247.

[73] Taufiqurrahmi N, Bhatia S. Catalytic cracking of edible and non-edible oils for the production of biofuels. *Energy Environ. Sci.* 2011;4:1087.

[74] Adam J, Blazsó M, Mészáros E, Stöcker M, Nilsen HM, Bouzga A, Hustad JE, Grønli M, Øye G. Pyrolysis of biomass in the presence of Al-MCM-41 type catalysts. *Fuel* 2005;84:1494.

[75] Iliopoulou EF, Antonakou EV, Karakoulia SA, Vasalos IA, Lappas AA, Triantafyllidis KS. Catalytic conversion of biomass pyrolysis products by mesoporous materials: Effect of steam stability and acidity of Al-MCM-41 catalysts. *Chem. Eng. J.* 2007;134:51.

[76] Valle B, Gayubo AG, Atutxa A, Alonso A, Bilbao J. Integration of thermal treatment and catalytic transformation for upgrading biomass pyrolysis oil. *Int. J. Chem. Reactor Eng.* 2007;5:A86.

[77] Lu Q, Zhu XF, Li WZ, Zhang Y, Chen DY. On-line catalytic upgrading of biomass fast pyrolysis products. *Chin. Sci. Bull.* 2009;54:11.

[78] Zhang HY, Xiao R, Wang DH, Zhong ZP ,Song M, Pan QW, He GY. Catalyt ic fast pyrolysis of biomass in a fluidized bed with fresh and spent fluidized catalytic cracking (FCC) cat alysts. *Energy Fuels* 2009;23(12):6199.

[79] Zhu XX, Mallinson RG, Resasco DE. Role of transalkylation reactions in the conversion of anisole over HZSM-5. *Appl. Catal. A* 2010;379:172.

[80] Ausavasukhi A, Sooknoi T, Resasco DE. Catalytic deoxygenation of benzaldehyde over gallium-modified ZSM-5 zeolite. *J. Catal.* 2009;268:68.

[81] Peralta MA, Sooknoi T, Danuthai T, Resasco DE. Deoxygenation of benzaldehyde over CsNaX zeolites. *J. Mol. Catal. A* 2009;312:78.

[82] Wang JJ, Chang J, Fuan J. Upgrading of bio-oil by catalytic esterification anddetermination of acid number for evaluating esterification degree. *Energy Fuels* 2010;24:3251.

[83] Moens L, Black SK, Myers MD, Czernik S. Study of the neutralization and stabilization of a mixed hardwood bio-oil. *Energy Fuels* 2009;23:2695.

[84] Hilten RN, Bibens BP, Kastner JR, Das KC. In-line esterification of pyrolysisvapor with ethanol improves bio-oil quality. *Energy Fuels* 2010;24:673.

[85] Miao S, Shanks BH. Esterification of biomass pyrolysis model acids over sul-fonic acid-functionalized mesoporous silicas. *Appl. Catal. A* 2009;359:113.

[86] Hino M, Arata K. Solid catalysts treated with anions. I catalytic activity of iron oxide treated with sulfate ion of dehydration of 2-propnaol and ethanol and polymerization of isobutyl vinylthe. *Chem. Letter.* 1979; 89(5):477.

[87] Xu JM, Jiang JC, Sun YJ, Lu YJ. Bio-oil upgrading by means of ethyl ester production in reactive distillation to remove water and toimprove storage and fuel characteristics. *Biomass Bioenergy* 2008;32:1056.

[88] Xiong WM, Fu Y, Lai DM, Guo QX. Upgrading of bio-oil via esterification catalysted with acidic ion-exchange resin. *Chem. J. Chinese U* 2009;30:1754.

[89] Wang JJ, Chang J, Fan J. Catalytic esterification of bio-oil by ion exchange resins. *J. Fuel Chem. Tech.* 2010;38:560.

[90] Hu X, Gunawan R, Mourant D, Lievens C, Li X, Zhang S, Chaiwat W, Li CZ. Acid-catalysed reactions between methanol and the bio-oil from the fast pyrolysis of mallee bark. *Fuel* 2012;97:512.

[91] Mahfud FH, Melian-Cabrera I, Manurung R, Heeres HJ. Biomass to fuels - Upgrading of flash pyrolysis oil by reactive distillation using a high boiling alcohol and acid catalysts. *Process Saf. Environ.* 2007;85:466.

[92] Xu JM, Jiang, JC, Lv W, Dai WD, Sun, YJ. Rice husk bio-oil upgrading by means of phase separation and the production of esters from the water phase, and novolac resins from the insoluble phase. *Biomass Bioenergy* 2010;34:1059.

[93] Xu JM, Jiang, JC, Dai WD, Zhang TJ, Xu Y. Bio-Oil Upgrading by Means of Ozone Oxidation and Esterification to Remove Water and to Improve Fuel Characteristics. *Energy Fuels* 2011;25:1798.

[94] Peng J, Chen P, Lou H, Zheng XM. Upgrading of bio-oil over aluminum silicate in supercritical ethanol. *Energy Fuels* 2008;22:3489.

[95] Peng J, Chen P, Lou H, Zheng XM. Catalytic upgrading of bio-oil by HZSM-5 in sub- and super-critical ethanol. *Bioresour Technol.* 2009;100:3415.

[96] Leadbeater NE, Stencel LM, Fast, easy preparation of biodiesel using microwave heating. *Energy Fuels* 2006;20:2281.

[97] Zhang Q, Chang J, Wang TJ, Xu Y. Upgrading bio-oil over different solid catalysts. *Energy Fuels* 2006;20:2717.

[98] Xiong WM, Zhu MZ, Deng Li, Fu Y, Guo QX. Esterification of Organic Acid in Bio-Oil using Acidic Ionic Liquid Catalysts. *Energy Fuels* 2009;23:2278.

[99] Tang Z, Lu Q, Zhang Y, Zhu X, Guo QX. One step bio-oil upgrading through hydrotreatment, esterification, and cracking. *Ind. Eng. Chem. Res.* 2009;48:6923.

[100] Li W, Pan CY, Sheng L, Liu Z, Chen P, Lou H, Zheng XM. Upgrading of high-boiling fraction of bio-oil in supercritical methanol. *Bioresour Technol.* 2011;102:9223.

[101] Zhang JX, Luo ZY, Dang Q, Wang J, Chen W. Upgrading of Bio-oil over Bifunctional Catalysts in Supercritical Monoalcohols. *Energy Fuels* 2012;26:2990.

[102] Dang Q, Luo ZY, Zhang JX, Wang J, Chen W, Yang Y. Experimental study on bio-oil upgrading over $Pt/SO_4^{2-}/ZrO_2/SBA-15$ catalyst in supercritical ethanol. *Fuel* 2012; In press.

[103] Xu Y, Wang TJ, Ma LL, Chen GY. Upgrading of fast pyrolysis liquid fuel from biomass over $Ru/\gamma-Al_2O_3$ catalyst. *Energy Convers Manage* 2012;55:172.

[104] Wu C, Liu RH. Sustainable hydrogen production from steam reforming of bio-oil model compound based on carbon deposition/elimination. *Int. J. Hydrogen Energy* 2011;36:2860.

[105] Rostrup-Nielsen JR, Sehested J, Nørskov JK. Hydrogen and synthesis gas by steam- and CO_2 reforming. *Adv. Catal.* 2002;47:65.

[106] Trane R, Dahl S, Skjøth-Rasmussen MS, Jensen AD. Catalytic steam reforming of bio-oil. *Int. J. Hydrogen Energy* 2012;37:6447.

[107] Aasberg-Petersen K, Dybkjaer I, Ovesen CV, Skjøth NC, Sehested J, Thomsen SG. Natural gas to synthesis gas - catalysts and catalytic processes. *J. Nat. Gas Sci. Eng.* 2011;3:423.

[108] Rioche C, Kulkarni S, Meunier FC, Breen JP, Burch R. Steam reforming of model compounds and fast pyrolysis bio-oil on supported noble metal catalysts. *Appl. Catal. A* 2005;61:130.

[109] Kechagiopoulos PN, Voutetakis SS, Lemonidou AA, Vasalos IA. Hydrogen production via steam reforming of the aqueous phase of bio-oil in a fixed bed reactor. *Energy Fuel* 2006;20:2155.

[110] Basagiannis AC, Verykios XE. Influence of the carrier on steam reforming of acetic acid over Ru-based catalysts. *Appl. Catal. B* 2008;82:77.

[111] Basagiannis AC, Veryki os XE. Catalytic steam reform ing of acetic acid for hydrogen production. *Int. J. Hydrogen Energy* 2007;32:334.

[112] Salehi E, Azad FS, Harding T, Abedi J. Production of hydrogen by steam reforming of bio-oil over Ni/Al_2O_3 catalysts: Effect of addition of promoter and preparation procedure. *Fuel Process Technol. 2011*;92:2203.

[113] Bimbela F, Chen D, Ruiz J, García L, Arauzo J. Ni/Al coprecipitated catalysts modified with magnesium and copper for the catalytic steam reforming of model compounds from biomass pyrolysis liquids. *Appl. Catal. B* 2012;119–120:1.

[114] Zhang YH, Li WZ, Zhang SP, Xu QL, Yan YJ. Steam Reforming of Bio-Oil for Hydrogen Production: Effect of Ni-Co Bimetallic Catalysts. *Chem. Eng. Technol.* 2012; 35:302.

[115] García L, French R, Czernik S, Chorn et E. Catalytic steam reform ing of bio-oil s for the production of hydroge n: effect s of cat alyst compositio n. *Appl. Catal. A* 2000; 201:225.

[116] Gong FY, Ye TQ, Yuan LX, Kan T, Torimoto Y, Yamamoto M, Li QX. Direct reduction of iron oxides based on steam reforming of bio-oil: a highly efficient approach for production of DRI from bio-oil and iron ores. *Green Chem.* 2009;11:2001.

[117] Ye QT, Yuan LX, Chen YQ, Kan T, Tu J, Zhu XF, Torimoto Y, Yamamoto M, Li QX. High Efficient Production of Hydrogen from Bio-oil Using Low-temperature Electrochemical Catalytic Reforming Approach Over NiCuZn–Al$_2$O$_3$ Catalyst. *Catal. Lett.* 2009;127:323.

[118] Chen YQ, Yuan LX, Ye TQ, Qiu SB, Zhu SF, Torimoto Y, Yamamoto M, Li QX. Effects of current upon hydrogen production from electrochemical catalytic reforming of acetic acid. *Int. J. Hydrogen Energy* 2009;34:1760.

[119] Xiong JX, Kan T, Li XL, Ye TQ, Li QX. Effects of Current upon Electrochemical Catalytic Reforming of Anisole. *Chin. J. Chem. Phys.* 2010;23:693.

[120] Lin SB, Ye TQ, Yuan LX, Hou T, Li QX. Production of Hydrogen from Bio-oil Using Low-temperature Electrochemical Catalytic Reforming Approach over CoZnAl Catalyst. *Chin. J. Chem. Phys.* 2010;23:451.

[121] Li XL, Ning S, Yuan LX, Li QX. Hydrogen Production From Crude Bio-oil and Biomass Char by Electrochemical Catalytic Reforming. *Chin. J. Chem. Phys.* 2011; 24:477.

[122] Davda R, Shabaker JW, Huber GW, Cortright RD, Dumesic JA. A review of catalytic issues and process conditions for renewable hydrogen and alkanes by aqueous-phase reforming of oxygenated hydrocarbons over supported metal catalysts. *Appl. Catal. B Environ.* 2005;56:171.

[123] Shabaker JW, Huber GW, Davda Rr, Cortright Rd, Dumesic JA. Aqueous-Phase reforming of ethylene glyeol over supported platinum catalysts. *Catal. Lett.* 2003; 88(1-2):1.

[124] Soares RR, Simonetti DA, Dumesic JA. Glycerol as a Source for Fuels and Chemicals by Low-Temperature Catalytic Processing. *Angew. Chem. Int. Ed.* 2006;45(24):3982.

[125] Davda RR, Shabaker JW, Huber GW, Cortright RD, Dumesic JA. Aqueous-phase r eforming of ethylene glycol on silica-supported metal catalysts. *Appl. Catal. B* 2003;43:13.

[126] Shabaker JW, Huber GW, Dumesic JA. Aqueous-phase reforming of oxygenated hydrocarbons over Sn-modified Ni catalysts. *J. Catal.* 2004;222(1):180.

[127] Dumesic JA, Huber GW. An overview of aqueous-phase catalytic processes for production of hydrogen and alkanes in a biorefinery. *Catal. Today* 2006;111:119.

[128] Luo N, Fu X, Cao F, Xiao T, Edwards PP. Glycerol aqueous phase reforming for hydrogen generation over Pt catalyst–effect of catalyst composition and reaction conditions. *Fuel* 2008;87:3483.

[129] Meryemoglu B, Hesenov A, Irmak S, Atanur O, Erbatur O. Aqueous-phase reforming of biomass using various types of supported precious metal and raney-nickel catalysts for hydrogen production. *Int. J. Hydrogen Energy* 2010;35(22):12580.

[130] King DL, Zhang L, Xia G, Karim AM, Heldebrant JD, Wang XQ, Peterson T, Wang Y. Aqueous phase reforming of glycerol for hydrogen production over Pt–Re supported on carbon. *Appl. Catal. B* 2010;99:206.

[131] Tokarev AV, Kirilin AV, Murz ina EV, Eranen K, Kustov LM, Murzin DY, et al. The role of bio-ethanol in aqueous phase reforming to sustainable hydrogen. *Int. J. Hydrogen Energy* 2010;35:12642.

[132] Tang Z, Monroe J, Dong J, Nenoff T, Weinkauf D. Platinum -loaded NaY zeolite for aq ueous-phase reforming of methanol and ethanol to hydrogen. *Ind. Eng. Chem. Res.* 2009; 48:2728.

[133] Vispute TP, Huber GW. Production of hydrogen, alkanes and polyols by aqueous phase processing of wood-derived pyrolysis oils. *Green Chem.* 2009;11:1433.

[134] Meryemoglu B, Kaya B, Irmak S, Hesenov A, Erbatur O. Comparison of batch aqueous-phase reforming of glycerol and lignocellulosic biomass hydrolysate. *Fuel* 2012;97: 241.

In: Biomass Processing, Conversion and Biorefinery ISBN: 978-1-62618-346-9
Editors: Bo Zhang and Yong Wang © 2013 Nova Science Publishers, Inc.

Chapter 13

BIOFUEL AND BIO-OIL UPGRADING

Changjun Liu[1] and Yong Wang[1,2]*
[1]The Gene and Linda Voiland School of Chemical Engineering
and Bioengineering, Washington State University, Pullman WA, US
[2]Institute for Integrated Catalysis,
Pacific Northwest National Laboratory, Richland WA, US

ABSTRACT

Biomass is the only renewable source of carbon-based fuels and chemicals. Lignocellulosic biomass contains about 45 wt% cellulose, 30 wt% hemicellulose and about 25 wt% lignin. Converting lignocellulosic biomass into valuable chemicals and liquid fuels has been extensively studied for the past decade due to the abundance of lignocellulosic biomass. Gasification, pyrolysis and hydrolysis are three major thermochemical routes to depolymerize lignocellulosic biomass. Unlike fossil carbon resources, biomass contains a large amount of oxygen which leads to the low heating value and other unfavorable properties of the resultant liquid products. Thus deoxygenation is particularly important in the case of fuels. Hydrodeoxygenation (HDO) is the most promising and efficient route to fulfill this purpose. Recent progresses in biofuel and bio-oil upgrading by hydrotreatment are summarized into three main catalytic routes with respect to their feedstock, namely, conventional hydrodeoxygenation of bio-oil, aqueous-phase hydrotreating of sugars, and hydrodeoxygenation of lignin-derived compounds.

1. INTRODUCTION

1.1. Bio-Oil, Sugars and Lignin-Derived Compounds

The global demand and consumption of energy are continuously increasing due to the increasing world population and the expanding human activity. Transportation fuel

---

* Corresponding author: Yong.Wang@pnnl.gov.

consumption constitutes about 20% of the total energy consumption of our society [1]. Traditionally the transportation fuel is derived from fossil resources like crude oil. The gradual depletion of fossil fuel resources coupled with the increasing deterioration of the environment make the use of alternative energy resources necessary [2, 3]. Biomass, as the only renewable carbon source, has been considered as one of the most promising sustainable natural resources with the potential to replace petroleum in the production of chemicals and liquid transportation fuels [4, 5]. The utilization of biofuels can also considerably help reduce overall life cycle greenhouse gas emissions [6]. The first generation biofuels (bio-ethanol and biodiesel) are derived from food grade biomass such as sugar, starch, fats and oils. However, the use of food grade biomass associates with constraint supply, competition to food, and low land energy efficiency [7]. Thus researches on second generation biofuels have been focused on nonedible biomass like agriculture wastes and forestry residues [6, 8, 9]. These nonedible biomasses comprise of lignocellulose including about 45 wt% cellulose, 30 wt% hemicellulose and 25 wt% lignin [9]. Current technologies for lignocellulosic biomass conversion include gasification to synthesis gas, hydrolysis into sugar, lignin fractions, and selective thermal processes such as pyrolysis and liquefaction [10]. A more specific strategy scheme is shown in Figure 1.

Fast pyrolysis is the most preferred process due to its high compatibility to a large variety of feedstocks and its potential to produce a liquid bio-oil without the initial need of hydrogen at high pressure [12]. It has been identified as one of the most feasible routes for bio-oil production which shows the highest liquid yield and retains most of the energy in the liquid products [13-15].

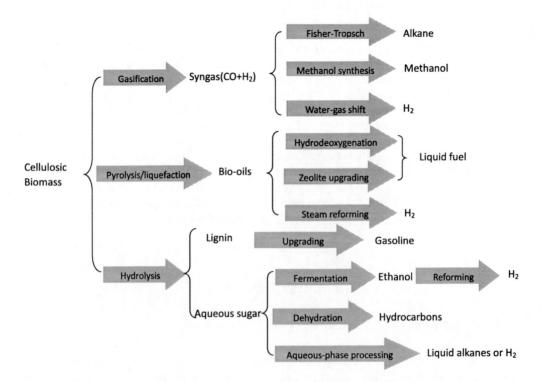

Figure 1. Strategies for fuel production from biomass (Adapted with permission from ref [11] Copyright 2006 American Chemical Society).

The primary liquid product of fast pyrolysis is generally called bio-oil, which is highly viscous and acidic. It has high oxygen content (around 35-40 wt%), low vapor pressure, low heating value. Thus bio-oil cannot be directly used as transportation fuels. High oxygen content of bio-oil is a major issue. A wide variety of oxygenates are found in bio-oils, including three main families of compounds: (i) small carbonyl compounds such as acetic acid, acetaldehyde, acetone, etc; (ii) sugar-derived compounds like furfural, levoglucosan, etc; and (iii) lignin-derived compounds [12, 16]. Conventional hydrotreating processes have been extensively investigated for bio-oil upgrading [17].

Lignocellulosic biomass can also be deconstructed into fermentable sugars and separated from lignin by mineral acid or enzyme catalyzed hydrolysis. Although the obtained fermentable sugars can be converted into bio-ethanol by the well-established fermentation industry, developing catalytic routes to more efficiently convert these water soluble sugars into liquid hydrocarbon appears more attractive. Several aqueous-phase hydrotreating processes have been developed in laboratory scale mainly by Dumesic's group [18-23].

Lignin in lignocellulosic biomass is generally separated from cellulose and hemicellulose in the hydrolysis process and pulping process in paper industry. Conversion of lignin into valuable chemicals and liquid fuel is another important aspect of biomass utilization. Either pyrolysis or hydrolysis can depolymerize lignin into its monomeric units. Both vapor phase hydrodeoxygenation (HDO) and aqueous-phase hydrotreating processes have been developed to obtain valuable aromatics and high-molecular-weight hydrocarbons from lignin-derived compounds [24-34].

Publications on novel catalysts and processes in biofuel and bio-oil upgrading greatly increased after 2002, with the great interest in developing more active, selective and low cost catalysts, and more efficient process. A universally applicable catalytic process does not exist for all crude bio-oils. Specific highly active and selective catalysts have been developed with respect to feeding derived from lignocellulose. This chapter focuses on the recent progresses in the fields of conventional HDO of pyrolysis oil, aqueous-phase hydrotreating of carbohydrate, and HDO of lignin-derived compounds.

1.2. Factors Used to Evaluate HDO Efficiency

The evaluation of the simple reactions of model compounds suggests that the thermodynamics does not appear to be a constraint of the HDO processes [14]. However, it is difficult to evaluate the conversion of each component due to the complex composition of bio-oils. Thus the oil yield and degree of deoxygenation are generally adopted for quantitative comparison as shown in Eq. 1.1 and 1.2 respectively.

$$Y_{oil} = \left(\frac{m_{oil}}{m_{feed}} \right) \times 100 \tag{1.1}$$

$$DOD = \left(1 - \frac{x_{O \text{ in product}}}{x_{O \text{ in feed}}} \right) \times 100 \tag{1.2}$$

where Y_{oil} is the yield of product oil, m_{oil} is the mass of product oil, m_{feed} is the mass of bio-oil feed. DOD is the degree of deoxygenation, and $x_{O\ in\ product}$ and $x_{O\ in\ feed}$ are the weight percent of oxygen in product oil and bio-oil feed respectively. The combination of these two parameters can give an overview of the reaction extent, in terms of the selectivity of oil product and the efficiency of oxygen removal. It is worth noting that these two parameters could be less descriptive when separated since 100% yield could be achieved with zero deoxygenation. In addition H/C ratio is a useful indicator of the aromatic character of the product [35].

2. HYDRODEOXYGENATION OF BIO-OIL

Fast pyrolysis is highly flexible to a wide variety of biomass feedstock and can maximize the bio-oil yield. Thus fast pyrolysis is considered as one of the most feasible routes for bio-oil production.

However, bio-oil from fast pyrolysis is a highly complex mixture of more than 300 identified oxygenated compounds [14, 36, 37]. The chemical composition classified by functional groups with relative abundance is shown in Figure 2. The distribution of these compounds mostly depends on the type of biomass used and the process severity [36, 38, 39].

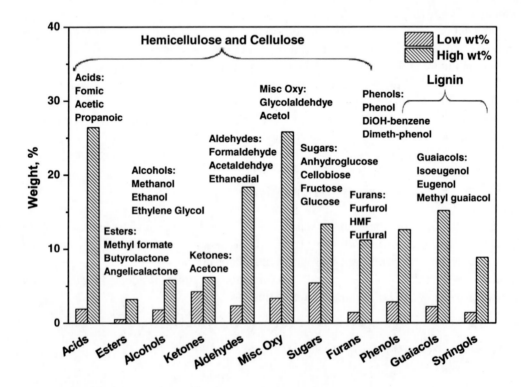

Figure 2. Chemical composition of bio-oil and the most abundant molecules of each of the components (Adapted with permission from ref [11] Copyright 2006 American Chemical Society).

Bio-oils are characterized by low vapor pressure, low heating value, high acidity, high viscosity and high reactivity [11, 38, 39]. These adverse characteristics are all related to the high oxygen content in the bio-oil [40]. Thus oxygen in bio-oil must be removed by upgrading before it can be used as a replacement for diesel and gasoline fuels [11, 14]. Generally, there are two routes for catalytic upgrading of bio-oil: hydrodeoxygenation (HDO) and zeolite cracking [14]. Compared to zeolite cracking, HDO produces high grade oil product equivalent to crude oil. HDO is considered as the most feasible and competitive route for the bio-oil upgrading to transportation fuels [11, 14, 41]. Traditionally the HDO technologies are mirrored from hydrodesulfuration using cobalt- or nickel-doped molybdenum sulfides as catalysts [42]. Oxygen is removed as H_2O and CO_2, and thus raises the energy density [43]. Removing oxygen is more difficult compared to HDS [44]. The historical developments in HDO since 1980s have been reviewed by Elliott [35] in 2007. More recently a review on catalytic upgrading of bio-oil to transportation fuels was published by Mortensen et al. [14].

2.1. Molybdenum Based Catalysts

Cobalt- and nickel-promoted molybdenum catalysts have been widely used in hydrodesulfuration (HDS) of petroleum products [45]. Hydrodeoxygenation, hydrodesulphurization, hydrodenitrogenation (HDN) and hydrogenation occur simultaneously during hydroprocessing of various feeds for fuel production [44]. Conventional hydroprocessing catalysts like $CoMo/Al_2O_3$ and $NiMo/Al_2O_3$ were most extensively studied for hydrodeoxygenation due to the similarity of HDS and HDO [35, 44]. Both the oxide and sulfided form of $CoMo/Al_2O_3$ and $NiMo/Al_2O_3$ were active HDO catalysts while the sulfided form of CoMo catalyst was much more active [46]. HDO is generally carried out at 250-450 °C with hydrogen pressure ranging from 7.5 to 30 MPa [36, 41, 47-49]. High hydrogen pressure is necessary to ensure a higher availability of hydrogen in the vicinity of catalyst since hydrogen solubility in oil increases with hydrogen pressure. Higher hydrogen availability also increases the hydrodeoxygenation rate and suppresses the coking [50].

Elliott et al. [51] tested bio-oil from a poplar wood over a sulfided CoMo catalyst at 355 °C and 13.8 MPa with a liquid hourly space velocity (LHSV) of 0.35 h^{-1}. The liquid product with 3.6% and 5.9% oxygen content was obtained with only 23% mass yield. The H/C atomic ratios of the upgraded bio-oil were about 1.45-1.55. Zhang et al. [52] separated bio-oil into water and oil phases. The hydrotreating of oil phase on sulfided $CoMo/Al_2O_3$ catalyst at 360 °C and 2 MPa of hydrogen pressure resulted in an oxygen content decreased from 41.8% of crude oil to 3% of the upgraded product. A theoretic yield is predicted to be 56-58 wt% for complete deoxygenation according to the purposed overall reaction (Eq. 1.3) [14, 53].

$$CH_{1.4}O_{0.4} + 0.7H_2 \rightarrow 1'CH_2' + 0.4H_2O \tag{1.3}$$

The conventional hydrotreating approach for hydrothermal product was proved to be inappropriate for bio-oil from fast pyrolysis due to the high water content and instability of bio-oil [35]. Carbonyl compounds such as aldehydes and ketones were found to be highly active and largely responsible for the instability of pyrolysis oils [48, 54, 55]. Heavy product tar, produced by the polymerization of carbonyl compounds under hydrotreating conditions, could plug the reactor system and encase the catalyst bed in a coke like product.

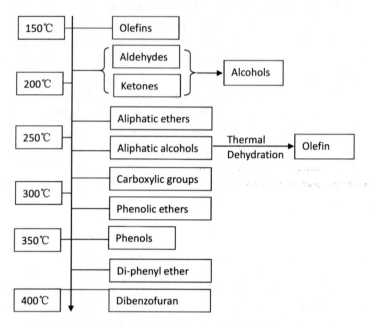

Figure 3. Reactivity scale of oxygenated groups under hydrotreatment conditions (Adapted with permission from ref [35] Copyright 2006 American Chemical Society).

Table 1. Two-Stage Hydroprocessing of Bio-oil [35]

	Experiment operating conditions		
Feed oil	Bio-oil	Stage 1 product	Combined
Catalyst	HT-400	HT-400	HT-400
Temperature, ℃	274	353	
Pressure, MPa	14.1	14.2	
LHSV,vol oil/(vol cat h)	0.62	0.11	
	Experimental results and product analyses		
H_2 consumption, L/L oil	60	576	457
Carbon conversion, wt%			
To aqueous	8.2	0.7	8.8
To gas (C1 to C4)	7.0	12.6	17
Oil product yield, L/L feed	0.69	0.62	0.43
Aqueous phase yield, L/L feed	0.34	0.38	0.61
C5-225℃, L/L feed	0.07	0.45	0.31
	Product inspections		
H/C atomic ratio	1.58	1.67	
Oxygen, wt%	32.7	2.3	
Density, g/mL	1.14	0.86	
C5-225℃, vol%	10	72	
Viscosity, cPs @ 60℃	14200	ND	
Moisture, wt%	14.1	ND	

Sheu et al. [56] suggested that some oxygen functional groups in bio-oil should be removed first to minimize the possibility of its polymerization at high temperature. Comparing the reactivity of various oxygenated groups over sulfided catalysts (Figure 3), it was found aldehydes and ketones are highly active in hydrotreatment at low temperature.

Thus it is possible to stabilize the bio-oil by reducing the amount of olefin and carbonyl compounds at a low temperature [35, 57]. Based on this observation, Elliott et al. [35] developed a two-stage hydrotreating process. Typical process parameters and results are shown in Table 1. The raw bio-oil was first hydrotreated at low temperature (<300 °C) in presence of either Ni or sulfided CoMo catalyst. Stability of the resulting bio-oil was then improved by removing the instable carbonyl components and olefinic side chains. Polymerization of the unstable species was still found in competition with hydrotreating reactions at temperature above 200-250 °C. However, lower temperature and enhanced hydrogen mass transfer favored the hydrotreating reactions over polymerization [50]. The composition of the low temperature hydrotreating product was found to be representative to the characteristics of bio-oil feedstock and the source biomass but the thermal stability was largely improved [41]. The elemental analysis showed that the oxygen content of the low-temperature hydrotreated product oil was 32 wt%. The stabilized oil was then processed at about 350 °C and 14.2 MPa in the presence of sulfided CoMo catalyst. Oil volumetric yield of 0.43 L/L feed was achieved in combined two-stage hydroteatment process. The final products after high temperature hydrocracking were all similar collections of cyclic hydrocarbons regardless the feedstock [41].

A nonisothermal catalyst bed was later developed to simplify the two-stage hydrotreating process and maximize liquid product yield [35]. Bio-oil from pine sawdust was hydrotreated in a nonisothermal fixed bed reactor using sulfided Ru/C for low temperature hydrotreating (~170 °C) and sulfide promoted Mo catalyst (CoMo or NiMo) for high temperature hydrotreating (~400 °C) at 13.8 MPa in large excess hydrogen flow. Oxygen content in the bio-oil after hydrotreating reduced from 40.5 wt% on a moisture-free basis to below 2.7% and the oil yield is in a range of 0.35 ~0.45g/g of dry feed [40].

The sulfur content of bio-oil is very low. It is generally accepted that a source of sulfur will be required to keep the sulfide phases active over long periods [58]. Although low concentration hydrogen sulfide was found to slightly promote the hydrogenation activity of CoMo catalyst, a high concentration of hydrogen sulfide had an inhibiting effect on the activity [58]. High concentration of hydrogen sulfide was confirmed to inhibit the dehydroxylation of guaiacol while favoring the demethylation to form catechol and also inhibit the hydrogenation of carbonyl groups [59]. Thus a careful control of the concentration of sulfide species in hydrodeoxygenation of bio-oil is necessary. Besides high concentration sulfide species, ammonia and water were also found of inhibiting effects [58].

Christensen et al. [60] analyzed the distillate fractions of hydrotreated bio-oil with different oxygen contents (8.2, 4.9 and 0.4 wt%) that were originally generated from lignocellulosic biomass. The results showed that carboxylic acids, carbonyls, aryl ethers, phenols, and alcohols were found in all the fractions of the highest oxygen content bio-oil. In addition, carboxylic acids and carbonyl compounds were concentrated in the light, naphtha and jet fractions. Phenolic compounds were almost exclusively found in the whole boiling range fractions of 4.9 wt% oxygen sample. Phenolic acidity was also found in the fractions from the lowest oxygen-content oil. Thus it is necessary to deepen not only the extent of deoxygenation but also the extent of removing carboxylic and phenolic functional groups in bio-oil hydroprocessing.

Economic analysis suggested that sending partial deoxygenated material to a refinery could lower the processing cost [61]. Reducing the severity of the hydrodeoxygenation so as to leave about 7 wt% oxygen in the partially upgraded bio-oil and avoid reducing aromatics could reduce hydrogen consumption, catalyst costs, and hydrotreater capital costs [61]. Mercader et al. [62, 63] showed in a laboratory scale FCC unit that the co-feeding of partially upgraded high oxygen content (17 - 28 wt%, on dry basis) bio-oil with a Long Residue can also produce near normal FCC gasoline and Light Cycle Oil products without excessive increase of undesired coke and dry gas.

Although sulfided Co- and Ni-promoted molybdenum catalysts showed good activity in HDO of bio-oil, some technical problems remain to be addressed.

1. Catalyst deactivation due to coking, alkali deposition, and sulfur loss[46].
2. Better supports with improved hydrothermal stability.
3. Improvement of the hydrodeoxygenation selectivity and suppression of the saturation of aromatic ring to produce high quality transportation fuels due to the poor octanes of cyclic hydrocarbons [35].
4. Minimization of hydrogen consumption and production of gasoline range fuels with a higher octane number.
5. Avoidance of potential sulfur contamination of the nearly sulfur-free bio-oil due to the co-feeding of low concentration sulfur species to maintain the long period activity of sulfided catalysts [42].

2.2. Transition Metal Catalysts

Sulfided catalysts were found to have a poor hydrostability in HDO of bio-oils due to the presence of high concentration water in bio-oil. In addition, sulfide species used to maintain the activity of sulfide catalysts could potentially contaminate the product oil. Thus the sulfide HDO catalysts are of limited application [64]. More stable catalysts and economically favorable clean processes are preferred for bio-oil upgrading.

Noble metals are selective hydrogenation catalysts with good low temperature activity [65-67]. Ru, Pd, Rh, and Pt catalysts exhibit high activity in partial hydrogenation of tricyclic aromatics [66]. They are more active than the sulfided molybdenum-based catalysts and can be used at lower temperature with high activity [68]. Pd, Pt, Ni, and Ru are capable of catalyzing the hydrogenation of C=C, C=O, and C-O-C bonds at moderate temperatures (97-147 °C) and moderate pressures (1-3 MPa) [20]. Sheu et al. [56] studied the kinetics of hydroteating pine pyrolytic oil on Pt/Al$_2$O$_3$/SiO$_2$, sulfided CoMo/γ-Al$_2$O$_3$, Ni-W/γ-Al$_2$O$_3$, and NiMo/γ-Al$_2$O$_3$. Pt/Al$_2$O$_3$/SiO$_2$ was found to exhibit the highest activity for oxygen removal among these four catalysts. Ru/C and Pd/C are also efficient catalysts used in low temperature stage hydrogenation of bio-oil to produce a partially upgraded bio-oil suitable to high temperature hydroprocessing [40, 41, 68].

Wildschut et al. [69] explored the catalytic activities of supported Ru, Pd, Rh, and Pt in hydrotreatment of bio-oil under both mild and severe reaction conditions. They found Ru/C showed superior activity and selectivity to classical Mo based catalysts. Their ^{1}H NMR and

2D GC analysis of the products further revealed lower contents of organic acids, aldehydes, ketones, and ethers than the corresponding feed while the amount of phenolics, aromatics and alkanes were considerably higher. Carbohydrate fraction in bio-oil was found to be readily hydrogenated to polyols on Ru/C at 250 °C and 10 MPa hydrogen pressure [70, 71] while phenol was hydrogenated to cyclohexane via cyclohexanol under the same conditions [72]. The performance of Ru/C was studied as a function of number of recycles. Severe clustering of metal particles and coke deposition were responsible for its deactivation [72]. Pd/C was found to be less active than Ru/C in hydrogenation using phenol, furfural, guaiacol and acetic acid as model compounds in a temperature range of 150-300 °C [68, 71]. Although Ru/C was found to more active than Pd/C, the capability of catalyzing aqueous-phase reforming and methanation limits its use to below 250 °C for efficient hydrogenation. Due to the high activity of Ru/C in aqueous-phase reforming, all the intermediates from hydrogenation of the model compounds were further gasified at 250 °C and above. Palladium is less active than ruthenium in hydrogenation but it does not catalyze aqueous-phase gasification which allows its operation at higher reaction temperatures to achieve a higher activity. A moderate yield to ethanol from acetic acid was obtained on palladium at 300 °C. For guaiacol, its aromatic ring was first saturated followed by subsequent HDO to cyclohexane via cyclohexanol [68]. For furans, Sitthisa et al. [73] revealed that Pd on Pd/SiO_2 readily interacts with the furan ring via the π bonds and forms the acyl surface species which leads to decarbonylation products and ring saturation products.

In addition to noble metal catalysts, other transition metals and bimetallic catalysts were also explored for hydrodeoxygenation [73-77]. By comparing the HDO of furfural over supported Cu, Pd and Ni catalysts at 210-290 °C under atmospheric pressure of hydrogen, Sitthisa et al. [73] found their catalytic behaviors are quite different. Cu/SiO_2 was found to facilitate the hydrogenation of carbonyl group of furfural due to the preferential adsorption of the carbonyl group on the Cu and the strong repulsion of Cu surface to the furan ring [73, 74]. Addition of Cu to Pd/SiO_2 results in the formation of Pd-Cu alloys, which in turn greatly suppresses the decarbonylation rate and increases the hydrogenation rate [76]. Decarbonylation products and the ring opening products were primarily formed on Ni/SiO_2 due to even stronger interactions between the Ni surface and furan ring [73, 75]. Ni-Fe/SiO_2 bimetallic catalysts were found to be highly selective in C=O hydrogenation and C—O hydrogenolysis and capable of converting furfural into methylfuran [75]. Ni-Fe/SiO_2 also showed high selectivity in converting benzyl alcohol into toluene while the aromatic ring remained unperturbed [75]. Ardiyanti et al. [77] found that the non-sulfided bimetallic 16Ni2Cu/δ-Al_2O_3 catalyst was more active than the monometallic Ni and Cu analogs. Although the activity of 16Ni2Cu is still not as active as the bench mark Ru/C catalyst, it at least suggested that higher activity and selectivity could be achieved by certain combination of non-precious transition metal catalysts.

Recently the transition metal phosphides have emerged as a promising group of hydroprocessing catalysts which are highly active and stable. Metal-rich phosphides combine the properties of metals and ceramics. They are good conductors of heat and electricity, and have high thermal and chemical stability. Zhao et al. [78] tested the activity of Ni_2P/SiO_2, Fe_2P/SiO_2, MoP/SiO_2, Co_2P/SiO_2, and WP/SiO_2 in gas phase HDO of guaiacol. Benzene and phenol were the main products while catechol and cresol were found as intermediates. Their activities were found to follow the order of: Ni_2P/SiO_2 > Co_2P/SiO_2 > Fe_2P/SiO_2, WP/SiO_2, MoP/SiO_2. Commercial 5%Pd/Al_2O_3 was found to be more active than these metal phosphides at lower contact time with catechol as main product. Bui et al. [79] later tested the HDO activity of a series of metal phosphides with 2-methyltetrahydrofuran as model compound at 300 °C and 0.1 MPa. The activity was found to decrease in the order of: Ni_2P/SiO_2 > WP/SiO_2 > MoP/SiO_2 > CoP/SiO_2 > FeP/SiO_2 > Pd/Al_2O_3. It suggested that the HDO activity of given metal phosphide differed from substrate to substrate. The selectivity of HDO products was affected by the preparation method of metal phosphides and it followed the trend of: MoP/SiO_2 > WP/SiO_2 > Ni_2P/SiO_2 > FeP/SiO_2 > CoP/SiO_2 (phosphite method) and MoP/SiO_2 ~ WP/SiO_2 > FeP/SiO_2 > Ni_2P/SiO_2 > CoP/SiO_2 (phosphate method). Li et al. [80] found that the activity of Ni_2P/SiO_2, MoP/SiO_2, and $NiMoP/SiO_2$ in HDO of anisole at 300 °C and 1.5 MPa was in the sequence of Ni_2P/SiO_2 > $NiMoP/SiO_2$ > MoP/SiO_2, and a higher Ni/Mo ratio led to a higher activity. Ni phosphide-containing catalysts were more active than $NiMo/\gamma$-Al_2O_3. However the deactivation of metal phosphides in the presence of water was observed [80].

Although metallic catalysts exhibit high activity towards HDO and avoid the contamination of product oil by sulfur species, there are also apparent issues with their use. Lowering the cost along with improving the stability and optimizing the selectivity towards more favorable less saturated HDO products are worth more efforts.

2.3. Effect of Supports

Support has comprehensive effects on the activity of HDO catalysts. The reaction pathway could even be altered by different supports. Alumina was most widely used as a support in traditional sulfided Mo-based catalysts. Despite its wide application, the observed high coking level and the instability of alumina supports were found under bio-oil hydrotreating conditions [81]. The formation of boehmite (γ-AlOOH) due to reaction with water led to a significant loss of surface area and physical strength [82]. Carbon, ZrO_2, and TiO_2 were tested as the replacement for alumina due to their better water tolerance [30, 68, 83, 84]. Other metal oxides SiO_2 [24, 28], SiO_2-Al_2O_3 [85-87], ZrO_2 [84], CeO_2-ZrO_2 [88], and SBA-15 [89] were also tested as supports in HDO catalysts. Ketchie et al. [90] found significant growth of the metal particle size on Ru/γ-Al_2O_3 and Ru/SiO_2 during the aqueous treatments while the Ru/TiO_2, Ru/C were quite stable. Kreuzer et al. [91] compared the hydrogenolysis activity of Pt supported on SiO_2, Al_2O_3, and TiO_2. It was found that the activity increased in the order of: Pt/TiO_2 > Pt/Al_2O_3 > Pt/SiO_2. Pt/SiO_2 showed almost 100% deoxygenation selectivity while unsupported Pt black showed zero.

3. AQUEOUS-PHASE HYDROTREATING OF CARBOHYDRATE

Current technologies for lignocellulosic biomass conversion include gasification to synthesis gas, hydrolysis (either using acids or enzymes) into sugar and lignin fractions, and selective thermal processes such as pyrolysis and liquefaction [10]. Hydrolysis of lignocellulose has been well-developed due to the demand of sugar for fermentation by the ethanol industry. Acid catalyzed hydrolysis can deconstruct the cellulose and hemicellulose into their corresponding sugars (e.g., glucose, fructose, xylose, and et al) and separate them from lignin [11, 92]. Selective removal of oxygen from the biomass-derived compounds is one of the key challenges in converting renewable biomass resources into fuels and chemicals [84]. Aqueous-phase C_5-C_6 sugars can be converted into mono-oxygenated fuels such as ethanol and butanol by fermentation. However, higher-molecular-weight hydrocarbons that more closely resemble current transportation fuels derived from oil (e.g., C_5-C_{12} for gasoline, C_9-C_{16} for jet fuel, and C_{10}-C_{20} for diesel applications) and thus could then be processed and distributed by existing petrochemical technologies and infrastructure are more preferable for existing transportation vehicles. Thus developing more efficient novel catalytic strategies to convert aqueous-phase sugars are necessary. Liquid hydrocarbon production from aqueous-phase sugars requires reduction steps to remove oxygen, combined with C—C bond-formation steps to increase the molecular weight. In addition, it is preferable to process these sugars in aqueous-phase since they are produced in aqueous-phase. Thus aqueous-phase processing could be of higher energy efficiency as it avoids vaporizing large amount of water and the hydrophobic products could spontaneously separate from water after HDO.

3.1. One-Pot Aqueous-Phase Reforming of Sugars into Alkanes

Glucose, fructose, xylose, and other monoses derived from cellulose and hemicellulose by the hydrolysis can be converted into alkanes via aqueous-phase reforming. Cortright et al. [23] first reported the concept of producing hydrogen and alkanes from sugars and alcohols by aqueous-phase reforming. In this process, sugars were converted into alkanes via repeated dehydration and hydrogenation steps. The required hydrogen was supplied by the aqueous-phase reforming portion. Bifucntional catalysts Pd/SiO_2-Al_2O_3, Pt/SiO_2-Al_2O_3 were found to be highly selective for aqueous-phase reforming of sorbitol to C_5 and C_6 alkanes [22]. Pt/Niobium-based solid acid was later found to have superior reactivity in aqueous-phase dehydration and hydrogenation of sorbitol to alkanes at temperatures near 530K and pressures of 5.4 MPa [93]. The higher reactivity of niobium-based catalysts was ascribed to the coordination environment of niobium acid center which involved hydroxyl groups associated with an adjacent Nb=O group [93]. Li et al. [94] found that Pt/zirconium phosphate showed a better selectivity to C_5 and C_6 species and improved stability under the hydrothermal conditions. The resulting gasoline-range molecules were produced with a high carbon yield of 73%. However the alkanes formed by this protocol are straight-chain compounds with only minor amounts of branched isomers. Also C_5-C_6 alkanes are too volatile to be used as gasoline additives.

3.2. Liquid Fuel via Various Platform Molecules

3.2.1. Aqueous-Phase Hydrodeoxygenation of Sugars

In aqueous-phase hydrodeoxygenation, the reforming reaction competes with hydrodeoxygenation pathway. The extent of C-C and C-O bond cleavage determines the selectivity of hydrodeoxygenation products [95]. Thus it is important to suppress the C-C cleavage and facilitate C-O cleavage by utilizing dehydration-hydrogenation process [20]. Bifunctional catalysts of both acid and metal sites are crucial for the dehydration/

hydrogenation in this process [21]. Without acid sites, glucose was found to be quantitatively hydrogenated into sorbitol over Ru/C at 100 °C and 8 MPa of hydrogen [96]. Direct hydrogenolysis of the C-O and C-C bonds in the alcohols does not occur under typical aqueous-phase reforming conditions [97]. Aqueous-phase dehydration rate was found to be much lower than in the gas phase due to the blocking of Lewis acid sites by abundant water [95]. Dehydration leads to the cleavage of C-O bonds in polyols, while the C-C bond cleavage of polyols with terminal hydroxyl groups results from sequential dehydrogenation, disproportionation and decarboxylation. The presence of terminal hydroxyl is critical for C-C cleavage. Both Brønsted and Lewis acid sites were found to be responsible for xylose dehydration [98]. Lewis sites are primarily accountable for the humins formed from xylose and furfural. In addition, the selectivity of the aqueous-phase dehydration of carbohydrates depends on the Brønsted to Lewis acid ratio. Zirconium phosphate was found to have 30 times higher furfural selectivity than catalysts with higher Lewis acid site concentration [98]. However, Lewis acid was found to catalyze the isomerization of aldose to ketose and ketose can be dehydrated on Brøsnted acids [99-101]. This xylose-xylulose-furfural route showed a lower activation energy than the direct dehydration of xylose [102]. High reaction temperature and high Ru loading facilitated the C-C bond cleavage in the aqueous-phase hydrodeoxygenation of carboxylic acids, while an acidic support in combination with a metal favored the hydrodeoxygenation of carboxyl due to its high metal-support interaction and surface acidity [84].

Ru/C was found to be the most active catalyst among Ru, Pt, Pd, Rh, Ir, Raney Ni, and Raney Cu for aqueous-phase hydrogenation of carboxyl acid to corresponding alcohol [103]. ReO$_x$-promoted Rh/C catalyst was found to be selective in the hydrogenolysis of secondary C-O bonds for a broad range of cyclic ethers and polyols [104]. Both experimental results and theoretical calculations based on density functional theory suggest that the selective hydrogenolysis of C-O bonds was facilitated by acid-catalyzed ring-opening and dehydration reactions coupled with metal-catalyzed hydrogenation. The hydroxyl groups on rhenium atoms associated with rhodium were found to be acidic due to the strong binding of oxygen atoms by rhenium. A properly chosen and tuned composition and structure of the bifunctional catalysts could optimize the ratio of C-O cleavage to C-C cleavage which in turn optimizes the yield of liquid alkanes [20, 105-108].

3.2.2. Mono-Functional Compounds as Intermediates

Conversion of carbohydrates into alkanes involves the removal of oxygen via hydrogenolysis of C-O or a dehydration-hydrogenation process. Kunkes et al. [109] developed a new strategy to convert sugars and polyols into high-octane components of gasoline instead of producing light alkanes by complete dehydration and hydrogenation of sugars in one pot (Figure 4).

First, sugars and polyols were converted into primary hydrophobic alcohols, ketones, carboxylic acids, and heterocyclic compounds over Pt-Re catalyst by controlled C-C and C-O cleavages.

This step removes more than 80% of the initial oxygen content of the sugars and polyols with the composition and yield of products tuned by adjusting operation parameters like temperature, pressure and space velocity [110]. Then, these mono-functional compounds can serve as intermediates for the conversion of renewable biomass resources to fuels and chemicals [109].

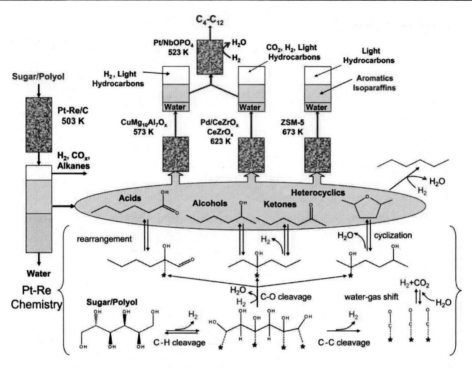

Figure 4. Liquid fuel from sugars using mono-functional compounds as intermediate (Reprinted from ref [109] with perisssion Copyright 2008 AAAS).

Ketonization and aldol condensation are important C-C coupling reactions to form large organic molecules from carbohydrate-derived carbonyl compounds [20]. Carbohydrate can be selectively converted into liquid alkanes of carbon atoms ranging from C_7 to C_{15} by coupling aqueous-phase processing with the aldol condensation [111]. The mono-functional compounds derived from first step can be converted into high-octane components of gasoline by ketonization and aldol condensation followed by final hydrodeoxygenation [110].

Gurbuz et al. [112] developed a dual-bed catalyst system which integrated the ketonization with aldol condensation to achieve a desirable carbon chain length for final liquid alkane products. $Ce_1Zr_1O_x$ was used to catalyze the ketonization of carboxylic acids and Pd/ZrO_2 was used to carry out the aldol condensation/hydrogenation of resulting ketones. After final hydrodeoxygenation over a Pt/SiO_2-Al_2O_3 catalyst, C_{7+} alkanes were obtained from sorbitol.

3.2.3. Furfural and 5-Hydroxymethylfurfural as Platform Molecules

Pentoses and hexoses from hydrolysis of lignocellulosic biomass can be dehydrated to furfural and 5-hydroxymethylfurfural (HMF) by acid catalyst, respectively. Furfural and HMF can then serve as platform molecules for both valuable chemicals and liquid fuels. HMF can be hydrogenated into dimethylfuran over CuRu/C catalyst, which was demonstrated as an alternative way to produce liquid fuels from carbohydrates [113]. In order to produce heavier molecule liquid fuels, Huber et al. [21] developed an aqueous-phase process (Figure 5) to convert biomass-derived carbohydrate into liquid alkanes by combining acid-catalyzed dehydration, aldol condensation, and aqueous-phase dehydration/hydrogena-tion (APD/H).

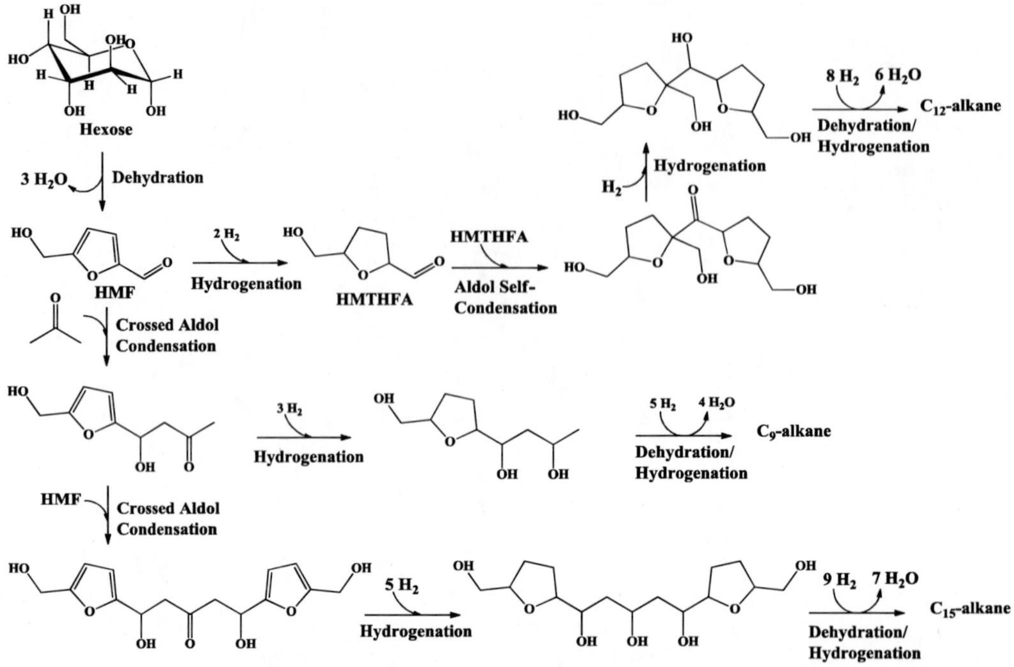

Figure 5. Reaction pathways for the conversion of biomass-derived glucose into liquid alkanes (Adapted from Ref [21] with perisssion Copyright 2005 AAAS).

In this process, dehydration of biomass-derived sugars into HMF or furfural is the initial step, which was accomplished in a biphasic reactor system [114, 115]. Then C_7 to C_{15} range liquid alkanes can be selectively produced by cross aldol condensation followed by hydrogenation/dehydration [111]. Pd/MgO-ZrO$_2$ catalyst [116] was found to be an efficient hydrothermal stable catalyst for aldol-condensation and sequential hydrogenation. The resulting alkanes carrying 90% of the energy of carbohydrate and H_2 feeds can be used as transportation fuel components [21].

However, HMF and furfural cannot undergo self-condensation reactions due to a lack in the necessary α-H atom for aldol-condensation. Thus other carbonyl compounds that form carbanion species are required. Hydroxymethyl tetrahydrofurfural (HMTHFA) and tetrahydrofurfural can be selectively produced by saturating the furan ring of HMF and furfural, respectively. HMTHFA and tetrahdyofurfural will then undergo self-condensation to form heavier molecules [117].

3.2.4. Levulinic Acid as a Platform Molecule

Aqueous-phase sugars from lignocellulose were produced by acid-catalyzed hydrolysis. Beside furfural and HMF, levulinic acid can also be directly produced via acid-catalyzed hydrolysis along with an equi-molar formic acid [118, 119]. Serrano-Ruiz et al. [19, 119] developed a cascade strategy to progressively remove oxygen from biomass with controlled reactivity and facilitated product separation (Figure 6). This process started with direct hydrolysis of cellulose into an equi-molar mixture of levulinic acid and formic acid using sulfuric acid as catalyst. Levulinic acid was then reduced to γ-valerolactone over Ru/C [119] or RuRe/C [120] catalysts in the presence of sulfuric acid at 150 °C and 3.5 MPa.

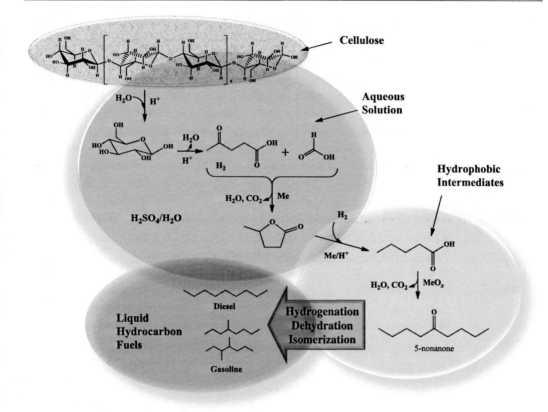

Figure 6. Strategy for the conversion of solid cellulose to liquid hydrocarbon fuels. H+: acid sites; Me: metal sites; MeOx: metal oxide sites (Adapted with permission from ref [19] Copyright 2010 Elsevier B.V.).

The sulfuric acid can be separated and recycled by extraction using ethyl acetate [19] or alkylphenol solvents [121]. To avoid using external solvent and energy-intensive distillation, Gurbuz et al. [18] developed an interesting reactive extraction process to separate the levulinic acid and formic acid from sulfuric acid. Butene from decarboxylation of γ-valerolactone was used as a reactive extraction solvent to form hydrophobic levulinate and formate esters with levulinic acid and formic acid in the presence of sulfuric acid at moderated temperature (< 100 °C) and short contact times (< 120 min). The product spontaneously separated from aqueous phase.

The obtained butyl esters were then converted to γ-valerolactone in a dual-catalyst-bed reactor packed with Pd/C and Ru/C catalysts. By-product 2-butanol can be dehydrated to butene over a SiO_2/Al_2O_3 catalyst. The resulting product, γ-valerolactone, undergoes ring-opening and hydrogenation to form pentanoic acid over Pd/niobia catalyst which can be subsequently converted to 5-nonanone over niobia and/or a ceria-zirconia catalyst by ketonization [119]. The activity and stability of Pd/niobia can be improved by adding 5 wt% silica [122]. Nonanones can then be converted to nonenes by hydrogenation/dehydration. Nonenes can be hydrogenated to n-nonane for use in diesel fuel. Oligomerization of nonenes over acid catalyst will lead to C_{18}-C_{27} alkenes. After hydrogenation, those C_{18}-C_{27} alkanes are suitable for jet and diesel fuel applications [123].

4. HYDRODEOXYGENATION OF LIGNIN-DERIVED COMPOUNDS

Lignin constitutes as much as 30 wt% of lignocellulosic biomass [124, 125]. Lignin is a complex polymer consisting of phenylpropane units with three building blocks, namely coniferyl alcohol, sinapyl alcohol and *p*-coumaryl alcohol (Figure 7).

Table 2. Bond dissociation energies (kJ/mol) [44]

RO–R	339
RO–Ar	422
R–OH	385
Ar–OH	468

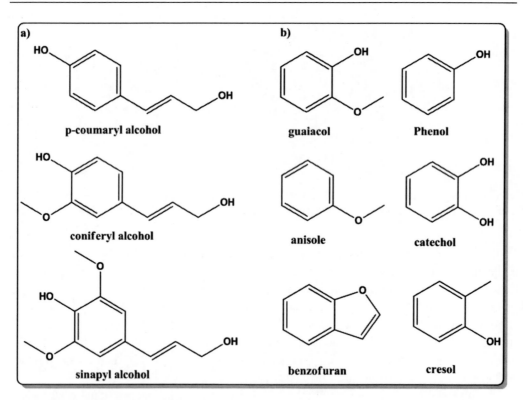

Figure 7. Monomeric blocks of lignin and typical model compounds.

It can be catalytically converted to renewable chemicals and liquid fuels via proper depolymerization and upgrading [126]. Other than pyrolysis and gasification, new technologies such as hydrolysis, hydrogenation, enzymatic reactions, and base-catalyzed depolymerization are also emerging and under development [34, 126-131]. Depolyerimzation of lignin leads to the production of phenolic molecules which represent an important part of the bio-oil.

It could account for up to 20-30 wt% of the organic phase [11, 44, 124, 125, 132]. Removal of oxygen from these phenolic molecules is one of the biggest challenges in HDO. It is also crucial for high quality liquid fuel production.

As shown in Table 2, the bond strength of the O attached to aromatic carbon ($C_{AR)}$ is about 84 kJ/mol greater than that of the O attached to aliphatic carbon (C_{AL}) [44]. This indicates that the HDO of aromatic ether and phenols are much more difficult. Generally more severe reaction conditions are required for hydrodeoxygenation of lignin-derived compounds [133].

4.1. Conventional HDO of Lignin-Derived Compounds

Conventional HDO of bio-oil has been widely studied due to the similarity between HDO and HDS. Special attention has been paid to the HDO of lignin-derived compounds due to its difficulty and importance [134, 135]. Generally, model compounds (Figure 7) such as phenol, anisole, cresol, guaiacol, catechol, and benzofuran are used for HDO mechanistic studies [12]. Two main reaction pathways were found for the oxygen removal from phenol [134]. One is the direct hydrogenolysis of C-O bond leading to the formation of aromatics, and another is hydrogenation of the aromatic ring followed by dehydration of resulting cyclic alcohol.

Conventional hydroprocessing catalysts based on sulfided CoMo/Al$_2$O$_3$ and NiMo/Al$_2$O$_3$ have been extensively studied [58, 85, 133-137]. Toluene and methylcyclohexane were produced as the major products from HDO of cresol over sulfided CoMo/Al$_2$O$_3$ [138]. The kinetic experiments suggested that the ring saturation proceeded in parallel with direct deoxygenation [138, 139]. The o-substitution was found to be important for direct hydrogenolysis of phenol. The reactivity of o-cresol is the least among phenol, m-cresol, p-cresol, and o-cresol regardless of the type of catalysts used [44, 136, 138-140]. Yoosuk et al. [141] studied the HDO of phenol over MoS$_2$ and CoMoS$_2$ at 350 °C and found that Co significantly enhanced the phenol conversion and benzene selectivity by promoting the direct hydrogenolysis reaction pathway. Jongerius et al. [31] found that the HDO of mono-oxygenated substrates like phenol, anisole, and cresol is more difficult than the HDO of catechol, guaiacol, and syringol with commercial sulfided CoMo/Al$_2$O$_3$ catalyst at 5.0 MPa hydrogen pressure and 300 °C. Commercial sulfided CoMo/Al$_2$O$_3$ catalysts are not active for aromatic ring saturation under these conditions.

Temperatures higher than 300 °C are required to convert guaiacyl groups on CoMo/Al$_2$O$_3$ and NiMo/Al$_2$O$_3$ catalysts [133]. The main reaction scheme of guaiacol involves the transformation to hydroxyphenol followed by HDO to phenol. Ferrari et al. [142] suggested that acid species located on the sulfide phase of CoMo/C are responsible for the demethylation of guaiacol to catechol. Centeno et al. [137] compared the activity of guaiacol over sulfided CoMo catalysts supported on alumina, carbon, and silica. They found that the catalysts supported on carbon led to direct elimination of the methoxyl group. Bui et al. [85] compared the MoS$_2$ and Co promoted MoS$_2$ in HDO of guaiacol and found that the presence of CoMoS phase facilitates the demethoxylation route and also promotes the direct deoxygenation of the resulting phenol intermediate to benzene. The Lewis acid-base sites of γ-Al$_2$O$_3$ support were also found to be responsible for demethylation of guaiacol over CoMo/γ-Al$_2$O$_3$ [58].

Noble metals are also interesting catalysts for HDO of lignin-derived compounds due to their well-known hydrogenation activities. Gutierrez et al. [83] showed that the activity and selectivity of ZrO_2-supported mono- and bimetallic noble metal (Rh, Pd, Pt) catalysts were similar or better than that of the conventional sulfided $CoMo/Al_2O_3$ catalyst in guaiacol hydrogenation at 100 °C and hydrodeoxygenation at 300 °C and 8 MPa hydrogen pressure. Lee et al. [30] compared the activity of Pt, Rh, Pd, and Ru supported on acidic matrices of Al_2O_3, SiO_2-Al_2O_3, and nitric-acid-treated carbon black in hydrodeoxygenation of guaiacol at 250 °C and 4.0 MPa H_2. Cyclohexane, cyclohexanol, cyclohexanone, and 2-methoxycyclohexanol were found as the main products. Their distributions were more dependent on the acid supports than on noble metals. The degree of deoxygenation appears to be dependent on the acidity of support. Wild et al. [143] found that Ru/C was very active for hydrodeoxygenation of pyrolysis lignin oil. Guaiacols, syringols, alkylphenols, and catechols can be converted into cycloalkanes, alkyl-substituted cyclohexanols, cyclohexanol and linear alkanes at 350 °C and 10.0 MPa H_2. Bejblova et al. [144] performed hydrodeoxygenation of benzophenone on a series of Pd catalysts supported on active carbon, alumina, ZSM-5, β-zeolite, and MCM-41 at 130 °C and 6 MPa. They found that Pd catalysts supported on both active carbon and acidic zeolite are very active in benzophenone HDO to diphenylmethane. Reaction rate was affected by the acidity of support with the highest reaction rate being obtained on the most acidic β-zeolite.

Non-noble metal catalysts and metal phosphides have also attracted much attention due to their potential of lowering catalyst cost. Bykova et al. [145] found Ni-based catalysts stabilized by SiO_2 and/or ZrO_2 were active in hydrodeoxygenation of guaiacol at 320 °C and 17 MPa H_2. High guaiacol conversion, above 85%, was achieved over SiO_2 supported Ni and NiCu bimetallic catalysts at 1 hour time-on-stream. For guaiacol conversion, $Ni36.5Cu2.3/ZrO_2$-SiO_2-La_2O_3 was considered as the most promising HDO catalyst when the deoxygenation degree, the amount of undesired gaseous products, and the coking extent on the catalyst surface were taken into account. Whiffen et al. [146] studied the activity of low-surface-area MoS_2, MoO_2, MoO_3, and MoP in the hydrogeoxygenation of p-cresol at 350 °C and 4.4 MPa H_2. The activity was found to decrease in the order of: MoP > MoS_2 > MoO_2 > MoO_3. Li et al. [80] studied the HDO of anisole over Ni_2P/SiO_2, MoP/SiO_2, and $NiMoP/SiO_2$ with different Ni/Mo ratios. Demethylation of anisole, hydrogenolysis of phenol and hydrogenation of benzene were found to be the main reactions. The HDO activity decreased in the order of Ni_2P/SiO_2 > $NiMoP/SiO_2$ > MoP/SiO_2. The superior activity of Ni_2P is attributed to the higher d electron density in Ni_2P.

4.2. Gas Phase HDO of Lignin-Derived Compounds

Conventional HDO processed under high hydrogen pressure requires a high capital investment and high operational costs. In this aspect, gas phase HDO at atmospheric pressure is preferable besides it could be more readily integrated with a pyrolysis unit. Transition metals supported on acidic supports have been exclusively studied in gas phase HDO due to their high activity. Zhu et al. [147] found that Pt/Hβ was capable of catalyzing the transalkylation and HDO of anisole at 400 °C and atmospheric pressure. Both transalkylation and HDO rates were higher than corresponding mono-functional catalysts. Benzene, toluene, and xylenes (BTX) were produced as main products. Similarly, Pt/Al_2O_3 was found to selectively catalyze the aromatic carbon-oxygen bond cleavage relative to the accompanying methyl group transfer reaction of guaiacol in the presence of H_2 at 300 °C [87]. Phenol and anisole were formed as primary HDO products while oxygen was removed as methanol and water. Runnebaum et al. [148] found that Pt/SiO_2-Al_2O_3 was three times more active than Pt/γ-Al_2O_3 in gas-phase HDO of 4-methylanisole due to the higher acidity of the SiO_2-Al_2O_3 support. Gonzalez-Borja et al. [149] found that bimetallic Pt-Sn/Inconel showed higher activity and stability than monometallic Pt and Sn catalysts in HDO of guaiacol and anisole at 400 °C and 0.1 MPa. Basic oxide, MgO, was also investigated as a support [29]. Compared to acidic supports (γ-Al_2O_3), Pt/MgO was more stable and selective in HDO of guaiacol at 300 °C and 0.14 MPa. Transalkylation was suppressed on Pt/MgO. Ni/SiO_2 was also tested in gas phase HDO of phenol [150, 151]. It was found that direct hydrogenolysis of phenol to benzene was favored by high Ni loading and elevated temperatures while the selectivity was also found to be strongly dependent on water content with higher water content favoring ring saturation. Fe/SiO_2 was tested in gas-phase hydrodeoxygenation of guaiacol [24, 28]. A yield of 38% to benzene and toluene was achieved at a guaiacol conversion of 74%. Zhao et al. [78] achieved 52-60% selectivity of benzene at guaiacol conversion of 54-80% over Ni_2P/SiO_2, Co_2P/SiO_2 and MoP/SiO_2 catalysts at 300 °C and atmospheric pressure. A high BTX yield was also achieved over Ga/Hβ catalyst in HDO of m-cresol at 450 °C [152]. Therefore, gas phase HDO seems to be more favorable for direct hydrogenolysis of lignin-derived compounds to benzene, toluene, and xylene without saturation of aromatic ring.

4.3. Aqueous-Phase HDO of Lignin-Derived Compounds

Water is a good medium for selective hydrogenation of bio-derived compounds under mild conditions [153]. Aqueous-phase processing of carbohydrate under mild conditions has been extensively studied [22, 23, 154, 155]. Recently, the aqueous-phase hydroprocessing has also been investigated in HDO of lignin-derived compounds.

Zhao et al. [153] selectively converted phenolic bio-oil components (phenols, guaiacols, and syringols) into cycloalkanes and methanol in the aqueous phase using a bifunctional catalytic system consisting of a carbon-supported noble-metal catalyst and a mineral acid. They reported that a series of reactions, including hydrogenation and dehydration, are involved in converting oxygenated compounds into hydrocarbons [156]. Ring saturation was found to be the initial step. Specifically, phenol was first hydrogenated to cyclohexanone and cyclohexanol, followed by dehydration of cyclohexanol to form cyclohexene on mineral acids and subsequent hydrogenation to form cyclohexane on metal site. Pd-, Pt-, Ru- and Rh-based catalysts favor the hydrogenation of phenol to cyclohexanol. The presence of both metal sites and acid sites are crucial for the overall HDO reaction. Kinetic studies suggest that the dehydration reaction is significantly slower than the hydrogenation reaction and keto/enol transformation.

Thus significantly larger concentrations of Brønsted acid sites are required to match the available metal sites for hydrogenation [157]. The combination of Pd/C catalyst and solid acids such as Nafion/SiO_2, HZSM-5, Amberlyst 15, sulfated zirconia and $Cs_{2.5}H_{0.5}PW_{12}O_{40}$ were later tested in order to replace mineral acids [26]. Pd/C combined with HZSM-5 showed an extremely high selectivity towards removal of oxygen-containing groups in lignin-derived substituted phenolic monomers and dimers. Raney® Ni combined with Nafion/SiO_2 are able to quantitatively covert substituted phenols into cycloalkanes by aqueous-phase HDO at 200 °C [158].

This opens a route to lower the catalyst cost. Recently, Zhao et al. [42] showed that C_5-C_9 alkanes, cycloalkanes, and aromatics with a high octane number were produced by aqueous-phase HDO of pyrolysis oil over a Ni/HZSM-5 catalyst at 350 °C and 5 MPa H_2. The proximity of metallic and acid sites is believed to be responsible for efficient HDO. Slight leaching of Ni was found after reaction. The stability of Ni/HZSM-5 catalyst was improved by adding the γ-Al_2O_3 binder. Ni leaching was found to be almost negligible even after 90 hours reaction [159].

In addition, Ni/Al_2O_3-HZSM-5 exhibited up to five times higher catalytic activity for phenol hydrogenation compared to Ni/HZSM-5. Kinetic studies indicate that phenol hydrogenation is the rate determining step on both catalysts. The reaction rates of phenol hydrogenation (r_1), cyclohexanone hydrogenation (r_2), cyclohexanol dehydration (r_3) and cyclohexene hydrogenation (r_4) follow the order of $r_1 < r_2 < r_3 \ll r_4$ [159].

It is worth noting that the hydrocarbon produced in above aqueous-phase hydrotreating processes is limited to the gasoline range products. Thus C-C coupling is necessary to produce jet fuel and diesel range products. Hong et al. [160] found that HY zeolite supported Pt can catalyze phenol HDO and ring-coupling in an aqueous-phase hydrotreating process. Zhao et al. [161] successfully converted phenol and substituted phenols into bicycloalkanes over a bifunctional Pd/Hβ catalyst by tandem reactions of hydroalkylation and hydrodeoxygenation at 200-250 °C. Detailed kinetic studies suggested that phenol selectively reacted with the *in situ* generated cyclohexanol or cyclohexene on Brønsted acid sites. Phenols were also found to preferentially alkylate with the *in situ* generated cyclohexanone. Phenol alkylation with alcohol intermediates was found to compete with alcohol dehydration on acid sites. A properly chosen solid acid catalyst and Pd loading can optimally balance the rates of hydrogenation, dehydration, and alkylation in the tandem reactions of phenols.

CONCLUSION

Lignocellulosic biomass can be deconstructed into small molecular platform intermediates via either fast pyrolysis or hydrolysis. Removal of excess oxygen from the resulting oxygenated intermediates and C-C coupling are necessary to upgrade it into high quality liquid hydrocarbon fuels.

HDO is the most promising and efficient route for biofuel and bio-oil upgrading. HDO of bio-oil over sulfided CoMo and NiMo catalysts could be co-processed with the conventional HDS in a FCC unit. The catalyst stability under hydrothermal conditions and the effect of oxygenates on the activity of HDO and HDS are worthy of further investigation. It is also desirable to minimize the hydrogen consumption by improving the HDO selectivity while suppressing the saturation of aromatic rings. Deep deoxygenation and elimination of phenolic residues remain to be a challenge. Non-noble metals and the emerging metal phosphides are promising catalysts for HDO of bio-oil.

Aqueous-phase HDO is advantageous for oxygen removal from carbohydrates produced by hydrolysis of lignocellulosic biomass. High quality liquid hydrocarbon fuel can be produced by properly coupling the HDO with ketonization and aldol condensation of carbonyl compounds. However, the tandem reactions involved in high-molecular-weight liquid alkane production lead to a long reaction process which in turn results in high capital investment and operational costs. Thus developing novel multifunctional catalysts which facilitate the integration of multiple reactions into one reactor are desired.

Multifunctional catalysts with high selectivity towards the cleavage of C_{AR}-O are desirable for the production of BTX from lignin-derived compounds. Alkylation of aromatic ring with aliphatic ketone/alcohol or alkene intermediates is a promising way to produce jet or diesel range liquid hydrocarbons from lignin-derived compounds.

REFERENCES

[1]   van Ruijven B, van Vuuren DP. Oil and natural gas prices and greenhouse gas emission mitigation. *Energy Policy* 2009;37:4797.

[2]   Balat M. Production of bioethanol from lignocellulosic materials via the biochemical pathway: A review. *Energy Conversion and Management* 2011;52:858.

[3]   Sorrell S, Speirs J, Bentley R, Brandt A, Miller R. Global oil depletion: A review of the evidence. *Energy Policy* 2010;38:5290.

[4]   Serrano-Ruiz JC, Luque R, Sepulveda-Escribano A. Transformations of biomass-derived platform molecules: from high added-value chemicals to fuels via aqueous-phase processing. *Chemical Society Reviews* 2011;40:5266.

[5]   van Leeuwen B, van der Wulp A, Duijnstee I, van Maris A, Straathof A. Fermentative production of isobutene. *Applied Microbiology and Biotechnology* 2012;93:1377.

[6]   Eisentraut A. sustainable production of second generation biofuels potential and perspectives in major economies and developing countries.: International Energy Agency; 2010.

[7]   McKendry P. Energy production from biomass (part 1): overview of biomass. *Bioresource Technology* 2002;83:37.

[8]   Carriquiry MA, Du X, Timilsina GR. Second-generation biofuels economics and policies. The world bank; 2010.

[9]   Sanderson K. Lignocellulose: A chewy problem. *Nature* 2011;474:S12.

[10]   Simonetti DA, Dumesic JA. Catalytic strategies for changing the energy content and achieving C-C coupling in biomass-derived oxygenated hydrocarbons. *ChemSusChem* 2008;1:725.

[11]   Huber GW, Iborra S, Corma A. Synthesis of transportation fuels from biomass: chemistry, catalysts, and engineering. *Chemical Reviews* 2006;106:4044.

[12]   Hicks JC. Advances in C-O bond transformations in lignin-derived compounds for biofuels production. *The Journal of Physical Chemistry Letters* 2011;2:2280.

[13]   Elliott DC, Baker EG, Beckman D, Solantausta Y, Tolenhiemo V, Gevert SB, et al. Technoeconomic assessment of direct biomass liquefaction to transportation fuels. *Biomass* 1990;22:251.

[14]   Mortensen PM, Grunwaldt JD, Jensen PA, Knudsen KG, Jensen AD. A review of catalytic upgrading of bio-oil to engine fuels. *Applied Catalysis A: General* 2011;407:1.

[15]   Agrawal R, Singh NR. Synergistic routes to liquid fuel for a petroleum-deprived future. *AIChE Journal* 2009;55:1898.

[16] Resasco DE. What should we demand from the catalysts responsible for upgrading biomass pyrolysis oil? *The Journal of Physical Chemistry Letters* 2011;2:2294.

[17] Elliott DC, Hart TR, Neuenschwander GG. Chemical processing in high-pressure aqueous environments. 8. Improved catalysts for hydrothermal gasification. *Industrial and Engineering Chemistry Research* 2006;45:3776.

[18] Gurbuz EI, Alonso DM, Bond JQ, Dumesic JA. Reactive extraction of levulinate esters and conversion to g-valerolactone for production of liquid fuels. *ChemSusChem* 2011;4:357.

[19] Serrano-Ruiz JC, Braden DJ, West RM, Dumesic JA. Conversion of cellulose to hydrocarbon fuels by progressive removal of oxygen. *Applied Catalysis B-Environmental* 2010;100:184.

[20] Chheda JN, Huber GW, Dumesic JA. Liquid-phase catalytic processing of biomass-derived oxygenated hydrocarbons to fuels and chemicals. *Angewandte Chemie International Edition* 2007;46:7164.

[21] Huber GW, Chheda JN, Barrett CJ, Dumesic JA. Production of liquid alkanes by aqueous-phase processing of biomass-derived carbohydrates. *Science* 2005;308:1446.

[22] Huber GW, Cortright RD, Dumesic JA. Renewable alkanes by aqueous-phase reforming of biomass-derived oxygenates. *Angewandte Chemie International Edition* 2004;43:1549.

[23] Cortright RD, Davda RR, Dumesic JA. Hydrogen from catalytic reforming of biomass-derived hydrocarbons in liquid water. *Nature* 2002;418:964.

[24] Olcese R, Bettahar MM, Malaman B, Ghanbaja J, Tibavizco L, Petitjean D, et al. Gas-phase hydrodeoxygenation of guaiacol over iron-based catalysts. Effect of gases composition, iron load and supports (silica and activated carbon). *Applied Catalysis B: Environmental* 2013;129:528.

[25] Ben H, Mu W, Deng Y, Ragauskas AJ. Production of renewable gasoline from aqueous phase hydrogenation of lignin pyrolysis oil. *Fuel* 2013;103:1148.

[26] Zhao C, Lercher JA. Selective hydrodeoxygenation of lignin-derived phenolic monomers and dimers to cycloalkanes on Pd/C and HZSM-5 catalysts. *ChemCatChem.* 2012;4:64.

[27] Zabeti M, Nguyen TS, Lefferts L, Heeres HJ, Seshan K. In situ catalytic pyrolysis of lignocellulose using alkali-modified amorphous silica alumina. *Bioresource Technology* 2012;118:374.

[28] Olcese RN, Bettahar M, Petitjean D, Malaman B, Giovanella F, Dufour A. Gas-phase hydrodeoxygenation of guaiacol over Fe/SiO2 catalyst. *Applied Catalysis B: Environmental* 2012;115–116:63.

[29] Nimmanwudipong T, Aydin C, Lu J, Runnebaum RC, Brodwater KC, Browning ND, et al. Selective hydrodeoxygenation of guaiacol catalyzed by platinum supported on magnesium oxide. *Catalysis Letters* 2012:1.

[30] Lee CR, Yoon JS, Suh Y-W, Choi J-W, Ha J-M, Suh DJ, et al. Catalytic roles of metals and supports on hydrodeoxygenation of lignin monomer guaiacol. *Catalysis Communications* 2012;17:54.

[31] Jongerius AL, Jastrzebski R, Bruijnincx PCA, Weckhuysen BM. CoMo sulfide-catalyzed hydrodeoxygenation of lignin model compounds: An extended reaction network for the conversion of monomeric and dimeric substrates. *Journal of Catalysis* 2012;285:315.

[32]   Horáček J, Homola F, Kubičková I, Kubička D. Lignin to liquids over sulfided catalysts. *Catalysis Today* 2012;179:191.

[33]   Bu Q, Lei H, Zacher AH, Wang L, Ren S, Liang J, et al. A review of catalytic hydrodeoxygenation of lignin-derived phenols from biomass pyrolysis. *Bioresource Technology* 2012;124:470.

[34]   Beauchet R, Monteil-Rivera F, Lavoie JM. Conversion of lignin to aromatic-based chemicals (L-chems) and biofuels (L-fuels). *Bioresource Technology* 2012;121:328.

[35]   Elliott DC. Historical developments in hydroprocessing bio-oils. *Energy and Fuels* 2007;21:1792.

[36]   Venderbosch RH, Prins W. Fast pyrolysis technology development. *Biofuels, Bioproducts and Biorefining* 2010;4:178.

[37]   Zhang J, Toghiani H, Mohan D, Pittman CU, Toghiani RK. Product analysis and thermodynamic simulations from the pyrolysis of several biomass feedstocks. *Energy and Fuels* 2007;21:2373.

[38]   Czernik S, Bridgwater AV. Overview of applications of biomass fast pyrolysis oil. *Energy and Fuels* 2004;18:590.

[39]   Mohan D, Pittman CU, Steele PH. Pyrolysis of wood/biomass for bio-oil: A critical review. *Energy and Fuels* 2006;20:848.

[40]   Elliott DC, Hart TR, Neuenschwander GG, Rotness LJ, Olarte MV, Zacher AH, et al. Catalytic hydroprocessing of fast pyrolysis bio-oil from pine sawdust. *Energy and Fuels* 2012;26:3891.

[41]   Elliott DC, Hart TR, Neuenschwander GG, Rotness LJ, Zacher AH. Catalytic hydroprocessing of biomass fast pyrolysis bio-oil to produce hydrocarbon products. *Environmental Progress and Sustainable Energy* 2009;28:441.

[42]   Zhao C, Lercher JA. Upgrading pyrolysis oil over Ni/HZSM-5 by cascade reactions. *Angewandte Chemie-International Edition* 2012;51:5935.

[43]   Zhang Q, Chang J, Wang T, Xu Y. Review of biomass pyrolysis oil properties and upgrading research. *Energy Conversion and Management* 2007;48:87.

[44]   Furimsky E. Catalytic hydrodeoxygenation. *Applied Catalysis A: General* 2000;199:147.

[45]   Song C. An overview of new approaches to deep desulfurization for ultra-clean gasoline, diesel fuel and jet fuel. *Catalysis Today* 2003;86:211.

[46]   Ryymin E-M, Honkela ML, Viljava T-R, Krause AOI. Insight to sulfur species in the hydrodeoxygenation of aliphatic esters over sulfided NiMo/γ-Al2O3 catalyst. *Applied Catalysis A: General* 2009;358:42.

[47]   de Miguel Mercader F, Groeneveld MJ, Kersten SRA, Way NWJ, Schaverien CJ, Hogendoorn JA. Production of advanced biofuels: Co-processing of upgraded pyrolysis oil in standard refinery units. *Applied Catalysis B: Environmental* 2010;96:57.

[48]   Venderbosch RH, Ardiyanti AR, Wildschut J, Oasmaa A, Heeres HJ. Stabilization of biomass-derived pyrolysis oils. *Journal of Chemical Technology and Biotechnology* 2010;85:674.

[49]   Huber GW, Corma A. Synergies between bio- and oil refineries for the production of fuels from biomass. *Angewandte Chemie International Edition* 2007;46:7184.

[50]   Mercader FDM, Koehorst PJJ, Heeres HJ, Kersten SRA, Hogendoorn JA. Competition between hydrotreating and polymerization reactions during pyrolysis oil hydrodeoxygenation. *AIChE Journal* 2011;57:3160.

[51] Marsman JH, Wildschut J, Mahfud F, Heeres HJ. Identification of components in fast pyrolysis oil and upgraded products by comprehensive two-dimensional gas chromatography and flame ionisation detection. *Journal of Chromatography A* 2007;1150:21.

[52] Zhang S, Yan Y, Li T, Ren Z. Upgrading of liquid fuel from the pyrolysis of biomass. *Bioresource Technology* 2005;96:545.

[53] Bridgwater AV. Production of high grade fuels and chemicals from catalytic pyrolysis of biomass. *Catalysis Today* 1996;29:285.

[54] Oasmaa A, Korhonen J, Kuoppala E. An approach for stability measurement of wood-based fast pyrolysis bio-oils. *Energy and Fuels* 2011;25:3307.

[55] Elliott DC, Oasmaa A, Preto F, Meier D, Bridgwater AV. Results of the IEA round robin on viscosity and stability of fast pyrolysis bio-oils. *Energy and Fuels* 2012;26:3769.

[56] Sheu Y-HE, Anthony RG, Soltes EJ. Kinetic studies of upgrading pine pyrolytic oil by hydrotreatment. *Fuel Processing Technology* 1988;19:31.

[57] Grange P, Laurent E, Maggi R, Centeno A, Delmon B. Hydrotreatment of pyrolysis oils from biomass: Reactivity of the various categories of oxygenated compounds and preliminary techno-economical study. *Catalysis Today* 1996;29:297.

[58] Laurent E, Delmon B. Study of the hydrodeoxygenation of carbonyl, carboxylic and guaiacyl groups over sulfided CoMo/γ-Al$_2$O$_3$ and NiMo/γ-Al$_2$O$_3$ catalyst: II. Influence of water, ammonia and hydrogen sulfide. *Applied Catalysis A: General* 1994;109:97.

[59] Ferrari M, Bosmans S, Maggi R, Delmon B, Grange P. Influence of the hydrogen sulfide partial pressure on the hydrodeoxygenation reactions over sulfided CoMo/Carbon catalysts. In: B. Delmon GFF, Grange P, editors. Studies in Surface Science and Catalysis: Elsevier; 1999, p. 85.

[60] Christensen ED, Chupka GM, Luecke J, Smurthwaite T, Alleman TL, Iisa K, et al. Analysis of oxygenated compounds in hydrotreated biomass fast pyrolysis oil distillate fractions. *Energy and Fuels* 2011;25:5462.

[61] French RJ, Hrdlicka J, Baldwin R. Mild hydrotreating of biomass pyrolysis oils to produce a suitable refinery feedstock. *Environmental Progress and Sustainable Energy* 2010;29:142.

[62] de Miguel Mercader F, Groeneveld MJ, Kersten SRA, Way NWJ, Schaverien CJ, Hogendoorn JA. Production of advanced biofuels: Co-processing of upgraded pyrolysis oil in standard refinery units. *Applied Catalysis B-Environmental* 2010;96:57.

[63] Mercader FdM, Groeneveld MJ, Kersten SRA, Geantet C, Toussaint G, Way NWJ, et al. Hydrodeoxygenation of pyrolysis oil fractions: process understanding and quality assessment through co-processing in refinery units. *Energy and Environmental Science* 2011;4:985.

[64] He Z, Wang X. Hydrodeoxygenation of model compounds and catalytic systems for pyrolysis bio-oils upgrading. *Catalysis for Sustainable Energy* 2012;1:28.

[65] Lin SD, Song C. Noble metal catalysts for low-temperature naphthalene hydrogenation in the presence of benzothiophene. *Catalysis Today* 1996;31:93.

[66] Sakanishi K, Ohira M, Mochida I, Okazaki H, Soeda M. The reactivities of polyaromatic hydrocarbons in catalytic hydrogenation over supported noble metals. *Bulletin of the Chemical Society of Japan* 1989;62:3994.

[67] Smith GV, Notheisz F. Heterogeneous catalysis in organic chemistry: Academic Press; 1999.

[68] Elliott DC, Hart TR. Catalytic hydroprocessing of chemical models for bio-oil. *Energy and Fuels* 2009;23:631.

[69] Wildschut J, Mahfud FH, Venderbosch RH, Heeres HJ. Hydrotreatment of fast pyrolysis oil using heterogeneous noble-metal catalysts. *Industrial and Engineering Chemistry Research* 2009;48:10324.

[70] Wildschut J, Iqbal M, Mahfud FH, Melian-Cabrera I, Venderbosch RH, Heeres HJ. Insights in the hydrotreatment of fast pyrolysis oil using a ruthenium on carbon catalyst. *Energy and Environmental Science* 2010;3:962.

[71] Wildschut J, Arentz J, Rasrendra CB, Venderbosch RH, Heeres HJ. Catalytic hydrotreatment of fast pyrolysis oil: Model studies on reaction pathways for the carbohydrate fraction. *Environmental Progress and Sustainable Energy* 2009;28:450.

[72] Wildschut J, Melián-Cabrera I, Heeres HJ. Catalyst studies on the hydrotreatment of fast pyrolysis oil. *Applied Catalysis B: Environmental* 2010;99:298.

[73] Sitthisa S, Resasco DE. Hydrodeoxygenation of furfural over supported metal catalysts: A comparative study of Cu, Pd and Ni. *Catalysis Letters* 2011;141:784.

[74] Sitthisa S, Sooknoi T, Ma Y, Balbuena PB, Resasco DE. Kinetics and mechanism of hydrogenation of furfural on Cu/SiO_2 catalysts. *Journal of Catalysis* 2011;277:1.

[75] Sitthisa S, An W, Resasco DE. Selective conversion of furfural to methylfuran over silica-supported Ni-Fe bimetallic catalysts. *Journal of Catalysis* 2011;284:90.

[76] Sitthisa S, Pham T, Prasomsri T, Sooknoi T, Mallinson RG, Resasco DE. Conversion of furfural and 2-methylpentanal on Pd/SiO_2 and $Pd-Cu/SiO_2$ catalysts. *Journal of Catalysis* 2011;280:17.

[77] Ardiyanti AR, Khromova SA, Venderbosch RH, Yakovlev VA, Heeres HJ. Catalytic hydrotreatment of fast-pyrolysis oil using non-sulfided bimetallic Ni-Cu catalysts on a $δ-Al_2O_3$ support. *Applied Catalysis B: Environmental* 2012;117–118:105.

[78] Zhao HY, Li D, Bui P, Oyama ST. Hydrodeoxygenation of guaiacol as model compound for pyrolysis oil on transition metal phosphide hydroprocessing catalysts. *Applied Catalysis a-General* 2011;391:305.

[79] Bui P, Cecilia JA, Oyama ST, Takagaki A, Infantes-Molina A, Zhao H, et al. Studies of the synthesis of transition metal phosphides and their activity in the hydrodeoxygenation of a biofuel model compound. *Journal of Catalysis* 2012;294:184.

[80] Li K, Wang R, Chen J. Hydrodeoxygenation of anisole over silica-supported Ni_2P, MoP, and NiMoP catalysts. *Energy and Fuels* 2011;25:854.

[81] Furimsky E, Massoth FE. Deactivation of hydroprocessing catalysts. *Catalysis Today* 1999;52:381.

[82] Elliott DC, Sealock LJ, Baker EG. Chemical-processing in high-pressure aqueous environments .2. Development of catalysts for gasification. *Industrial and Engineering Chemistry Research* 1993;32:1542.

[83] Gutierrez A, Kaila RK, Honkela ML, Slioor R, Krause AOI. Hydrodeoxygenation of guaiacol on noble metal catalysts. *Catalysis Today* 2009;147:239.

[84] Chen L, Zhu Y, Zheng H, Zhang C, Zhang B, Li Y. Aqueous-phase hydrodeoxygenation of carboxylic acids to alcohols or alkanes over supported Ru catalysts. *Journal of Molecular Catalysis A: Chemical* 2011;351:217.

[85]   Bui VN, Laurenti D, Afanasiev P, Geantet C. Hydrodeoxygenation of guaiacol with CoMo catalysts. Part I: Promoting effect of cobalt on HDO selectivity and activity. *Applied Catalysis B: Environmental* 2011;101:239.

[86]   Bui VN, Toussaint G, Laurenti D, Mirodatos C, Geantet C. Co-processing of pyrolisis bio oils and gas oil for new generation of bio-fuels: Hydrodeoxygenation of guaïacol and SRGO mixed feed. *Catalysis Today* 2009;143:172.

[87]   Nimmanwudipong T, Runnebaum R, Block D, Gates B. Catalytic reactions of guaiacol: reaction network and evidence of oxygen removal in reactions with hydrogen. *Catalysis Letters* 2011;141:779.

[88]   Bykova MV, Ermakov DY, Kaichev VV, Bulavchenko OA, Saraev AA, Lebedev MY, et al. Ni-based sol–gel catalysts as promising systems for crude bio-oil upgrading: Guaiacol hydrodeoxygenation study. *Applied Catalysis B: Environmental* 2012;113–114:296.

[89]   Tyrone Ghampson I, Sepúlveda C, Garcia R, García Fierro JL, Escalona N, DeSisto WJ. Comparison of alumina- and SBA-15-supported molybdenum nitride catalysts for hydrodeoxygenation of guaiacol. *Applied Catalysis A: General* 2012;435–436:51.

[90]   Ketchie WC, Maris EP, Davis RJ. In-situ X-ray absorption spectroscopy of supported ru catalysts in the aqueous phase. *Chemistry of Materials* 2007;19:3406.

[91]   Kreuzer K, Kramer R. Support effects in the hydrogenolysis of tetrahydrofuran on platinum catalysts. *Journal of Catalysis* 1997;167:391.

[92]   Sun Y, Cheng J. Hydrolysis of lignocellulosic materials for ethanol production: a review. *Bioresource Technology* 2002;83:1.

[93]   West RM, Tucker MH, Braden DJ, Dumesic JA. Production of alkanes from biomass derived carbohydrates on bi-functional catalysts employing niobium-based supports. *Catalysis Communications* 2009;10:1743.

[94]   Li N, Tompsett GA, Huber GW. Renewable high-octane gasoline by aqueous-phase hydrodeoxygenation of C_5 and C_6 carbohydrates over Pt/zirconium phosphate catalysts. *ChemSusChem* 2010;3:1154.

[95]   Peng BX, Zhao C, Mejia-Centeno I, Fuentes GA, Jentys A, Lercher JA. Comparison of kinetics and reaction pathways for hydrodeoxygenation of C_3 alcohols on Pt/Al_2O_3. *Catalysis Today* 2012;183:3.

[96]   Gallezot P, Nicolaus N, Flèche G, Fuertes P, Perrard A. Glucose hydrogenation on ruthenium catalysts in a trickle-bed reactor. *Journal of Catalysis* 1998;180:51.

[97]   Wawrzetz A, Peng B, Hrabar A, Jentys A, Lemonidou AA, Lercher JA. Towards understanding the bifunctional hydrodeoxygenation and aqueous phase reforming of glycerol. *Journal of Catalysis* 2010;269:411.

[98]   Weingarten R, Tompsett GA, Conner Jr WC, Huber GW. Design of solid acid catalysts for aqueous-phase dehydration of carbohydrates: The role of Lewis and Brønsted acid sites. *Journal of Catalysis* 2011;279:174.

[99]   Choudhary V, Pinar AB, Sandler SI, Vlachos DG, Lobo RF. Xylose isomerization to xylulose and its dehydration to furfural in aqueous media. *ACS Catalysis* 2011;1:1724.

[100] Yang Y, Hu C-W, Abu-Omar MM. Synthesis of furfural from ylose, xylan, and biomass using $AlCl_3 \cdot 6\,H_2O$ in biphasic media via xylose isomerization to xylulose. *ChemSusChem* 2012;5:405.

[101] Wang TF, Pagan-Torres YJ, Combs EJ, Dumesic JA, Shanks BH. Water-compatible lewis acid-catalyzed conversion of carbohydrates to 5-hydroxymethylfurfural in a biphasic solvent system. *Topics in Catalysis* 2012;55:657.

[102] Kruger JS, Nikolakis V, Vlachos DG. Carbohydrate dehydration using porous catalysts. *Current Opinion in Chemical Engineering* 2012;1:312.

[103] Olcay H, Xu L, Xu Y, Huber GW. Aqueous-phase hydrogenation of acetic acid over transition metal catalysts. *ChemCatChem* 2010;2:1420.

[104] Chia M, Pagan-Torres YJ, Hibbitts D, Tan QH, Pham HN, Datye AK, et al. Selective hydrogenolysis of polyols and cyclic ethers over bifunctional surface sites on rhodium-rhenium catalysts. *Journal of the American Chemical Society* 2011;133:12675.

[105] Li N, Huber GW. Aqueous-phase hydrodeoxygenation of sorbitol with Pt/SiO$_2$–Al$_2$O$_3$: Identification of reaction intermediates. *Journal of Catalysis* 2010;270:48.

[106] Maris EP, Davis RJ. Hydrogenolysis of glycerol over carbon-supported Ru and Pt catalysts. *Journal of Catalysis* 2007;249:328.

[107] Maris EP, Ketchie WC, Murayama M, Davis RJ. Glycerol hydrogenolysis on carbon-supported PtRu and AuRu bimetallic catalysts. *Journal of Catalysis* 2007;251:281.

[108] Lin YC, Huber GW. The critical role of heterogeneous catalysis in lignocellulosic biomass conversion. *Energy and Environmental Science* 2009;2:68.

[109] Kunkes E, Simonetti D, West R, Serrano-Ruiz J, Gartner C, Dumesic J. Catalytic conversion of biomass to monofunctional hydrocarbons and targeted liquid-fuel classes. *Science* 2008;322:417.

[110] West RM, Kunkes EL, Simonetti DA, Dumesic JA. Catalytic conversion of biomass-derived carbohydrates to fuels and chemicals by formation and upgrading of mono-functional hydrocarbon intermediates. *Catalysis Today* 2009;147:115.

[111] Huber GW, Dumesic JA. An overview of aqueous-phase catalytic processes for production of hydrogen and alkanes in a biorefinery. *Catalysis Today* 2006;111:119.

[112] Gurbuz EI, Kunkes EL, Dumesic JA. Dual-bed catalyst system for C-C coupling of biomass-derived oxygenated hydrocarbons to fuel-grade compounds. *Green Chemistry* 2010;12:223.

[113] Roman-Leshkov Y, Barrett CJ, Liu ZY, Dumesic JA. Production of dimethylfuran for liquid fuels from biomass-derived carbohydrates. *Nature* 2007;447:982.

[114] Chheda JN, Roman-Leshkov Y, Dumesic JA. Production of 5-hydroxymethylfurfural and furfural by dehydration of biomass-derived mono- and poly-saccharides. *Green Chemistry* 2007;9:342.

[115] West RM, Liu ZY, Peter M, Gaertner CA, Dumesic JA. Carbon-carbon bond formation for biomass-derived furfurals and ketones by aldol condensation in a biphasic system. *Journal of Molecular Catalysis a-Chemical* 2008;296:18.

[116] Barrett CJ, Chheda JN, Huber GW, Dumesic JA. Single-reactor process for sequential aldol-condensation and hydrogenation of biomass-derived compounds in water. *Applied Catalysis B-Environmental* 2006;66:111.

[117] Chheda JN, Dumesic JA. An overview of dehydration, aldol-condensation and hydrogenation processes for production of liquid alkanes from biomass-derived carbohydrates. *Catalysis Today* 2007;123:59.

[118] Lange J-P. Lignocellulose conversion: An introduction to chemistry, process and economics. *Biofuels, Bioproducts and Biorefining* 2007;1:39.

[119] Serrano-Ruiz JC, Wang D, Dumesic JA. Catalytic upgrading of levulinic acid to 5-nonanone. *Green Chemistry* 2010;12:574.

[120] Braden DJ, Henao CA, Heltzel J, Maravelias CC, Dumesic JA. Production of liquid hydrocarbon fuels by catalytic conversion of biomass-derived levulinic acid. *Green Chemistry* 2011;13:3505.

[121] Alonso DM, Wettstein SG, Bond JQ, Root TW, Dumesic JA. Production of biofuels from cellulose and corn stover using alkylphenol solvents. *ChemSusChem* 2011;4:1078.

[122] Pham HN, Pagan-Torres YJ, Serrano-Ruiz JC, Wang D, Dumesic JA, Datye AK. Improved hydrothermal stability of niobia-supported Pd catalysts. *Applied Catalysis A: General* 2011;397:153.

[123] Alonso DM, Bond JQ, Serrano-Ruiz JC, Dumesic JA. Production of liquid hydrocarbon transportation fuels by oligomerization of biomass-derived C-9 alkenes. *Green Chemistry* 2010;12:992.

[124] Stöcker M. Biofuels and biomass-to-liquid fuels in the biorefinery: Catalytic conversion of lignocellulosic biomass using porous materials. *Angewandte Chemie International Edition* 2008;47:9200.

[125] Ragauskas AJ, Williams CK, Davison BH, Britovsek G, Cairney J, Eckert CA, et al. The path forward for biofuels and biomaterials. *Science* 2006;311:484.

[126] Zakzeski J, Bruijnincx PCA, Jongerius AL, Weckhuysen BM. The catalytic valorization of lignin for the production of renewable chemicals. *Chemical Reviews* 2010;110:3552.

[127] Yoshikawa T, Yagi T, Shinohara S, Fukunaga T, Nakasaka Y, Tago T, et al. Production of phenols from lignin via depolymerization and catalytic cracking. *Fuel Processing Technology*.

[128] Toledano A, Serrano L, Labidi J. Organosolv lignin depolymerization with different base catalysts. *Journal of Chemical Technology and Biotechnology* 2012;87:1593.

[129] Roberts VM, Stein V, Reiner T, Lemonidou A, Li X, Lercher JA. Towards quantitative catalytic lignin depolymerization. *Chemistry – A European Journal* 2011;17:5939.

[130] Miller JE, Evans L, Littlewolf A, Trudell DE. Batch microreactor studies of lignin and lignin model compound depolymerization by bases in alcohol solvents. *Fuel* 1999;78:1363.

[131] Pandey MP, Kim CS. Lignin depolymerization and conversion: A review of thermochemical methods. *Chemical Engineering and Technology* 2011;34:29.

[132] Vitolo S, Seggiani M, Frediani P, Ambrosini G, Politi L. Catalytic upgrading of pyrolytic oils to fuel over different zeolites. *Fuel* 1999;78:1147.

[133] Laurent E, Delmon B. Study of the hydrodeoxygenation of carbonyl, carboxylic and guaiacyl groups over sulfided $CoMo/\gamma$-Al_2O_3 and $NiMo/\gamma$-Al_2O_3 catalysts: I. Catalytic reaction schemes. *Applied Catalysis A: General* 1994;109:77.

[134] Laurent E, Delmon B. Influence of oxygen-, nitrogen-, and sulfur-containing compounds on the hydrodeoxygenation of phenols over sulfided $CoMo/\gamma$-Al_2O_3 and $NiMo/\gamma$-Al_2O_3 catalysts. *Industrial and Engineering Chemistry Research* 1993;32:2516.

[135] Şenol Oİ, Ryymin EM, Viljava TR, Krause AOI. Effect of hydrogen sulphide on the hydrodeoxygenation of aromatic and aliphatic oxygenates on sulphided catalysts. *Journal of Molecular Catalysis A: Chemical* 2007;277:107.

[136] Weigold H. Behaviour of Co-Mo-Al_2O_3 catalysts in the hydrodeoxygenation of phenols. *Fuel* 1982;61:1021.

[137] Centeno A, Laurent E, Delmon B. Influence of the support of CoMo sulfide catalysts and of the addition of potassium and platinum on the catalytic performances for the hydrodeoxygenation of carbonyl, carboxyl, and guaiacol-type molecules. *Journal of Catalysis* 1995;154:288.

[138] Odebunmi EO, Ollis DF. Catalytic hydrodeoxygenation: I. Conversions of o-, p-, and m-cresols. *Journal of Catalysis* 1983;80:56.

[139] Wandas R, Surygala J, Śliwka E. Conversion of cresols and naphthalene in the hydroprocessing of three-component model mixtures simulating fast pyrolysis tars. *Fuel* 1996;75:687.

[140] Shin E-J, Keane MA. Catalytic hydrogen treatment of aromatic alcohols. *Journal of Catalysis* 1998;173:450.

[141] Yoosuk B, Tumnantong D, Prasassarakich P. Unsupported MoS_2 and $CoMoS_2$ catalysts for hydrodeoxygenation of phenol. *Chemical Engineering Science* 2012;79:1.

[142] Ferrari M, Bosmans S, Maggi R, Delmon B, Grange P. CoMo/carbon hydrodeoxygenation catalysts: influence of the hydrogen sulfide partial pressure and of the sulfidation temperature. *Catalysis Today* 2001;65:257.

[143] de Wild P, Van der Laan R, Kloekhorst A, Heeres E. Lignin valorisation for chemicals and (transportation) fuels via (catalytic) pyrolysis and hydrodeoxygenation. *Environmental Progress and Sustainable Energy* 2009;28:461.

[144] Bejblová M, Zámostný P, Červený L, Čejka J. Hydrodeoxygenation of benzophenone on Pd catalysts. *Applied Catalysis A: General* 2005;296:169.

[145] Bykova MV, Ermakov DY, Kaichev VV, Bulavchenko OA, Saraev AA, Lebedev MY, et al. Ni-based sol-gel catalysts as promising systems for crude bio-oil upgrading: Guaiacol hydrodeoxygenation study. *Applied Catalysis B-Environmental* 2012;113:296.

[146] Whiffen VML, Smith KJ. Hydrodeoxygenation of 4-methylphenol over unsupported MoP, MoS_2, and MoO_x Catalysts *Energy and Fuels* 2010;24:4728.

[147] Zhu XL, Lobban LL, Mallinson RG, Resasco DE. Bifunctional transalkylation and hydrodeoxygenation of anisole over a Pt/HBeta catalyst. *Journal of Catalysis* 2011;281:21.

[148] Runnebaum RC, Nimmanwudipong T, Limbo RR, Block DE, Gates BC. Conversion of 4-methylanisole catalyzed by Pt/γ-Al_2O_3 and by Pt/SiO_2-Al_2O_3: Reaction networks and evidence of oxygen removal. *Catalysis Letters* 2012;142:7.

[149] Gonzalez-Borja MA, Resasco DE. Anisole and guaiacol hydrodeoxygenation over monolithic Pt-Sn Catalysts. *Energy and Fuels* 2011;25:4155.

[150] Shin E-J, Keane MA. Gas-phase hydrogenation/hydrogenolysis of phenol over supported nickel catalysts. *Industrial and Engineering Chemistry Research* 2000;39:883.

[151] Nimmanwudipong T, Runnebaum RC, Block DE, Gates BC. Catalytic conversion of guaiacol catalyzed by platinum supported on alumina: Reaction network including hydrodeoxygenation reactions. *Energy and Fuels* 2011;25:3417.

[152] Ausavasukhi A, Huang Y, To AT, Sooknoi T, Resasco DE. Hydrodeoxygenation of m-cresol over gallium-modified beta zeolite catalysts. *Journal of Catalysis* 2012;290:90.

[153] Zhao C, Kou Y, Lemonidou AA, Li XB, Lercher JA. Highly selective catalytic conversion of phenolic bio-oil to alkanes. *Angewandte Chemie-International Edition* 2009;48:3987.

[154] Huber GW, Shabaker JW, Dumesic JA. Raney Ni-Sn catalyst for H_2 production from biomass-derived hydrocarbons. *Science* 2003;300:2075.

[155] Soares RR, Simonetti DA, Dumesic JA. Glycerol as a source for fuels and chemicals by low-temperature catalytic processing. *Angewandte Chemie* 2006;118:4086.

[156] Zhao C, He J, Lemonidou AA, Li X, Lercher JA. Aqueous-phase hydrodeoxygenation of bio-derived phenols to cycloalkanes. *Journal of Catalysis* 2011;280:8.

[157] Zhao C, He JY, Lemonidou AA, Li XB, Lercher JA. Aqueous-phase hydrodeoxygenation of bio-derived phenols to cycloalkanes. *Journal of Catalysis* 2011;280:8.

[158] Zhao C, Kou Y, Lemonidou AA, Li XB, Lercher JA. Hydrodeoxygenation of bio-derived phenols to hydrocarbons using RANEY®Ni and Nafion/SiO_2 catalysts. *Chemical Communications* 2010;46:412.

[159] Zhao C, Kasakov S, He J, Lercher JA. Comparison of kinetics, activity and stability of Ni/HZSM-5 and Ni/Al2O3-HZSM-5 for phenol hydrodeoxygenation. *Journal of Catalysis* 2012;296:12.

[160] Hong DY, Miller SJ, Agrawal PK, Jones CW. Hydrodeoxygenation and coupling of aqueous phenolics over bifunctional zeolite-supported metal catalysts. *Chemical Communications* 2010;46:1038.

[161] Zhao C, Camaioni DM, Lercher JA. Selective catalytic hydroalkylation and deoxygenation of substituted phenols to bicycloalkanes. *Journal of Catalysis* 2012;288:92.

In: Biomass Processing, Conversion and Biorefinery   ISBN: 978-1-62618-346-9
Editors: Bo Zhang and Yong Wang   © 2013 Nova Science Publishers, Inc.

Chapter 14

BIODIESEL PRODUCTION

Oscar Marin-Flores[1], Anna Lee Tonkovich[2]
and Yong Wang[1,3]*
[1]Voiland School of Chemical Engineering and Bioengineering,
Washington State University, Pullman, WA
[2]Heliae, LLC, Gilbert, AZ
[3]Institute for Integrated Catalysis,
Pacific Northwest National Laboratory, Richland, WA

ABSTRACT

Biodiesel has attracted a great deal of attention due to not only its environmental benefits but also the fact that it is made from renewable resources. However, the cost of production of biodiesel still appears to be prohibitive which makes it commercially unviable. Some of the solutions proposed to overcome this drawback involve the use of waste cooking oils as raw material, the implementation of continuous transesterification processes and the recovery of high quality glycerol as by-product. The feedstock used defines whether the synthesized biodiesel is of first, second or third generation. The most widely used method to produce biodiesel is the transesterification of vegetable oils and animal fats. The rates of transesterification appear to be affected by certain variables such as molar ratio of glycerides to alcohol, catalysts, reaction temperature, reaction time and free fatty acids and water content of oils or fats. The mechanism and kinetics of the transesterification are analyzed in the present article to show how the reaction occurs and progresses with time.

1. INTRODUCTION

The generally accepted definition is based on the ASTM biodiesel standard D6751, which defines biodiesel as a "fuel comprised of mono-alkyl esters of long-chain fatty acids derived from vegetable oils or animal fats". Biodiesel can be mixed with petroleum diesel in

---

* Phone 509-371-6273. Fax 509-371-6498. E-mail: wang42@wsu.edu.

any percentage, which is represented by a number following a B. Thus, B10 is 10 percent biodiesel with 90 percent petroleum, or B100 is 100 percent biodiesel, with no petroleum addition [1].

1.1. Biodiesel History

The diesel engine was developed by inventor Rudolph Diesel in the 1890s to overcome the inefficiencies detected on the steam engines of the late 1800s. The early diesel engines were designed to run on many different fuels such as kerosene or coal although with time it was clear that, because of their high energy content, vegetable oils would become excellent fuels. The use of vegetable oil to produce diesel was first attributed to E. Duffy and J. Patrick in 1853, many years before the first diesel engine became functional. Shortly after Diesel's death in 1913 petroleum became widely available in a variety of forms, including the class of fuel we know today as "diesel fuel". Once petroleum became available and inexpensive, the diesel engine was redesigned to make it more suitable to petroleum diesel fuel. On 1937, G. Chavanne of the University of Brussels was granted a patent for the transformation of vegetable oils to be used as fuels. This patent described the transesterification of vegetable oils using either ethanol or methanol to separate the fatty acids from the glycerol by replacing the glycerol with short linear alcohols. However, today's biodiesel industry was not established in Europe until the late 1980s. The 1970s' energy crisis and the increasing interest in environment re-launched the study of biodiesel and boomed it rapidly. During the period between 2001 and 2009, the world production of biodiesel increased by more than 16 times (from 9.57 billion tons to 157.6 billion tons). With the exception of traditional biodiesel production countries (i.e., EU and US), a significant increase of market has been expected in countries such as China, Brazil, Japan, Indonesia, and Malaysia [2].

1.2. Biodiesel Properties

Many industrial devices to produce biodiesel have been set up in Europe and USA; for that reason, certain standards for biodiesel have been formulated in these countries. Such standards are listed in Table 1.

1.3. Advantages and Disadvantages of Biodiesel

Compared to fossil diesel, biodiesel has the following advantages [3–6]:

- Biodiesel is a renewable energy source as opposed to oil-derived diesel, the reserves of which are finite.
- Biodiesel can easily decompose under natural conditions. Over 90% of pure biodiesel can be degraded in matter of few weeks.
- Compared with common diesel and petrol, biodiesel has higher combustible value that makes it relatively safe to be stored and transport.

- Engine modification is not required as it has similar properties to those of diesel fuel.
- Biodiesel contains much less sulfur which not only reduces the amount of pollutant emissions but also enables an effective lubrication of movable parts during the work of the engine. Pure or blend biodiesel also could suppress the net production of carbon dioxide.
- Biodiesel would make an area become independent of its need for energy as it can be produced locally.

Although biodiesel is more environmentally friendly than diesel fuel, it still has some disadvantages:

- High viscosity and surface stress would lead to bigger drops which may cause problems with the system of fuel injection.
- Vegetable oil contains much more unsaturated compounds than diesel, so biodiesel from it is much easier subjected to oxidation.
- It is more expensive due to the raw material utilized in its production.

Table 1. Biodiesel standards [6]

Properties	EN14214	ASTM D6751
FAME content	$\geq 96.5\%$ (m/m)	–
Density at 15°C	$\geq 860, \leq 900$ $(kg/m)^3$	–
Viscosity at 40°C	$\geq 3.5, \leq 5.0$ (mm^2/s)	$\geq 1.9, \leq 6.0$ (mm^2/s)
Flash point	$\geq 101°C$	$\geq 130°C$
Sulfur content	≤ 10 mg/kg	≤ 50 mg/kg
Carbon residue remnant (at 10% distillation remnant)	$\leq 0.3\%$ (m/m)	$\leq 0.05\%$ (m/m)
Cetane number	≥ 51.0	≥ 47
Sulfated ash content	– $\leq 0.02\%$ (m/m)	– $\leq 0.02\%$ (m/m)
Water content	≤ 500 mg/kg	$\leq 0.05\%$ (v/v)
Total contamination	≤ 24 mg/kg	–
Copper band corrosion (3 hours at 50°C)	Class 1 max	No. 3 max
Oxidation stability, 100°C	≥ 6 hours	≥ 3 hours
Acid value	≤ 0.5	≤ 0.8
Iodine value	≤ 120	–
Linolenic Acid Methylester	$\leq 12\%$ (m/m)	–
Polyunsaturated (>4 Double bonds) Methylester	$\leq 1\%$ (m/m)	–
Methanol content	$\leq 0.2\%$ (m/m)	–
Monoglyceride content	$\leq 0.8\%$ (m/m)	–
Diglyceride content	$\leq 0.2\%$ (m/m)	–
Triglyceride content	$\leq 0.2\%$ (m/m)	–
Free Glycerine	$\leq 0.02\%$ (m/m)	≤ 0.02
Total Glycerine	$\leq 0.25\%$ (m/m)	≤ 0.25
Group I metals (Na+K)	≤ 5 mg/kg	≤ 5
Group II metals (Ca+Mg)	≤ 5 mg/kg	–
Phosphorous content	≤ 4 mg/kg	$\leq 0.001\%$ (m/m)

2. PRODUCTION OF BIODIESEL

2.1. Biodiesel Sources

Different vegetable oils and fats can be used to produce biodiesel [7]. However, given that the price of edible oils is higher than that of diesel fuel, alternative sources such as waste vegetable oils and non-edible crude vegetable oils are preferred as potential inexpensive biodiesel sources. Thus, depending on the source, the synthesized biodiesel can be classified as follows:

- First Generation Biodiesel: produced from oil crops (rapeseed, soybeans, sunflower, palm, coconut, used cooking oil, animal fats, etc.
- Second Generation Biodiesel: produced from novel starch, oil and sugar crops such as Jathropa or Miscanthus, which are considered more suitable biodiesel sources since they do not compromise the food industry.
- Third generation Biodiesel: produced from microalgae, which due to its high lipid content appears to possess a high potential for biodiesel production.

2.2. Methods of Biodiesel Production

A number of well-established methods are currently available for the production of biodiesel fuel [8]. These methods are intended to reduce the viscosity of the crude oils in order to deliver a product with suitable properties to be used as engine fuels. This can be accomplished using the approaches described below.

2.2.1. Blending of Crude Oils or Dilution

Crude oils can be mixed directly or diluted with diesel fuel to improve the viscosity so as to solve the problems associated with the use of pure vegetable oils with high viscosities in compression ignition engines.

2.2.2. Micro-Emulsification

Microemulsions are clear, stable isotropic fluids with three components: an oil phase, an aqueous phase and a surfactant. This ternary phase can improve spray characteristics by explosive vaporization of the low boiling constituents in the micelles. All micro-emulsions with butanol, hexanol and octanol can meet the maximum viscosity limitation for diesel engines.

2.2.3. Pyrolysis

This process involves the conversion of one substance into another using heat or with the aid of catalytic materials in the absence of air or oxygen. The materials to be pyrolyzed can be vegetable oils, animal fats, natural fatty acids and methyl ester of fatty acids. For instance, the viscosity of pyrolyzed soybean oil distillate is 10.2 cSt at 37.8 °C, which is higher than the ASTM specified range for No. 2 diesel fuel but acceptable as still well below the viscosity of soybean oil.

2.2.4. Transesterification

Transesterification or alcoholysis is a reversible process where an alcohol is displaced from an ester by another. The transesterification reaction is represented by the general equation as Figure 1. If methanol is used in this process it is called methanolysis. Methanolysis of triglycerides is represented in Figure 2. Figure 3 shows a schematic diagram of the steps involved in biodiesel production. Alcohol, catalyst, and oil are combined in a reactor and agitated for approximately 1 h at 60°C. Smaller plants often use batch reactors, but larger plants use continuous stirred-tank reactors (CSTR) or plug flow reactors.

$$RCOOR^1 + R^2OH \overset{\text{Catalyst}}{\rightleftharpoons} RCOOR^2 + R^1OH$$

$$\text{Ester} \qquad \text{Alcohol} \qquad\qquad \text{Ester} \qquad \text{Alcohol}$$

Figure 1. General equation for transesterification.

$$
\begin{array}{ccccc}
CH_2-OCOR^1 & & & CH_2-OH & R^1COOCH_3 \\
| & & \text{Catalyst} & | & | \\
CH-OCOR^2 & + 3CH_3OH & \rightleftharpoons & CH-OH & + \quad R^2COOCH_3 \\
| & & & | & | \\
CH_2-OCOR^3 & & & CH_2-OH & R^3COOCH_3 \\
\text{Triglyceride} & \text{Methanol} & & \text{Glycerol} & \text{Methyl esters}
\end{array}
$$

Figure 2. General equation for transesterification of triglycerides with methanol

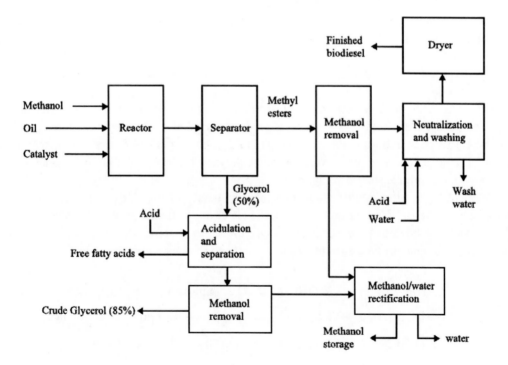

Figure 3. Process flow schematic for biodiesel production [9].

The reaction is usually carried out in two steps. In this system, approximately 80% of the alcohol and catalyst is added to the oil in a first stage CSTR. Then the reacted stream from this reactor goes through a glycerol removal step before entering a second CSTR. The remaining 20% of the alcohol and catalyst are added in this reactor. This system provides a very complete reaction with the potential of using less alcohol than single-step systems.

Following the reaction, the glycerol by-product is removed from the methyl esters. Due to the low solubility of glycerol in the esters, this separation generally occurs quickly and may be accomplished with either a settling tank or a centrifuge. Excess methanol can make the separation difficult and it is usually removed from the reaction stream after the glycerol and methyl esters are separated to prevent reversing the transesterification reaction. Water may be added to the reaction mixture after the transesterification is complete to improve the separation of glycerol.

2.3. Transesterification Kinetics and Mechanism

The transesterification of triglycerides produce fatty acid alkyl esters and glycerol. The mechanism of transesterification is described in Figure 4.

Each step in the mechanism is reversible and a little excess of alcohol is used to shift the equilibrium towards the formation of esters. In the presence of excess alcohol, the forward reaction is pseudo-first order while the reverse reaction is found to be second order. Alkali-catalyzed transesterification is about 4000 times faster than the acid-catalyzed one. For this reason, the alkali-catalyzed process is most frequently used commercially.

The mechanism of alkali-catalyzed transesterification is shown in Figure 5. The first step involves the attack of the alkoxide ion to the carbonyl carbon of the triglyceride molecule, which results in the formation of a tetrahedral intermediate.

The reaction of this intermediate with an alcohol produces the alkoxide ion in the second step. In the last step the rearrangement of the tetrahedral intermediate gives rise to an ester and a diglyceride.

Brönsted acids such as sulfonic and sulfuric acids can be used to achieve an acid-catalyzed transesterification. These catalysts produce very high yields in alkyl esters but with low rates of reaction, requiring typically temperatures higher than 100°C and more than 3 h to attain full conversion. The mechanism of acid-catalyzed transesterification of vegetable oil (for a monoglyceride) is shown in Figure 6.

Acid-catalyzed transesterification can be extended to di- and tri-glycerides. The protonation of the carbonyl group of the ester leads to the formation of a carbocation, which after a nucleophilic attack of the alcohol gives rise to a tetrahedral intermediate. This intermediate eliminates glycerol to form a new ester and to regenerate the catalyst.

$$\text{Triglyceride} + R^1OH \rightleftharpoons \text{Diglyceride} + RCOOR^1$$

$$\text{Diglyceride} + R^1OH \rightleftharpoons \text{Monoglyceride} + RCOOR^1$$

$$\text{Monoglyceride} + R^1OH \rightleftharpoons \text{Glycerol} + RCOOR^1$$

Figure 4. General equation for transesterification of triglycerides.

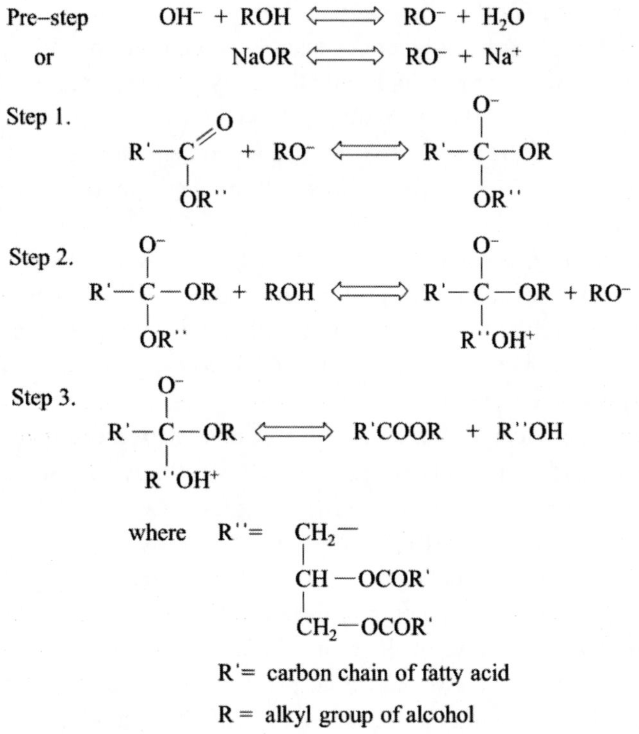

Pre–step $OH^- + ROH \rightleftharpoons RO^- + H_2O$

or $NaOR \rightleftharpoons RO^- + Na^+$

Step 1.

$$R'-\underset{\underset{OR''}{|}}{C}\overset{\overset{O}{\diagup}}{} + RO^- \rightleftharpoons R'-\underset{\underset{OR''}{|}}{\overset{\overset{O^-}{|}}{C}}-OR$$

Step 2.

$$R'-\underset{\underset{OR''}{|}}{\overset{\overset{O^-}{|}}{C}}-OR + ROH \rightleftharpoons R'-\underset{\underset{R''OH^+}{|}}{\overset{\overset{O^-}{|}}{C}}-OR + RO^-$$

Step 3.

$$R'-\underset{\underset{R''OH^+}{|}}{\overset{\overset{O^-}{|}}{C}}-OR \rightleftharpoons R'COOR + R''OH$$

where $R'' = $ $\underset{\underset{CH_2-OCOR'}{\overset{\overset{CH_2-}{|}}{\underset{|}{CH-OCOR'}}}}{}$

R' = carbon chain of fatty acid

R = alkyl group of alcohol

Figure 5. Mechanism of base-catalyzed transesterification.

where $R'' = \begin{bmatrix} OH \\ OH \end{bmatrix}$: glyceride

R' = carbon chain of fatty acid

R = alkyl group of the alcohol

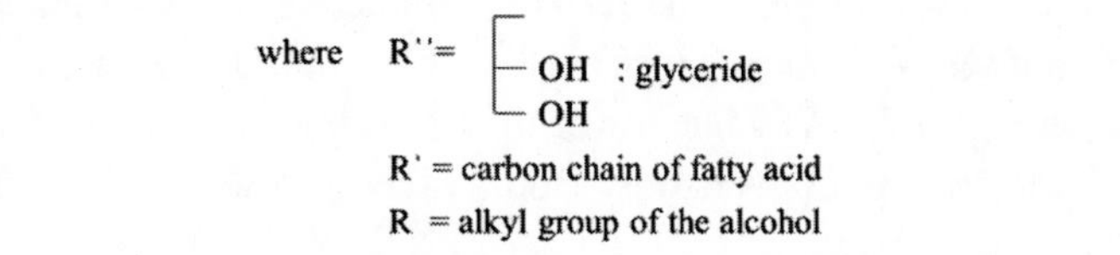

Figure 6. Mechanism of acid-catalyzed transesterification.

2.4. Variables Affecting the Transesterification Reaction

The process of transesterification is affected by various factors depending upon the reaction conditions used [3, 10–14]. The effects of such factors are discussed below.

2.4.1. Catalyst Type and Concentration

The transesterification of triglycerides can be catalyzed by alkali, acid, enzyme or heterogeneous catalysts, among which alkali catalysts like sodium hydroxide, sodium methoxide, potassium hydroxide, potassium methoxide are considered the most effective ones. If the oil has high free fatty acid content and more water, acid catalyzed transesterification is favorable. Acids such as sulfuric, phosphoric, hydrochloric or organic sulfonic are suitable for this process.

As a catalyst in the process of alkaline methanolysis, sodium hydroxide or potassium hydroxide has been extensively used. Refined and crude oils with either sodium hydroxide or potassium hydroxide catalyst resulted in appreciable conversions. Attempts have been made to use basic alkaline-earth metal compounds in the transesterification of rapeseed oil for production of fatty acid methyl esters.

The reaction appears to proceed if methoxide ions are present in the reaction medium. Alkaline-earth metal hydroxides, alkoxides and oxides catalyze this process but produce low reaction rates as the reaction mixture forms a three-phase system oil–methanol-catalyst, which becomes highly limited by diffusion reasons.

Although chemical transesterification using alkaline catalysts gives high conversion levels of triglycerides to their corresponding methyl esters in short reaction times, the reaction exhibits several drawbacks. Some of them are: high energy consumption, difficult glycerol recovery, hard separation of catalyst from the product, required treatment of alkaline waste water, and interference of free fatty acids and water with the reaction.

Enzymatic catalysts are able to effectively catalyze the transesterification of triglycerides in either aqueous or non-aqueous systems, which can overcome the problems mentioned above.

The glycerol formed can be easily removed without any complex process, and also the free fatty acids contained in waste oils and fats can be completely converted to alkyl esters. Unfortunately, the production cost of this type of catalyst is in general significantly greater than that of an alkaline one.

2.4.2. Free Fatty Acids and Moisture

Free fatty acid and moisture content are key parameters for determining the viability of the vegetable oil transesterification process. To carry out the base-catalyzed reaction to completion a free fatty acid (FFA) value lower than 3% is needed. It has been found that high oil acidity levels lead to low conversions. Both, excess as well as insufficient amount of catalyst may cause soap formation.

The starting materials used for base-catalyzed alcoholysis should meet certain specifications. Thus, triglycerides should have lower acid value and all materials are expected to be substantially anhydrous. The excess of sodium hydroxide catalyst compensates for higher acidity, but the resulting soap leads to an increase in the viscosity or the formation of gels that interferes with both the reaction and the separation of glycerol. When the reaction conditions are not appropriate, ester yields are significantly reduced. The methoxide and hydroxide of sodium or potassium should be maintained in anhydrous state. Prolonged contact with air will affect the effectiveness of these catalysts due to their interaction with moisture and carbon dioxide.

2.4.3. Molar Ratio of Alcohol to Oil and Type of Alcohol

The stoichiometric ratio for transesterification requires three moles of alcohol and one mole of triglyceride to yield three moles of fatty acid alkyl esters and one mole of glycerol. However, transesterification is an equilibrium reaction in which a large excess of alcohol is required to promote the forward reaction. To maximize the conversion to the ester, a molar ratio of 6:1 should be used. The molar ratio has no effect on acid, peroxide, saponification and iodine value of methyl esters. However, the high molar ratio of alcohol to vegetable oil interferes with the separation of glycerin because there is an increase in solubility. When glycerin remains in solution, the reverse reaction is favored lowering the yield of esters.

The base catalyzed formation of ethyl ester is difficult compared to the formation of methyl esters. Specifically the formation of stable emulsion during ethanolysis is a problem. Methanol and ethanol are not miscible with triglycerides at ambient temperature, and the reaction mixtures are usually mechanically stirred to enhance mass transfer.

2.4.4. Effect of Reaction Time and Temperature

The conversion increases with reaction time. Some investigators studied the effect of reaction time on transesterification of beef tallow with methanol. The reaction was very slow during the first minute due to mixing and dispersion of methanol into beef tallow. A significant increase in the rate of reaction was observed from 1 to 5 min. The production of beef tallow methyl esters reached the maximum value at about 15 min.

Transesterification can occur at different temperatures, depending on the oil used. For the transesterification of refined oil with methanol, the reaction was studied with three different temperatures. After 0.1 h, the measured ester yields were 94, 87 and 64% for 60, 45 and 32°C, respectively. After 1 h, the formation of esters was identical at 60 and 45°C runs and only slightly lower at 32°C.

2.4.5. Mixing Intensity

Mixing is very important in the transesterification process, as oils or fats are immiscible with sodium hydroxide–methanol solution. Once the two phases are mixed and the reaction is started, stirring is no longer needed. The effect of mixing on the transesterification of beef tallow has been matter of research. The experimental findings showed no reaction in absence of mixing. However, when NaOH–MeOH was added to the melted beef tallow in the reactor while stirring, the stirring speed was insignificant. Hence, the reaction time was the controlling factor in determining the yield of methyl esters. This suggested that the stirring speeds investigated exceeded the threshold requirement of mixing.

2.5. Transesterification under Supercritical Conditions

The direct methanolysis of triglycerides with enzymatic catalyst in flowing supercritical carbon dioxide has been also investigated. Corn oil and methanol were pumped into the carbon dioxide stream to yield fatty acid methyl ester >98% [8]. Direct methanolysis of soy flakes gives FAME at similar yield. Transesterification of rapeseed oil in supercritical methanol was investigated although without the use of catalyst. It was demonstrated that, in a preheating temperature of 350°C, 450 seconds of supercritical treatment of methanol was sufficient to convert the rapeseed oil to methyl esters and that, although the prepared methyl esters were basically the same as those of the common method with a basic catalyst, the yield of methyl esters by the former was found to be higher than that by the later [3]. In addition, it was found that this new supercritical methanol process requires the shorter reaction time and simpler purification procedure because of the unused catalyst.

CONCLUSION

Biodiesel has recently received much attention because of its environmental benefits and the fact that it can be obtained from renewable sources. Most of the investigations focused on the production of biodiesel have been performed on edible oils and only very few on non-edible sources. Transesterification appears to be the most widely used method to synthesize biodiesel. The process of transesterification seems to be affected by several factors such as reaction time, temperature, type of alcohol, among others.

REFERENCES

[1]   Balat M, Potential alternatives to edible oils for biodiesel production–A review. *Energy Conversion and Management* 2011; 52:1479–1492.

[2]   Singh S, Singh D, Biodiesel production through the use of different sources and charcaterization of oils and their esters as the substitute of diesel: A review. *Renewable and Sustainable Energy Reviews* 2010; 14:200–216.

[3]   Meher L, Vidya Sagar D, Naik S, Technical aspects of biodiesel production by transesterification–a review. *Renewable and Sustainable Energy Reviews* 2006; 10:248–268.

[4]   Ching Juan J, Agung Kartika D, Yeong Wu T, Yun Hin T, Biodiesel production from jathropa oil by catalytic and non-catalytic approaches: *An overview. Bioresource Technology* 2011;102:452–460.

[5]   Ying Koh M, Idaty Mohd.Gazhi T, A review of biodiesel production from Jathropa curcas L. oil, *Renewable and Sustainable Energy Reviews* 2011; 15:2240–2251.

[6]   Xiao G, Gao L, *Biofuel Production–Recent Developments and Prospects,* DOI:10.5772/959.

[7]   Huang G, Chen F, Wei D, Zhang X, Chen G, Biodiesel production by microalgal biotechnology. *Applied Energy* 2010; 87:38–46.

[8]   Ma F, Hanna M, Biodiesel production: a review, *Bioresource Technology* 1999; 70:1–15.

[9]   Van Gerpen J, Biodiesel processing and production, *Fuel Processing Technology* 2005; 86:1097–1107.

[10] Ameer Basha S, Raja Gopal K, Jebaraj S, A review on biodiesel production, combustion, emissions and performance, *Renewable and Sustainable Energy Reviews* 2009; 136:1628–1634.

[11] Phan A, Phan T, Biodiesel production from waste cooking oils, *Fuel* 2008; 87:3490–3496.

[12] Bhatti H, Hanif M, Qasim M, Rehman A, Biodiesel production from waste tallow, *Fuel* 2008; 87:2961–2966.

[13] Bajaj A, Lohan P, Jha P, Mehrotra, Biodiesel production through lipase catalyzed transesterification: An overview, *Journal of Molecular Catalysis B: Enzymatic* 2010; 62:9–14.

[14] Gog A, Roman M, Tosa M, Paizs C, Irimie F, Biodiesel production using enzymatic transesterification–Current state and perspectives, *Renewable Energy* 2012; 39:10–16.

In: Biomass Processing, Conversion and Biorefinery   ISBN: 978-1-62618-346-9
Editors: Bo Zhang and Yong Wang   © 2013 Nova Science Publishers, Inc.

Chapter 15

UTILIZATION OF *CEIBA PENTANDRA* SEED OIL AS POTENTIAL FEEDSTOCK FOR BIODIESEL PRODUCTION

S. Yusup[1,], M. M. Ahmad[1], Y. Uemura[1], S. Abu Bakar[1], R. Nik Mohamad Kamil[1], A. T. Quitain[2] and S. Shari[1]*

[1]Biomass Processing Laboratory, Centre for Biofuel and Biochemical Research, Green Technology Mission Oriented Research, Universiti Teknologi Petronas, Bandar Seri Iskandar, Tronoh, Perak, Malaysia
[2]Graduate School of Science and Technology, Kumamoto University, Kurokami Kumamoto, Japan

ABSTRACT

The utilization of *Ceiba Pentandra* seed oil as potential feedstock for biodiesel production has been explored. About 28% wt/wt of *Ceiba Pentandra* seed oil was obtained from the extraction process. *Ceiba Pentandra* seed oil which contains high free fatty acid (FFA) was pre-treated using acid-catalyzed esterification process. Transesterification process had been carried out using L_9 Taguchi method. It was found that, the highest yield of *Ceiba Pentandra* biodiesel which is 90.52% was obtained at reaction temperature of 70°C and methanol-to-oil molar ratio of 10:1. The biodiesel property of *Ceiba Pentandra* seed oil Methyl Ester has been characterized and it meets the properties of biodiesel as stated in standard method of ASTM D6751 and EN14214.

Keywords: *Ceiba Pentandra*, Biodiesel, Free Fatty Acid, Esterification, Transesterification, Taguchi

---

* Phone +605368 8208. Fax +6053688205. E-mail: drsuzana_yusuf@petronas.com.my.

1. INTRODUCTION

The current energy supply originates mainly from renewable fossil fuels such as petroleum, coal and natural gas. Their consumption are more rapid than their production. Due to the environmental concerns and its limited supply, researchers are exploring alternative sources such as diesel fuels. "Biodiesel" which is scientifically known as Fatty Acid Methyl Esters (FAME) is one of the alternatives for fossil fuels. Biodiesel is becoming more favorable nowadays as an alternative fuel, due to the possible depletion of fossil oil. Studies on biodiesel have been conducted since decades ago [1, 2]. Unlike fossil fuel, available sources for biodiesel production are unlimited, since it is produced from renewable sources. Hence, they are able to cope with energy demand consumption that is increasing rapidly. It is biodegradable, non-toxic and has low emission profiles as well as environmental friendly [3]. Few studies have been conducted on biodiesel production based on edible oil and non edible oil [4]. Palm oil and sunflower oil are examples of common edible oil that are widely used as the feedstock for biodiesel. However, in order to prevent the food versus fuel crisis, more sources of non edible oil are studied as potential feedstock for biodiesel [5]. There are four methods to produce biodiesel that include the direct use and blending, microemulsions, thermal cracking (pyrolysis), and transesterification [6]. In most cases, transesterification process has become the most preferable method [7]. The purpose of transesterification is to lower the viscosity of the oil. This process reduces the viscosity to a value that is comparable to that of diesel, hence improves combustion [8]. Transesterification is a chemical reaction in which triglyceride reacts with alcohol in the presence of catalyst [9]. During the reaction, the long chain and branched chain fatty acid molecules are converted to Mono Ester by transesterification [1]. The transesterification reaction consists of reversible steps; the conversion of Triglycerides to Diglycerides, followed by Diglycerides to Monoglycerides. The reaction is as shown in the following equation (1) [10].

$$
\begin{array}{ll}
CH_2O-\overset{\overset{O}{\|}}{C}-R & \\
CH-O-\overset{\overset{O}{\|}}{C}-R \quad +CH_3OH \quad \underrightarrow{OH^-} \quad 3CH_3O-\overset{\overset{O}{\|}}{C}-R \quad + & CH_2OH \\
CH_2O-\overset{\overset{O}{\|}}{C}-R & CH-OH \\
\quad\quad\quad\quad\quad\quad\quad\quad \text{Alcohol} \quad \text{Catalyst} \quad \text{Esters} & CH_2OH \\
\text{Glyceride} & \text{Glycerol}
\end{array}
$$

(1)

Source: A. Demirbas, *Biodiesel: A Realistic Fuel Alternative for Diesel Engines*. Springer-Verlag London Limited, 2008 [10].

BIODIESEL FEEDSTOCK

Biodiesel feedstock can originate from an unlimited amount of sources such as vegetable oil, animal fat, algae and oil from different type of seeds. In western countries such as US and Europe, edible oils such as soybean oil, rapeseed oil and sunflower oil are utilized for biodiesel production [11].

More than 95% of the biodiesel feedstock comes from edible oil which will contribute towards global imbalance in the food supply and market demand. This world scenario is called the "food versus fuel" in which the consumption of edible oil for food supply competes with biodiesel production. The increase in edible oil price and the perception that the utilization of edible oil cause for the price increase have prompted utilization of non-edible oil as the feedstock for biodiesel production [5]. Non edible feedstocks such as Moringa Oleifera seed oil, Tobacco seed oil, Jatropha Curcas seed oil and Castor seed oil have been researched [12-14].

It is found that *Ceiba Pentandra* seed oil has the potential to be one of biodiesel feedstock [15]. *Ceiba Pentandra* seed oil is locally known as *Kapok* or *Kekabu,* and is abundance in Western India, Malaysia, Vietnam, Indonesia and Philippines. *Ceiba Pentandra* is a humid tree of the Malvales and a family of Malvaceae. In Malaysia, *Ceiba Pentandra* seed oil is commonly found in the northern parts of Peninsular Malaysia as shown in Figure 1 [17]. The tree can produce between 500 to 4000 fruits at one time, with each fruit containing 200 seeds. [18].

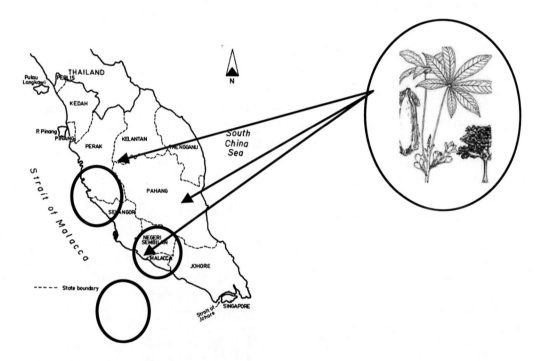

Figure 1. *Ceiba Pentandra* trees plantations in Peninsular of Malaysia [17].

The fiber contains almost 64% cellulose [18]. The seeds are brownish black in color and contain 25-28% of oil [19].

In 1964 up to 1966, several researches had studied about the *Ceiba Pentandra* seed oil content as well as the fatty acid composition inside the oil [20]. They reported that the *Ceiba Pentandra* seed oil contains Cyclopropenoid Fatty Acids (CPFA) which can cause various types of physiological disorders in a farm. Therefore, the seed of *Ceiba Pentandra* is not suitable for consumption.

Figure 2. *Ceiba Pentandra* seeds surrounded by its lint [16].

3. PHYSIOCHEMICAL PROPERTIES OF *CEIBA PENTANDRA*

Physically, *Ceiba Pentandra* seed oil has similar properties as cotton seed oil [19]. Table 1 shows the physiochemical properties of *Ceiba Pentandra* seed oil which has been analyzed by Berry.K, [18], Kurniawan, W [15] and Salimon, J et al., [16]. The *Ceiba Pentandra* seed oil is characterized for its density, water content, refractive index, saponification value and iodine value. Feedstock characterization is performed to determine its potential as viable feedstock for biodiesel production.

Table 1. Physiochemical Properties of *Ceiba Pentandra* seed oil

Property	Average Value		
	Berry, K (1979) [18]	Kurniawan, W (2004) [15]	Salimon, J et al. (2005) [16]
Density at 15°C (kg/L)	-	0.910-0.912	-
Water content (%)	3.4	-	0.04
Refractive Index (at 25°C)	1.4656	1.466-1.472	1.46
Saponification value (mg/g)	183	189-225	208
Iodine value	94.98	85-110	104

Density value is required in order to estimate the cetane number of biodiesel [21]. Density specification for biodiesel is included in the European standard (EN 14214) and American Standard (ASTM D5002-94). Yusup.S et al. [22] reported that the density of Rubber Seed Oil from 910 kg/m^3 was reduced to 874 kg/ m^3 when converted to rubber seed oil methyl ester (RSOME) during transesterification process [23]. High water content in biodiesel can cause many problems such as microbial growth in the transportation equipment and corrosion in fuel tank [5]. Therefore, water content value in biodiesel needs to be determined and reduced to avoid all the problems. There common method to determine the water content of biodiesel is using Karl-Fischer method which follows EN 14214 standard method [10]. Saponification value belongs to all fatty acids that are present in oil. It depends on the molecular weight and fatty acid concentration percentage present in FAMEs. Saponification value is required in order to determine the average molecular mass of oils and fats [24]. Iodine value determination is also important since it can determine the percentage of unsaturated fatty acid components of the oil. It is also relate to the oxidative stability and reflects the tendency of oil to oxidise and form engine deposits [24].

3.1. Fatty Acid Composition of *Ceiba Pentandra* Seed Oil Methyl Esters

Fatty acid composition is important in order to identify the percentage of saturated and unsaturated fatty acid in the oil. Gas Chromatography - Mass Spectroscopy (GC-MS) can analyze the percentage of fatty acid of the oil. GC system is equipped with a Triple Axis inert XL EI/CI mass selective detector and capillary column RT-2500 (100 x 0.25 mm; film thickness 0.20 μmeter). 1.0 μL of sample was injected into the column using a split mode (split ratio 1:100). Helium was used as a carrier gas at 1.2 ml/min. The column oven temperature was adjusted from 150°C to 250°C at a linear ramp rate of 4°C/min. The initial and final hold up time was 1 and 5 minute respectively. For GC/MS detection, an electron ionization method was used with the ionization energy of 70 eV.

When *Ceiba Pentandra* seed oil is converted to biodiesel through the transesterification reaction, the free fatty acid of the oil is also converted to Fatty Acid Methyl Esters (FAME). The fatty acid composition of Ceiba Pentandra seed oil has been analyzed by Berry, K in 1979 [18] and Rossell and Pritchard in 1991 [25] and is shown in Table 2. The value of fatty acid was determined in weight percentage. It shows that the Ceiba Pentandra seed oil contains high linoleic acid which is unsaturated fatty acid. It is strongly proven the analysis showed a high percentage of Linoleic acid in Ceiba Pentandra seed oil. Thus this oil property makes it the excellent oil for engine performance during cold weather after its conversion to biodiesel [22, 26, 27].

Table 2. Fatty acid composition of Ceiba Pentandra seed oil [18]

Ceiba Pentandra seed oil	Fatty Acid Composition (wt %)		
	Berry, K (1979) [18]	Rossell & Pritchard (1991) [25]	This work (2011)
Caproic acid (C6)	-	-	9.42
Myristic acid (C14)	0.25	-	-
Palmitic acid (C16)	24.31	20.90	23.17
Palmitoleic acid (C16:1)	0.40	-	-
Stearic acic (C18)	2.65	13.70	4.73
Oleic acid (C18:1)	21.88	20.80	22.88
Linoleic acid (C18:2)	38.92	42.60	30.00
Linolenic acid (C18:3)	1.00	1.9	-
Arachidic acid (C20)	-	-	1.18

4. PRE-TREATMENT OF HIGH FFA *CEIBA PENTANDRA* SEED OIL USING ACID-CATALYZED ESTERIFICATION PROCESS

Generally, a one step process can be applied using acid as the catalyst. H_2SO_4 is the acid catalyst which is commonly used for the reaction. However the reaction is relatively slow even though it can give higher biodiesel yield compared to transesterification reaction using base [28]. Base-catalyzed transesterification is much more effective in terms of reaction rate [29]. Esterification reaction using acid as catalyst is one of the techniques to neutralize the high FFA content in the biodiesel feedstock [30]. When base catalyst is introduced in the transesterification of feedstock which contains high FFA, the FFA will be converted into soap and it will further inhibit the separation of the Alkyl Esters and Glycerol [31]. These soap and emulsion formations make the separation of product and by-product of biodiesel difficult [32]. In order to convert vegetable oil to ester effectively, the FFA content of the feedstock for the alkaline-catalyzed transesterification reaction should not exceed 1% [33]. Otherwise, saponification reaction of oil and fat with base catalyst will take part in the transesterification reaction.

To avoid the saponification reaction from occurring, esterification reaction is applied using acid catalyst to reduce the FFA content in the oil below 1%. This is followed by base catalyzed transesterification reaction using the treated oil [27, 34]. The determination of acid value, as well as the FFA content before pre-treatment is essential for efficient transesterification process. The acid esterification reaction of *Ceiba Pentandra* seed oil was carried out at 65°C for 3 hr in a sohxlet extractor using 10 to 1 methanol-to-oil ratio as well as 0.5 and 1.0 wt/wt of sulfuric acid as catalyst. This process was repeated until the acid value of the sample was less than 2 mg KOH/ g of oil or below 1% of FFA content.

5. TRANSESTERIFICATION OF *CEIBA PENTANDRA* SEED OIL USING L₉ TAGUCHI APPROACH

The transesterification experiment of *Ceiba Pentandra* seed oil has been designed using Taguchi method. [35] The standard L_9 orthogonal array was chosen to determine the optimum condition for this process. The values of methanol-to-oil ratio, amount of catalyst, reaction temperature, and reaction time are varied as presented in Table 3. Each run is verified by repetition.

50 g of *Ceiba Pentandra* seed oil was heated in a round bottom flask at a desired temperature using heating plate. The methanol and potassium hydroxide (CH_4O-KOH) solution of a specific amount was then added into the round bottom flask and stirred at the desired temperature. Once the transesterification reaction was completed, the mixture was left overnight for complete separation. Two layers of immiscible phase were obtained. The upper layer consisted of *Ceiba Pentandra* seed oil Methyl Esters and the lower layer contained Glycerol, excess CH_4O, and unreacted catalyst. The *Ceiba Pentandra* biodiesel was purified with warm deionized water to remove the residual catalyst. Anhydrous Na_2SO_4 was then added to the Methyl Esters to absorb the residual water. The mixture was then filtered to obtain pure biodiesel. The by-product at the lower layer was removed.

Table 3. Transesterification parameters using L₉ Taguchi method

Run	Methanol-to-oil molar ratio	Catalyst amount (wt %)	Temperature (°C)	Time (hr)
1	6	0.5	60	1
2	6	1.0	65	2
3	8	0.5	65	3
4	8	2.0	60	2
5	10	2.0	65	1
6	10	1.0	60	3
7	10	0.5	70	2
8	8	1.0	70	1
9	6	2.0	70	3

Table 4. *Ceiba Pentandra* biodiesel yield percentage for 9 runs

Run	Methanol-to-oil molar ratio	Catalyst amount (wt %)	Temperature (°C)	Time (hr)	Yield (%)
1	6	0.5	60	1	74.43
2	6	1.0	65	2	26.74
3	8	0.5	65	3	64.86
4	8	2.0	60	2	72.91
5	10	2.0	65	1	75.00
6	10	1.0	60	3	64.83
7	10	0.5	70	2	90.52
8	8	1.0	70	1	74.83
9	6	2.0	70	3	76.23

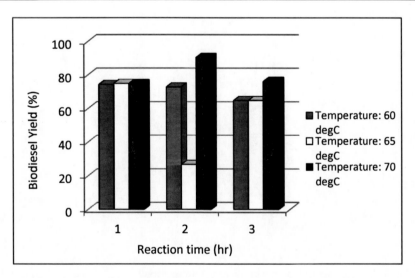

Figure 3. Graph of *Ceiba Pentandra* biodiesel versus reaction time at different temperature (Methanol-to-oil molar ratio of 10 to 1 and Catalyst amount at 0.5 wt %).

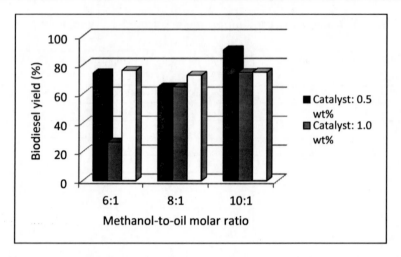

Figure 4. Graph of *Ceiba Pentandra* biodiesel versus methanol to oil molar ratio at different catalyst weight (Temperature of 70°C and reaction time at 2 hr).

Table 4 lists the *Ceiba Pentandra* biodiesel yield for all the trials. Based on the result, the highest yield of *Ceiba Pentandra* biodiesel was obtained in run number 7, where 90.52% of biodiesel was produced. The reaction occurs at 70°C, using 10:1 methanol-to-oil ratio and 0.5wt% of KOH as catalyst, for 2 hrs.

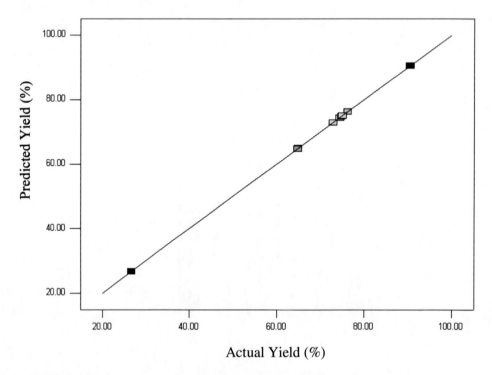

Figure 5. Relationship between actual and predicted yield biodiesel.

Figure 3 and 4 shows the result of the analysis of variance (ANNOVA). They show the yield obtained under the effect of temperature, catalyst weight, methanol-to-oil molar ratio and reaction time. Figure 3 shows that at 70°C, the yield increases to 90.52 within 2 hrs of reaction time. Increasing the reaction time to 3 hrs does not increase the yield further. In Figure 4, by increasing the methanol-to-oil molar ratio and the catalyst weight, the yield increases but increasing more catalyst amount to 2 wt% does not increase the yield further. At 0.5 wt% of catalyst weight and 10 to 1 of methanol-to-oil molar ratio, the maximum yield obtained is 90.52%.

Figure 5 shows the relation between the predicted *Ceiba Pentandra* biodiesel yield and its actual yield. The data shows good agreement with each other as represented by the straight line.

Based on Table 5, temperature has the highest effect compared to other parameters. Therefore, it can be concluded that temperature has the most significant influence on the reaction, while time gives the least effect.

Table 5. Average signal to noise ratio

Level	Methanol-to-oil molar ratio	Catalyst Weight (wt%)	Temperature (°C)	Time (min)
1	34.54	37.60	36.98	37.47
2	36.99	34.09	34.09	34.98
3	37.62	37.47	38.08	36.71
Rank	3	2	1	4

6. FUEL PROPERTIES OF *CEIBA PENTANDRA* SEED OIL METHYL ESTERS FOLLOWING INTERNATIONAL STANDARD (ASTM D 6751 AND EN 14214)

The important part in producing the biodiesel is that its quality must conform to the biodiesel standard methods of ASTM D 6751 and EN 14214. *Ceiba Pentandra* biodiesel properties such as density, cetane number, oxidative stability and kinematic viscosity have been characterized following the biodiesel standard method listed in Table 6. The biodiesel cetane number expresses the ignition property of biodiesel similar to cetane number in fossil base diesel. It is related to the ignition delay time a fuel experiences upon injection into the combustion chamber. The cetane number decreases with decreasing chain length and increasing degree of un-saturation and branching. Higher cetane number results in shorter ignition delay which makes the fuel easier to ignite. In this study, the cetane number for *Ceiba Pentandra* seed oil methyl esters is 57.52. The minimum cetane number prescribed in ASTM D6751 and EN 14214 is 47 and 51 respectively.

Higher heating value (HHV) or calorific value is the measurement of the energy produced during biodiesel burning. This property is related to the stability of biodiesel. In this analysis, HHV of *Ceiba Pentandra* seed oil Methyl Ester which was measured in a bomb calorimeter was 39.4 MJ/kg as prescribed in the ASTM D2015. The oxidative stability of biodiesel was analyzed following EN 14112 which is 6 and 3hr (minimum). The oxidation occurs due to the presence of air, eminent temperature and presence of metal content which promote the oxidation of the unsaturated bonds. The induction time of *Ceiba Pentandra* seed oil methyl esters was determined to be 12.3 hr.

The flow of biodiesel consists of cloud point (CP), pour point (PP) and cloud filter plugging point (CFPP). These properties are temperature dependant and the limits of cloud point are not specified in both the ASTM and EN standards since the value is associated with the weather condition and the place where the test is conducted. Furthermore, the biodiesel blend standard of ASTM D7467 does not have any temperature limits. The EN 14214 standard also does not mention a low temperature parameter in its specifications. However, it discusses the use of a low-temperature filterability test which is cold filter plugging point (CFPP). CP, CFPP and PP values of *Ceiba Pentandra* seed oil Methyl Esters are 4.4, 4 and 1°C respectively. From previous research work which has been done on rapeseed biodiesel, the rapeseed biodiesel has lower values of CP and PP which are CP -3°C and PP 9°C [36]. Jatropha, pongamia and palm biodiesel have been tested on its cloud point and pour point by Sarin et al. [37]and it shows that Palm, Jatropha, and Pongamia biodiesels have cloud points of 16.0, 4.0, and −1.0 °C and pour points of 12.0,−3.0, and −6.0 °C respectively. The low temperature properties of *Ceiba Pentandra* seed oil Methyl Esters can be improved by using different types of additives.

Kinematic viscosity is a measurement of fluid's thickness or resistance to flow over time. In determining the biodiesel quality, kinematic viscosity is one of the important elements which need to be monitored since it continuously increases with decreasing biodiesel's oxidative stability [5]. Higher oil viscosity causes operational problem such as the formation of large droplets on injection which results in poor combustion, engine deposits as well as increase of the exhaust smoke and gas emission.

Table 6. Properties of *Ceiba Pentandra* biodiesel following the standard method [38]

Property	Unit	Ceiba Pentandra Biodiesel	Biodiesel Standards (ASTM D6751)	Biodiesel Standards (EN 14214) [38]
Density at 15°C	Kg/m^3	880.7	-	-
Cetane number	-	57.52	47 min	51 min
Higher heating value	(MJ/kg)	39.4	-	-
Water content	%	0.05	<0.03	<0/.05
Oxidative stability	hr	12.3	3 min	6 min
Cloud point	°C	4.4	5	a
Pour point	°C	1	-15	b
Cloud filter plugging point	°C	4	-	b
Flash point	°C	145	93 min	120 min
Kinematic Viscosity at 40°C	mm^2/s	1.8	1.9-6	3.5-5
Sulphur content	wt%	0.024	0.05 max	-
Acid value	mg KOH/ mg oil	0.4	<0.5	<0.8

CONCLUSION

Fossil fuel, which is the main source for energy supply is depleting. Thus, biodiesel has the potential to substitute the fossil fuel in order to cope with the energy demand. Biodiesel is produced from renewable sources, and there are numerous sources available worldwide. Biodiesel is environmental friendly since it emits low carbon pollutant, biodegradable and non-toxic. *Ceiba Pentandra* seed oil has been identified as a potential source of biodiesel since its properties are comparable and satisfy the standard for biodiesel (EN 14214 and ASTM D6751). Parametric studies have been conducted for its utilization in both esterification and transesterification process. The highest yield of *Ceiba Pentandra* biodiesel is obtained using a methanol-oil ratio of 10:1, 0.5 wt% of catalyst, at 70°C, and reaction time of 2 hr.

REFERENCES

[1]   F. Ma and M. A. Hanna, "Biodiesel production: a review," *Bioresour. Technol.,* vol. 70, pp. 1-15, 10, 1999.

[2]   E. M. Shahid and Y. Jamal, "Production of biodiesel: A technical review," *Renewable and Sustainable Energy Reviews,* vol. 15, pp. 4732-4745, 12, 2011.

[3]   A. Demirbas, "Importance of biodiesel as transportation fuel," *Energy Policy,* vol. 35, pp. 4661-4670, 9, 2007.

[4]   H. C. Ong, T. M. I. Mahlia, H. H. Masjuki and R. S. Norhasyima, "Comparison of palm oil, Jatropha curcas and Calophyllum inophyllum for biodiesel: A review," *Renewable and Sustainable Energy Reviews,* vol. 15, pp. 3501-3515, 10, 2011.

[5]   G. Knothe, J. Krahl and J. V. and Gerpen, Eds., *The Biodiesel Hanbook.* United States of America: AOCS Press, 2010.

[6]   M. Balat and H. Balat, "Progress in biodiesel processing," *Appl. Energy,* vol. 87, pp. 1815-1835, 6, 2010.

[7]   D. Y. C. Leung, X. Wu and M. K. H. Leung, "A review on biodiesel production using catalyzed transesterification," *Appl. Energy,* vol. 87, pp. 1083-1095, 4, 2010.

[8]   M. C. Navindgi, M. Dutta and B. Sudheer, "Performance Evaluation, Emission Characteristics and Economic Analysis of Four Non-Edible Straight Vegetable Oils on a Single Cylinder CI Engine," *ARPN Journal of Engineering and Applied Sciences,* vol. 7, pp. 173-179, 2012.

[9]   D. Ayhan, "Biodiesel from waste cooking oil via base-catalytic and supercritical methanol transesterification," *Energy Conversion and Management,* vol. 50, pp. 923-927, 4, 2009.

[10]   A. Demirbas, *Biodiesel: A Realistic Fuel Alternative for Diesel Engines.* Springer-Verlag London Limited, 2008.

[11]   A. S. Ramadhas, S. Jayaraj and C. Muraleedharan, "Use of vegetable oils as I.C. engine fuels—A review," *Renewable Energy,* vol. 29, pp. 727-742, 4, 2004.

[12]   U. Rashid, F. Anwar, M. Ashraf, M. Saleem and S. Yusup, "Application of response surface methodology for optimizing transesterification of Moringa oleifera oil:

Biodiesel production," *Energy Conversion and Management,* vol. 52, pp. 3034-3042, 8, 2011.

[13] P. Berman, S. Nizri and Z. Wiesman, "Castor oil biodiesel and its blends as alternative fuel," *Biomass Bioenergy,* vol. 35, pp. 2861-2866, 7, 2011.

[14] N. Usta, B. Aydoğan, A. H. Çon, E. Uğuzdoğan and S. G. Özkal, "Properties and quality verification of biodiesel produced from tobacco seed oil," *Energy Conversion and Management,* vol. 52, pp. 2031-2039, 5, 2011.

[15] W. Kurniawan, "Optimasi rasio berat kalium hidroksida terhadap minyak biji kapuk (ceiba pentandra L.gaertn) dalam reaksi transesterifikasi menggunakan etanol," Fakultas Matematika Dan Ilmu Pengetahuan Alam, Universitas Diponegoro, 2004.

[16] J. Salimon and K. A. A. and Kadir, "Fatty Acid Composition and Physicochemical Properties in Kekabu Seed Oil," vol. 2, pp. 117-120, 2005.

[17] D. Zina, "The Kapok Connection: Study Explains Rainforest Similarities," *National Science Foundation,* vol. 2012, Retrieved November 7, 2012, from http://www. sciencedaily.com /releases/2007/06/070615152639.htm, 2007.

[18] S. Berry K., "The Characteristics of the Kapok (Ceiba pentadra,Gaertn.) Seed Oil," *Pertanika,* vol. 2, pp. 1-4, 1979.

[19] Q. Shu, Q. Zhang, G. Xu, Z. Nawaz, D. Wang and J. Wang, "Synthesis of biodiesel from cottonseed oil and methanol using a carbon-based solid acid catalyst," *Fuel Process Technol,* vol. 90, pp. 1002-1008, 0, 2009.

[20] S. F. Dias, D. G. Valente and J. M. F. Abreu, "Comparison between ethanol and hexane for oil extraction from Quercus suber L. Fruits," *Grasas y Aceites,* vol. 54, pp. 378-383, 2003.

[21] A. Srivastava and R. Prasad, "Triglycerides-based diesel fuels," *Renewable and Sustainable Energy Reviews,* vol. 4, pp. 111-133, 6, 2000.

[22] S. Yusup and M. Khan, "Basic properties of crude rubber seed oil and crude palm oil blend as a potential feedstock for biodiesel production with enhanced cold flow characteristics," *Biomass Bioenergy,* vol. 34, pp. 1523-1526, 10, 2010.

[23] S. Yusup and M. A. Khan, "Base catalyzed transesterification of acid treated vegetable oil blend for biodiesel production," *Biomass Bioenergy,* vol. 34, pp. 1500-1504, 10, 2010.

[24] C. J. Jan, P. Natale and C. Stefano, *Biodiesel Science and Technology: From Soil to Oil.* Washington, DC: Woodhead publishing limited, 2010.

[25] J. B. Rossell and J. L. R. and Pritchard, Eds., *Analysis of Oilseeds, Fats and Fatty Foods.* Northern Ireland: The Universities Press (Belfast) Ltd, 1991.

[26] M. Çetinkaya, Y. Ulusoy, Y. Tekìn and F. Karaosmanoğlu, "Engine and winter road test performances of used cooking oil originated biodiesel," *Energy Conversion and Management,* vol. 46, pp. 1279-1291, 5, 2005.

[27] G. Jon Van, "Biodiesel processing and production," *Fuel Process Technol,* vol. 86, pp. 1097-1107, 6/25, 2005.

[28] M. A. Dubé, A. Y. Tremblay and J. Liu, "Biodiesel production using a membrane reactor," *Bioresour. Technol.,* vol. 98, pp. 639-647, 2, 2007.

[29] J. Ye, S. Tu and Y. Sha, "Investigation to biodiesel production by the two-step homogeneous base-catalyzed transesterification," *Bioresour. Technol.,* vol. 101, pp. 7368-7374, 10, 2010.

[30] A. Hayyan, F. S. Mjalli, M. A. Hashim, M. Hayyan, I. M. AlNashef, S. M. Al-Zahrani and M. A. Al-Saadi, "Ethanesulfonic acid-based esterification of industrial acidic crude palm oil for biodiesel production," *Bioresour. Technol.,* vol. 102, pp. 9564-9570, 10, 2011.

[31] M. A. Khan, S. Yusup and M. M. Ahmad, "Acid esterification of a high free fatty acid crude palm oil and crude rubber seed oil blend: Optimization and parametric analysis," *Biomass Bioenergy,* vol. 34, pp. 1751-1756, 12, 2010.

[32] M. Di Serio, R. Tesser, L. Pengmei and Santacesaria., "Heterogenous catalyst for biodiesel production," *Energy and Fuels,* pp. 207-217, 2008.

[33] A. Hayyan, M. Z. Alam, M. E. S. Mirghani, N. A. Kabbashi, N. I. N. M. Hakimi, Y. M. Siran and S. Tahiruddin, "Reduction of high content of free fatty acid in sludge palm oil via acid catalyst for biodiesel production," *Fuel Process Technol,* vol. 92, pp. 920-924, 5, 2011.

[34] P. Felizardo, M. J. Neiva Correia, I. Raposo, J. F. Mendes, R. Berkemeier and J. M. Bordado, "Production of biodiesel from waste frying oils," *Waste Manage.,* vol. 26, pp. 487-494, 2006.

[35] P. J. Ross, Ed., *Taguchi Technqiues for Quality Engineering.* McGraw-Hill Professional, 1995.

[36] S. A. Biktashev, R. A. Usmanov, R. R. Gabitov, R. A. Gazizov, F. M. Gumerov, F. R. Gabitov, I. M. Abdulagatov, R. S. Yarullin and I. A. Yakushev, "Transesterification of rapeseed and palm oils in supercritical methanol and ethanol," *Biomass Bioenergy,* vol. 35, pp. 2999-3011, 7, 2011.

[37] A. Sarin, R. Arora, N. P. Singh, R. Sarin, R. K. Malhotra and K. Kundu, "Effect of blends of Palm-Jatropha-Pongamia biodiesels on cloud point and pour point," *Energy,* vol. 34, pp. 2016-2021, 11, 2009.

[38] A. Kumar Tiwari, A. Kumar and H. Raheman, "Biodiesel production from jatropha oil (Jatropha curcas) with high free fatty acids: An optimized process," *Biomass Bioenergy,* vol. 31, pp. 569-575, 8, 2007.

In: Biomass Processing, Conversion and Biorefinery   ISBN: 978-1-62618-346-9
Editors: Bo Zhang and Yong Wang   © 2013 Nova Science Publishers, Inc.

Chapter 16

MICROALGAE FOR BIODIESEL PRODUCTION AND WASTEWATER TREATMENT

*Rifat Hasan[1], Lijun Wang[2] and Bo Zhang[2],**
[1]Department of Chemical Engineering,
North Carolina A and T State University, Greensboro, NC, US
[2]Biological Engineering Program, Department of Natural Resources and Environmental
Design, North Carolina A&T State University, Greensboro, NC, US

ABSTRACT

Microalgae are promising third-generation biofuel feedstocks that offer many potential technical and economic advantages. Algae can use and sequester CO_2 from many sources and may be processed into a broad spectrum of products including biodiesel, green diesel and gasoline replacements, bioethanol, methane, heat, bio-oil, biochar, animal feed and biomaterials, etc. This chapter reviews the microalgae studies for wastewater treatment and biodiesel production. Under suitable conditions, microalgae can be cultured in wastewater to reduce nitrate, phosphate and organic matter. These algae can then be used in biofuel production reducing the load on the limited arable lands, so that the food production does not get disrupted to meet the demand of feedstock for biofuel production. With the current requirement for renewable fuels, especially in the transportation sector, there is a need to develop a range of sustainable biofuels resources and that will be a significant step towards the replacement of fossil fuels.

1. INTRODUCTION

Microalgae are prokaryotic or eukaryotic photosynthetic microorganisms, which have unicellular or simple multicellular structure, and perform oxygenic photosynthesis like higher plants. Examples of prokaryotic microorganisms are Cyanobacteria (*Cyanophyceae*) and eukaryotic microalgae are green algae (*Chlorophyta*) and diatoms (*Bacillariophyta*) [1].

* Phone 336-334-7787. Fax 336-334-7270. E-mail: bzhang@ ncat.edu.

Algae are essential to global carbon, nitrogen and sulfur cycling. Approximately 45% of photosynthetic carbon assimilation is by algae. Algae are present in all existing habitat where light is available. Algae have close associations with many other organisms (e.g. lichens, coral, numerous protozoans, etc.). It is estimated that more than 50,000 species exist, but only a limited number of around 30,000, have been studied and analyzed [2]. Among those, most widely used microalgae for wastewater treatment as well as for biofuel production are *Chlorella* sp. The first use of microalgae by humans dates back 2000 years to the Chinese, who used Nostoc to survive during famine. However, microalgal biotechnology only really began to develop in the middle of the last century. Nowadays, there are numerous commercial applications of microalgae. For example, (i) microalgae can be used to enhance the nutritional value of food and animal feed owing to their chemical composition; (ii) they play a crucial role in aquaculture; (iii) they can be incorporated into cosmetics; (iv) they can be used in wastewater treatment, and (v) biofuel production.

Generally, they are cultivated as a source of highly valuable molecules. Microalgae in human nutrition are currently marketed in different forms such as tablets, capsules and liquids. They can also be incorporated into pastas, snack foods, candy bars or gums, and beverages. Owing to their diverse chemical properties, they can act as a nutritional supplement or represent a source of natural food colorants [3]. In addition to its use in human nutrition, microalgae can be incorporated into the feed for a wide variety of animals ranging from fish (aquaculture) to pets and farm animals. In fact, 30% of the current world algal production is sold for animal feed applications. Microalgae are also used to refine the products of aquaculture. Some microalgal species are established in the skin care market, the main ones being *Arthrospira* and *Chlorella*. Some cosmeticians have even invested in their own microalgal production system (LVMH, Paris, France and Daniel Jouvance, Carnac, France). Microalgae extracts can be mainly found in face and skin care products (e.g., anti-aging cream, refreshing or regenerant care products, emollient and as an anti-irritant in peelers). Microalgae are also represented in sun protection and hair care products. However, pure molecules can also be extracted when their concentrations are sufficiently high. This leads to valuable products like fatty acids, pigments and stable isotope biochemicals [3].

Microalgae have the ability to mitigate CO_2 emission and produce oil with a high productivity, thereby having the potential for applications in producing the third-generation of biofuels. The key technologies for producing microalgal biofuels include identification of preferable culture conditions for high oil productivity, development of effective and economical microalgae cultivation systems, as well as separation and harvesting of microalgal biomass and oil [4]. In this chapter, we will review these key technologies.

2. MICROALGAE CULTIVATION

2.1. Microalgae Culture Conditions

The growth characteristics and composition of microalgae are known to significantly depend on the cultivation conditions. There are four major types of cultivation conditions for microalgae: photoautotrophic, heterotrophic, mixotrophic and photoheterotrophic cultivation [5].

2.1.1. Phototrophic Cultivation

Phototrophic cultivation occurs when the microalgae use light, such as sunlight, as the energy source, and inorganic carbon (e.g., carbon dioxide) as the carbon source to form chemical energy through photosynthesis [6].

Phototrophic cultivation is the most commonly used cultivation condition for microalgae growth [7]. It is found that under phototrophic cultivation, there is a large variation in the lipid content of microalgae, ranging from 5% to 68%, depending on the type of microalgae species. Normally a nitrogen-limiting or nutrient-limiting condition was used to increase the lipid content in microalgae [8].

As a result, achieving higher lipid content is usually at the expense of lower biomass productivity. Thus, lipid content is not the sole factor determining the oil-producing ability of microalgae. Instead, both lipid content and biomass production need to be considered simultaneously.

Hence, lipid productivity, representing the combined effects of oil content and biomass production, is a more suitable performance index to indicate the ability of microalgae with regard to oil production.

The highest lipid productivity reported in the literature is about 179 mg/L/d by *Chlorella* sp. under phototrophic cultivation using 2% CO_2 with 0.25 vvm aeration [9]. The major advantage of using autotrophic cultivation to produce microalgal oil is the consumption of CO_2 as carbon source for the cell growth and oil production. However, when CO_2 is the only carbon source, the microalgae cultivation site should be close to factories or power plants which can supply a large quantity of CO_2 for microalgal growth. Moreover, compared to other types of cultivation, the contamination problem is less severe when using autotrophic growth. Therefore, outdoor scale-up microalgae cultivation systems (such as open ponds and raceway ponds) are usually operated under phototrophic cultivation conditions [8].

2.1.2. Heterotrophic Cultivation

Some microalgae species can not only grow under phototrophic conditions, but also use organic carbon under dark conditions, just like bacteria. The situation when microalgae use organic carbon as both the energy and carbon source is called heterotrophic cultivation [5]. This type of cultivation could avoid the problems associated with limited light that hinder high cell density in large scale photobioreactors during phototrophic cultivation [6]. As higher biomass production and productivity could be obtained from using heterotrophic cultivation. Some microalgae species show higher lipid content during heterotrophic growth, and a 40% increase in lipid content was obtained in *Chlorella protothecoides* by changing the cultivation condition from phototrophic to heterotrophic [10].

2.1.3. Mixotrophic Cultivation

Mixotrophic cultivation is when microalgae undergo photosynthesis and use both organic compounds and inorganic carbon (CO_2) as a carbon source for growth. This means that the microalgae are able to live under either phototrophic or heterotrophic conditions, or both. Microalgae assimilate organic compounds and CO_2 as carbon sources, and the CO_2 released by microalgae via respiration will be trapped and reused under phototrophic cultivation [8]. Compared with phototrophic and heterotrophic cultivation, mixotrophic cultivation is rarely used in microalgal oil production.

2.1.4 Photoheterotrophic Cultivation

Photoheterotrophic cultivation is when the microalgae require light when using organic compounds as the carbon source. The main difference between mixotrophic and photoheterotrophic cultivation is that the latter requires light as the energy source, while mixotrophic cultivation can use organic compounds to serve this purpose. Hence, photoheterotrophic cultivation needs both sugars and light at the same time [5]. Although the production of some light-regulated useful metabolites can be enhanced by using photoheterotrophic cultivation [11], using this approach to produce biodiesel is very rare, as is the case with mixotrophic cultivation.

2.2. Parameters Involved

The most important parameters that have large affect on the growth and lipid content of microalgae are temperature, wastewater concentration (nutrient concentration), light intensity and CO_2 concentration. Several studies have been accomplished on these parameters [12]. It is found that temperature 15-25°C is good for the growth of microalgae. Widely used temperature of for microalgae culture is 20 °C. Light intensity 50-250 μmol m^{-2}s^{-1} is being used in different studies to culture microalgae [13]. It is found that higher light intensity can cause microalgae to die. Smaller light intensity is found to be good for high lipid content in microalgae. Green microalgae contain chlorophylls that use light to absorb CO_2 from air. CO_2 flow rate also influence the growth of microalgae. These four different parameters have been studied separately several times for different microalgae sp. So far these parameters are not studied altogether. To optimize the growth and lipid content of microalgae it is required to combine all the parameters so that the efficiency can be maximized. Combining all these parameters growth and oil content of different types microalgae can be optimized which can be used for further studies.

3. WASTEWATER TREATMENT USING MICROALGAE

Nowadays, it is truism to recognize that the pollution problem is a major concern of society. Environmental laws are given general applicability and their enforcement has been gradually stricter. So, in terms of health, environment and economy, the battle against pollution has become a major concern [14]. Today, the strategic importance of fresh water and air is universally recognized more than ever before. Issues concerning sustainable water management can be found almost in every agenda all over the world. There are fewer things invented where using one item both water and air pollution can be minimized. Microalgae are one of them which can be used to reduce these crises as it ensures sustainable management of both air and water.

Without proper treatment, excess nitrogen and phosphorus discharged in wastewater can lead to ecosystem damage [15]. The negative effects of such nutrient overloading of receiver systems include low dissolved oxygen concentrations and fish kills, undesirable pH shifts, and cyanotoxin production. Chemical and physical based technologies are on hand to remove these nutrients, but they consume significant amounts of energy and chemicals, making them costly processes. Chemical treatment often leads to secondary contamination of the sludge byproduct as well, creating additional troubles of safe disposal. The energy and cost required for treatment of wastewater remain a problem for industries and municipalities. Compared to physical and chemical treatment processes, algae based treatment can potentially achieve nutrient deduction in a less expensive and ecologically safer way with the added benefits of resource recovery and recycling [16].

The history of the commercial use of algal cultures spans about 75 years with application to wastewater treatment and mass production of different strains such as *Chlorella* and *Dunaliella* [14]. Since the land-space requirements of microalgal wastewater treatment systems are substantial, several efforts are being made to develop wastewater treatment systems based on the use of hyper concentrated algal cultures. Microalgae can treat human sewage, livestock wastes, agro-industrial wastes and industrial wastes. Microalgal systems can also be used for the treatment of other wastes such as piggery effluent, the effluent from food processing factories and other agricultural wastes [14]. Therefore algae can be used in wastewater treatment for a range of purposes, some of which are used for the removal of coliform bacteria, reduction of chemical and biochemical oxygen demand, removal of N and/or P, and also for the removal of heavy metals [14]. Microalgae produced for wastewater treatment can further be used as the feedstock for biofuel production.

3.1. Composition of Typical Wastewater

Wastewater is a complex mixture of natural organic and inorganic materials as well as man-made compounds. Three quarters of organic carbon in sewage are present as carbohydrates, fats, proteins, amino acids, and volatile acids. The inorganic constituents include large concentrations of sodium, calcium, potassium, magnesium, chlorine, sulfur, phosphate, bicarbonate, ammonium salts and heavy metals [17]. As wastewater contains the highest amount of ammonia nitrogen and active phosphorus, that could be a suitable growth medium for microalgae for the dual purposes of removing nutrients and obtaining a feedstock for biofuel production.

3.2. Microalgae Culture in Wastewater

Growing algae requires consideration of three primary nutrients: carbon, nitrogen, and phosphorus. Micronutrients required in trace amounts include silica, calcium, magnesium, potassium, iron, manganese, sulfur, zinc, copper, and cobalt, although the supply of these essential micronutrients rarely limits algal growth when wastewater is used. If not already available in the water source, the addition of commercial fertilizers can significantly increase production costs; making the price of algae derived fuel cost prohibitive. For this reason, wastewater is an attractive resource for algae production [16].

Several studies have been conducted to culture different types of microalgae in different types of wastewater to remove the nutrients. The feasibility of growing *Chlorella* sp. in the centrate, a highly concentrated municipal wastewater stream generated from activated sludge thickening process, for simultaneous wastewater treatment and energy production was tested [18]. The results showed that by the end of a 14-day batch culture, algae could remove ammonia, total nitrogen, total phosphorus, and chemical oxygen demand (COD) by 93.9%, 89.1%, 80.9%, and 90.8%, respectively from raw centrate, and the fatty acid methyl ester (FAME) content was 11.04% of dry biomass providing a biodiesel yield of 0.12 g-biodiesel/L-algae culture solution. In another study, *Chlamydomonas reinhardtii* was grown in wastewaters taken from three different stages of the treatment process [19].

4. BIODIESEL PRODUCTION FROM MICROALGAE

4.1. Biofuel Production Potential

Since the last few decades, fossil fuels have become an integral part of human daily lives. Specifically, fossil fuels are burned to produce energy for transportation and electricity generation, in which these two sectors have played a vital role in improving human living standard and accelerating advance technological development. However, fossil fuels are non-renewable source that are limited in supply and will one day be exhausted. In addition, burning fossil fuels have raised numerous environmental concerns, including green house gas (GHG) effects which significantly contribute towards global warming. Apart from that, as energy crisis is beginning to hit almost every part of the world due to rapid industrialization and population growth, the search for renewable energy sources has become the key challenge in this century in order to stimulate a more sustainable energy development for the future [20]. Therefore, discovering and constructing renewable, carbon neutral transportation fuel systems are possibly two of the most vital issues for current society [21].

With growing concerns surrounding the continued use of fossil fuels, renewable biofuels have received a large amount of recent attention. Biodiesel, a promising substitute for petroleum fuels, has the potential to address sustainability and energy security issues because it is derived from plant oils or animal fats, and has much lower green-house gas emission. Currently, soybean oil is the major feedstock for commercial biodiesel production; other oil feedstock including canola, corn, jatropha, waste cooking oil, and animal fats are also being tested. While biofuels produced using oil crops and waste oils cannot alone meet the existing demand for fuel, microalgae appear to be a more promising feedstock option [21]. Microalgae-based biofuels are an appealing choice [22] to meet these mandates because of microalgae's (1) Rapid growth rate (cell doubling time of 1–10 days [23]), (2) High lipid content (more than 50% by cell dry weight [24]), (3) Smaller land usage (15–300 times more oil production than conventional crops on a per-area basis [25]), and (4) High carbon dioxide (CO_2) absorption and uptake rate [26]. Given these advantages, microalgae-based biofuels have been recognized as the "third-generation of biomass energy" [27] and the "only current renewable source of oil that could meet the global demand for transport fuels" [23].

In the recent years, the potential and prospect of microalgae for sustainable energy development have been extensively reviewed and microalgae are foreseen to be the fuel of the future. In fact, microalgae biofuels have been placed globally as one of the leading research fields which can bring enormous benefits to human beings and the environment [20]. Under suitable culture conditions, some microalgal species are able to accumulate up to 50–70% of oil/lipid per dry weight [21]. The fatty acid profile of microalgal oil is suitable for the synthesis of biodiesel [7]. The major attraction of using microalgal oil for biodiesel is the tremendous oil production capacity by microalgae, as they could produce up to 58,700 L oil per hectare, which is one or two magnitudes higher than that of any other energy crop [21]. However, mass production of microalgal oil faces a number of technical hurdles that render the current development of the algal industry economically unfit. In addition, it is also necessary, but very difficult, to develop cost-effective technologies that would permit efficient biomass harvesting and oil extraction. Nevertheless, since microalgae production is regarded a feasible approach to mitigate global warming, it is clear that producing oil from microalgal biomass would provide significant benefits, in addition to the fuel [4].

4.2. Biodiesel Production

Biodiesel is a potential substitute for conventional diesel fuel. One of the biotechnological processes that have received increasing interest from companies and researchers is the cultivation of microalgae, which are an excellent source of organic compounds such as fatty acids [28].

The fatty acids that are produced by microalgae can be extracted and converted into biodiesel (Figure 1) [29]. Over the past 30 years, microalgal biotechnology for lipids production has developed extensively [30]. Microalgae exhibit a great variability in lipid content. Among microalgae species, oil contents can reach up to 80%, and levels of 20–50% are quite common [31].

The microalgae *Chlorella* has up to 50% lipids and *Botryococcus* has 80%. The variations are due to different growing conditions and methods of extraction of lipids and fatty acids. One of the main factors that influences the lipid and fatty acid content of microalgae in terms of cultivation is the CO_2 concentration. In areas where microalgae are grown for biodiesel production alongside fossil fuel power stations, CO_2 release can be significantly reduced and the lipid content increased [29, 32] .

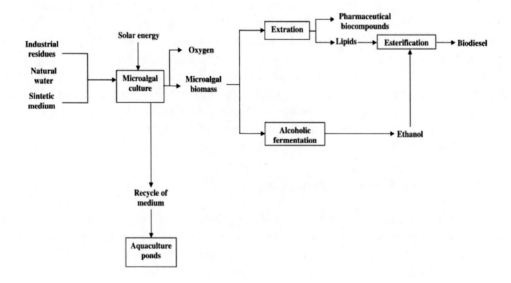

Figure 1. Flow diagram of biomass microalgae for biodiesel production.

The carbon and hydrogen content of microalgal biofuel is greater than in plant biofuel, even though the oxygen content is lower. The H/C and O/C mean molar ratio of microalgal biofuel was 1.72 and 0.26, while the H/C and O/C molar ratio of plant biofuel was 1.38 and 0.37, respectively [33].

Microalgal biofuel is characterized by lower oxygen content and a higher H/C ratio than biofuel from plants, sunflower and cotton [34].

The high hydrogen content of microalgal biofuel is due to chlorophyll and proteins. Microalgal biofuel has a high calorific value, low viscosity and low density than plant biofuel. These physical properties of microalgae make them more appropriate for biofuel than lignocellulosic materials [34].

The mean annual productivity of microalgal biomass in a tropical climate region is 1.53 kgm^{-3} day^{-1}, with a mean 30.0% of lipids extracted from the biomass, the concentration per hectare of total area is around 123.0 m^3 for 90.0% of the 365 days of a year, since the remaining 10.0% are used for maintenance and cleaning of the bioreactors [33]. Thus, the yield of biodiesel from microalgae is 98.4 m^3ha^{-1}, for the production of 5.4 billion m^3 of biodiesel per year, the area requirement is approximately 5.4 Mha. This represents only 3.0% of the area currently used for cultivation of plants for biodiesel production. This would be a possible scenario even if the concentration of lipids in the microalgal biomass was 15.0% of dry weight [35].

4.3. Transesterification Technologies in the Production of Biodiesel

Widely used process to produce biodiesel from microalgae is transesterification. The viscosities of vegetable oils and microalgal oils are usually higher than that of diesel oils [36]. Hence, they cannot be applied to engines directly.

The transesterification of microalgal oils will greatly reduce the original viscosity and increase the fluidity [6]. Few reports on the production of biodiesel from microalgal oils are available [21]. Nevertheless the technologies of the biodiesel production for vegetable oils can be applied to the biodiesel production of microalgal oils because of the similar physical and chemical properties. In the process of transesterification, alcohols are key substrates in transesterification.

The commonly used alcohols are methanol, ethanol, propanol, butanol, and amyl alcohol, but methanol is applied more widely because of its low-cost and physical advantages [6]. Alkali, acid, or enzyme catalyzed processes can be applied in transesterification though these processes have their own advantages and disadvantages (Table 1) [6].

4.4. Economics of Biodiesel Production

Cost of producing microalgal biodiesel can be reduced substantially by using a biorefinery based production strategy, improving capabilities of microalgae through genetic engineering and advances in engineering of photobioreactors [21]. Microalgal oils can potentially substitute petroleum as a source of hydrocarbon feedstock for the petrochemical industry.

For this to happen, microalgal oil will need to be sourced at a price that is roughly related to the price of crude oil, as follows:

$$C_{algal\ oil} = 0.0069\ C_{petroleum} \tag{1}$$

where $C_{algal\ oil}$ ($ per liter) is the price of microalgal oil and $C_{petroleum}$ is the price of crude oil in $ per barrel [21]. Eq. (1) assumes that algal oil has roughly 80% of the energy content of crude petroleum.

Table 1. Application of transesterification technologies

Type of transesterification	Advantages	Disadvantages
Chemical catalysis	1) Reaction condition can be well controlled	1) Reaction temperature is relatively High and process is complex
	2) Large scale production	2) The later disposal process is complex
	3) The cost of production process is cheap	3) The process need much energy
	4) Methanol produced can be recycled	4) Need an installation for methanol recycle
	5) High conversion of the production	5) The wastewater pollutes the environment
Enzyme catalysis	1) Moderate reaction condition	1) Limitation of enzyme in the conversion of short chain of fatty acids
	2) The small amount of methanol is required	2) Chemical exist in the process of production are poisonous to enzyme
	3) Have no pollution to natural environment	
Supercritical fluid techniques	1) Easy to be controlled	1) High temperature and pressure in the reaction condition leads to high cost of production
	2) It is safe and fast	
	3) Friendly to environment	

CONCLUSION

As demonstrated, it is promising to use microalgae for wastewater treatment and at the same time for biodiesel production. With the favorable conditions and present advanced technology it is economically feasible to reduce greenhouse gas emission by growing algae in wastewater and processing it into biodiesel. Thus the improved climate change will affect the basic elements of human life: water, food, health and the environment and will affect millions of people all the way through famine, drought and floods.

ACKNOWLEDGMENTS

This publication was made possible by Grant Number NC.X-269-5-12-130-1 from the National Institute of Food and Agriculture. Its contents are solely the responsibility of the authors and do not necessarily represent the official views of the National Institute of Food and Agriculture.

REFERENCES

[1]   Li Y, Horsman M, Wu N, Lan CQ, Dubois-Calero N. Biofuels from microalgae. *Biotechnology Progress* 2008;24(4):815.

[2]   Richmond A. Handbook of microalgal culture: biotechnology and applied phycology. *Blackwell Science Ltd* 2004.

[3]   Spolaore P, Joannis-Cassan C, Duran E, Isambert A. Commercial applications of microalgae. *Journal of Bioscience and Bioengineering* 2006;101:87.

[4]   Chen C-Y, Yeh K-L, Aisyah R, Lee D-J, Chang J-S. Cultivation, photobioreactor design and harvesting of microalgae for biodiesel production: A critical review. *Bioresource Technology 2011*;102:71.

[5]   Chojnacka K, Marquez-Rocha FJ. Kinetic and stoichiometric relationships of the energy and carbon metabolism in the culture of microalgae. Biotechnology 2004;3:21.

[6]   Huang G, Chen F, Wei D, Zhang X, Chen G. Biodiesel production by microalgal biotechnology. *Applied Energy* 2010;87:38.

[7]   Gouveia L, Oliveira AC. Microalgae as a raw material for biofuels production. *Biotechnol.* 2009;36:269.

[8]   Mata TM, Martins AA, Caetano NS. Microalgae for biodiesel production and other applications: A review. *Renewable and Sustainable Energy Reviews* 2010;14:217.

[9]   Chiu S-Y, Kao C-Y, Chen C-H, Kuan T-C, Ong S-C, Lin C-S. Reduction of CO2 by a high-density culture of Chlorella sp. in a semicontinuous photobioreactor. *Bioresource Technology* 2008;99:3389.

[10] Xu H, Miao X, Wu Q. High quality biodiesel production from a microalga Chlorella protothecoides by heterotrophic growth in fermenters. *Journal of Biotechnology* 2006;126:499.

[11] Ogbonna JC, Ichige E, Tanaka H. Regulating the ratio of photoautotrophic to heterotrophic metabolic activities in photoheterotrophic culture of Euglena gracilis and its application to alpha-tocopherol production. *Biotechnol.* 2002;24:953.

[12] Roleda MY, Slocombe SP, Leakey RJG, Day JG, Bell EM, Stanley MS. Effects of temperature and nutrient regimes on biomass and lipid production by six oleaginous microalgae in batch culture employing a two-phase cultivation strategy. *Bioresource Technology* 2013;129:439.

[13] Perrine Z, Negi S, Sayre RT. Optimization of photosynthetic light energy utilization by microalgae. *Algal. Research* 2012;1:134.

[14] Abdel-Raouf N, Al-Homaidan AA, Ibraheem IBM. Microalgae and wastewater treatment. *Saudi Journal of Biological Sciences* 2012;19:257.

[15] Correll DL. Role of phosphorus in the eutrophication of receiving waters: a review. *Environ Qual.* 1998;27:261.

[16] Christenson L, Sims R. Production and harvesting of microalgae for wastewater treatment, biofuels, and bioproducts. *Biotechnology Advances* 2011;29:686.

[17] Lim S-L, Chu W-L, Phang S-M. Use of Chlorella vulgaris for bioremediation of textile wastewater. *Bioresource Technology* 2010;101:7314.

[18] Li Y, Chen Y-F, Chen P, Min M, Zhou W, Martinez B, et al. Characterization of a microalga Chlorella sp. well adapted to highly concentrated municipal wastewater for nutrient removal and biodiesel production. *Bioresource Technology* 2011;102:5138.

[19] Kong Q, Li L, Martinez B, Chen P, Ruan R. Culture of Microalgae Chlamydomonas reinhardtii in Wastewater for Biomass Feedstock Production. *Appl Biochem Biotechnol* 2009;160:9.

[20] Lam MK, Lee KT. Microalgae biofuels: A critical review of issues, problems and the way forward. *Biotechnology Advances* 2012;30:673.

[21] Chisti Y. Biodiesel from microalgae. *Biotechnology Advances* 2007;25:294.

[22] Zhang X, Hu Q, Sommerfeld M, Puruhito E, Chen Y. Harvesting algal biomass for biofuels using ultrafiltration membranes. *Bioresource Technology* 2010;101:5297.

[23] Schenk PM, Thomas-Hall SR, Stephens E, Marx UC, Mussgnug JH, Posten C, et al. Second generation biofuels: High-efficiency microalgae for biodiesel production. *Bioenerg. Res.* 2008;1:20.

[24] Hu Q, Sommerfeld, M., Jarvis, E., Ghirardi, M., Posewitz, M., Seibert, M., Darzins, A. Microalgal triacylglycerols as feedstocks for biofuel production: Perspectives and advances. *Plant J.* 2008;54:621.

[25] Xin L, Hong-ying H, Ke G, Ying-xue S. Effects of different nitrogen and phosphorus concentrations on the growth, nutrient uptake, and lipid accumulation of a freshwater microalga Scenedesmus sp. *Bioresource Technology* 2010;101:5494.

[26] Jorquera O, Kiperstok A, Sales EA, Embiruçu M, Ghirardi ML. Comparative energy life-cycle analyses of microalgal biomass production in open ponds and photobioreactors. *Bioresource Technology* 2010;101:1406.

[27] Gressel J. Transgenics are imperative for biofuel crops. *Plant Science* 2008;174:246.

[28] Colla LM, Bertolin TE, Costa JAV. Fatty acids profile of Spirulina platensis grown under different temperatures and nitrogen concentrations. *Z Naturforsch* 2004;59:55.

[29] Brown LM, Zeiler KG. Aquatic biomass and carbon dioxide trapping. *Energy Conversion and Management* 1993;34:1005.

[30] Scragg AH, Morrison J, Shales SW. The use of a fuel containing Chlorella vulgaris in a diesel engine. *Enzyme and Microbial. Technology* 2003;33:884.

[31] Powell EE, Hill GA. Economic assessment of an integrated bioethanol–biodiesel–microbial fuel cell facility utilizing yeast and photosynthetic algae. *Chemical Engineering Research and Design* 2009;87:1340.

[32] Sawayama S, Inoue S, Dote Y, Yokoyama S-Y. CO_2 fixation and oil production through microalga. *Energy Conversion and Management* 1995;36:729.

[33] Costa JAV, de Morais MG. The role of biochemical engineering in the production of biofuels from microalgae. *Bioresource Technology* 2011;102:2.

[34] Miao X, Wu Q, Yang C. Fast pyrolysis of microalgae to produce renewable fuels. *Journal of Analytical and Applied Pyrolysis* 2004;71:855.

[35] Chisti Y. Biodiesel from microalgae beats bioethanol. *Trends in Biotechnology* 2008;26:126.

[36] Fuls J, Hawkins CS, Hugo FJC. Tractor engine performance on sunflower oil fuel. *Journal of Agricultural Engineering Research* 1984;30:29.

In: Biomass Processing, Conversion and Biorefinery
Editors: Bo Zhang and Yong Wang

ISBN: 978-1-62618-346-9
© 2013 Nova Science Publishers, Inc.

Chapter 17

THERMOCHEMICAL CONVERSION OF FERMENTATION-DERIVED OXYGENATES TO FUELS

Karthikeyan K. Ramasamy[*1,2] *and Yong Wang*[1,2]
[1]Pacific Northwest National Laboratory, Richland, WA
[2]Voiland School of Chemical Engineering and Bioengineering,
Washington State University, Pullman, WA

ABSTRACT

Though ethanol is currently the dominant biofuel as a blend component to gasoline (generated from renewable resources through fermentation), several properties make ethanol undesirable as standalone transportation fuel (e.g., low energy density and high water solubility). In the near future, the production capacity of fermentation-derived ethanol/oxygenates are projected to rise beyond the blending needs. A viable solution for next-generation biofuels is through conversion of oxygenates to hydrocarbon compounds similar to those in gasoline, diesel, and jet fuel. This chapter discusses the thermochemical conversion of fermentation-derived alcohols/oxygenates to fuel-range hydrocarbons.

1. INTRODUCTION

In recent years, much research has focused on the development of renewable energy resources in the wake of projected crude oil depletion and the negative environmental effects of crude oil usage.

Figure 1. Simplified block diagram of the biological conversion of biomass to oxygenates.

* Phone 509-372-6976. Fax 509-372-4732. E-mail: karthi@pnnl.gov.

Renewable fuels (e.g., methanol, ethanol, butanol, and biodiesel) are generated through thermochemical or biological routes. The primary biological route in renewable fuel production is the fermentation of sugars to alcohols/oxygenates. For centuries, biological fermentation processes have been used to generate food and beverages (e.g., cheese, yogurt, wine, and beer).

Materials containing sugar (e.g., sugar cane and sugar beets) and materials that can be converted to sugars (e.g., corn, grains, agricultural wastes, and wood) can be used to produce alcohol and other oxygenates via fermentation. Unlike sugar cane and corn, lignocellulose material such as agricultural wastes and wood consists of cellulose, hemicellulose, and lignin. At present, the recalcitrant nature of the lignin allows only the cellulose and hemicellulose portion of the lignocellulose material to be converted to oxygenates through fermentation. Figure 1 shows a simplified block diagram of the biological conversion of biomass to oxygenate compounds. First, the starch, cellulose, and hemicellulose components of the biomass material are converted to simple sugars through hydrolysis and then the fermentation process produces alcohols and other oxygenates [1]. Conventional alcohol fermentation occurs close to room temperature in anaerobic conditions.

2. BIOETHANOL

Prior to the late 1970s, lead additives were used in gasoline as an octane enhancer. In the late 1970s, ethanol replaced lead as an octane enhancer due to its environmentally preferable properties. In the late 1980s, ethanol consumption was increased when some U.S. states mandated that ethanol and other oxygenates be used to reduce carbon emissions. In 2011, the United States produced 14 billion gallons of fermentation-derived ethanol. Currently, most gasoline available in United States contains up to 10 percent ethanol as a blend component. Recently, the Energy Independence and Security Act of 2007 mandated that the United States produce 36 billion gallons of biofuel (e.g., ethanol, butanol and other renewable fuels) annually by 2022. However, due to its low energy content, incompatibility with existing automobile engines, and corrosion capacity, ethanol cannot yet be used to its fullest extent. For this reason, the amount of ethanol/oxygenates blended with gasoline is not expected to grow beyond 15 percent, which allows the opportunity to convert the oxygenates generated from biomass to high-energy-content olefins, aromatics, and fuel-range hydrocarbons [2].

In recent years, biological production methods have almost completely replaced petrochemical production methods for ethanol. Current research efforts are focused on efficiently and economically converting lignocellulosic biomass to alcohol. Figure 2 shows the generalized reaction mechanism for biologically converting cellulose to ethanol. In ethanol fermentation using the Saccharomyces cerevisiae microorganism, each mole of glucose converts to two moles of ethanol and two moles of carbon dioxide [3].

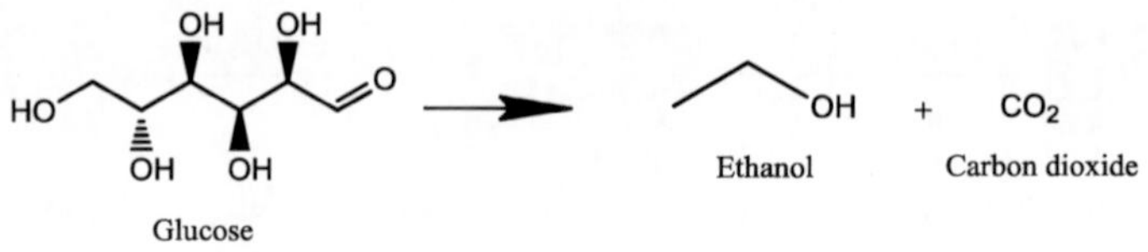

Figure 2. Biological ethanol generation from glucose using *Saccharomyces cerevisiae [3]*.

2.1. Ethanol Conversion

In the 1970s, ExxonMobil developed a methanol-to-gasoline process (MTG) using the HZSM-5 zeolite catalyst and, in the 1980s, commercialized the technology in New Zealand [4]. In the MTG process, the conversion of methanol to hydrocarbons and water is complete and stoichiometric when the catalyst is active. The shape-selective HZSM-5 catalyst limits the product composition to around a chain length of 12 carbons. Using the HZSM-5 catalyst, and varying the catalyst compositions and process conditions, methanol can be converted into light olefins, gasoline, and diesel-range hydrocarbons [4]. For several reasons, when converting alcohol to fuel-range hydrocarbons using the HZSM-5 catalyst, higher alcohols (e.g., ethanol and butanol) are preferable over methanol [8]. Methanol conversion creates high percentages of durene (1,2,4,5-tetramethylbenzene) and benzene. Durene is undesirable in the transportation fuel due to its high melting point (79.2°C), which causes it to solidify at room temperature. Benzene is an environmentally controlled substance due to its carcinogenic nature. Thus, benzene and durene are not tolerated in the transportation fuels [6-7]. In addition, higher alcohols release less water through dehydration than methanol because they have a higher ratio of hydrogen to oxygen.

Between 175 and 250°C, ethanol over HZSM-5 catalyst primarily forms diethyl ether through intermolecular dehydration; as the temperature increases to 280°C, intramolecular dehydration forms ethylene. Above 300°C, heavier hydrocarbon tends to form. In a simplified reaction mechanism (See Figure 3), over HZSM-5 catalyst and above 300°C ethanol forms ethylene and water through dehydration reaction. Then, ethylene goes through number of reactions (e.g., oligomerization, dehydrocyclization, hydrogenation, and cracking) to form a complex mixture of hydrocarbon products (i.e., paraffins, olefins, saturated cyclics, aromatics, and naphthalene) with a carbon number between C_2 and C_{12} [8].

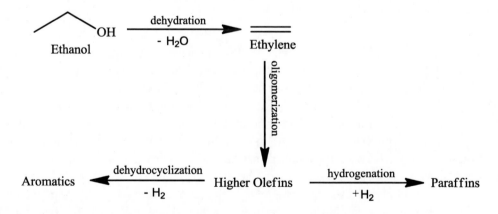

Figure 3. Simplified reaction mechanism ethanol to fuel over HZSM-5 catalyst.

Figure 4 shows the product class distribution and liquid hydrocarbon yield generated at various temperatures ≥300°C for experiments conducted using HZSM-5 catalyst with a silicon to aluminum (Si to Al) ratio of 15 to 1 at 300 psig pressure and $2h^{-1}$ weight hourly space velocity (WHSV). The optimum liquid hydrocarbon yield occurs around 360°C. Beyond this temperature, the liquid hydrocarbon yield drops due to increased activity of cracking reactions, which result in the formation of more gaseous products [8].

The fermentation-derived alcohol is produced as a dilute solution in water; the typical ethanol concentration is between 8 and 14 percent. Because the removal of all the water from the diluted ethanol is expensive, understanding the impact of feed dilution on process economics is imperative. After a series of experiments using samples with dilution ranges up to 30 percent water, Costa et al. [7] concluded that dilution up to 4 percent water level does not influence product distribution or catalyst activity. In an experiment with 15 percent water, the catalyst activity periods were reduced by half and catalyst crystalline structure (as observed by x-ray diffraction) was irreversibly lost. Similar results were indicated after an investigation on ethanol dilution by Derouane et al. [7, 9, 10].

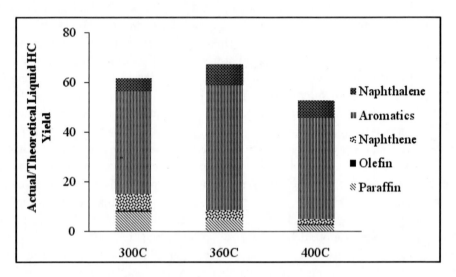

Figure 4. Product class distribution and liquid hydrocarbon yield generated at various temperatures for the experiment conducted on HZSM-5 catalyst with Si to Al ratio of 15 to 1 at 300 psig and 2 h^{-1} WHSV [8].

The activity of the HZSM-5 catalyst can be differentiated on the basis of its Si to Al ratio, which varies from tens to thousands. The number of acid sites and the catalyst properties (e.g, acid strength, acid density and hydrothermal stability) are highly dependent on the catalyst Si to Al ratio. Maderia et al. [11] investigated the influence of HZSM-5 Si to Al ratio on catalytic performance and the deactivation rate of ethanol transformation into hydrocarbons. A HZSM-5 catalyst with Si to Al ratio of 40 to 1 was the most stable and selective variant due to an optimum balance between the number of Brønsted acid sites and the amount of radicals (i.e., active sites for ethanol conversion into higher hydrocarbons).

Figure 5 shows the conversion of ethanol, liquid hydrocarbon yield, and ethylene selectivity over time-on-stream for the experiment conducted with a Si to Al ratio of 15 to 1 at 360°C, 300 psig, and 2 h^{-1} WHSV. Throughout the duration of the experiment, ethanol conversion remained close to 100 percent [12]. Initially the liquid hydrocarbon yield was lower and the ethylene selectivity higher before the reaction reached a steady state. This phenomenon, termed the hydrocarbon pool mechanism by Dahl et al. [13, 14], results from the required initial induction period wherein radicals or co-catalyst are formed to sustain continued higher hydrocarbon generation. In the later stage of the catalyst's lifetime, liquid hydrocarbon starts to decline while ethylene selectivity increases. The decrease in liquid hydrocarbon is due to the combination of active site poisoning and pore blockage from high molecular weight products. Close to complete conversion of ethanol to ethylene is attained even after the catalyst deactivation for liquid hydrocarbon. This phenomenon shows that the primary dehydration reaction and the secondary reaction require different active sites and/or catalyst acid strength. The deactivation of the HZSM-5 catalyst due to site poisoning and/or the pore blockage can be easily regenerated by burning off the carbon at 550°C in air atmosphere with virtually no effect on the catalyst activity in subsequent cycles [15]. The catalyst activity can be maintained at a desired level for a long duration using a fluidized or multi-bed reactor and alternating active and regeneration cycles.

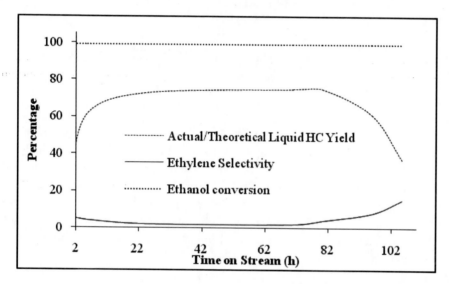

Figure 5. Ethanol conversion, liquid hydrocarbon yield, and ethylene selectivity over catalyst time-on-stream for the experiment conducted on HZSM-5 catalyst with Si to Al ratio of 15 to 1, 360°C, 300 psig, 2 h^{-1} WHSV [12].

2.2. Ethylene Oligomerization

Ethylene can be produced via ethanol dehydration, which is easily accomplished using water-stable solid catalysts (e.g., titanium, alumina, amorphous silica-alumina, zeolites, and other catalyst with acidic properties) at temperatures between 150 and 300°C under atmospheric pressure conditions [16]. Ethanol dehydration to ethylene is a well-understood, commercially proven process. Using an ethylene oligomerization process to produce distillate-range hydrocarbons is important for achieving high-energy-content renewable fuel. The generation of via olefin oligomerization has several advantages over an MTG-based process because olefins are easier to transform through various processes (e.g., isomerization and hydrogenation) to meet current jet and diesel fuel requirements. Olefin chain growth process (commonly called Ziegler process), primarily uses homogenous organic transition metal catalysts to produce polyethylene with a higher carbon number. Recent research has focused on developing a modified Ziegler process to selectively form olefin with carbon numbers between C_4 and C_8 [17].

Heterogeneous ethylene oligomerization processes with improved process economics and ease of operation [18] compared to the homogenous catalysis are the subject of ongoing research. Figure 6 shows the simplified reaction pathway from ethanol to higher olefins. Activity of solid acid catalyst (e.g., zeolites) to convert ethylene to fuel-range olefins through oligomerization reaction is very low. Kiyosi et al. [19] found that nickel deposited on Kieselguhr could catalyze dimerization of ethylene at room temperature. This discovery led to additional research on Ni-exchanged solid acid catalysts with various pore dimensions [18, 20].

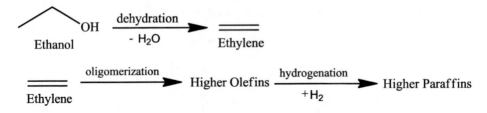

Figure 6. Simplified reaction mechanism of ethanol to higher hydrocarbons.

Heydenrych et al. [21] conducted experiments on Ni-exchanged amorphous silica-alumina and showed the amount of oligomers to be ~11.5g/g_{cat} h at 120°C and 507 psi. Lallemand et al. [22] conducted experiments on Ni-exchanged dealuminated Y zeolites and showed the amount of oligomers to be ~30g/g_{cat} h at 50°C and 580 psi. Hulea et al. [23] conducted experiments on Ni-exchanged AlMCM-41 mesoporous materials and showed the amount of oligomers to be ~63.2g/g_{cat} h at 150°C and 507 psi. Research is ongoing to understand the influence of metal content, metal oxidation state, porosity, and acidity on the ethylene oligomerization catalyst.

3. BIOBUTANOL

Biobutanol can be produced by acetone-butanol-ethanol (ABE) fermentation. Figure 7 shows the generalized reaction mechanism for biologically converting glucose to butanol. The use of ABE fermentation was second only to ethanol fermentation in the early part of the twentieth century, but later disappeared due to the availability of the cheap fossil fuel. As fossil fuel reserves are depleted, ABE fermentation could see resurgence.

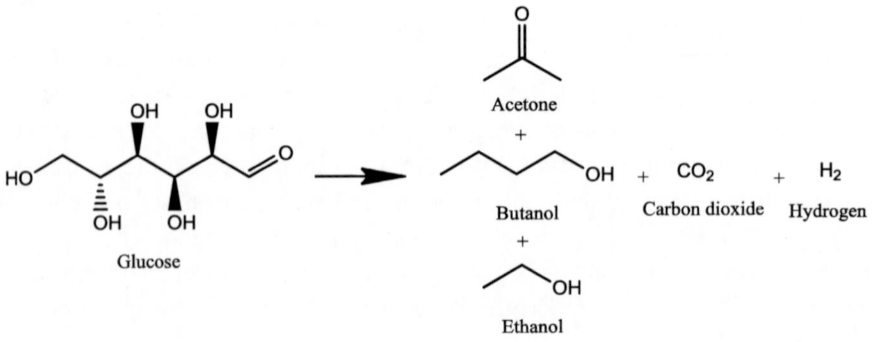

Figure 7. Biological butanol generation from glucose by *Clostridia acetobutylicum* or *Clostridia beijerinckii*.

Butanol has traditionally been produced by anaerobic fermentation of sugar substrates using various species of solventogenic clostridia. ABE fermentation process performance has been limited by factors such as substrate inhibition and low butanol concentration due to low solvent tolerance. These limitations lead to low butanol productivity and a high downstream processing cost for butanol recovery [24, 25].

Steady progress has been made in research into efficient production of butanol from sustainable and renewable carbon sources. Recent advancement in strain development, process engineering, and in situ product recovery to avoid the butanol poisoning allow for lower cost butanol production [24, 25], reviving butanol as a promising renewable fuel intermediate. In butanol fermentation using *Clostridia acetobutylicum* or *Clostridia beijerinckii*, each mole of glucose converts to various ratios of acetone, butanol, ethanol, carbon dioxide, and hydrogen. The typical ratio between acetone, butanol, and ethanol is 3 to 6 to 1. Recent strain and the process developments have increased butanol concentration. Butanol has several advantages over ethanol (e.g., higher energy content, low vapor pressure, and lower water solubility) when blended with gasoline. Thus, butanol has applications in higher blend transportation fuel or, in some cases, as a standalone fuel for existing internal combustion engines. The following sections describe different thermochemical processes that can be used to convert butanol to higher value hydrocarbon that can fit in to jet fuel and distillate fuel.

3.1. Butanol Conversion

Similar to ethanol, butanol can be converted to hydrocarbon transportation fuel through various routes. Costa et al. [26] investigated a butanol/acetone mixture over HZSM-5 catalyst at MTG-based operating conditions and concluded the mixture can be converted to gasoline-range, hydrocarbon-rich aromatics. Converting high-energy butanol to jet fuel and distillate-range hydrocarbon would be more profitable than making gasoline-range highly aromatic hydrocarbons. Like ethanol, butanol is easy to dehydrate using water-stable catalysts in mild operating conditions [27]. Butene oligomerization is comparatively easier than ethylene oligomerization, because in ethylene oligomerization a primary cation is involved in every consecutive step. However, in olefin with carbon number greater than two, linear paraffins can be formed by consecutive oligomerization and stretching reactions involving only secondary and tertiary carbenium ions, which are at least 45 to 55 kJ/mol lower in heat of formation than the primary cations [28].

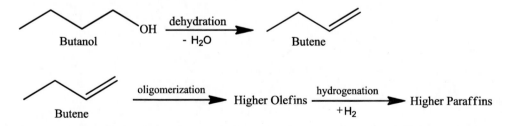

Figure 8. Simplified reaction pathway of butanol to higher hydrocarbons.

Maximizing the concentration of linear butenes during isobutanol dehydration is preferable because diesel fuel generally consists of lightly branched high-molecular-weight C_{10} to C_{16} hydrocarbons. Highly pure isobutene is undesirable as a starting molecule because oligomerization tends to lead to branched products. Figure 8 shows the simplified reaction pathway to convert butanol to higher olefins. Research is available on butane-oligomerization formed distillate-range olefins using various solid catalysts (e.g., amberlyst 15, HZSM-5, Mordenite, and MCM-41) at low to mild operating temperatures with pressures ranging up to 1000 psig [29, 30]. As mentioned in the earlier section olefins with higher carbon number can be hydrogenated and isomerized to generate paraffinic compounds that can fit in to jet fuel and distillate fuel.

3.2. Acetone Conversion

Acetone, along with butanol and ethanol, is one of the main products of the biobutanol (ABE) fermentation process [25]. Acetone can be converted into fuel-range hydrocarbon through acid and basic catalysts [31, 32]. Hutchings et al. compared H beta and HZSM-5 catalytic activity to convert acetone into hydrocarbons and showed that higher isobutene selectivity can be achieved at high acetone conversion using H beta [33]. Figure 9 shows that the conversion of acetone into hydrocarbons is a complex process. Initially, diacetone alcohol is formed through the aldol condensation (dimerization) process. The diacetone alcohol then undergoes cracking to form butene and acetic acid. The acetic acid undergoes ketonization to form acetone. In addition, in the presence of acid catalyst, diacetone alcohol forms mesityl oxide through dehydration, which, in turn, reacts with acetone and release water to form mesitylene.

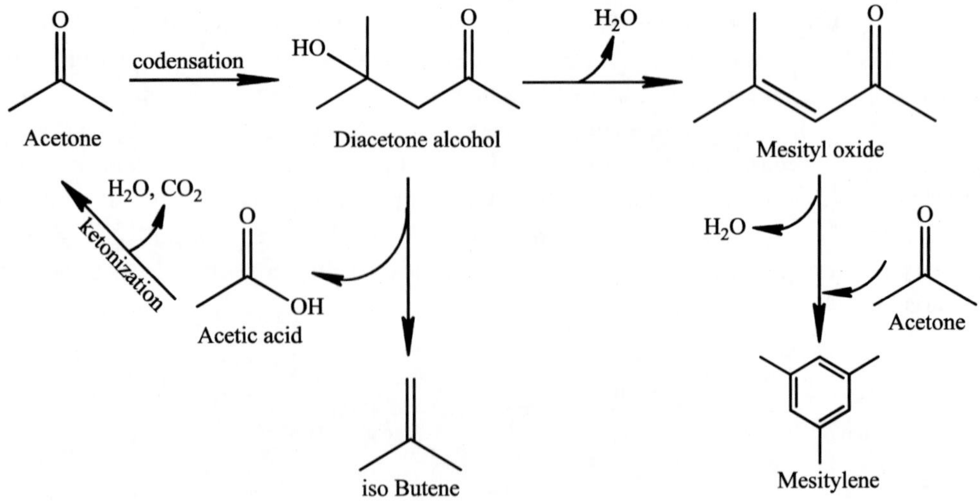

Figure 9. Simplified reaction mechanism of acetone to hydrocarbons.

Under acid-catalyzed conditions, isobutene and mesitylene undergo further reactions (e.g., isomerization, oligomerizarion, and transalkylation) to form a very complex hydrocarbon mixture [34]. Selectively producing isobutene has advantage over the complex mixture of aromatic, straight, and branched chain compounds. Selective isobutene production allows oligomerization and the potential for high-value hydrocarbon to be used in jet fuel and distillate. Alkaline and alkaline earth exchanged beta zeolite were recently studied with the goal to improve isobutene selectivity over the higher aromatic hydrocarbons. In particular, potassium and barium exchanged catalysts showed higher selectivity to isobutene [35, 36].

CONCLUSION

Generating liquid hydrocarbons from renewable biomass resources that meet current and future requirements for transportation fuels is important to reduce dependence on fossil-fuel resources and alleviate the effects of global warming. Conversion of fermentation-derived oxygenates into infrastructure compatible hydrocarbon fuels is a potential pathways to meeting renewable fuel goals. Simple modifications to commercially proven MTG technology allow for the generation of gasoline-range hydrocarbons from fermentation-derived alcohols such as ethanol and butanol. Further research and development is required to optimize the catalyst and process conditions needed to convert alcohols and other oxygenates to distillate-range hydrocarbons.

REFERENCES

[1]   Zhang Y.H, Ding S.Y, Mielenz J.R, Cui J.B, Elander R.T, Laser M, Himmel M.E, McMillan J.R, Lynd L.R, Fractionating recalcitrant lignocellulose at modest reaction conditions, *Biotechnology and Bioengineering,* 2007; 97:214.

[2]   Energy Information Administration, Annual Energy Outlook (2012) http://www.eia.gov/forecasts/aeo/pdf/0383%282012%29.pdf. Accessed on 2nd Dec 2012.

[3]   Van Maris A.J.A, Abbott D.A, Bellissimi E, van den Brink J, Kuyper M, Luttik M.A.H, Wisselink H.W, Scheffers W.A, van Dijken J.P, Pronk J.T, Alcoholic fermentation of carbon sources in biomass hydrolysates by Saccharomyces cerevisiae: current status, Antonie van Leeuwenhoe, 2006;90:391.

[4]   Touhami M, Mike S, Gas Conversion to Liquid Fuels and Chemicals: *The Methanol Route-Catalysis and Processes Development*, Catalysis Reviews, 2009; 51:1.

[5]   Kokotailo G.T, Meier W.M, In the Properties and Applications of Zeolites; Townsend, R.P., Ed, *The Chemical Society London*, 1979; 133.

[6]   Tret'yakov V. F, Makarfi Y.I, Tret'yakov K. V, Frantsuzova N. A, Talyshinskii R. M, The Catalytic Conversion of Bioethanol to Hydrocarbon Fuel: A Review and Study, *Catalysis in Industry,* 2010; 2:402.

[7]   Costa E, Ugulna A, Aguado J, Herndndez P.J, Ethanol to Gasoline Process: Effect of Variables, Mechanism, and Kinetics. *Industrial and Engineering Chemistry* Process *Design and Development,* 1985; 24:239.

[8]   Unpublished information generated by authors at *Pacific Northwest National Laboratory under Laboratory Directed Research and Development Funding.*

[9]   Derouane E.G, Nagy J.B, Dejaifve P, Van Hooff J.H.C, Spekman B.P, Vjzdrine J.C, Naccache C, Elucidation of the Mechanism of Conversion of Methanol and Ethanol to Hydrocarbons on a New Type of Synthetic Zeolite, *Journal Of Catalysis*, 1978;53:40.

[10]   Csicsery S.M, Catalysis by Shape Selective Zeolites - Science and Technology, Pure *and Applied* Chemistry, *1986;* 58:841.

[11]   Madeira F.F, Tayeb K.B, Pinard L, Vezinb H, Mauryc S, Cadran N, Ethanol Transformation into Hydrocarbons on ZSM-5 Zeolites: Influence of Si/Al Ratio on Catalytic Performances and Deactivation Rate. Study of the Radical Species Role, Applied Catalysis A: General 2012; 444:171.

[12]   Ramasamy K.K, Wang Y, Catalyst Activity Comparison of Alcohols over Zeolites, *Journal of Energy Chemistry*, 2013;22

[13]   Dahl M.I, Kolboe S, On the reaction mechanism for propene formation in the MTO reaction over SAPO-34, *Catalysis Letters,* 1993; 20:329.

[14]   Olsbye U, Bjørgen M, Svelle S, Lillerud K, Kolboe S, Mechanistic Insight into the Methanol-to-Hydrocarbons reaction, *Catalysis Today*, 2005;106:108.

[15]   Aguayo A, Gayubo A, Atutxa A, Valle B, Bilbao J, Regeneration of a HZSM-5 zeolite catalyst deactivated in the transformation of aqueous ethanol into hydrocarbons, *Catalysis Today,* 2005;107:410.

[16]   Takahara I, Saito R, Inaba M, Murata K. Dehydration of ethanol into ethylene over solid acid catalysts. *Catalysis Letter* 2005; 105:249.

[17]   Agapie T. Selective ethylene oligomerization: Recent advances in chromium catalysis and mechanistic investigations. *Coordination Chemistry Reviews* 2011;255:861

[18]   Keil F.J, Methanol-to-hydrocarbons: process technology. *Microporous and Mesoporous Materials*. 1999;29:49.

[19]   Kiyosi M, Reactivity of lower hydrocarbons. VI. Activation of specific bonds in ethylene on catalytic surface, *Kogyo Kagaku Zasshi*. 1938; 41:350.

[20]   Chi T. Studies of a new alkene oligomerization catalyst derived from nickel sulfate. *Catalysis Today*. 1999; 51:153.

[21]   Heydenrych M.D, Nicolaides C.P, Scurrell M.S, Oligomerization of ethene in a slurry reactor using a Nickel (II)-exchanged silica–alumina catalyst. *Journal of Catalysis*. 2001; 197:49.

[22]   Lallemand M, Finiels A, Fajula F, Hulea V, Catalytic oligomerization of ethylene over Ni-containing dealuminated Y zeolites. *Applied Catalysis A: General*. 2006;301:196.

[23]   Hulea V, Fajula F. Ni-exchanged AlMCM-41 – An efficient bifunctional catalyst for ethylene oligomerization. *Journal of Catalysis*. 2004; 225:213.

[24]   Jang Y, Butanol production from renewable biomass by clostridia, *Bioresource Technology*. 2012; 123: 653.

[25]   Claassen P.A.M, van Lier J.B, Lopez Contreras A.M, van Niel E.W.J, Sijtsma L, Stams A.J.M, de Vries S.S, Weusthuis R.A, Utilisation of biomass for the supply of energy carriers. Applied Microbiology *and* Biotechnology. 1999; 52:741.

[26]   Costa E, Aguado J, Ovejero G, Cafiizares P, Conversion of n -Butanol-Acetone Mixtures to C1-Cl0 Hydrocarbons on HZSM-5 Type Zeolites, *Industrial Engineering Chemical Research*, 1992; 31: 21

[27]   Ramesh K, Borgna A, Modified Catalyst Composition for Conversion of Alcohol to Alkene. World patent number:WO2009022990A1

[28] Van den berg J.P, Wolthuizen J. P, Clague A. D. H, Hays G. R, Huis R, Van hooff J. H. C, Low-Temperature Oligomerization of Small Olefins on Zeolite H-ZSM-5. An Investigation with High-Resolution Solid-State 13C-NMR, *Journal of Catalysis*, 1983; 80:130.

[29] Catani R, Mandreoli M, Rossini S, Vaccari A, Mesoporous Catalysts for the Synthesis of Clean Diesel Fuels by Oligomerisation of Olefins. *Catalysis Today* 2002;75:125.

[30] Hauge K, Bergene E, Chen D, Fredriksen G. R, Holmen A, Oligomerization of isobutene over solid acid catalysts. *Catalysis Today,* 2005;100:463.

[31] Veloso C.O, Monteiro J.L.F, Sousa-Aguiar E.F, Aldol condensation of acetone over alkali cation exchanged zeolites. *Studies in Surface Science and Catalysis,* 1994; 84:1913.

[32] Thotla S, Agarwal V, Mahajani S.M, Aldol Condensation of Acetone with Reactive Distillation Using Water as a Selectivity Enhancer, *Industrial Engineering Chemical Research.* 2007;46;8371

[33] Hutchings G.J, Johnston P, Lee F.D, Williams C.D, Acetone conversion to isobutene in high selectivity using zeolite catalyst, *Catalysis Letters.* 1993; 21:49

[34] Cruz-Cabeza A.J, Esquivel D, Jiménez-Sanchidrián C, Romero-Salguero F.J, Metal-Exchanged β zeolites as catalysts for the conversion of acetone to hydrocarbons, *Materials.* 2012; 5:121.

[35] Tago T, Konno H, Ikeda S, Yamazaki S, Ninomiya W, Nakasaka Y, Masuda T, Selective production of isobutylene from acetone over alkali metal ion-exchanged BEA zeolites. *Catalysis Today.* 2011; 164:158.

[36] Esquivel D, Cruz-Cabeza A.J, Jiménez-Sanchidrián C, Romero-Salguero F.J, Local environment and acidity in alkaline and alkaline-earth exchanged β zeolite: Structural analysis and catalytic properties. *Microporous and* Mesoporous *Materials.* 2011; 142:672.

In: Biomass Processing, Conversion and Biorefinery   ISBN: 978-1-62618-346-9
Editors: Bo Zhang and Yong Wang   © 2013 Nova Science Publishers, Inc.

Chapter 18

MICROBIAL CONVERSION OF BIO-BASED CHEMICALS: PRESENT AND FUTURE PROSPECTS

Huibin Zou, Guang Zhao and Mo Xian[*]

Qingdao Institute of Bioenergy and Bioprocess Technology,
Chinese Academy of Sciences, Qingdao, Shandong, China

ABSTRACT

The fast development of microbial biotechnology has enabled innovative bioconversion processes for the production of an extensive range of bio-based chemicals from renewable biomass. However, in order to economically and technically compete with petroleum-based chemicals, it is essential to develop robust microbes and efficient processing techniques that are capable of producing bio-chemicals with high performance and low production cost. Recently, the strategies of developing microbial strains using metabolic engineering techniques have been successfully applied to the industrial production of several value-added chemicals. This review summarizes the key economic factors and relative advanced biotechnologies for developing bio-based chemicals, and gives several successful commodity examples. In addition, the review addresses the current trends in developing value-added bioproducts, which represent the near-term opportunities of robust microbial strains for industrial applications.

1. INTRODUCTION

There are widespread concerns over the crude oil – especially for the vulnerable and increasing oil price [1]. The production of fuels and chemicals from renewable biomass is becoming increasingly attractive as the alternative bio-based products can impact on the chemical market for healthy prices. Also, there has been a global movement toward reducing the use of fossil resources for environmental issues. In addition to the reduced consumption of petroleum, many bio-processes actually absorb CO_2, directly reducing the greenhouse gas

[*] Tel/Fax: +86 532 80662765   E-mail: xianmo@qibebt.ac.cn

emissions [2]. The bio-processes are often cleaner processes, using less energy and generating less waste.

Microbial processes have been used for the production of natural products for a long time, and recently the development of synthetic biology to manipulate microbial genetic background has revolutionized the field of industrial biotechnology. In most cases, industrial biotechnology tools allow the industry to develop novel processes that are cleaner and better for the environment with reduced economic costs.

The market potential of microbes and microbial products has been estimated by a market forecasting firm (BCC research, 2011). The global market value for microbial products, in bulk chemical production, is about $156 billion in 2011 and will be increased to $259 billion in 2016, with annual growth rate of 10.7%. The underlying market for microbes is estimated to reach $6.8 billion in 2016 from nearly $4.9 billion in 2011.

However, the transition from traditional petro-economy to bio-economy will be neither quick nor smooth, and the research and development process is in need of considerable capital investment and policy support.

Also the technological challenges are immense: most of the technologies are just out of the laboratories and need considerable efforts in scale up and downstream processing. The present study is a state-of-the-art review discussing the present status of bio-based commodity chemicals with their necessary techno-economics, and several successive stories of commercialized bio-chemicals (succinic acid, lactic acid, isobutanol, and 1,3 propanediol). In addition, the review addresses the near term opportunities of industrial biotechnology and recent advances in value-added bioproduct candidates, including their metabolic engineering approaches and fermentation technologies.

2. CURRENT STATUS OF COMMERCIALIZED BIO-BASED PRODUCTS

After the "golden years of synthetic chemistry revolution" in the last century, the petrochemicals dominate today's chemical industry. With the rapid development of experimental and industrial biotechnology in this century, we are entering into "synthetic biology revolution" [3] and bio-based products represent a growing market with an extensive range of fuels, chemicals and plastics. Early commercialized bio-based products like cellulose, starch, sugars, oils and proteins were directly processed from biomass by physical or chemical treatments, while the next wave of bio-based products should be biochemically converted by microbial and enzymatical bioprocesses from fermentable biomass (usually starch and sugars).

According to the 2010 report from the World Economic Forum, the estimated market for biofuels, bio-based bulk chemicals and bioprocessing enzymes will grow to $95 billion by 2020 [4]. However, bio-based products are currently still at an early development stage. Nearly 87% of global chemical production still relies on fossil resources, with mature, well understood technologies and significant advantages in economy and scale, while many bio-refineries still remain in the small-sized operation, immature technologies with low production efficiency.

To achieve economical and technical competitiveness, more sustainable models must be developed to maximize the use of biomass, increase the processing efficiency, and explore new products and value streams.

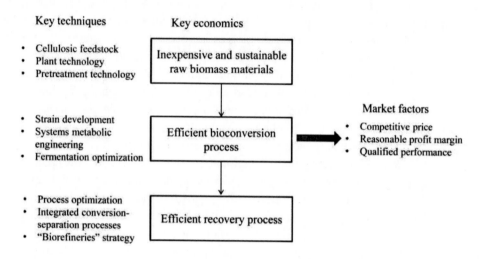

Figure 1. Key techno-economics of commercialized bio-based products.

2.1. General Issues of Bio-Based Chemicals

The success of commodity bio-based products largely depends on the inexpensive and sustainable raw biomass materials and efficient conversion of these materials to value-added products. In Figure 1, we summarized several key economic factors (inexpensive and sustainable supply of fermentable raw material, efficiency of bioconversion and recovery process) and relative advanced biotechnologies aiming to reduce the unit cost of bio-based products.

2.1.1. Supply of Biomass Feedstock

Most biobased industrial products typically started with bulk and labile biomass feedstock like sugars and starch. United States has well established supply chains for pricing and delivery of corn, mostly for corn starch derived bio-ethanol. The production capacity of corn ethanol has been significantly improved to about 14 billion gallons per year in 2011 [2], resulting in the increased corn price in the global market. For the healthy and sustainable development, bio-industries are seeking for sufficient feedstock at a stable price. Cellulosic biomass is the most abundant source of carbon; current biotechnology has focused on consolidated pretreatment and enzymatic conversion of cellulose to sugars, however the commercialization in the present scenario is not economically viable [5]. Plant biotechnology has been well utilized in increasing the productivity for many biomass feedstock like corn and soy. Moreover, plant genetic engineering technology was used in modifying plant structure to reduce the cost of the expensive pretreatment process or increase the polysaccharides content [6]. For example, genetically modified switchgrass has reduced recalcitrance to thermal, chemical, enzymatic and microbial treatments [7].

2.1.2. Bioconversion Process

Bioconversion is the central part of whole production process for the bio-based products. It selectively needs suitable and inexpensive raw material from the upstream process, and the productivity and yield of the catalytic strains or enzymes towards objective products largely affect the downstream recovery processes. In order to efficiently produce bio-based products from renewable resources, advanced biotechnologies are needed to find and improve natural enzymes and microorganisms which have superior metabolic capabilities. Industrial strains have been traditionally developed by the random mutagenesis and screening processes, however, this approach has many limitations and rational metabolic engineering has become popular for strain development [8]. Recently, genetically engineered microbes are commonly used to carry out the fermentation and systems metabolic engineering has been developed at the systems-level for developing strains for a variety of bio-products. The advanced techniques include *in silico* modeling, omics (genome, transcriptome, proteome, metabolome, and fluxome), gene synthesis, synthetic regulatory circuits, and enzyme/pathway engineering [9-13]. Other than stain development, fermentation strategy development also improves the efficiency of the bioconversion process to a large degree by experimentally or statistically optimizing the fermentation process (medium, conditions, process control). For example, recent fermentation strategies like simultaneous saccharification and co-fermentation (SSCF) improved the conversion efficiency of xylose [14], and co-fermentation strategy using two strains improved the conversion efficiency of sugar mixtures [15].

2.1.3. Separation and Recovery Process

Separation of objective products from the bioconversion system is where major costs are generated. The difficulties in developing an efficient separating process are associated with the complexity of the fermentation broth; the recovery process usually needs multiple steps with extensive energy and material consumption; also the waste management for environmental issues needs to be considered. Effective recovery techniques play vital roles in the commercialization of bio-based products. A variety of separation techniques, such as precipitation, ion-exchange adsorption, solvent extraction, crystallization, membrane separation, and chromatography, have been successfully established for the downstream processing of commercialized bio-based chemicals. However, the immature recovery techniques are the main economic barriers against the commercialization of many underdeveloped bio-based chemicals. To reduce the cost and improve the recovery efficiency, many new recovery techniques have been proposed. For example, the integrated fermentation and *in situ* product recovery processes were used in bio-butanol production [16], which reduced butanol toxicity to the microbes, and improved substrate utilization and overall efficiency of the integrated conversion-separation processes. Similar recovery strategy was also proposed in fumaric acid fermentation and separation [17]. Because one of the difficulties in developing an efficient separating process is associated with the complexity of the fermentation broth, the integration of recovery techniques for different chemicals for the complete utilization of the fermentation broth should lead to the development of "biorefineries" strategy [5]. This strategy allows the production of objective products and many other valuable co-products at smaller volumes, improves the overall effectiveness of the process, and makes it techno-economically competitive. In summary, commercialization of bio-based products is expected to follow common business imperatives. The bio-based products must offer equal or better market price than petroleum-based products. And RandD efforts may focus on optimizing the process, increasing the efficiency, developing innovate advanced techniques, improving the performance of the final products, and developing new value-added products that are acceptable by the consumers.

2.2. Representative Bio-Based Commodity Chemicals

The first generation biofuels, bio-ethanol from corn and biodiesel from oils and fats, are typical examples of early bio-based commodity chemicals. Now it is met with general acceptance that more value-added bio-based chemicals need to be developed for a sustainable bio-economy. In 2004, the US Department of Energy (DOE) conducted an initial screening and categorization of renewable chemicals that could be co-produced as side streams of biofuels and bioenergy. The analysis yielded a list of 12 sugar-derived chemicals as top targets for further research and development within industrial biotechnology [18]. Since 2004, with the speeding research and development of biotechnology, biotech companies have developed economic bioprocess pathways for more value-added bio-products. Early examples of commercial or near term value-added chemicals included several alternative biofuels, key building blocks and bio-based monomers with broad applications ranging from fuels, cosmetics, home cleaning products, food/drink packaging, and car industry.

2.2.1. Succinic Acid

As one of the top 12 sugar-derived chemicals [18], succinic acid is a key C4 building block for a wide range of secondary chemicals used in the chemical industries (e.g., synthesis of 1,4-butanediol, tetrahydrofuran, butyrolactone, maleic succimide, itaconic acid and N-methylpyrrolidinone) [25]. Succinic acid could also be used as a monomer for the production of biodegradable polymers [26, 27]. Until recently, nearly all succinic acids were produced from petroleum feedstocks. With cost-competitiveness and superior performance, bio-based succinic acid has successfully reached production quantities in a variety of companies [2, 3]. Myriant and BioAmber have provided breakthrough biotechnologies in commercial production of succinic acid from glucose using genetically engineered E. coli [19]. As the cost of downstream processing can make up a very high portion (50–70 %) in the total production cost for bio-based succinic acid, in situ product removal and biorefinery strategies were proposed to minimize the production cost [28].

Table 1. Representative value-added bio-based commodity chemicals

Chemical	Status	Production strains	Companies	Summarized by
Succinic acid	Commercialized	Engineered Escherichia coli	Myriant, BioAmber, DSM	[2, 3, 19]
Lactic acid and polylactic acid	Commercialized	Lactobacilli strains; Engineered E. coli	NatureWorks	[2, 20, 21]
1,3 Propanediol	Commercialized	Engineered E. coli	DuPont Tate and Lyle	[2, 22]
Isobutanol	Commercialized	Engineered E. coli; Engineered yeast	Gevo	[2, 23]
Isoprene	Near-term commercialized	Engineered E. coli; Engineered yeast	Genencor; Amyris	[24]

2.2.2. Lactic Acid and Polylactic Acid (PLA)

Lactic acid is one of the most successful bio-based commodity chemicals, which can be fermented by a variety of bacteria and fungi [21]. *Lactobacilli spp.* is commonly used for the industrial production of lactic acid with high yield (nearly 1g/g glucose) and matured techniques. PLA was synthesized from the lactic acid by ring opening polymerization (ROP) of lactide [29]. PLA has been considered to be an environmentally friendly alternative to petroleum-based plastic, because of its biodegradable, biocompatible, and compostable properties. Compared to other biodegradable polyesters, PLA has the highest potential due to its availability on the market and its low price. Currently, NatureWorks has established global market channels for its PLA biopolymer (marketed under the Ingeo trademark) in more than 70,000 stores globally and over 100 million pounds in annual sales [2]. To further reduce the production cost, engineered *E. coli* strains were developed to improve the optical purity of lactic acid [30], and to employ direct sugar-PLA pathways [20].

2.2.3. 1,3-Propanediol (PDO)

Bio-derived PDO can be produced from renewable biomass by robust microbes. In nature, many microorganisms including *Klebsiella*, *Citrobacters*, and *Clostridium* were found to produce 1,3-PDO from glycerol. DuPont constructed engineered *E. coli* strains, which integrated the pathways for glycerol and 1,3-PDO production, can directly produce 1,3-PDO from sugar substrates [22].

The PDO can be used as a key building block for a wide range of secondary chemicals or as a monomer ingredient in polymers. Over a dozen products can be made using bio-derived PDO as a key ingredient. Zemea® and Susterra® propanediol are two grades of 100% renewably sourced Bio-PDO™, manufactured by DuPont Tate and Lyle Bio Products. On a pound-for-pound basis, the bio-based PDO consumes 38% less energy and emits 42% fewer greenhouse gas emissions compared to petroleum-based PDO [2].

2.2.4. Isobutanol

Isobutanol can be used as an ingredient in nearly 40% of traditional chemicals (such as butenes, toluenes and xylenes) or transportation fuels. The solvent, rubber and fuel ingredients markets of related chemicals are each worth several billion dollars. Current isobutanol based chemicals are subject to volatile oil prices and supply, so bio-derived isobutanol can serve as a price-stable replacement for petroleum-derived isobutanol [2]. Gevo (http://www.gevo.com) is now producing bio-based isobutanol from sugars by an Integrated Fermentation Technology (GIFT) that combines genetically engineered yeast with a continuous separation process to recover the isobutanol from the fermentation broth. Gevo also engineered *E. coli* to improve the yield of isobutanol by blocking competitive pathways for ethanol and acetic acid [23].

2.2.5. Isoprene

Isoprene can be used as a key chemical to produce a diverse range of industrial products used in surgical gloves, motor mounts and fittings, rubber bands, golf balls, and shoes. Until recently, most of isoprene products are produced from petroleum-derived feedstocks. The research collaboration between Genencor and Goodyear Tire has resulted in the production of BioIsoprene®, a synthetic cis-polyisoprene [24].

This technology has resulted in the production, recovery, polymerization of, and manufacture of tires with the isoprene produced via fermentation. Genencor and Amyris also successfully explored techniques for bio-isoprene fermentation. In this process, the isoprene monomer is firstly produced from sugar substrates by engineered strains (*E. coli* for Genencor and yeast for Amyris) that carry the entire isoprene synthetic pathways; then the isoprene produced in the gas-phase is recovered from the fermentation off-gas in a continuous process. Although the technology will not be full-scale within several years, it has been proved that isoprene is a representative example for obtaining renewable chemicals to replace the scarce natural resources.

3. METABOLIC ENGINEERING FOR NEW BIO-CHEMICALS

Several bio-based chemicals have been successfully commercialized in the past years, and a wide range of new bio-chemicals are being developed by manipulating metabolic networks in the engineered microorganisms. Many of the native or engineered metabolic pathways in microorganisms are cross linked, which offer great opportunities to design novel bio-based molecules for a variety of applications.

3.1. Fatty Acids and Derivatives

Fatty acids are carboxylic acids with a long aliphatic chain and have diverse uses in different fields, including food industry, biofuel, cosmetics, rubber, and surfactant, *etc.* Fatty acids usually come from two main processes: chemical synthesis derived from fossil fuel and biosynthesis by organisms using renewable materials [31]. In recent years, because of the gradual and inescapable exhaustion of the Earth's fossil energy resources, biosynthesis of fatty acids in microbial cells as an alternative to traditional synthetic routes has become a research hotspot [32].

In all organisms, fatty acids are synthesized in high flux and converted into phospholipids to build up the cell membranes. Fatty acid biosynthesis and regulation have been extensively characterized (Figure 2) [33, 34], which provides rich information for metabolic engineering to increase the fatty acids production. Organisms create fatty acids from acetyl-coenzyme A (CoA) precursor, which is converted into malonyl-CoA and malonyl-acyl carrier protein (ACP) by acetyl-CoA carboxylase (ACC) and malonyl-CoA:ACP transacylase (FabD), respectively. The β-ketoacyl-ACP synthase III (FabH) initiates the fatty acyl elongation by condensing malonyl-ACP and acetyl-CoA to generate acetoacetyl-ACP. Then, a series of β-keto-reduction, dehydration, and enoyl-reduction reactions, catalyzed by β-ketoacyl-ACP reductase (FabG), (3R)-hydroxyacyl-ACP dehydratase (FabZ), and enoyl-ACP reductase (FabI) respectively, take place in turn to transform the acetoacetyl-ACP (a β-ketoacyl-ACP) into acyl-ACP. The same cycle can be repeated for several times to elongate the growing acyl chain after addition of two carbon atoms from malonyl-ACP as catalyzed by β-ketoacyl-ACP synthase I (FabB). On the other side, organisms also have the β-oxidation pathway to degrade the excess fatty acid, in which fatty acid is activated to acyl-CoA by acyl-CoA synthetase (FadD) and broken down to generate acetyl-CoA, the entry molecule for the citric acid cycle.

Engineering *E. coli* metabolic pathway to produce fatty acids in high yields has been reported by several groups. Steen *et al.* overexpressed a truncated version of the endogenous thioesterase TesA, which lacks the membrane insertion domain and catalyzes the hydrolysis of acyl-ACP, releasing free fatty acids from the fatty acid synthesis pathway. In order to block the fatty acids degradation, they also deleted the *fadE* gene, which encodes the acyl-CoA dehydrogenase for the first step of β-oxidation [35]. The resultant strain had the improved ability to produce free fatty acids, and accumulated 1.2 g/L fatty acids from 2% glucose in shake flasks, reaching 14% of the theoretical limit. Although a range of free fatty acids (C8-C18) could be detected, the C14:0 fatty acid was the major type, indicating that TesA has a substrate preference for C14 fatty acyl-ACP. In another study, Liu *et al.* overexpressed the endogenous acetyl-CoA carboxylase genes *accBCDA* to elevate the intracellular level of malonyl-CoA, overexpressed both the endogenous *tesA* gene and a plant thioesterase gene from *Cinnamomum camphorum*, and deleted the *fadD* gene [36]. This engineered *E. coli* strain produced 4.5 g/L/day total fatty acid in a fed-batch fermentation with 20% of theoretical yield. Moreover, Liu *et al.* provided a high intracellular level of NADPH by overexpressing the endogenous malic enzyme gene *maeB* or adding malate to the culture medium, which showed a positive effect on the fatty acids production in the engineered *E. coli*. Coexpression of acetyl-CoA carboxylase genes from *Acinetobacter calcoaceticus* and *E. coli maeB* gene resulted in a 5.6-fold higher fatty acids production compared with the wild-type strain [37]. As unsaturated fatty acids are preferred for biodiesel production, Cao *et al.* overexpressed two fatty acid synthesis genes (*fabA* and *fabB*) in engineered *E. coli* to elevate the unsaturated fatty acids proportion in the total fatty acids [38].

Metabolic engineering for improved extracellular fatty acid production in microalgae was also reported. In cyanobacterium *Synechocystis* sp. PCC6803, a codon-optimized *E. coli tesA* gene was overexpressed and the *slr2132* gene coding for phosphotransacetylase was knocked out to weaken the polar cell wall layers for fatty acid secretion. In this engineered strain, the fatty acid secretion yield was increased to 197 mg/L at a cell density of 1.0×10^9 cells/mL [39]. To further increase the fatty acid yield and simplify the fatty acid separation process, a CO_2-limitation-inducible green recovery system was constructed. The lipase genes from *Staphylococcus hyicus* and *Fusarium oxysporum* were inserted into the *Synechocystis* chromosome under the control of *cmpA* promoter corresponding to CO_2 concentration. Under CO_2 limitation, the diacylglycerols in cell membrane were degraded, leading to the release of fatty acids into the culture medium [40].

Fatty acids also can be used as platform chemicals to synthesize a series of derivatives that have broad applications, such as alkanes, alkenes, unsaturated fatty acids, fatty alcohols, fatty acid ethyl esters (FAEEs), and hydroxyl fatty acids (HFAs). All these chemicals can be synthesized through the fatty acid synthesis pathway from renewable biomass by metabolic engineered microorganisms.

Alkanes and alkenes are the predominant components of petroleum. Recently, an alkane biosynthesis pathway from cyanobacteria was discovered. In this pathway, the acyl-ACP was reduced to aldehydes by an acyl-ACP reductase (AAR), and the aldehydes were converted into alkanes by an aldehyde decarboxylase (ADC) (Figure 2). An engineered *E. coli* strain carrying those two genes from *Synechococcus* sp. produced 300 mg/L of alkanes (C13-C17), and more than 80% of the produced alkanes were found outside the cells [41]. Furthermore, a three-gene cluster *oleACD* responsible for long-chain alkene biosynthesis from *Micrococcus leteus* was discovered.

Heterologous expression of these genes in a fatty acid-producing *E. coli* led to the production of long-chain internal alkenes, predominantly C27:3 and C29:3 [42]. In the process of alkene biosynthesis, OleA catalyzed a non-decarboxylative Claisen condensation reaction and converted two molecules of fatty acyl-CoA into a β-ketoacid, which was further transformed into olefins by OleC and OleD with unknown mechanisms [43]. In addition, a cytochrome P450 enzyme OleT$_{JE}$ from *Jeotgalicoccus* spp. can catalyze the decarboxylation of free fatty acids to terminal alkenes, and expression of the OleT$_{JE}$ in *E. coli* resulted in terminal alkenes (mostly C18-C20) accumulation [44].

Fatty alcohols are mainly used in the production of detergents and surfactants. They are also components of cosmetics, foods, and industrial solvents. Due to their amphipathic nature, fatty alcohols behave as the nonionic surfactants. To achieve the fatty alcohol production in *E. coli*, the *fadD* gene was overexpressed together with an acyl-CoA reductase gene *acr1* from *Acinetobacter calcoaceticus*, which catalyzed the conversion of acyl-CoA into fatty aldehyde. Then the fatty aldehyde was further reduced to fatty alcohol by the endogenous aldehyde reductase of *E. coli* [35]. Fatty alcohols with different chain length have diverse applications: the C12/14 alcohols can be used as lubricant additives, and the C16/18 alcohols can be employed as defoamers, solubility retarders, and consistency giving factors. For selective production of desired fatty alcohols, the substrate specificity of synthesis enzymes in this pathway was investigated, and appropriate enzymes were selected to construct the pathways for C12/14 and C16/18 fatty alcohol synthesis. After optimization of gene expression level, the C12/14 alcohol production was increased to 449.2 mg/L, accounting for 75.0% of the total fatty alcohol production [45].

Long chain hydroxyfatty acids are widely used in the production of biodegradable plastics, cyclic lactones, and pharmaceutical agents. Although the biosynthesis of hydroxyfatty acid was reported in various organisms, the exogenous addition of fatty acids as substrate was required [46-48]. Recently, Wang *et al.* introduced the fatty acid hydroxylase gene P450$_{BM3}$ from *Bacillus megaterium* into a fatty acid producing *E. coli* strain. This engineered *E. coli* strain produced 117 mg/L hydroxyfatty acids directly from renewable carbohydrates resources in shake flasks [49].

Fatty acid ethyl esters (FAEEs) are currently used as biodiesel with qualified performance for diesel engines. A FAEE-producing *E. coli* strain was constructed based on a fatty acid production strain as following: A pyruvate decarboxylase gene *pdc* and an alcohol dehydrogenase gene *adhB* were overexpressed to convert pyruvate into ethanol. At the same time, FadD was overexpressed to activate free fatty acids to acyl-CoA, and then a wax-ester synthase gene *atfA* was introduced to esterify the acyl-CoA to FAEEs. The resultant strain accumulated 427 mg/L FAEEs with 9.4% of theoretical yield [35].

3.2. 3-Hydroxypropionate

As an important platform chemical, 3-hydroxypropionate (3HP) has been listed as one of the top 12 value-added chemicals from biomass [18]. 3HP can be readily converted to commodity chemicals including acrylic acid, methyl acrylate, acrylamide, PDO, malonic acid, and acrylonitrile. The combined market of those chemicals is well over $1 billion/year.

Traditional chemical synthesis of 3HP mainly derives from acrylic acid and 3-hydroxypropylene nitrile [50-52], however the chemical routes were not commercially feasible because of the high production cost and environmental issues.

3HP was found to be a characteristic metabolic intermediate in *Chloroflexus aurantiacus*, a phototrophic green non sulfur bacterium [53, 54]. The autotrophic CO_2 fixation pathway, termed 3HP cycle, starts with acetyl-CoA carboxylation to malonyl-CoA, which not only serves as a precursor for fatty acid biosynthesis, also can be reduced with NADPH by malonyl-COA reductase (MCR) to 3HP [55]. 3HP is converted to succinyl-CoA with the fixation of another CO_2 molecule [56], and succinyl-CoA is used to regenerate the initial CO_2 acceptor molecule acetyl-CoA, releasing glyoxylate as the primary CO_2 fixation product [57]. Although similar pathways have been discovered in various bacteria like *Metallosphaera sedula*, *Sulfolobus* spp. [58], and *Cenarchaeum* [59], no known organism produces 3HP as a major metabolic end product. Several organisms are able to produce 3HP from acrylic acid, 1,3-PDO, or propionate [60].

Unfortunately, these routes are commercially inefficient because the starting compounds required are even more expensive than 3HP. In view of this, genetically modifying metabolic pathways is required for efficient 3HP biosynthesis.

Two main carbon resources, glycerol and glucose, were often used in the 3HP biosynthesis. Glycerol is the main by-product of biodiesel production, and the price of crude glycerol decreased to only $0.011/kg [61]. For 3HP biosynthesis from glycerol, Suthers *et al.* introduced glycerol dehydratase gene *dhaB* from *Klebsiella pneumoniae* and aldehyde dehydrogenase gene *ald4* from *Saccharomyces cerevisiae* into *E. coli* strain (Figure 3A). DhaB converted glycerol into 3-hydroxypropionic aldehyde (3HPA) in a coenzyme B_{12}-dependent manner, and Ald4 further oxidized 3HPA to 3HP. Co-overexpression of these two enzymes resulted in 0.17 g/L 3HP production using glycerol as sole carbon resource under anaerobic conditions [62].

To improve the 3HP production, a series of researches have been done during the past few years. Along with the oxidation of 3HPA to 3HP, a mass of NADH was produced, leading to the imbalance of cellular reducing power. Shinzo overexpressed the propanol dehydrogenase catalyzing 3HPA to 1,3-PDO to consume the excess NADH (Figure 3A), and replaced the *ald4* gene with *pdu* operon from *Klebsiella*. The engineered strain produced 3HP and 1,3-PDO *simultaneously with 40% of theoretical yield [63]. Raj et al. replaced ald4 gene with E. coli aldH gene, resulting in 2.5 g/L 3HP production [64, 65]. However the 3HP titer was still too low to be produced commercially. The low titer was mainly attributed to the instability of DhaB and the imbalanced activities between DhaB and AldH [64]. To stabilize the DhaB enzyme, glycerol dehydratase reactivase (GDR) genes gdrAB were co-overexpressed along with dhaB.*

Also the α-ketoglutaric semialdehyde dehydrogenase KGSADH from Azospirillum brasilense, which had a better activity to produce 3HP, was used as an alternative to AldH. The recombinant strain produced 38.7 g/L 3HP with 35% of the theoretical yield under aerobic fed-batch conditions [65].

In order to further improve the 3HP production, enzymes from diverse organisms were characterized and used in 3HP biosynthesis. For example, the CoA-dependent propionaldehyde dehydrogenase PduP from Lacobacillus exhibited broad substrate specificity and can utilize both NAD^+ and $NADP^+$ as cofactor [66], and the aldehyde dehydrogenase PuuC from K. pneumoniae showed higher efficiency [67].

Besides *E. coli*, *K. pneumoniae* was also manipulated as the host of extraneous metabolic pathway to produce the 3HP. Compared with *E. coli*, *K. pneumoniae* has some expected advantages: *K. pneumoniae* has endogenous glycerol dehydratase and glycerol dehydratase activitase, which increase the strain tolerance to the high concentration of glycerol and dissolved oxygen [64, 68].

Furthermore, *K. pneumoniae* can synthesize coenzyme B_{12}, avoiding addition of the expensive coenzyme B_{12} [69]. The 3HP production in two-stage oxidoreduction potential control fermentation reached 2.8g/L, accompanied with 9.8 g/L PDO (1,3-propanediol) [70]. Deletion of *dhaT* gene, encoding the main propanol dehydrogenase of *K. pneumoniae*, increased 3HP production to 3.6 g/L along with 3.0 g/L PDO in shake flasks and 16 g/L 3HP with the co-production of 16.8 g/L PDO under microaerobic fed-batch fermentation in 24h [71]. In a glycerol fed-batch bioreactor experiment under a constant DO (dissolved oxygen) of 5%, the strain *K. pneumoniae* Δ*dhaT*Δ*yqhD* overexpressing both PuuC and DhaB could produce >28 g/L 3HP in 48 h with a yield of 40% on glycerol [69].

Seven 3HP synthetic pathways have been patented by Cargill *et al.* [72-75], in which 3HP could be produced at a theoretical yield of 100% from glucose. All these seven routes started from common metabolic intermediates such as pyruvate, acetyl-CoA, and aspartate, so various C5 and C6 sugars derived from lignocellulosic biomass can be used as raw material for 3-HP production. Among these pathways, the most representative one used lactyl-CoA as the intermediate (Figure 3B).

In this pathway, lactic acid was activated into lactyl-CoA and then converted into acryloyl-CoA by propionyl-CoA transferase (PCT) and lactyl-CoA dehydratase (LCD) from *Megasphaera elsdenii*, respectively. Then acryloyl-CoA was transformed into 3HP-CoA by 3HP-CoA dehydratase (OS19) from *C. aurantiacus*. After the improvement of key enzyme activities and the optimization of culture conditions, the recombinant strain can accumulate 3HP up to 25 g/L. However, LCD was easily inactivated by oxygen, so the fermentation had to be performed under strictly anaerobic conditions. In addition, the overexpression of PCT affected the cellular pool of acetyl-CoA and inhibited the cell growth. Those disadvantages severely limited the application of this pathway.

Another interesting route from glucose was the malonyl-CoA pathway, in which the common intermediate malonyl-CoA was directly transformed into 3HP by malocyl-CoA reductase (MCR) from *C. aurantiacus* (Figure 3B).

A high conversion yield of 3HP from glucose was expected since the malonyl-CoA route was energetically well balanced. The conversion of glucose to each mol of acetyl-CoA generates 2 mol of NADH, 1 mol of ATP and 1 mol of CO_2; while carboxylation of each mol of acetyl-CoA to malonyl-CoA utilizes 1 mol of CO_2 along with 1 mol of ATP, and subsequent reduction of malonyl-CoA to 3-HP utilizes 2 mol of NADPH. The recombinant *E. coli* strain carrying *mcr* gene produced 0.064 g/L 3HP in 24 h in the shake flask cultivation under aerobic conditions with glucose as the sole carbon source [76]. Co-overexpression of the genes *accADBC* and *pntAB*, encoding acetyl-CoA carboxylase and nicotinamide nucleotide transhydrogenase that converts NADH to NADPH respectively, increased 3-HP production to 0.19 g/L [76]. Low activity of malonyl-CoA reductase should be responsible for the low 3HP titer. Under fermentation conditions, the remained activity of MCR at 37°C was only about 50% of the maximal-activity at 57°C [77]. Furthermore, MCR activity was strictly regulated by its postulated physiological requirement.

It was reported that only about 37% of the maximal rate could be measured in heterotrophically grown *C. aurantiacus* cells, although these cells grew five times faster than cells under autotrophic conditions [55]. When expressed in *E. coli*, MCR enzymatic activity could be further down-regulated due to the change of physiological environment.

3.3. Advanced Alcohols

Bioethonal fermented from corn and biodiesel esterified from edible vegetable oils and animal fats were termed as the first generation biofuels, which have captured 90% of the current biofuel market [78]. However, ethanol is not an ideal fuel because it has a lower energy density than gasoline. Additionally, ethanol is miscible with water, making it difficult to distill from culture broth, and corrosive to storage and distribution infrastructures [79]. On the other hand, longer chain alcohols are non-hygroscopic, less volatile, and have higher energy densities compared with ethanol. These advantages allow them to be used in existing gasoline, diesel, and jet engines. Except for 1-butanol [80, 81], there is no report showing longer chain alcohols production from renewable sources with high yield. Although some alcohols such as isobutanol, 2-methyl-1-butanol, and 1-hexanol have been identified as microbial by-products [82-84], none of them can be produced from renewable sources at industrial levels. Due to the energy and environmental concerns, metabolic engineering for microbial production of longer chain alcohols has already become a research hotpoint.

Unlike the fermentative pathways for ethanol and 1-butanol, longer chain alcohols are mainly derived from acyl-CoA and 2-keto acids. As mentioned above, acyl-CoA, the activated form of long-chain fatty acids, can be converted to corresponding alcohols by appropriate acyl-CoA reductase and aldehyde reductase (Figure 4A) [35, 45]. 2-keto acids are intermediates in amino acid biosynthesis in all organisms, and can be converted to aldehyde by 2-keto acid decarboxylases (KDCs) and then to alcohols by alcohol dehydrogenase (ADH) or aldehyde reductase (Figure 4A). A 2-keto acid pathway was constructed in *E. coli* by introducing the *kivd* gene from *Lactococcus lactis* [85] and the *adh2* or *adh6* gene from *S. cerevisiae* [86, 87]. A wide range of 2-keto acids derived from various amino acid biosynthesis pathways can be converted to their corresponding alcohols. For example, 2-ketobutyrate and 2-keto-3-methyl-valerate in isoleucine biosynthesis pathway can be converted to 1-propanol and 2-methyl-1-butanol; 2-keto-isovalerate in valine biosynthesis pathway is the substrate for isobutanol; 2-keto-4-methyl-pentanoate in leucine biosynthesis pathway is the precursor for 3-methyl-1-butanol (3MB); phenylpyruvate in phenylalanine biosynthesis pathway can be converted into 2-phenylethanol [88]. The alcohol production is completely determined by the 2-keto acid pool of the engineered *E. coli*, and overexpression of pathway genes to increase the flux toward a specific 2-keto acid can elevate the production level of corresponding alcohol. For example, isobutanol production at 22 g/L with 86% of theoretical yield was achieved when the *alsS* gene from *Bacillus subtilis* and endogenous genes *ilvCD* were overexpressed to enhance the 2-ketoisovalerate biosynthesis [88]. To further improve the 3MB production in the engineered *E. coli*, random mutagenesis was performed. Under selective pressure toward leucine biosynthesis, a new *E. coli* strain with increased 3MB production was selected. Using the two-phase fermentation technique to continuously remove the toxic 3MB from the aqueous cellular environment, this strain produced 9.5 g/L 3MB with a yield of 0.11 g/g glucose in 60 h [89].

Different with short chain (C3-C5) alcohols which can be directly derived from intermediates in amino acid biosynthesis, longer chain alcohols synthesis involves two types of 2-keto acid chain elongation pathways: the 2-isopropylmalate synthase (IPMS) pathway and the acetohydroxy acid synthase (AHAS) pathway (Figure 4B). The non-recursive nature IPMS pathway is for the ketoisocaproate production in leucine biosynthesis, which employs one acetyl-CoA to extend the 2-keto acid chain by one carbon unit.

In IPMS pathway, LeuA is the first enzyme and determines the substrate specificity of this elongation pathway. Through structure-based enzymatic engineering of LeuA, several LeuA mutants were selected which preferentially selected longer-chain substrates for catalysis and can recursively catalyze up to five elongation cycles [90]. Using this engineered *leuA*BCD* operon, the simultaneous production of C3-C8 *n*-alcohols was achieved in a threonine-hyperproduction strain carrying the *thrABC* operon (driving the carbon flux toward threonine), *ilvA* gene (converting threonine into 2-ketobutyrate), *leuA* gene (G462D mutant, providing elongation activity for small substrates), and *kivd-adh6* gene (catalyzing 2-keto acids into alcohols) [90]. The nature role of AHAS pathway is to produce 2-keto-isovalerate in valine/leucine biosynthesis. In long chain alcohols biosynthesis, this pathway is used to increase the 2-keto acid carbon number by two and generate a branch point at the expense of one pyruvate molecule. Furthermore, these two elongation pathways can be applied repetitively in hybrid fashion to generate a series of 2-keto acids with different carbon numbers and structures [91].

For the acyl-CoA chain elongation, besides the fatty acid biosynthesis pathway discussed above (Figure 2), a CoA-dependent reverse β-oxidation cycle was reported recently in *E. coli* (Figure 4B) [92, 93]. This pathway elongates the chain by two carbon units with consumption of one acetyl-CoA, and enables the elongation at maximum carbon and energy efficiency, better than 2-keto acid elongation pathway and fatty acid synthesis pathway which requires ATP-dependent activation of acetyl-CoA to malonyl-CoA. Combined with appropriate thioesterases and dehydrogenases, this pathway was successfully used to produce 1-hexanol [92] and 1-octanol [93] in *E. coli*. This pathway can also be used in long-chain fatty acids (C>10) production at higher efficiency.

1,4-Butanediol (BDO) is an important commodity alcohol, which is industrially used to produce over 2.5 million tons of plastics, polyesters and spandex fibers annually. BDO can not be produced naturally by any known organism, and is manufactured entirely from petroleum-based feedstocks. To design an artificial BDO synthetic pathway, Simpheny Biopathway Predictor software was used to elucidate all potential pathways based on transformation of functional groups by known chemistry [94].

From over 10,000 identified pathways for BDO synthesis from common metabolites, the pathway proceeding through the 4-hydroxybutyrate intermediate was selected as the highest priority using the following criteria: theoretical BDO yield, pathway length, thermodynamic feasibility, number of non-native steps, and number of steps without currently identified enzymes. To construct this artificial route in *E. coli*, candidate genes from various bacteria were expressed and tested *in vivo* to determine their effects on BDO production, and appropriate enzymes were elected to assemble the artificial BDO synthetic pathway. In the engineered *E. coli* strain, the intermediate succinate from the TCA cycle was activated to succinyl-CoA by the entogenous enzyme succinyl-CoA synthetase SucCD, and then the succinyl-CoA was converted into succinyl semialdehyde by CoA-dependent succinate semialdehyde dehydrogenase SucD from *Porphyromonas gingivalis* [95].

The succinyl semialdehyde can also be derived from another TCA cycle intermediate α-ketoglutarate by α-keto acid decarboxylase SucA from *Mycobacterium bovis* [96]. Catalyzed by 4-hydroxybutyrate (4HB) dehydrogenase 4HBd from *P gingivalis* [95], succinyl semialdehyde was transformed to 4HB, which is activated to CoA derivative by 4HB-CoA transferase Cat2 from *P gingivalis* [97]. With two sequential reduction reactions catalyzed by CoA-dependent aldehyde dehydrogenase encoded by *Clostridium beijerinckii* gene 025B [98] and *E. coli* endogenous alcohol dehydrogenases respectively, 4HB-CoA was converted into BDO via 4-hydroxybutyraldehyde. To achieve a high level BDO production, the competitive metabolic pathways for ethanol, formate, lactate and succinate were blocked by deletion of *adhE* (alcohol dehydrogenase gene), *pfl* (pyruvate formate lyase gene), *ldh* (lactate dehydrogenase gene) and *mdh* (malate dehydrogenase gene). To improve the cell growth under microaerobic condition, *K. pneumoniae lpdA* gene, which was proved functional anaerobically [99], was used to replace the *E. coli* native *lpdA* gene encoding lipoamide dehydrogenase, an E3 component of the pyruvate dehydrogenase. For better expression of endogenous enzymes for succinate synthesis, *arcA* gene encoding a transcriptional repressor [100] was knocked out. In addition, a mutation was introduced into the citrate synthase gene *gltA* to improve TCA cycle flux [101]. The resultant *E. coli* strain produced over 18 g/L BDO from glucose in 5 days [94]. Furthermore, this stain can produce BDO from xylose, sucrose and biomass-derived mixed sugar streams. This work presented a systems-based metabolic engineering approach to design, construct and optimize the artificial pathway which is not naturally existed in any known organisms.

3.4. Polyhydroxyalkanoates

Over the past 20 years, exploration into alternative plastics derived from biologically renewable resources has never ceased. Polyhydroxyalkanoates (PHAs) represent one class of biopolymers that are currently under extensive investigation. PHAs are a family of bacterial polyesters, which are synthesized by various microorganisms as intracellular storage compounds for energy and carbon under unfavorable growth conditions due to the imbalanced nutrient supply. They are generally biodegradable, biocompatible, and have sustainable properties. Since polyhydroxybutyrate (PHB) was discovered by Lemoigne in 1926, more than 150 possible monomer units have been identified, which offer a broad range of PHAs to compete with petrochemically derived polymers such as polypropylene and polyethylene [102]. PHAs have been developed into a wide array of uses ranging from single-use bulk, commodity plastics, to specialized medical applications [103, 104]. Recent studies have also demonstrated that PHAs can be used as biofuels or fuel additives [105].

PHA polymers are produced via a series of enzymatic reactions in both native and recombinant organisms. In the native PHA-producing organisms, PHAs are accumulated as granules that are surrounded by specific lipids and proteins [106]. Recombinant PHA-producing organisms are engineered to produce PHAs with various physical properties depending on the types of PHA synthases, the starting carbon feedstocks and the PHA synthesis processes [107, 108]. Based on the structures of the monomers, PHA are classified into short-chain-length (scl) PHAs, which consist of monomers with 3-5 carbon units such as 3HP, 3-hydroxybutyrate (3HB), 4-hydroxybutyrate (4HB), and 3-hydroxyvalerate (3HV); medium-chain-length (mcl) polymers, which consist of monomers with 6-14 carbon units such as 3-hydroxyhexanoate (3HHx), 3-hydroxyheptanoate (3HHp) and 3-hydrotetradecanoate (3HTD). On the basis of types and arrangement of the monomers, PHAs are divided into homopolymers PHAs including P3HP, P3HB, P4HB and PHV; random copolymers PHAs like PHBV, PHBHHx, and P3HB4HB; block copolymer PHAs such as PHB-*b*-PHVHHp, PHB-*b*-PHBV and PHB-b-PHHx. So far, PHB, PHBV, PHBHHx, P3HB4HB and mcl PHA have been applied in large-scale production [109].

PHB is the most representative homopolymers in PHA family for its high contents both in wild type bacteria *Ralstonioa eutropha* and recombinant *E. coli*. As the most common wild type strain for the industrial production of PHB, *Ralstonia eutropha* could reach over 200g/L cell density containing over 80% PHB in a one cubic meter fermentor after 60h of fermentation (Tianjin Northern Food Co. Ltd). Another wild strain *Alcaligenes latus* was reported to reach a cell density of 142 g/L with 50% PHB after just 18 h of growth. When using recombinant *E. coli* for PHB production, the cell density could reach 206 g/L containing 73% PHB with a productivity of 3.4 g/L/h. There were also a large number of reports about other homopolymers, including poly(3-hydroxypropionate) (P3HP), poly(4-hydroxybutyrate) (P4HB), poly(3-hydroxyvalerate) (PHV), poly(3-hydroxyhexanoate) (PHHx), poly(3-hydroxyheptanoate) (PHHp), poly(3-hydroxyoctanoate) (PHO), poly(3-hydroxynonanoate) (PHN), poly(3-hydroxydodecanoate) (PHD), and poly(3-hydroxydodecanoate) (PHDD).

Polylactic acid (PLA) was normally synthesized by a two-step process including lactic acid fermentation and chemical polymerization. Recently, engineered *E. coli* strain for PLA production was developed. In this strain, lactyl-CoA was supplied by *Clostridium propionicum* propionate CoA transferase, polymerized by the mutants of PHA synthetase from *Pseudomonas* sp. MBEL 6-19 processing broad substrate specificities (Figure 5) [110, 111]. Then these two key enzymes were further engineered for higher efficiency. Though the yield was still too low, this technology provides the possibility to biosynthesize different PHA polymers.

While PHB has limited applications due to its poor thermal and mechanical properties, the random incorporation of a second monomer exhibits significant improvements [103, 112]. Owing to the improvement in physical properties, 3HB based copolymers have attracted industrial interests. Typically, copolymers PHBV, P3HB4HB and PHBHHx have been developed into large-scale application [109]. However, the copolymer production needs addition of corresponding precursors, which added extra expenses for copolymer production. Therefore, to reduce the cost of production, a great deal of efforts has been made to design new engineered pathways using low cost substrate for copolymers production (Figure 5). Qiu *et al.* engineered *Aeromonas hydrophila* 4AK4 and *Pseudomonas putida* GPp104 to synthesize poly(3-hydroxybutyrate-*co*-3-hydroxyhexanoate) (PHBHHx) using gluconate and glucose as carbon sources [113]. In this study, a truncated *tesA* gene was introduced into *A. hydrophila* 4AK4. When additional PHBHHx synthesis genes, *phaPCJ*, were co-overexpressed, the PHBHHx content reached 15% (wt/wt cell dry weight) with 19% (mol/mol) 3-hydroxyhexanoate. The recombinant strain *Pseudomonas putida* GPp104 harboring *phaC* gene encoding PHBHHx synthase of *A. hydrophila*, *phaB* gene encoding acetoacetyl-CoA reductase of *Ralstonia eutropha* and *phaG* gene encoding 3-hydroxyacyl-ACP-CoA transferase of *P. putida*, can synthesize 19% (wt/wt cell dry weight) PHBHHx containing 5% (mol/mol) 3-hydroxyhexanoate from glucose [114].

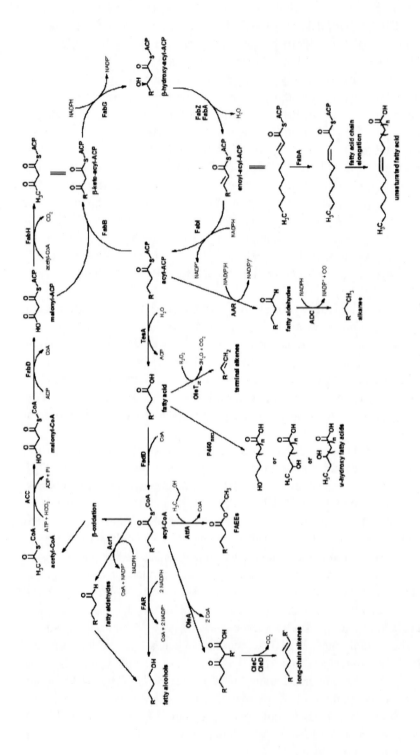

Figure 2. Pathways for the production of fatty acids and derivatives. ACC, acetyl-CoA carboxylase; FabD, malonyl-CoA:ACP transacylase; FabH, β-keto-acyl-ACP synthase III; FabB, β-keto-acyl-ACP synthase I; FabG, β-keto-acyl-ACP reductase; FabZ, β-hydroxyacyl-ACP dehydratase; FabI, enoyl-acyl-ACP reductase; TesA, acyl-ACP thioesterase; FadD, acyl-CoA synthase; Acr1, acyl-CoA reductase; FAR, acyl-CoA reductase; OleA, enzyme promoting condensation of two long-chain fatty acids; OleC and OleD, enzymes in long chain alkenes synthesis pathway with unknown intermediates and mechanism; AtfA, wax-ester synthase; P450$_{BM3}$, a cytochrome P450 enzyme catalyzing fatty acid hydroxylation; AAR, acyl-ACP reductase; ADC, aldehyde decarboxylase; OleT$_{JE}$, a cytochrome P450 enzyme that reduces fatty acids to alkenes; FabA, β-hydroxyacyl-ACP dehydratase with *trans*-2-decenoyl-ACP isomerase activity.

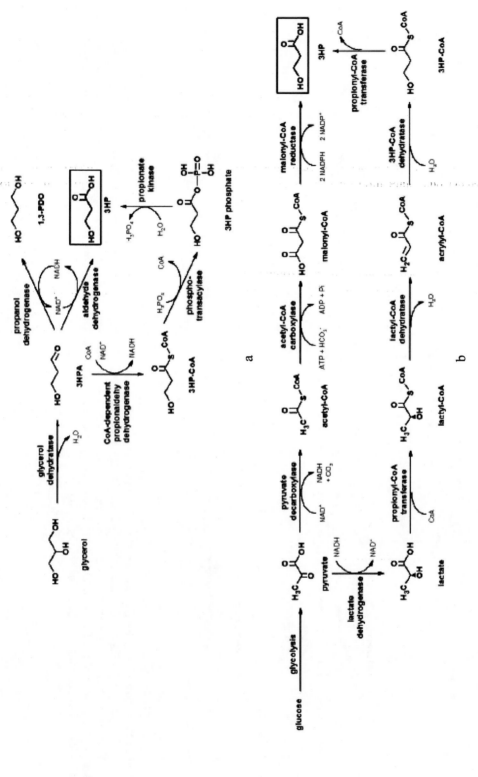

Figure 3. Biosynthetic pathways for 3HP from glycerol (A) and glucose (B). 3HP, 3-hydroxypropionate; 3HPA, 3-hydroxypropionaldehyde; 1,3-PDO, 1,3-propanediol.

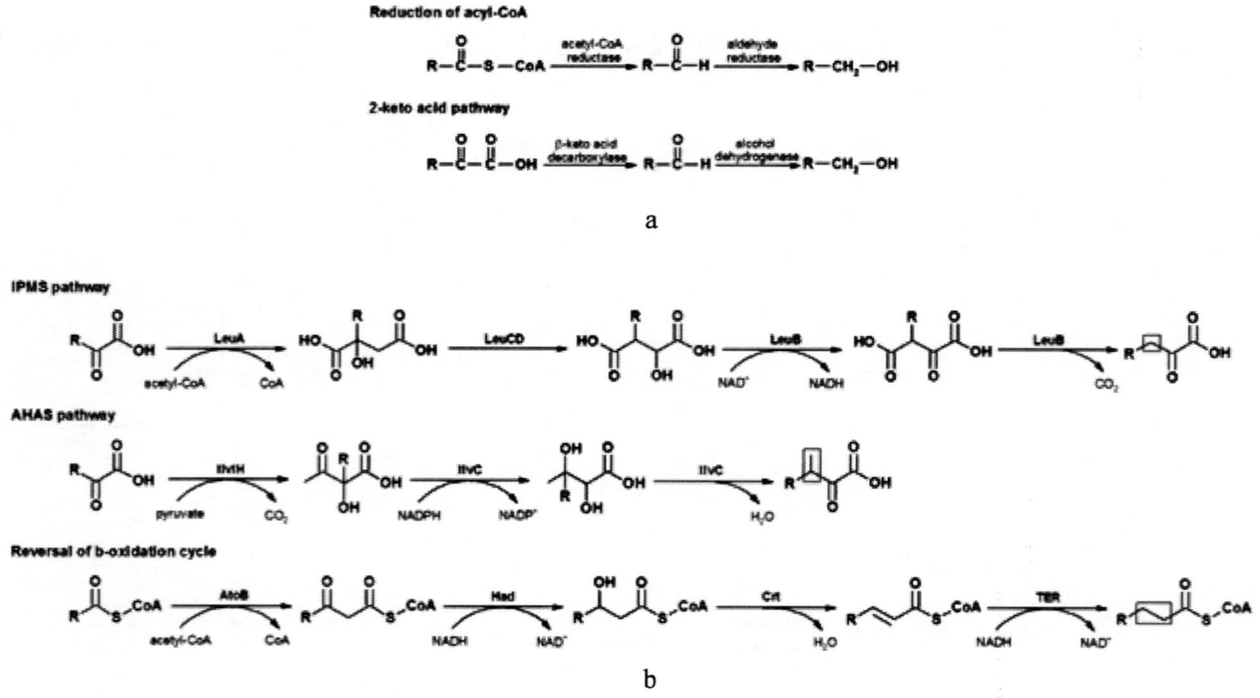

Figure 4. Schematic biosynthetic pathways for longer-chain or branch alcohols. (A) Pathways used to introduce hydroxyl group into the desirable product. (B) Carbon chain elongation pathways used to compose desirable hydrocarbon chains. Fatty acid synthesis pathway, which is mentioned in Fig 2, is also used for chain elongation. IPMS, isopropylmalate synthase in leucine biosynthesis; LeuA, 2-isopropylmalate synthase; LeuCD, isopropylmalate isomerase complex; LeuB, isopropylmalate dehydrogenase; AHAS, acetohydroxy acid synthase in valine/isoleucine biosynthesis; IlvIH, acetohydroxybutanol synthase; IlvC, 2-aceto-2-hydrpxybutyrate reductase; IlvD, 2,3-dihydroxy-3-methylvalerate hydrolysase; AtoB, acetyl-CoA acetyltransferase; Had, hydroxyacetyl-CoA dehydrogenase; Crt, crotonase; Ter, trans-enoyl-CoA reductase.

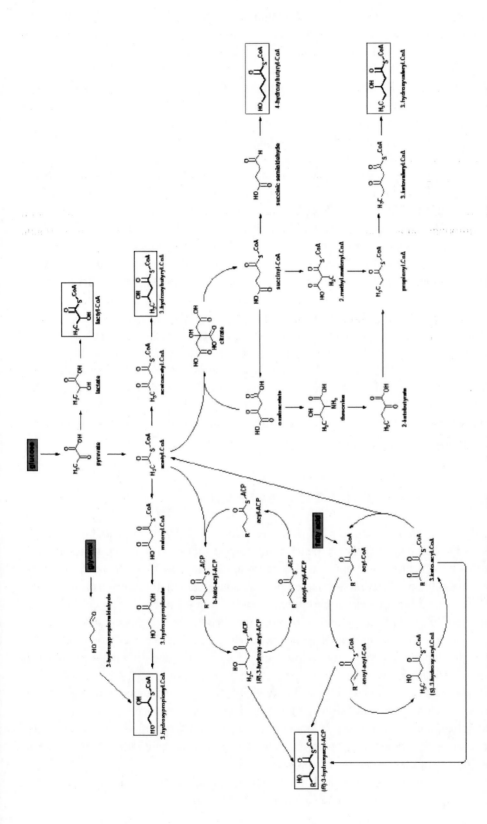

Figure 5. Metabolic pathways for precursors of polyhydroxalkanoates (PHAs) from carbon sources including glucose, glycerol, and fatty acids. The precursors with different carbon unit numbers (shown in boxes) can be polymerized into PHA homopolymers, random copolymers, and block copolymers. Polylactic acid (PLA) and poly(4-hydroxybutyrate) are also included as special polyesters produced by PHA related pathways.

In addition, genes involved in succinate degradation in *Clostridium kluyveri* and PHB accumulation pathway of *Ralstonia eutropha* were co-expressed for the synthesis of [P(3HB-co-4HB)]. After the *E. coli* native succinate semialdehyde dehydrogenase genes *sad* and *gabD* were both deleted to enhance the carbon flux to 4HB biosynthesis, the recombinant mutant produced 9.4g/L cell dry weight containing 65.5% P(3HB-*co*-11.1mol%4HB) using glucose as sole carbon source.

Though the physical properties of PHA copolymers has been remarkably enhanced, they still suffered from property detrimental aging effects as native PHB [115]. To overcome the material property defects of copolymers, block copolymers (consisting of polymer chains which have two or more unique polymer regions covalently bonded together) have been found to be a feasible way [116]. So far, various types of block copolymers such as PHB-*b*-PHBV, PHB-*b*-PHVHHp, P3HB-*b*-P4HB and PHB-*b*-PHHx have been successfully produced, which showed improved thermal and mechanical properties [116-119]. The first report about biosynthesis of block polymer appeared in 2006, and in this study PHB-*b*-PHBV block copolymer was synthesized by *Ralstonia eutropha*. The block polymer materials were against polymer aging, while it also showed some limited characteristics due to the inherited brittleness of scl PHA [116]. To further improve the properties, a β-oxidation weakened *P. putida* KTOY06ΔC strain was constructed to produce the scl and mcl PHA block copolymer. The *P. putida* produced copolymer had two blocks, with poly-3-hydroxybutyrate (PHB) as one block and random copolymer of 3-hydroxyvalerate (3HV) and 3-hydroxyheptanoate (3HHp) as another block. The resulting block copolymer showed the highest tensile strength of 7.5 MPa and Young's modulus of 366 MPa, much better than random copolymer (7MPa, 127MPa), and homopolymers of PHB and PHVHHp (5.3Mpa, 257MPa) [117]. Besides random copolymers, block copolymers provided a new strategy to gain novel materials with improved or specific properties.

4. FUTURE PROSPECTS

Although various bio-based products have been developed by new technologies and a wide range of bio-based chemicals are penetrating diverse industrial markets, there are still challenges that need further developments. Overcoming the economic and technical barriers for the bioprocessing of different bio-based products is essential for the development of economically competitive processes based on modern biological and technical sciences. Availability, variability and sustainability of non-food feedstocks (lignocellulose or algae feedstock) for bio-production, are major issues to be addressed. For lignocellulose feedstock, one major challenging area of research is to develop inexpensive and effective enzymes for lignocellulose saccharification.

Another challenging task is to utilize the synthetic biology and metabolic engineering approaches to develop more efficient and robust microbes for the bioconversion processes. Correspondingly, improvements in fermentation strategy, media optimization and recovery strategy have to be performed along with the genetic engineering techniques to improve the yield and efficiency of the bio-production process.

Also, reuse of process streams and integration of processes for reducing the energy demands can be applied to make the conversion process economical.

The development of highly efficient "biorefinery" strategies that integrate the production of several bio-based products in one process could reduce the costs and allow bio-based products to economically compete with the petroleum-based products.

A particularly encouraging phenomenon is the extent of industry's efforts in pulling the research and development of bio-based products. There are now plenty of examples showing that the industry not only employed products and production methods from laboratories, but also was involved in RandD and created more demands for new candidate products. For example, the maturation of bio-isoprene from the laboratory bench to large-scale production was under the cooperation between Genencor and Goodyear in the automotive and tire industries, in which the industrial part offered a clear and practical proposition. In summary, cost-effective solutions for the techno-eco barriers of bio-based products will depend on continued research and development in laboratories, capital investment, policy support from the government, technical and financial support from industries, and market acceptance with existing business infrastructure. Fortunately, we are on the way to build a worldwide bio-based economy which is already well known for the scientists, industries, business, consumers and our governments.

REFERENCES

[1]   Kerr, R. A. Oil resources - The looming oil crisis could arrive uncomfortably soon. *Science* 2007;316:351.

[2]   Erickson, B., Nelson, J. E., Winters, P. Perspective on opportunities in industrial biotechnology in renewable chemicals. *Biotechnol. J.* 2012;7:176.

[3]   Singh, R. Facts, Growth, and Opportunities in Industrial Biotechnology. *Org. Process Res. Dev.* 2011;15:175.

[4]   King, D., Inderwildi, O. R., Williams, A. *The Future of Industrial Biorefineries*. World Economic Forum 2010.

[5]   Menon, V., Rao, M. Trends in bioconversion of lignocellulose: Biofuels, platform chemicals and biorefinery concept. *Prog. Energ. Combust.* 2012;38:522.

[6]   Sticklen, M. B. Plant genetic engineering for biofuel production: towards affordable cellulosic ethanol. *Nat. Rev. Genet.* 2008;9:433.

[7]   Fu, C. X., Mielenz, J. R., Xiao, X. R., Ge, Y. X., Hamilton, C. Y., Rodriguez, M., et al. Genetic manipulation of lignin reduces recalcitrance and improves ethanol production from switchgrass. *Proc. Natl. Acad. Sci. US* 2011;108:3803.

[8]   Park, J. H., Lee, S. Y., Kim, T. Y., Kim, H. U. Application of systems biology for bioprocess development. *Trends Biotechnol.* 2008;26:404.

[9]   Jang, Y. S., Park, J. M., Choi, S., Choi, Y. J., Seung, D. Y., Cho, J. H., et al. Engineering of microorganisms for the production of biofuels and perspectives based on systems metabolic engineering approaches. *Biotechnol. Adv.* 2012;30:989.

[10]   Kim, H. U., Kim, T. Y., Lee, S. Y. Metabolic flux analysis and metabolic engineering of microorganisms. *Mol. Biosyst.* 2008;4:113.

[11]   Lee, S. Y., Lee, D. Y., Kim, T. Y. Systems biotechnology for strain improvement. *Trends Biotechnol.* 2005;23:349.

[12] Palsson, B., Zengler, K. The challenges of integrating multi-omic data sets. *Nat. Chem. Biol.* 2010;6:787.

[13] Park, J. M., Kim, T. Y., Lee, S. Y. Constraints-based genome-scale metabolic simulation for systems metabolic engineering. *Biotechnol. Adv.* 2009;27:979.

[14] Olofsson, K., Wiman, M., Liden, G. Controlled feeding of cellulases improves conversion of xylose in simultaneous saccharification and co-fermentation for bioethanol production. *J. Biotechnol.* 2010;145:168.

[15] Eiteman, M. A., Lee, S. A., Altman, E. A co-fermentation strategy to consume sugar mixtures effectively. *J. Biol. Eng.* 2008;2:3.

[16] Ezeji, T. C., Qureshi, N., Blaschek, H. P. Bioproduction of butanol from biomass: from genes to bioreactors. *Curr. Opin. Biotechnol.* 2007;18:220.

[17] Xu, Q., Li, S., Huang, H., Wen, J. Key technologies for the industrial production of fumaric acid by fermentation. *Biotechnol. Adv.* 2012;30:1685.

[18] Werpy, T., Petersen, G. Top value added chemicals from biomass. Washington D.C.: Office of Energy Efficiency and Renewable Energy, US Department of Energy; 2004.

[19] Lee, S. J., Lee, D. Y., Kim, T. Y., Kim, B. H., Lee, J. W., Lee, S. Y. Metabolic engineering of *Escherichia coli* for enhanced production of succinic acid, based on genome comparison and in silico gene knockout simulation. *Appl. Environ. Microbiol.* 2005;71:7880.

[20] Park, S. J., Lee, S. Y., Kim, T. W., Jung, Y. K., Yang, T. H. Biosynthesis of lactate-containing polyesters by metabolically engineered bacteria. *Biotechnol. J.* 2012;7:199.

[21] Wee, Y. J., Kim, J. N., Ryu, H. W. Biotechnological production of lactic acid and its recent applications. *Food Technol. Biotechnol.* 2006;44:163.

[22] Nakamura, C. E., Whited, G. M. Metabolic engineering for the microbial production of 1,3-propanediol. *Curr. Opin. Biotechnol.* 2003;14:454.

[23] Atsumi, S., Wu, T. Y., Eckl, E. M., Hawkins, S. D., Buelter, T., Liao, J. C. Engineering the isobutanol biosynthetic pathway in *Escherichia coli* by comparison of three aldehyde reductase/alcohol dehydrogenase genes. *Appl. Microbiol. Biotechnol.* 2010; 85:651.

[24] Whited, G. M., Feher, F. J., Benko, D. A., Cervin, M. A., Chotani, G. K., McAuliffe, J. C., et al. Development of a gas-phase bioprocess for isoprene-monomer production using metabolic pathway engineering. *Ind. Biotechnol.* 2010;6:152.

[25] Delhomme, C., Weuster-Botz, D., Kuhn, F. E. Succinic acid from renewable resources as a C-4 building-block chemical-a review of the catalytic possibilities in aqueous media. *Green Chem.* 2009;11:13.

[26] Lu, S. F., Chen, M., Chen, C. H. Mechanisms and kinetics of thermal degradation of poly(butylene succinate-co-propylene succinate)s. *J. Appl. Polym. Sci.* 2012;123:3610.

[27] Zheng, L. C., Li, C. C., Guan, G. H., Zhang, D., Xiao, Y. N., Wang, D. J. Investigation on Isothermal Crystallization, Melting Behaviors, and Spherulitic Morphologies of Multiblock Copolymers Containing Poly(butylene succinate) and Poly(1,2-propylene succinate). *J. Appl. Polym. Sci.* 2011;119:2124.

[28] Cheng, K. K., Zhao, X. B., Zeng, J., Wu, R. C., Xu, Y. Z., Liu, D. H., et al. Downstream processing of biotechnological produced succinic acid. *Appl. Microbiol. Biotechnol.* 2012;95:841.

[29] Vink, E. T. H., Rabago, K. R., Glassner, D. A., Springs, B., O'Connor, R. P., Kolstad, J., et al. The sustainability of NatureWorks™ polylactide polymers and Ingeo™ polylactide fibers: an update of the future. *Macromol. Biosci.* 2004;4:551.

[30] Yu, C., Cao, Y. J., Zou, H. B., Xian, M. Metabolic engineering of *Escherichia coli* for biotechnological production of high-value organic acids and alcohols. *Appl. Microbiol. Biotechnol.* 2011;89:573.

[31] Chisti, Y. Biodiesel from microalgae. *Biotechnol. Adv.* 2007;25:294.

[32] Chisti, Y. Biodiesel from microalgae beats bioethanol. *Trends Biotechnol.* 2008;26:126.

[33] Chan, D. I., Vogel, H. J. Current understanding of fatty acid biosynthesis and the acyl carrier protein. *Biochem. J.* 2010;430:1.

[34] Fujita, Y., Matsuoka, H., Hirooka, K. Regulation of fatty acid metabolism in bacteria. *Mol. Microbiol.* 2007;66:829.

[35] Steen, E. J., Kang, Y., Bokinsky, G., Hu, Z., Schirmer, A., McClure, A., et al. Microbial production of fatty-acid-derived fuels and chemicals from plant biomass. *Nature* 2010;463:559.

[36] Liu, T., Vora, H., Khosla, C. Quantitative analysis and engineering of fatty acid biosynthesis in *E. coli. Metab. Eng.* 2010;12:378.

[37] Meng, X., Yang, J., Cao, Y., Li, L., Jiang, X., Xu, X., et al. Increasing fatty acid production in *E. coli* by simulating the lipid accumulation of oleaginous microorganisms. *J. Ind. Microbiol. Biotechnol.* 2011;38:919.

[38] Cao, Y. J., Yang, J. M., Xu, X., Liu, W., Xian, M. Increasing unsaturated fatty acid contents in *Escherichia coli* by coexpression of three different genes. *Appl. Microbiol. Biotechnol.* 2010;87:271.

[39] Liu, X., Sheng, J., Curtiss, R., 3rd. Fatty acid production in genetically modified cyanobacteria. *Proc. Natl. Acad. Sci. US* 2011;108:6899.

[40] Liu, X., Fallon, S., Sheng, J., Curtiss, R., 3rd. CO_2-limitation-inducible Green Recovery of fatty acids from cyanobacterial biomass. *Proc. Natl. Acad. Sci. US* 2011;108:6905.

[41] Schirmer, A., Rude, M. A., Li, X., Popova, E., del Cardayre, S. B. Microbial biosynthesis of alkanes. *Science* 2010;329:559.

[42] Beller, H. R., Goh, E. B., Keasling, J. D. Genes involved in long-chain alkene biosynthesis in *Micrococcus luteus. Appl. Environ. Microbiol.* 2010;76:1212.

[43] Frias, J. A., Richman, J. E., Erickson, J. S., Wackett, L. P. Purification and characterization of OleA from *Xanthomonas campestris* and demonstration of a non-decarboxylative Claisen condensation reaction. *J. Biol. Chem.* 2011;286:10930.

[44] Rude, M. A., Baron, T. S., Brubaker, S., Alibhai, M., Del Cardayre, S. B., Schirmer, A. Terminal olefin (1-alkene) biosynthesis by a novel p450 fatty acid decarboxylase from *Jeotgalicoccus* species. *Appl. Environ. Microbiol.* 2011;77:1718.

[45] Zheng, Y. N., Li, L. L., Liu, Q., Yang, J. M., Wang, X. W., Liu, W., et al. Optimization of fatty alcohol biosynthesis pathway for selectively enhanced production of C12/14 and C16/18 fatty alcohols in engineered *Escherichia coli. Microb. Cell Fact.* 2012;11: 65.

[46] Lu, W., Ness, J. E., Xie, W., Zhang, X., Minshull, J., Gross, R. A. Biosynthesis of monomers for plastics from renewable oils. *J. Am. Chem. Soc.* 2010;132:15451.

[47] Kuo, T. M., Nakamura, L. K., Lanser, A. C. Conversion of fatty acids by *Bacillus sphaericus*-like organisms. *Curr. Microbiol.* 2002;45:265.

[48] Boddupalli, S. S., Pramanik, B. C., Slaughter, C. A., Estabrook, R. W., Peterson, J. A. Fatty acid monooxygenation by P450BM-3: product identification and proposed mechanisms for the sequential hydroxylation reactions. *Arch. Biochem. Biophy.* 1992; 292:20.

[49] Wang, X., Li, L., Zheng, Y., Zou, H., Cao, Y., Liu, H., et al. Biosynthesis of long chain hydroxyfatty acids from glucose by engineered *Escherichia coli*. *Bioresour. Technol.* 2012; 114:561.

[50] Shen, C., You, S. A new method for synthesis of beta-hydroxypropionic acid. *Hebei. Chem. Eng. Ind.* 1995:10.

[51] Haas, T., Meier, M., Brossmer, C. Method and platinum group-metal catalysts for preparing malonic acid or its salts by the oxidation of 3-hydroxypropion aldehyde or 3-hydroxypropionic acid or its water-soluble salts. DE 19629372 A1 1998:2.

[52] Ishida, H., Ueno, E. Manufacture of 3-hydroxypropionic acid with good selectivity and yield. JP 2000159724 A2 2000:1.

[53] Eisenreich, W., Strauss, G., Werz, U., Fuchs, G., Bacher, A. Retrobiosynthetic analysis of carbon fixation in the phototrophic eubacterium *Chloroflexus aurantiacus*. *Eur. J. Biochem.* 1993;215:619.

[54] Herter, S., Farfsing, J., Gad'On, N., Rieder, C., Eisenreich, W., Bacher, A., et al. Autotrophic CO_2 fixation by *Chloroflexus aurantiacus*: study of glyoxylate formation and assimilation via the 3-hydroxypropionate cycle. *J. Bacteriol.* 2001;183:4305.

[55] Hugler, M., Menendez, C., Schagger, H., Fuchs, G. Malonyl-coenzyme A reductase from *Chloroflexus aurantiacus*, a key enzyme of the 3-hydroxypropionate cycle for autotrophic CO_2 fixation. *J. Bacteriol.* 2002;184:2404.

[56] Alber, B. E., Fuchs, G. Propionyl-coenzyme A synthase from *Chloroflexus aurantiacus*, a key enzyme of the 3-hydroxypropionate cycle for autotrophic CO_2 fixation. *J. Biol. Chem.* 2002;277:12137.

[57] Herter, S., Busch, A., Fuchs, G. L-malyl-coenzyme A Lyase/beta-methylmalyl-coenzyme A lyase from *Chloroflexus aurantiacus*, a bifunctional enzyme involved in autotrophic CO_2 fixation. *J. Bacteriol.* 2002;184:5999.

[58] Alber, B., Olinger, M., Rieder, A., Kockelkorn, D., Jobst, B., Hugler, M., et al. Malonyl-coenzyme A reductase in the modified 3-hydroxypropionate cycle for autotrophic carbon fixation in archaeal *Metallosphaera* and *Sulfolobus* spp. *J. Bacteriol.* 2006;188:8551.

[59] Berg, I. A., Kockelkorn, D., Buckel, W., Fuchs, G. A 3-hydroxypropionate/4-hydroxybutyrate autotrophic carbon dioxide assimilation pathway in Archaea. *Science* 2007; 318:1782.

[60] Hasegawa, J., Ogura, M., Kanema, H., Kawaharada, H., Watanabe, K. Production of β-hydroxypropionic acid from propionic acid by a *Candida rugosa* mutant unable to assimilate propionic acid. *J. Ferment. Technol.* 1982;60:591.

[61] Yazdani, S. S., Gonzalez, R. Anaerobic fermentation of glycerol: a path to economic viability for the biofuels industry. *Curr. Opin. Biotechnol.* 2007;18:213.

[62] Suthers, P. F., Cameron, D. C. Production of 3-hydroxypropionic acid in recombinant organisms. *PCT Int. Appl.*, WO2001016346 A1 2001:1.

[63] Shinzo, T. Process for producing 1,3-propanediol and/or 3-hydroxypropionic acid. WO20051093060A1 2004.

[64] Raj, S., Rathnasingh, C., Jo, J., Park, S. Production of 3-hydroxypropionic acid from glycerol by a novel recombinant *Escherichia coli* BL21 strain. *Proc. Biochem.* 2008; 43:1440.

[65] Rathnasingh, C., Raj, S. M., Jo, J. E., Park, S. Development and evaluation of efficient recombinant *Escherichia coli* strains for the production of 3-hydroxypropionic acid from glycerol. *Biotechnol. Bioeng.* 2009;104:729.

[66] Luo, L. H., Seo, J. W., Baek, J. O., Oh, B. R., Heo, S. Y., Hong, W. K., et al. Identification and characterization of the propanediol utilization protein PduP of *Lactobacillus reuteri* for 3-hydroxypropionic acid production from glycerol. *Appl. Microbiol. Biotechnol.* 2011;89:697.

[67] Raj, S. M., Rathnasingh, C., Jung, W. C., Selvakumar, E., Park, S. A novel NAD$^+$-dependent aldehyde dehydrogenase rncoded by the *puuC* gene of *Klebsiella pneumoniae* DSM 2026 that utilizes 3-hydroxypropionaldehyde as a substrate. *Biotechnol. Bioproc. E* 2010;15:131.

[68] Zhang, G. L., Xu, X. L., Li, C., Ma, B. B. Cloning, expression and reactivating characterization of glycerol dehydratase reactivation factor from *Klebsiella pneumoniae* XJPD-Li. *World J. Microb. Biot.* 2009;25:1947.

[69] Ashok, S., Sankaranarayanan, M., Ko, Y., Jae, K. E., Ainala, S. K., Kumar, V., et al. Production of 3-hydroxypropionic acid from glycerol by recombinant *Klebsiella pneumoniae* Δ*dhaT*Δ*yqhD* which can produce vitamin B$_{12}$ naturally. *Biotechnol. Bioeng.* 2012:DOI: 10.1002/bit.24726.

[70] Zhu, J. G., Ji, X. J., Huang, H., Du, J., Li, S., Ding, Y. Y. Production of 3-hydroxypropionic acid by recombinant *Klebsiella pneumoniae* based on aeration and ORP controlled strategy. *Korean J. Chem. Eng.* 2009;26:1679.

[71] Ashok, S., Raj, S. M., Rathnasingh, C., Park, S. Development of recombinant *Klebsiella pneumoniae dhaT* strain for the co-production of 3-hydroxypropionic acid and 1,3-propanediol from glycerol. *Appl. Microbiol. Biotechnol.* 2011;90:1253.

[72] Gokarn, R. R., Selifonova, O. V., Jessen, H. J., Steven, J. G., Selmer, T., Buckel, W. *3-hydroxypropionic acid and other organic compounds.* Patent application no PCT/US 2001/043607 2001.

[73] Liao, H. H., Gokarn, R. R., Gort, S. J., Jessen, H. J., Selifonova, O. V. *Alanine 2,3-aminomutase.* Patent application no EP1575881 2005.

[74] Liao, H., Gokarn, R. R., Gort, S. J., Jessen, H. J., Selifonova, O. V. *Production of 3-hydropropionic acid using β -alanine/pyruvate aminotransferase.* Patent application no 20070107080 2007.

[75] Marx, A., Wendisch, V. F., Rittmann, D., Buchholz, S. *Microbiological production of 3-hydroxypropionic acid.* Patent application no WO/2007/042494 2007.

[76] Rathnasingh, C., Raj, S. M., Lee, Y., Catherine, C., Ashok, S., Park, S. Production of 3-hydroxypropionic acid via malonyl-CoA pathway using recombinant *Escherichia coli* strains. *J. Biotechnol.* 2012;157:633.

[77] Kroeger, J. K., Zarzycki, J., Fuchs, G. A spectrophotometric assay for measuring acetyl-coenzyme A carboxylase. *Anal. Biochem.* 2011;411:100.

[78] Antoni, D., Zverlov, V. V., Schwarz, W. H. Biofuels from microbes. *Appl. Microbiol. Biotechnol.* 2007;77:23.

[79] Lee, S. K., Chou, H., Ham, T. S., Lee, T. S., Keasling, J. D. Metabolic engineering of microorganisms for biofuels production: from bugs to synthetic biology to fuels. *Curr. Opin. Biotechnol.* 2008;19:556.

[80] Lin, Y. L., Blaschek, H. P. Butanol production by a butanol-tolerant strain of *Clostridium acetobutylicum* in extruded corn broth. *Appl. Environ. Microbiol.* 1983;45:966.

[81] Nair, R. V., Bennett, G. N., Papoutsakis, E. T. Molecular characterization of an aldehyde/alcohol dehydrogenase gene from *Clostridium acetobutylicum* ATCC 824. *J. Bacteriol.* 1994;176:871.

[82] Dickinson, J. R., Harrison, S. J., Hewlins, M. J. An investigation of the metabolism of valine to isobutyl alcohol in *Saccharomyces cerevisiae*. *J. Biol. Chem.* 1998;273:25751.

[83] Dickinson, J. R., Harrison, S. J., Dickinson, J. A., Hewlins, M. J. An investigation of the metabolism of isoleucine to active Amyl alcohol in *Saccharomyces cerevisiae*. *J. Biol. Chem.* 2000;275:10937.

[84] Thauer, R. K., Jungermann, K., Henninger, H., Wenning, J., Decker, K. The energy metabolism of *Clostridium kluyveri*. *Eur. J. Biochem.* 1968;4:173.

[85] De la Plaza, M., Fernandez de Palencia, P., Pelaez, C., Requena, T. Biochemical and molecular characterization of α -ketoisovalerate decarboxylase, an enzyme involved in the formation of aldehydes from amino acids by *Lactococcus lactis*. *FEMS Microbiol. Lett.* 2004;238:367.

[86] Russell, D. W., Smith, M., Williamson, V. M., Young, E. T. Nucleotide sequence of the yeast alcohol dehydrogenase II gene. *J. Biol. Chem.* 1983;258:2674.

[87] Larroy, C., Rosario Fernandez, M., Gonzalez, E., Pares, X., Biosca, J. A. Properties and functional significance of *Saccharomyces cerevisiae* ADHVI. *Chem-Biol. Interact.* 2003;143-144:229.

[88] Atsumi, S., Hanai, T., Liao, J. C. Non-fermentative pathways for synthesis of branched-chain higher alcohols as biofuels. *Nature* 2008;451:86.

[89] Conno, M. R., Cann, A. F., Liao, J. C. 3-Methyl-1-butanol production in *Escherichia coli*: random mutagenesis and two-phase fermentation. *Appl. Microbiol. Biotechnol.* 2010; 86:1155.

[90] Marcheschi, R. J., Li, H., Zhang, K., Noey, E. L., Kim, S., Chaubey, A., et al. A synthetic recursive "+1" pathway for carbon chain elongation. *ACS Chemical Biology* 2012;7:689.

[91] Zhang, K., Sawaya, M. R., Eisenberg, D. S., Liao, J. C. Expanding metabolism for biosynthesis of nonnatural alcohols. *Proc. Natl. Acad. Sci. US* 2008;105:20653.

[92] Dekishima, Y., Lan, E. I., Shen, C. R., Cho, K. M., Liao, J. C. Extending carbon chain length of 1-butanol pathway for 1-hexanol synthesis from glucose by engineered *Escherichia coli*. *J. Am. Chem. Soc.* 2011;133:11399.

[93] Dellomonaco, C., Clomburg, J. M., Miller, E. N., Gonzalez, R. Engineered reversal of the β -oxidation cycle for the synthesis of fuels and chemicals. *Nature* 2011;476:355.

[94] Yim, H., Haselbeck, R., Niu, W., Pujol-Baxley, C., Burgard, A., Boldt, J., et al. Metabolic engineering of *Escherichia coli* for direct production of 1,4-butanediol. *Nat. Chem. Biol.* 2011;7:445.

[95] Nelson, K. E., Fleischmann, R. D., DeBoy, R. T., Paulsen, I. T., Fouts, D. E., Eisen, J. A., et al. Complete genome sequence of the oral pathogenic Bacterium *porphyromonas gingivalis* strain W83. *J. Bacteriol.* 2003;185:5591.

[96] Tian, J., Bryk, R., Itoh, M., Suematsu, M., Nathan, C. Variant tricarboxylic acid cycle in *Mycobacterium tuberculosis*: identification of α -ketoglutarate decarboxylase. *Proc. Natl. Acad. Sci. US* 2005;102:10670.

[97] Scherf, U., Buckel, W. Purification and properties of 4-hydroxybutyrate coenzyme A transferase from *Clostridium aminobutyricum*. *Appl. Environ. Microbiol.* 1991;57:2699.

[98] Toth, J., Ismaiel, A. A., Chen, J. S. The *ald* gene, encoding a coenzyme A-acylating aldehyde dehydrogenase, distinguishes *Clostridium beijerinckii* and two other solvent-producing clostridia from *Clostridium acetobutylicum*. *Appl. Environ. Microbiol.* 1999; 65:4973.

[99] Menzel, K., Zeng, A. P., Deckwer, W. D. Enzymatic evidence for an involvement of pyruvate dehydrogenase in the anaerobic glycerol metabolism of *Klebsiella pneumoniae*. *J. Biotechnol.* 1997;56:135.

[100] Iuchi, S., Lin, E. C. arcA (*dye*), a global regulatory gene in *Escherichia coli* mediating repression of enzymes in aerobic pathways. *Proc. Natl. Acad. Sci. US* 1988;85:1888.

[101] Stokell, D. J., Donald, L. J., Maurus, R., Nguyen, N. T., Sadler, G., Choudhary, K., et al. Probing the roles of key residues in the unique regulatory NADH binding site of type II citrate synthase of *Escherichia coli*. *J. Biol. Chem.* 2003;278:35435.

[102] Steinbuchel, A., Lutke-Eversloh, T. Metabolic engineering and pathway construction for biotechnological production of relevant polyhydroxyalkanoates in microorganisms. *Biochem. Eng. J.* 2003;16:81.

[103] Sudesh, K., Abe, H., Doi, Y. Synthesis, structure and properties of polyhydroxyalkanoates: biological polyesters. *Prog. Polym. Sci.* 2000;25:1503.

[104] Philip, S., Keshavarz, T., Roy, I. Polyhydroxyalkanoates: biodegradable polymers with a range of applications. *J. Chem. Technol. Biot.* 2007;82:233.

[105] Chen, G. Q. Biofunctionalization of polymers and their applications. *Adv. Biochem. Eng. Biotechnol.* 2011;125:29.

[106] Uchino, K., Saito, T., Gebauer, B., Jendrossek, D. Isolated poly(3-hydroxybutyrate) (PHB) granules are complex bacterial organelles catalyzing formation of PHB from acetyl coenzyme A (CoA) and degradation of PHB to acetyl-CoA. *J. Bacteriol.* 2007; 189:8250.

[107] Simon-Colin, C., Alain, K., Raguenes, G., Schmitt, S., Kervarec, N., Gouin, C., et al. Biosynthesis of medium chain length poly(3-hydroxyalkanoates) (mcl PHAs) from cosmetic co-products by *Pseudomonas raguenesii* sp nov., isolated from Tetiaroa, French Polynesia. *Bioresour. Technol.* 2009;100:6033.

[108] Gao, X., Chen, J. C., Wu, Q., Chen, G. Q. Polyhydroxyalkanoates as a source of chemicals, polymers, and biofuels. *Curr. Opin. Biotechnol.* 2011;22:768.

[109] Chen, G. Q. A microbial polyhydroxyalkanoates (PHA) based bio- and materials industry. *Chem. Soc. Rev.* 2009;38:2434.

[110] Jung, Y. K., Kim, T. Y., Park, S. J., Lee, S. Y. Metabolic engineering of *Escherichia coli* for the production of polylactic acid and its copolymers. *Biotechnol. Bioeng.* 2010; 105:161.

[111] Yang, T. H., Kim, T. W., Kang, H. O., Lee, S. H., Lee, E. J., Lim, S. C., et al. Biosynthesis of polylactic acid and its copolymers using evolved propionate CoA transferase and PHA synthase. *Biotechnol. Bioeng.* 2010;105:150.

[112] Chee, J. W., Amirul, A. A., Muhammad, T. S. T., Majid, M. I. A., Mansor, S. M. The influence of copolymer ratio and drug loading level on the biocompatibility of P(3HB-co-4HB) synthesized by *Cupriavidus* sp (USMAA2-4). *Biochem. Eng. J.* 2008;38:314.

[113] Qiu, Y. Z., Ouyang, S. P., Shen, Z., Wu, Q., Chen, G. Q. Metabolic engineering for the production of copolyesters consisting of 3-hydroxybutyrate and 3-hydroxyhexanoate by *Aeromonas hydrophila. Macromol. Biosci.* 2004;4:255.

[114] Qiu, Y. Z., Han, J., Guo, J. J., Chen, G. Q. Production of poly(3-hydroxybutyrate-co-3-hydroxyhexanoate) from gluconate and glucose by recombinant *Aeromonas hydrophila* and *Pseudomonas putida. Biotechnol. Lett.* 2005;27:1381.

[115] Shen, L., Worrell, E., Patel, M. Present and future development in plastics from biomass. *Biofuel. Bioprod. Bior.* 2010;4:25.

[116] Pederson, E. N., McChalicher, C. W., Srienc, F. Bacterial synthesis of PHA block copolymers. *Biomacromolecules* 2006;7:1904.

[117] Li, S. Y., Dong, C. L., Wang, S. Y., Ye, H. M., Chen, G. Q. Microbial production of polyhydroxyalkanoate block copolymer by recombinant *Pseudomonas putida. Appl. Microbiol. Biotechnol.* 2011;90:659.

[118] Hu, D., Chung, A. L., Wu, L. P., Zhang, X., Wu, Q., Chen, J. C ., et al. Biosynthesis and characterization of polyhydroxyalkanoate block copolymer P3HB-b-P4HB. *Biomacromolecules* 2011;12:3166.

[119] Tripathi, L., Wu, L. P., Chen, J., Chen, G. Q. Synthesis of Diblock copolymer poly-3-hydroxybutyrate -block-poly-3-hydroxyhexanoate [PHB-b-PHHx] by a β -oxidation weakened *Pseudomonas putida* KT2442. *Microb. Cell Fact.* 2012;11:44.

In: Biomass Processing, Conversion and Biorefinery ISBN: 978-1-62618-346-9
Editors: Bo Zhang and Yong Wang © 2013 Nova Science Publishers, Inc.

Chapter 19

PRETREATMENT TECHNOLOGIES FOR PRODUCTION OF LIGNOCELLULOSIC BIOFUELS

Bo Zhang[1,2,]*

[1]School of Chemical Engineering and Pharmacy,
Wuhan Institute of Technology, Hubei, China
[2]Biological Engineering Program, Department of Natural
Resources and Environmental Design, North Carolina
A and T State University, Greensboro, NC, US

ABSTRACT

Rising oil prices and uncertainty over the security of existing fossil fuel reserves, combined with concerns over global climate change, have created the need for new transportation fuels and bioproducts to substitute for fossil carbon-based materials. Ethanol is considered to be the next generation transportation fuel with the most potential, and significant quantities of ethanol are currently being produced from corn and sugar cane via a fermentation process. Utilizing lignocellulosic biomass as a feedstock is seen as the next step towards significantly expanding ethanol production. The biological conversion of cellulosic biomass into bioethanol is based on the breakdown of biomass into aqueous sugars using chemical and biological means, including the use of hydrolotic enzymes. From that point, the fermentable sugars can be further processed into ethanol or other advanced biofuels. Therefore, pretreatment is required to increase the surface accessibility of carbohydrate polymers to hydrolytic enzymes. This chapter reviews recent developments of several widely used pretreatment technologies, including alkali, hot-water, acid and inorganic salt (ionic liquid and Lewis acid) pretreatments. Recent advancements in the pretreatment field include: 1) application of novel chemicals or processes on biomass fractionation; 2) the use of new enzyme mixtures such as combinations of purified xylan related enzymes, appreciation of soluble inhibitors of cellulases, and especially the evaluation of pretreated biomass at varying cellulase loading; 3) application of a wide variety of sophisticated techniques for analyzing native and pretreated biomass solids, especially microscopic techniques and

* Phone 336-334-7787. Fax 336-334-7270. E-mail: bzhang@ ncat.edu.

methods for measuring surface area; and 4) greater efforts at scale up and commercialization of biomass processes for biofuels and chemicals.

1. INTRODUCTION

Rising oil prices and uncertainty over the security of existing fossil fuel reserves, combined with concerns over global climate change, have created the need for new transportation fuels and bioproducts to substitute for fossil carbon-based materials. Ethanol is considered next generation transportation fuel with the most potential, and significant quantities of ethanol are currently being produced from corn and sugar cane via a fermentation process. The use of lignocellulosic biomass as a feedstock is seen as the next step towards significantly expanding ethanol production capacity. Several biorefinery processes have been developed to produce biofuels and chemicals from biomass feedstock (Figure 1). There are two primary biorefinery platforms: the biological conversion route and the thermochemical route. In the thermochemical route, biomass is converted into syngas through gasification or into bio-oils through pyrolysis and catalytic hydrothermal treatment, which can be further upgraded to liquid fuels and other chemicals, such as menthol, gasoline, diesel fuel, and biodegradable plastics. While the biological route is based on the breakdown of biomass into aqueous sugars using chemical and biological means. The fermentable sugars can be further processed to ethanol or other advanced biofuels. However, in order to efficiently convert lignocellulosic biomass into bioethanol, technological barriers that include pretreatment, saccharification of the cellulose and hemicellulose matrixes, and simultaneous fermentationof hexoses and pentoses, still need to be addressed.

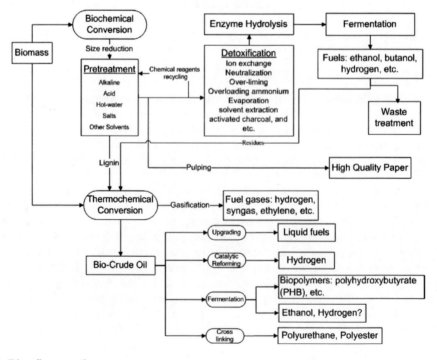

Figure 1. Biorefinery options.

Pretreatment has been considered as the most expensive processing step in cellulosic ethanol processes, representing about 18% of the total cost [1-3]. Therefore, developing a cost-effective and efficient biomass pretreatment technology is the most critical need for lignocellulosic biofuels. Pretreatment is required to increase the surface accessibility of carbohydrate polymers to the hydrolytic enzymes, which is a key step toward efficient utilization of biomass for ethanol or other advanced biofuels production. Pretreatment enables enzymes to attach to the carbohydrate polymers, and then hydrolyze these polysaccharides into fermentable sugars. The goal of pretreatment is to pre-extract hemicellulose, disrupt the lignin seal and liberate the cellulose from the plant cell wall matrix (Figure 2)[4]. Depending on the application and type of pretreatment catalyst, pretreatment techniques have generally been divided into three categories: physical, chemical, and biological pretreatment. Each pretreatment method has its own advantages and disadvantages, and no single pretreatment approach is suitable for all biomass species. Recent advancements in the pretreatment field include: 1) application of novel chemicals or processes on biomass fractionation; 2) the use of new enzyme mixtures such as combinations of purified xylan related enzymes, appreciation of soluble inhibitors of cellulases, and especially the evaluation of pretreated biomass at varying cellulase loading; 3) application of a wide variety of sophisticated techniques for analyzing native and pretreated biomass solids, especially microscopic techniques and methods for measuring surface area; and 4) greater efforts at scale up and commercialization of biomass processes for biofuels and chemicals. The processes of pretreatment have previously been reviewed [2, 3, 5, 6], and this chapter reviews recent developments of several widely used pretreatment technologies, including alkali, hot-water, acid and inorganic salt pretreatments.

2. ALKALINE PRETREATMENT

Using alkaline chemicals such as dilute sodium hydroxide, aqueous ammonia and lime to remove lignin has long been known to improve cellulose digestibility [7]. Commercially, DuPont-Danisco has developed a proprietary mild alkaline pretreatment process through its collaboration with the US. Department of Energy National Renewable Energy Lab (NREL) [8].

2.1. Sodium Hydroxide (NaOH)

Among the alkaline reagents, sodium hydroxide (NaOH) was widely used for pretreatment because its alkalinity is much higher than others. Zhang et al. [9] reported that 54.8% of cattail lignin and 43.7% of the hemicellulose were removed when applying a 4% NaOH solution. The overall effectiveness of alkali pretreatment was found to be a function of NaOH concentration. Nearly 78% of the cellulose from raw cattails was converted to fermentable glucose in 48 h using a cellulase loading of 60 FPU/g glucan.

Adding additional chemicals with NaOH could improve the pretreatment performance. For instance, by combining 5% NaOH with a 5% H_2O_2 solution that helps in additional lignin removal by oxidative action on lignin, the maximum overall sugar yield obtained from high lignin hybrid poplar pretreated at 80°C was 80% [10]. When spruce wood chips were pretreated with a NaOH/urea mixture solution,a 70% glucose yield was obtained at the low temperature of (-15°C) using 7% NaOH/12% urea solution, however, only 20% and 24% glucose yields were obtained at temperatures of 23°C and 60°C, respectively [11]. SEM confirmed that the NaOH/urea pretreatment could disrupt the bonds between hemicelluloses, cellulose, and lignin, and thus change the structure of cellulosic material. Partial lignin, hemicelluloses and cellulose was removed from spruce, enhancing the enzymatic hydrolysis efficiency of the lignocellulosic biomass.

The enzyme kinetics of sodium hydroxide pretreated biomass (wheat straw) were studied via two theoretical approaches, which describe the influence of enzyme concentration (6.25–75 g/L) on the production of reducing sugars [12]. The first approach used a modified Michaelis–Menten equation to determine the hydrolysis model and kinetic parameters (maximal velocity and half-saturation constant). The second approach, the Chrastil approach, was used to study all the time values from the rate of product formation. This approach considers that the product formation reactions are diffusion limited, and the time curves are dependent on the structure of the heterogeneous enzyme systems.

Compared to other pretreatment technologies, alkali pretreatment usually does not require high temperatures and high pressures, so the cost of pretreatment equipment is significantly low. However, pretreatment time is in the range of hours to days. Sodium hydroxide is expensive, and the recovery process is complex. Another challenge of alkaline pretreatment is the neutralization of alkali, which forms significant amount of salts.

2.2. Lime

Lime pretreatment has the following advantages: lime is the least expensive alkali at $0.06/kg; is safe to handle; and can be simply recovered [13]. The mechanism is similar to the NaOH pretreatment. When using over lime (0.5g $Ca(OH)_2$/g biomass) to pretreat corn stover at 25-55°C, lignin and hemicellulose were selectively removed, the degree of crystallinity slightly increased from 43% to 60% with delignification, but cellulose was not affected [14].

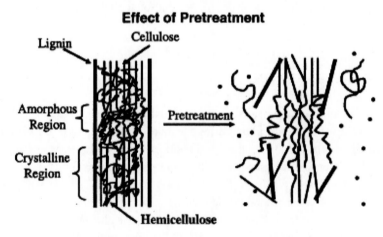

Figure 2. Schematic of goals of pretreatment on lignocellulosic material [4].

Lime pretreatment technology has been thoroughly studied on various biomass sources, such as sugarcane bagasse [15], switchgrass [16] and poplar [17].Sierra et al. combined the lime pretreatment with high-pressure oxygen, and found that the digestibility of poplar wood was significantly increased [18, 19]. Poplar was treated with 0.18-0.39 g lime/g biomass at 140-160°C for 2 hr under 14.8-21.7 bar absolute. The overall glucose and xylose yields obtained under these recommended conditions were ~95% and ~73.1%, respectively.

Lime provides a lower cost alternative than other alkalis, and lime pretreatment also requires lower temperatures and pressures. As with the NaOH pretreatment, the drawbacks of the lime pretreatment are long residence time, a large amount of wash water, or time consuming neutralization process and the formation of large quantities of salts, such as $Ca(SO)_4$.

Terrabon, Inc. (www.terrabon.com), who is the winner of the 2011 transformative technologies contest, licensed the MixAlco technology from Texas AandM University. The MixAlco process consists of a lime pretreatment followed by fermentation by microorganisms, producing a mixture of carboxylic acids, such as acetic, propionic or butyric acids. Calcium carbonate is added to neutralize the acids and form their corresponding carboxylate salts, such as calcium acetate, propionate and butyrate. The salts can be dewatered, concentrated, dried and thermally converted to ketones such as acetone, which can be hydrogenated to produce secondary alcohols such as isopropanol, propanol and butanol; or the carboxylic acids can be recovered from the fermentation solution by reacting with tertiary amines to form tertiary amine carboxylates and calcium carbonate that are then cracked to regenerate the tertiary amine and produce the carboxylic acids. According to the company, those primary or secondary alcohols can then be oligomerized to produce gasoline, diesel or jet fuel. The tertiary amine and calcium carbonate are recycled within the process so no chemicals are consumed. But the cost of recycling 25% of calcium carbonate via lime kiln is expensive. Unfortunately, due to short of funds, Terrabon filed for bankruptcy protection in September 2012, and the company's operations will cease.

2.3. Ammonia Fiber Expansion (AFEX)

During an ammonia fiber expansion/explosion, biomass particles are rapidly heated by high-pressure ammonia for a period time, then the pretreatment process is terminated by rapidly releasing pressure, which causes the biomass to undergo an explosive decompression. Ninety-nine percent of the ammonia is recovered and reused, while the remainder serves as a downstream nitrogen source for fermentation. The AFEX process decrystallizes cellulose, depolymerizes hemicellulose, and depolymerizes and removes lignin from cellulose/hemicellulose. The ammonia fiber expansion/explosion (AFEX) process has been found to be an effective pretreatment for promoting enzymatic hydrolysis of a wide variety of biomass sources, including corn stover [20], switchgrass [21], coastal Bermuda grass [22], forage and sweet sorghum bagasse [23]. The AFEX-treated hemicellulose is in oligomeric form, which can usually be hydrolyzed to fermentable pentoses by most commercial cellulase and xylanase mixtures.

AFEX treatment parameters include temperature, moisture content, ammonia loading and residence time. Teymouri et al. [20] optimized the AFEX pretreatment process for corn stover. The reactivity of pretreated biomass was affected by all operation variables including temperature, moisture content, ammonia loading and treatment time. Under the optimal conditions the glucose yield was close to the theoretical value, and the xylose yield was ~80%,using 60 FPU of cellulase enzyme/g of glucan. They concluded that AFEX is a promising pretreatment technology because of low enzyme requirements, the absence of degradation products, and much higher solids loadings.

Recently, Lee et al. [22] compared the autohydrolysis with AFEX pretreatments using coastal Bermuda grass as the feedstock. An AFEX pretreatment of (100°C for 30 min) and an autohydrolysis pretreatment (170°C for 60 min) resulted in a sugar yield of 94.8% and 55.4%, respectively, when applying a enzymatic loading of 30 FPU/g. AFEX pretreatment did not change the chemical composition of coastal Bermuda grass, but caused re-localization of lignin components. Bals et al. [24] studied the effectiveness of ammonia fiber expansion pretreatment on two harvest times and locations for switchgrass, and suggested that both harvest date and ecotype/location would determine the optimal pretreatment conditions.

2.4. Ammonia Recycle Percolation

Ammonia Recycle Percolation (ARP) is a process that pretreats biomass with 5-15 wt% aqueous ammonia in a flow-through column reactor. Ammonia, being a selective reagent for lignin, noncorrosive, and a relatively less expensive chemical, is an appropriate choice for pretreatment. SEM and FTIR spectra showed that the aqueous ammonia causes swelling and delignification of biomass at high temperatures. The ARP process could solubilize ~50% xylan from corn stover, but retains over 92% of the cellulose. Enzymatic digestibility of ARP-treated corn stover is 93% with 10 FPU/g-glucan enzyme loading [25]. The X-ray crystallography data showed that the basic cellulosic crystalline structure was not altered. By further minimizing the liquid throughput and optimizing the operation conditions for corn stover, 59–70% of lignin removal and 48–57% of xylan retention were achieved. Enzymatic digestibilities were 95, 90, and 86% with 60, 15, and 7.5 FPU/g of glucan, respectively [26]. ARP also showed great success in the delignification of sorghum [27], poplar [28], switchgrass [29] etc.

In order to reduce liquid loadings of the ARP process and make the process more cost effective, the soaking in aqueous ammonia (SAA) at low temperature – which retains the hemicellulose in the solids by minimizing the interaction with hemicellulose during treatment – was reported as a feasible approach to increase the fermentation yield and simplify the bioconversion scheme [30, 31].

3. HOT-WATER PRETREATMENT

Hot water pretreatment is often called "autohydrolysis." The major advantages of this method are less expense, lower corrosion to equipment, less xylose degradation and thus fewer byproducts including inhibitory compounds in the extracts [32]. Hot water under pressure can penetrate the cell structure of biomass, hydrate cellulose, and remove hemicellulose, thus effectively improving the enzymatic digestibility of biomass cellulose.

A pretreatment at 190°C for more than 10 minutes could effectively dissolve the xylan fraction of aquatic plant cattail as soluble oligomers. Both the glucose yield and xylose yield obtained from the pretreated cattail increased with the escalation of the final pretreatment temperature and treatment time. When cattails were pretreated at 190°C for 15 minutes, the highest glucose yield of 77.6% from the cellulose was achieved in 48 h using a cellulase loading of 60 FPU/g glucan[33].

At optimal conditions, 90% of the cellulose from corn stover pretreated in hot-water can be hydrolyzed to glucose[34]. The pretreatment process of bagasse was studied over a temperature range of 170-203°C, and time range of 1- 46 min. An 80% conversion yield was achieved, and no hydrolysis inhibitors were detected [35]. Hot water pretreatment was also reported to improve enzymatic digestibility of switchgrass, resulting in an 80% glucose yield [36]. The optimal hot-water pretreatment conditions for hybrid poplar consisting of 15% solids (wt/vol) were 200°C for 10 min, which resulted in the highest fermentable sugar yield, in the range of 54% and 67% [37].

Inbicon Biomass Refinery at Danmark established a demonstration facility based on hot-water pretreatment technology. Their goal is to demonstrate 4 ton/hr of continuous operation at industrial scale by 2013 [38].

4. ACID PRETREATMENT

The use of acid hydrolysis for the conversion of cellulose to glucose is a process that has been studied for the last 100 years. Generally, there are two types of acid hydrolysis: dilute and concentrated, each having unique properties and effects on biomass, and each having advantages and disadvantages in terms of economics.

4.1. Dilute-Acid Pretreatment

Dilute acid (0.5-1.0% sulfuric acid) pretreatment at moderate temperatures (140-190°C) can effectively remove and recover most of the hemicellulose as dissolved sugars. In the process, lignin is disrupted and partially dissolved, thus increasing cellulose susceptibility to enzymes [39]. Under this method, glucose yields from cellulose increase with hemicellulose removal to almost 100% [40]. Dilute acid hydrolysis consists of two chemical reactions. One reaction converts cellulosic materials to sugar and the other converts sugars into other chemicals, many of which inhibit the growth of downstream fermentation microbes. The same conditions that cause the first reaction to occur, simultaneously cause over-degradation of sugars and lignin, creating inhibitory compounds such as organic acids, furans, and phenols.

Partial cellulose may be degraded as oligomers or monomers during the acid pretreatment process. Sugar (glucose and xylose) yields were often reported for the pretreatment and enzyme hydrolysis stage separately, and as the total for both stages. Lloyd and Wyman [41] reported that up to 92% of the total sugars originally available in corn stover could be recovered via coupled dilute acid pretreatment and enzymatic hydrolysis. Conditions achieving maximum individual sugar yields were often not the same as those that maximized the total sugar yields, demonstrating the importance of clearly defining pretreatment goals when optimizing the process.

Dilute sulfuric acid pretreatment has been applied to a number of feedstocks including aquatic plants [42], switchgrass [43], hardwood [44], softwood [45], etc. The dilute-sulfuric acid pretreatment is an effective pretreatment approach that results in a high sugar yield from most biomass feedstocks. The disadvantages of the dilute sulfuric acid pretreatment are 1) corrosivity, requiring expensive construction materials, 2) formation of inhibitory compounds, and 3) the requirement for acid neutralization.

Verenium (www.verenium.com) is using a combination of acid pretreatments, enzymes, and two types of bacteria to make ethanol from bagasse that's left over from processing sugarcane to make sugar. It will also process "energy cane", a relative of sugarcane that's lower in sugar and higher in fiber.

4.2. SO$_2$-Catalyzed Steam Explosion

SO$_2$ impregnated steam-explosion has been considered as one of the most cost-effective pretreatment processes. Similar to the AFEX pretreatment, an SO$_2$ catalyst is used to presoak the biomass before a steam-explosion pretreatment. SO$_2$ steam-explosion promotes hemicellulose hydrolysis and increases the reactivity of various biomasses.

For example, when corn fiber was exposed to 3% SO$_2$, and then pretreated at 190°C for 5 minutes, coupled SO$_2$-catalyzed steam explosion and enzymatic hydrolysis resulted in a maximum conversion efficiency of 81% of the combined original hemicellulose and cellulose [46, 47]. An SO$_2$-catalyzed steam explosion could effectively increase the saccharification efficiencies of softwoods [48]. However, even with optimized steam pretreatment conditions, delignification after pretreatment was necessary in order to obtain sufficient hydrolytic conversion and subsequent fermentability [49].

An improved saccharification of SO$_2$ catalyzed steam-exploded corn stover was observed by adding polyethylene glycol (PEG6000). Adding PEG6000 could lower the enzyme loading and facilitate desorption of enzyme protein from lignocellulose. With 20% solid loading, the highest glucose concentrations of 102 g/L and 91.3% sugar yield were obtained [50].

In summary, SO$_2$ catalyzed steam-explosions has been tested at pilot-scale [51], and showed high efficiency on the pretreatment of soft woods. However, in order to obtain complete hydrolysis of the cellulosic component at reduced enzyme loadings, a delignification step is still required. At the same time, SO$_2$ is highly toxic and may cause unsafe impacts on health and environment. Also, some inhibitors are derived from the degradation of carbohydrates during the catalyzed steam-explosion process.

In 2010, Mascoma (www.mascoma.com), which has spent the past five years developing its consolidated bioprocessing (CBP) technology, acquired Canada's SunOpta BioProcess Inc., a division of SunOpta Inc. SunOpta's steam explosion technology is a first-step pretreatment process which exposes the cellulosic fibers of various materials, including woody biomass, switchgrass or agricultural waste, for further conversion to biofuels. Another commercial cellulosic ethanol company Abengoa Bioenergy/Iogen, is also using an acid steam explosion pretreatment process [52].

4.3. Concentrated Acid Hydrolysis

The concentrated acid process for producing sugars and ethanol from lignocellulosic biomass has a long history that goes back to 1883. Concentrated acid hydrolysis (about 70% acid content) uses a low temperature (100°F/38°C) and low pressure. The rate of cellulose recovery from the initial pre-treatment process and the conversion rate of cellulose to glucose under this process are much higher (90%) than with dilute acid hydrolysis. One concentrated acid hydrolysis model was developed by USDA and further refined by Purdue University and the Tennessee Valley Authority (TVA) [53]. Among the improvements added by these researchers were 1) recycling of dilute acid from the hydrolysis step and reusing it in the hemicellulose pretreatment step and 2) improved recycling of sulfuric acid by the use of a chromatographic column. Minimizing the use of sulfuric acid and recycling the acid cost-effectively are critical factors in the economic feasibility of the process.

The primary advantage of the concentrated process is the high sugar recovery efficiency, which can be on the order of more than 90% of both hemicellulose and cellulose sugars [54]. The low temperatures and pressures employed also allow the use of relatively low cost materials, such as fiberglass tanks and piping. The weaknesses, compared to other processes, are its relatively slow rate of conversion, and the fact that more economical and efficient acid recovery systems need to be developed. Unless the acid is removed, large quantities of lime must be used to neutralize the sugar solution, which requires the disposal of salts. This increases the cost and makes the end product more expensive.BlueFire Ethanol Incorporated (www.bluefireethanol.com) in the United States is currently working with DOE to commercialize this technology. BlueFire Ethanol Incorporated uses the Arkenol patented process, and it's a viable, world-wide cellulose-to-ethanol company with demonstrated production experience with ethanol from wood wastes, urban trash (post-sorted MSW), rice and wheat straws and other agricultural residues. To demonstrate the efficacy of the technology, the company has constructed and operated a pilot plant near its southern California offices for roughly eight years since 2003. BlueFire is building its second biorefinery plant at Fulton, MS. Meanwhile, HCL CleanTech (www.hclcleantech.com) uses concentrated hydrochloric acid (HCl) technology, which efficiently hydrolyzes all cellulosic materials and so allows a large variety of feedstock to be used with minimal configuration. HCL CleanTech announced that cost of the sugars produced by HCL CleanTech is more than 17% lower than the cost of corn mill sugars, while their quality is very similar. HCL CleanTech's process is more than 80% environmentally friendlier than corn mill processes by a Life Cycle Analysis comparison. In June 2010, HCL CleanTech has begun running its first pilot plant at the Southern Research Institute, NC.

5. INORGANIC SALTS PRETREATMENT

5.1. Ionic Liquid

An ionic liquid (IL) is a salt in the liquid state with a melting point typically below 100°C (212 °F). ILs are largely composed of ions and short-lived ion pairs, and developed as environmentally friendly alternative to organic solvents. Because of their extremely low-volatility, ILs are expected to have minimal environmental impact as a pretreatment reagent. ILs have been shown to be highly effective at solvating cellulose to technically useful concentrations [55]. During IL pretreatment, the microcrystalline cellulose is first dissolved and then recovered as essentially amorphous or as a mixture of amorphous and partially crystalline cellulose by rapidly quenching the solution with an antisolvent.

Graenacher first suggested in 1934 that molten N-ethylpyridinium chloride in the presence of nitrogen-containing bases, which have a relatively high melting point of 118 °C, could be used to dissolve cellulose, followed by subsequent chemical and mechanical processing [56].

Several ionic liquids containing 1-butyl-3-methylimidazolium cations ([C4mim]$^+$) that would dissolve cellulose have been screened [55]. It has been shown that these ILs can be used as non-derivatizing solvents for cellulose, and ILs incorporating anions as hydrogen bond acceptors were the most effective, whereas ILs containing 'non coordinating' anions, including [BF4]$^-$ and [PF6]$^-$ were nonsolvents. Chloride containing ILs appear to be the most effective solvents, presumably solubilizing cellulose through hydrogen-bonding from hydroxyl functions to the anions of the solvent.

Li et al. [57] compared dilute acid pretreatment of switchgrass with ionic liquid [C2mim][OAc] pretreatment in terms of delignification, saccharification efficiency and reducing sugar yields. During ionic liquid pretreatment, switchgrass cellulose undergoes dissolution and precipitation by an anti-solvent, resulting in reduced cellulose crystallinity and increased surface area, and a glucan yield of 96.0%. Lignin removal by IL is more effective than that of acid pretreatment.

Dadi et al. [58] studied the saccharification kinetics of cellulose pretreated using an ionic liquid: 1-n-butyl-3-methylimidazolium chloride. The initial enzymatic hydrolysis rates were approximately 50-fold higher for regenerated cellulose as compared to untreated Avicel PH-101 cellulose. They [59] further compared the effect of two ILs: 1-n-butyl3-methylimidazolium chloride (BMIMCl) and 1-allyl-3-methylimidazolium chloride (AMIMCl). By optimizing the IL treatment conditions, the digestibility of the IL-treated cellulose is significantly enhanced, and the initial hydrolysis rates were up to 90 times greater than those of untreated cellulose.

In order to select inexpensive, efficient and environmentally sound solvents for processing cellulosic biomass, Zhao et al. [60] studied a number of chloride- and acetate-based ILs for cellulose regeneration. Their data suggested that all regenerated celluloses are less crystalline (58–75% lower), and the initial hydrolysis rates were 2–10 times faster than the respective untreated celluloses.

The mechanism of the IL pretreatment was visualized using different technologies. Poplar wood was swollen by ionic liquid (1-ethyl-3-methylimidazolium acetate) pretreatment at room temperature [61], and silver and gold nano-particles of diameters ranging from 20 to 100 nm were able to be incorporated at depths up to 4 μm. Confocal surface-enhanced Raman images and Quantitative X-ray fluorescence microanalyses confirmed the incorporation of these nano-particles.

Auto-fluorescent mapping of switchgrass cell walls was used to visualize the mechanisms of biomass dissolution during ionic liquid pretreatment [62]. Treating switchgrass in the ionic liquid of 1-n-ethyl-3-methylimidazolium acetate resulted in the disruption and solubilization of the plant cell wall at mild temperatures. The results showed that swelling of the plant cell wall, attributed to disruption of inter- and intramolecular hydrogen bonding between cellulose fibrils and lignin, was followed by complete dissolution of the biomass.

Application of ionic liquids for the lignocellulosic biomass fractionation or pretreatment is still in its infancy. Most studies were done using cellulose. There are still multiple challenges needed to be overcome before applying this new concept in the field. Some of these challenges include the high cost of ILs; induction of cellulase inactivation; effective regeneration; causticity and toxicity of ILs; selection of stable ILs with suitable physical properties; and process scale-up.

5.2. Lewis Acid

Lewis acid is defined as a molecular entity that is an electron-pair acceptor and therefore able to react with a Lewis base to form a Lewis adduct by sharing the electron pair furnished by the Lewis base. A typical example of a Lewis acid in action is in the Friedel–Crafts alkylation reaction [63]. The key step is the acceptance by $AlCl_3$ of a chloride ion lone-pair, forming $AlCl_4^-$ and creating the strongly acidic, that is to say, electrophilic, carbonium ion.

Recently, it has been discovered that Lewis acid, specially chloride salts, can react with cellulosic biomass directly, in an aqueous phase or in ionic liquid with excellent selectivity [64, 65]. Liu and Wyman [66] evaluated the effect of several inorganic salts KCl, NaCl, $CaCl_2$, $MgCl_2$, and $FeCl_3$ on xylose monomer and xylotriose degradation. $FeCl_3$ was found to significantly increase xylose monomer and xylotriose degradation, resulting in degradation ratio of 65% and 78% for xylose and xylotriose, respectively. Also, losses of xylose and xylotriose were described using first order homogeneous kinetics. Yu et al. [67] investigated the effect of the metal salts NaCl, KCl, $CaCl_2$, $MgCl_2$, $FeCl_3$, $FeCl_2$, and $CuCl_2$ solutions on the decomposition of sweet sorghum bagasse. The hemicellulose removal by using transition metal chlorides is higher than that of using the alkaline earth metal chlorides or alkaline metal chlorides.

Chloride salts may react with biomass, forming a saccharide-metal cation intermediate complex. The total sugar yield from sweet sorghum bagasse undergoing a 0.1% $CuCl_2$ solution pretreatment reached 90.4%.

Liu et al. [68] used NaCl, KCl, $CaCl_2$, $MgCl_2$, $FeCl_2$, $FeSO_4$, $FeCl_3$, and $Fe_2(SO_4)_3$, as catalysts for the degradation of corn stover hemicellulose. Under optimal conditions of 0.1 M $FeCl_3$ at 140°C for 20 min, the xylose recovery yield and cellulose removal amount were 90% and <10%, respectively. They further optimized a $FeCl_3$ pretreatment process of corn stover for cellulose conversion [69].

The optimum yield of 98.0% was obtained. FTIR, SEM and XRD analysis indicated that $FeCl_3$ pretreatment may damage the surface of corn stover and disrupt almost all the ether linkages and some ester linkages between lignin and carbohydrates, but have no effect on delignification by analysis.

The advantages of the Lewis acid pretreatment include the lower corrosion to equipment and lower xylanase demand for hydrolysis. However, the mechanism of Lewis acid pretreatment is not clear, the overall efficiency of the pretreatment needs to be improved, and the effect of Lewis acid residues on the downstream processing needs to be evaluated.

6. OTHER COMMERCIAL PRETREATMENT TECHNOLOGIES

Several lignocellulose-based ethanol demonstration plants are operating in the US, though there still exists technical, economical, and commercial barriers. The pretreatment technology used by some of these cellulosic ethanol companies were reviewed above. Some of them do not use pretreatment or do not talk about their methods for pretreatment of cellulose. The following two commercial cellulosic ethanol companies use different pretreatment approaches. PureVision (www.purevisiontechnology.com) has developed a biomass fractionator to pretreat lignocellulose to yield a highly pure cellulose fraction. The biomass fractionator is based on sequentially treating biomass with hot water, hot alkaline solutions, and polishing the cellulose fraction with a wet alkaline oxidation step. PureVision now carries out fractionation and rapid hydrolysis testing at the bench and half-ton per day scales, with a focus on scaling up to a 20-ton per day. In January 2013, the company is undertaking shakedown of its first-of-a-kind biomass-to-sugars pilot plant.

Lignol's solvent-based pre-treatment technology (www.lignol.ca) was originally developed by a former affiliate of General Electric ("GE"), and then further developed and commercialized for wood-pulp applications by a subsidiary of Repap Enterprises Inc. Lignol has modified the pre-treatment process and integrated it with proprietary capabilities to convert cellulose to ethanol. Lignol's process includes an expensive pre-treatment step that fractionates biomass into separate streams of cellulose, hemicellulose and lignin, enabling the company to produce a variety of high value products, including furfural, acetic acid and a trademarked lignin, known as HP-L. The successful test of the effectiveness of Novozymes' enzymes and Lignol's substrate is an important step towards establishing Lignol's first commercial project, which is to be located in the United States.

CONCLUSION

Although pretreatment technologies have been extensively studied, it has become unclear who will be the pretreatment technology 'winners'. The advantages and disadvantages of pretreatment technologies reviewed in the article are summarized in Table 1.

Table 1. Simple comparison of different pretreatments

Methods	Features and Advantages	Disadvantages
Sodium hydroxide (NaOH)	remove lignin, low temperature and pressure,	inhibitory, long pretreatment time, expensive chemical, complex recovery process
Lime	the least expensive alkali; safe to handle; can be simply recovered	inhibitory, long residence time, neutralization or washing process required
Ammonia fiber expansion (AFEX)	low enzyme requirement, minimized degradation products, high solids loadings	formation of oligomeric form of hemicellulose degradation products

Methods	Features and Advantages	Disadvantages
Ammonia recycle percolation (ARP)	high efficiency for delignification,	low solids loadings
Hot-water	low cost, less inhibition, low corrosion, less residues, high simplicity	high temperature, particle size reduction required to obtain high yield
Dilute sulfuric acid	high yield	inhibitory, causticity reagent, acid neutralization required
SO_2-catalyzed steam explosion	high solids loadings, rapid penetration of biomass	highly toxic, inhibitory
Concentrated acid hydrolysis	high yield	slow rate of conversion, acid recovery systems required
Ionic liquid	effective at solvating cellulose, minimal environmental impact	high cost, induction of cellulase inactivation; effective regeneration; causticity and toxicity of ILs; selection of stable ILs
Lewis acid	lower corrosion to equipment, lower enzyme demands	the effect of Lewis acid residues on the downstream processing need to be evaluated

To develop a cost-effective pretreatment process, the following criteria should be considered: 1) the process requires minimized size reduction; 2) the requirement of pretreatment reagents is minimized; 3) the hemicellulose recovery is maximized; 4) the process does not form significant amount of inhibitors for enzyme hydrolysis and fermentation steps; and 5) the process is designed in a way to simplify the downstream processing.

REFERENCES

[1] Lynd, L., Elamder, R., Wyman, C. (1996) Likely features and costs of mature biomass ethanol technology. *Applied Biochemistry and Biotechnology* 57-58: 741-761.

[2] Mosier, N., Wyman, C., Dale, B., Elander, R., Lee, Y. Y., et al. (2005) Features of promising technologies for pretreatment of lignocellulosic biomass. *Bioresource Technology* 96: 673-686.

[3] Yang, B., Wyman, C. E. (2008) Pretreatment: the key to unlocking low-cost cellulosic ethanol. *Biofuels, Bioproducts and Biorefining* 2: 26-40.

[4] Hsu, T. A., Ladisch, M. R., Tsao, G. T. (1980) Alcohol from cellulose. *Chemical Technology* 10: 315–319.

[5] Zheng, Y., Pan, Z., Zhang, R. (2009) Biomass pretreatment for cellulosic ethanol production. *International Journal of Agricultural and Biological Engineering* 2: 1-18.

[6] Elander, R. T., Dale, B. E., Holtzapple, M., Ladisch, M. R., Lee, Y. Y., et al. (2009) Summary of findings from the Biomass Refining Consortium for Applied Fundamentals and Innovation (CAFI): corn stover pretreatment. *Cellulose* 16: 649-659.

[7]   Li, Y., Ruan, R., Chen, P. L., Liu, Z., Pan, X., et al. (2004) Enzymatic hydrolysis of corn stover pretreated by combined dilute alkaline treatment and homogenization. *Transactions of the ASAE* 47: 821-825.

[8]   DuPont-Danisco, Fact Sheet. http://www.tennessee.edu/media/kits/biorefinery/docs/ ddce_fact_sheet.pdf accessed on May 31, 2011

[9]   Zhang, B., Shahbazi, A., Wang, L., Diallo, O., Whitmore, A. (2010) Alkali Pretreatment and Enzymatic Hydrolysis of Cattails from Constructed Wetlands. *Am. J. Eng. Applied Sci.* 3: 328-332.

[10] Gupta, R. (2008) Alkaline pretreatment of biomass for ethanol production and understanding the factors influencing the cellulose. *Ph.D. dissertation*, Auburn University.

[11] Zhao, Y., Wang, Y., Zhu, J. Y., Ragauskas, A., Deng, Y. (2008) Enhanced enzymatic hydrolysis of spruce by alkaline pretreatment at low temperature. *Biotechnology and Bioengineering* 99: 1320-1328.

[12] Carrillo, F., Lis, M. J., Colom, X., López-Mesas, M., Valldeperas, J. (2005) Effect of alkali pretreatment on cellulase hydrolysis of wheat straw: Kinetic study. *Process Biochemistry* 40: 3360-3364.

[13] Kim, S. H. (2004) Lime pretreatment and enzymatic hydrolysis of corn stover. *Ph. D. dissertation*, Texas AandM University.

[14] Kim, S., Holtzapple, M. (2006) Effect of structural features on enzyme digestibility of corn stover. *Bioresource Technology* 97: 583-591.

[15] Rabelo, S. C., Filho, R. M., Costa, A. C. (2008) Lime Pretreatment of Sugarcane Bagasse for Bioethanol Production. *Applied Biochemistry and Biotechnology* 153: 139-150.

[16] Xu, J., Cheng, J. J., Sharma-Shivappa, R. R., Burns, J. C. (2010) Lime pretreatment of switchgrass at mild temperatures for ethanol production. *Bioresource Technology* 101: 2900-2903.

[17] Sierra, R., Garcia, L. A., Holtzapple, M. T. (2010) Selectivity and delignification kinetics for oxidative and nonoxidative lime pretreatment of poplar wood, part III: Long-term. *Biotechnology Progress* 26: 1685-1694.

[18] Sierra, R., Granda, C., Holtzapple, M. T. (2009) Short-term lime pretreatment of poplar wood. *Biotechnology Progress* 25: 323-332.

[19] Sierra, R., Holtzapple, M. T., Granda, C. B. (2010) Long-term lime pretreatment of poplar wood. *Journal: DOI 10.1002/aic.12350*.

[20] Teymouri, F., Laureanoperez, L., Alizadeh, H., Dale, B. (2005) Optimization of the ammonia fiber explosion (AFEX) treatment parameters for enzymatic hydrolysis of corn stover. *Bioresource Technology* 96: 2014-2018.

[21] Alizadeh, H., Teymouri, F., Gilbert, T., Dale, B. (2005) Pretreatment of switchgrass by ammonia fiber explosion (AFEX). *Applied Biochemistry and Biotechnology* 124: 1133-1141.

[22] Lee, J. M., Jameel, H., Venditti, R. A. (2010) A comparison of the autohydrolysis and ammonia fiber explosion (AFEX) pretreatments on the subsequent enzymatic hydrolysis of coastal Bermuda grass. *Bioresource Technology* 101: 5449-5458.

[23] Li, B.-Z., Balan, V., Yuan, Y.-J., Dale, B. E. (2010) Process optimization to convert forage and sweet sorghum bagasse to ethanol based on ammonia fiber expansion (AFEX) pretreatment. *Bioresource Technology* 101: 1285-1292.

[24] Bals, B., Rogers, C., Jin, M., Balan, V., Dale, B. (2010) Evaluation of ammonia fibre expansion (AFEX) pretreatment for enzymatic hydrolysis of switchgrass harvested in different seasons and locations. *Biotechnology for Biofuels* 3: 1.

[25] Kim, T., Lee, Y. (2005) Pretreatment and fractionation of corn stover by ammonia recycle percolation process. *Bioresource Technology* 96: 2007-2013.

[26] Kim, T., Lee, Y., Sunwoo, C., Kim, J. (2006) Pretreatment of corn stover by low-liquid ammonia recycle percolation process. *Applied Biochemistry and Biotechnology* 133: 41-57.

[27] Salvi, D., Aita, G., Robert, D., Bazan, V. (2010) Ethanol production from sorghum by a dilute ammonia pretreatment. *Journal of Industrial Microbiology and Biotechnology* 37: 27-34.

[28] Gupta, R., Lee, Y. Y. (2009) Pretreatment of hybrid poplar by aqueous ammonia. *Biotechnology Progress* 25: 357-364.

[29] Wyman, C., Dale, B., Elander, R. T. H., M. T., Ladisch, M. R., Lee, Y. Y., et al. (2009) Glucose and xylose yields from switchgrass for ammonia fiber expansion, ammonia recycle percolation, dilute sulfuric acid, hot water, lime, and sulfur dioxide pretreatments followed by enzymatic hydrolysis. In:*The 31st Symposium on Biotechnology for Fuels and Chemicals* San Francisco, CA.

[30] Kim, T., Lee, Y. (2005) Pretreatment of corn stover by soaking in aqueous ammonia. *Applied Biochemistry and Biotechnology* 124: 1119-1131.

[31] Kim, T., Lee, Y. (2007) Pretreatment of corn stover by soaking in aqueous ammonia at moderate temperatures. *Applied Biochemistry and Biotechnology* 137-140: 81-92.

[32] Huang, H., Ramaswamy, S., Tschirner, U., Ramarao, B. (2008) A review of separation technologies in current and future biorefineries. *Separation and Purification Technology* 62: 1-21.

[33] Zhang, B., Shahbazi, A., Wang, L., Diallo, O., Whitmore, A. (2010) Hot-water pretreatment of cattails for extraction of cellulose. *Journal of Industrial Microbiologyand Biotechnology* 3(7): 819-824.

[34] Mosier, N., Hendrickson, R., Ho, N., Sedlak, M., Ladisch, M. (2005) Optimization of pH controlled liquid hot water pretreatment of corn stover. *Bioresource Technology* 96: 1986-1993.

[35] Laser, M., Schulman, D., Allen, S. G., Lichwa, J., Antal, M. J., et al. (2002) A comparison of liquid hot water and steam pretreatments of sugar cane bagasse for bioconversion to ethanol. *Bioresource Technology* 81: 33-44.

[36] Kim, Y., Mosier, N. S., Ladisch, M. R. (2008) Effect of Liquid Hot Water Pretreatment on Switchgrass Hydrolysis. In:*AIChE meeting*, Philadelphia, PA.

[37] Kim, Y., Mosier, N. S., Ladisch, M. R. (2009) Enzymatic digestion of liquid hot water pretreated hybrid poplar. *Biotechnology Progress* 25: 340-348.

[38] Energy, D.*Ethanol from straw - the INBICON biorefinery demonstration facility.* In.

[39] Yang, B., Wyman, C. E. (2004) Effect of xylan and lignin removal by batch and flowthrough pretreatment on the enzymatic digestibility of corn stover cellulose. *Biotechnology and Bioengineering* 86: 88-98.

[40] Knappert, D. R., Grethlein, H. E., Converse, A. O. (1981) Partial acid hydrolysis of poplar wood as a pretreatment for enzymatic hydrolysis. *Biotechnol. Bioengin.Symp.* 11: 67–77.

[41] Lloyd, T., Wyman, C. (2005) Combined sugar yields for dilute sulfuric acid pretreatment of corn stover followed by enzymatic hydrolysis of the remaining solids. *Bioresource Technology* 96: 1967-1977.

[42] Zhang, B., Shahbazi, A., Wang, L., Diallo, O., Whitmore, A. (2011) Dilute-sulfuric acid pretreatment of cattails for cellulose conversion, *Bioresource Technology*, 102: 9308–9312.

[43] Yang, Y., Sharma-Shivappa, R., Burns, J. C., Cheng, J. J. (2009) Dilute Acid Pretreatment of Oven-dried Switchgrass Germplasms for Bioethanol Production†. *Energy and Fuels* 23: 3759-3766.

[44] Torget, R., Himmel, M. E., Grohmann, K. (1991) Dilute sulfuric acid pretreatment of hardwood bark. *Bioresource Technology* 35: 239-246.

[45] Nguyen, Q., Tucker, M., Keller, F., Eddy, F. (2000) Two-stage dilute-acid pretreatment of softwoods. *Applied Biochemistry and Biotechnology* 84-86: 561-576.

[46] Bura, R., Mansfield, S., Saddler, J., Bothast, R. (2002) SO2-catalyzed steam explosion of corn fiber for ethanol production. *Applied Biochemistry and Biotechnology* 98-100: 59-72.

[47] Bura, R., Bothast, R., Mansfield, S., Saddler, J. (2003) Optimization of SO2-catalyzed steam pretreatment of corn fiber for ethanol production. *Applied Biochemistry and Biotechnology* 106: 319-335.

[48] Ewanick, S. M., Bura, R., Saddler, J. N. (2007) Acid-catalyzed steam pretreatment of lodgepole pine and subsequent enzymatic hydrolysis and fermentation to ethanol. *Biotechnology and Bioengineering* 98: 737-746.

[49] Kumar, L., Chandra, R., Chung, P. A., Saddler, J. (2010) Can the same steam pretreatment conditions be used for most softwoods to achieve good, enzymatic hydrolysis and sugar yields? *Bioresource Technology* 101: 7827-7833.

[50] Ouyang, J., Ma, R., Huang, W., Li, X., Chen, M., et al. (2011) Enhanced saccharification of SO2 catalyzed steam-exploded corn stover by polyethylene glycol addition. *Biomass and Bioenergy* DOI: 10.1016/j.biombioe.2011.01.047.

[51] De Bari, I., Nanna, F., Braccio, G. (2007) SO2-Catalyzed Steam Fractionation of Aspen Chips for Bioethanol Production: Optimization of the Catalyst Impregnation. *Industrial and Engineering Chemistry Research* 46: 7711-7720.

[52] ABENGOA BIOENERGY, Abengoa Bioenergy Hybrid of Kansas, LLC (ABHK). http://www.kcc.state.ks.us/energy/kwrec_09/presentations/B1_Robb.pdf accessed on May 31, 2011

[53] US Department of Energy - Energy Efficiency and Renewable EnergyBiomass Program, Concentrated Acid Hydrolysis. http://www1.eere.energy.gov/biomass/printable_versions/concentrated_acid.html accessed on May 31, 2011

[54] Badger, P. C. (2002) Ethanol from cellulose: A general review. In:*Trends in new crops and new uses* (Janick, J.and Whipkey, A., eds), pp. 17-21. ASHS Press, Alexandria, VA.

[55] Swatloski, R. P., Spear, S. K., Holbrey, J. D., Rogers, R. D. (2002) Dissolution of Cellose with Ionic Liquids. *Journal of the American Chemical Society* 124: 4974-4975.

[56] Graenacher, C., *Manufacture and Application of New Cellulose Solutions and Cellulose Derivatives Produced therefrom*, US 1934/1943176.

[57] Li, C., Knierim, B., Manisseri, C., Arora, R., Scheller, H. V., et al. (2010) Comparison of dilute acid and ionic liquid pretreatment of switchgrass: Biomass recalcitrance,

delignification and enzymatic saccharification. *Bioresource Technology* 101: 4900-4906.

[58] Dadi, A. P., Varanasi, S., Schall, C. A. (2006) Enhancement of cellulose saccharification kinetics using an ionic liquid pretreatment step. *Biotechnology and Bioengineering* 95: 904-910.

[59] Dadi, A., Schall, C., Varanasi, S. (2007) Mitigation of cellulose recalcitrance to enzymatic hydrolysis by ionic liquid pretreatment. *Applied Biochemistry and Biotechnology* 137-140: 407-421.

[60] Zhao, H., Jones, C., Baker, G., Xia, S., Olubajo, O., et al. (2009) Regenerating cellulose from ionic liquids for an accelerated enzymatic hydrolysis. *Journal of Biotechnology* 139: 47-54.

[61] Lucas, M., Macdonald, B. A., Wagner, G. L., Joyce, S. A., Rector, K. D. (2010) Ionic Liquid Pretreatment of Poplar Wood at Room Temperature: Swelling and Incorporation of Nanoparticles. *ACS Applied Materials and Interfaces* 2: 2198-2205.

[62] Singh, S., Simmons, B. A., Vogel, K. P. (2009) Visualization of biomass solubilization and cellulose regeneration during ionic liquid pretreatment of switchgrass. *Biotechnology and Bioengineering* 104: 68-75.

[63] March, J. (1992) *Advanced Organic Chemistry.* 4th edn. J. Wiley and Sons, New York.

[64] Wan, Y., Chen, P., Zhang, B., Yang, C., Liu, Y., et al. (2009) Microwave-assisted pyrolysis of biomass: Catalysts to improve product selectivity. *Journal of Analytical and Applied Pyrolysis* 86: 161-167.

[65] Zhao, H., Holladay, J. E., Brown, H., Zhang, Z. C. (2007) Metal Chlorides in Ionic Liquid Solvents Convert Sugars to 5-Hydroxymethylfurfural. *Science* 316: 1597-1600.

[66] Liu, C., Wyman, C. (2006) The enhancement of xylose monomer and xylotriose degradation by inorganic salts in aqueous solutions at 180°C. *Carbohydrate Research* 341: 2550-2556.

[67] Yu, Q., Zhuang, X., Yuan, Z., Qi, W., Wang, Q., et al. (2011) The effect of metal salts on the decomposition of sweet sorghum bagasse in flow-through liquid hot water. *Bioresource Technology* 102: 3445-3450.

[68] Liu, L., Sun, J., Cai, C., Wang, S., Pei, H., et al. (2009) Corn stover pretreatment by inorganic salts and its effects on hemicellulose and cellulose degradation. *Bioresource Technology* 100: 5865-5871.

[69] Liu, L., Sun, J., Li, M., Wang, S., Pei, H., et al. (2009) Enhanced enzymatic hydrolysis and structural features of corn stover by FeCl3 pretreatment. *Bioresource Technology* 100: 5853-5858.

In: Biomass Processing, Conversion and Biorefinery
Editors: Bo Zhang and Yong Wang

ISBN: 978-1-62618-346-9
© 2013 Nova Science Publishers, Inc.

Chapter 20

BIOCHEMICAL CONVERSION OF ETHANOL FROM LIGNOCELLULOSE: PRETREATMENT, ENZYMES, CO-FERMENTATION, AND SEPARATION

Xian-Bao Zhang and Ming-Jun Zhu[*]

School of Bioscience & Bioengineering, South China University of Technology,
Guangzhou Higher Education Mega Center, Panyu, GuangzhouGuangdong province,
P.R. China

ABSTRACT

Bioethanol produced from lignocellulosic materials, such as agricultural and forest residues, has the potential to replace the limited fossil oil due to its high efficiency and renewability. The conversion of ethanol from cellulosic biomass is mainly involved in four steps: pretreatment, holocellulose (cellulose and hemicellulose) hydrolysis, fermentation and ethanol separation. The purpose of pretreatment is to improve the hydrolysis of cellulose by diminishing the physicochemical structural and compositional factors hindering the hydrolysis, so that the loading and mixtures of enzymes can be optimized and the hydrolysis and fermentation can be maximized. Separation technology, just like the other steps, has a great impact on ethanol yield. In the work, we reviewed the current popular pretreatment technologies, enzymatic hydrolysis, fermentation strains, separation techniques and recent advances for ethanol production from lignocellulose.

1. INTRODUCTION

With the depletion of the fossil fuel, the price of oil keeps increasing and the effect on climate change caused by the over emissions of greenhouse gas-CO_2 from the use of oil and coal is becoming more and more severe. In addition, the roaring development of highly oil-dependent global economies and the growing population conspire to increase the consumption of fossil fuels and oil availability could become more and more difficult in the

---

[*] Author to whom correspondence should be addressed; E-Mail: mjzhu@scut.edu.cn; Tel.: +86-020-39380623.

future. Therefore, the demand for alternative energy resources is increasing rapidly. Some renewable sources, such as wind, water, solar, geothermal heat can be utilized for the energy industry while biomass may be an alternative source that is more suitable for fuel production and the chemical industry in the near future [1]. Due to its chemical structure and relatively high energy value, the renewable green biomass can be utilized for liquid and gas fuel production (bioethanol, biogas), and for heat generation and power production [2, 3]. Bioethanol from cellulosic biomass has become a promising alternative fuel for fossil fuels due to the characteristics of being high efficient, renewable and capable of reducing the negative environmental impacts generated by the fossil fuels used all over the world. The ethanol produced from renewable materials like corn, wheat and sugarcane has been exploited in many countries such as America and Brazil [4], and it has been added to the gasoline by up to 10% in volume as partial replacement of petroleum since the 1980s [5]; however, the cost of ethanol fuel is higher than that of fossil fuels and it is an important factor impeding the large scale expansion of bioethanol production. In the long run, the conversion of bioethanol from low-cost lignocellulosic material is better than that from grains that human feed on, thereby gaining more and more attentions worldwide for many years [4, 5], as the green fuel from lignocellulosic biomass avoids the competition between food and fuel caused by the grain-based bioethanol production [6]. As the most abundant renewable resource in the world, lignocellulose is generally utilized in fuel, biomaterial and energy [7]. Lignocellulosic material is widely distributed and generated mainly from forestry residues, agricultural residues, grasses, municipal solid waste and waste paper [8]. Therefore, it is a potential resource to produce low-cost ethanol [5].

Lignocelluloses consist mainly of cellulose, hemicellulose and lignin, accounting for about 90% of dry matter in lignocelluloses [9]. The cell walls of plant biomass are composed primarily of cellulose which exists in both amorphous and crystalline forms. Hemicellulose occupies the space around the cellulose fibrils and binds the lignin structure to the cellulose fibrils in an extensively cross-linked coating of lignin. Cellulose and hemicellulose together are referred to as holocellulose. Lignin acts as a shield to reinforce the overall structure of lignocellulose and protects the holocellulose from being hydrolyzed by enzymes or microorganisms. Ethanol production from lignocellulosic biomass involves two processes: hydrolysis of cellulose in the lignocellulosic materials to fermentable reducing sugars, and fermentation of the sugars to ethanol [5]. The hydrolysis of cellulose present in biomass to sugars and other organic compounds that can be converted to fuels is hindered by many factors from physicochemical structure and composition. Owing to these structural characteristics, pretreatment is an essential step to improve the accessibility of homocellulose to enzymes or microorganisms and to decrease the cost of the process [10, 11]. Extensive pretreatment approaches, including physical, chemical, and biological treatments, have been investigated on a wide range of feedstocks. Although the raw material of lignocellulose is cheap, the hydrolysis processes for gaining fermentable sugars are cost-intensive [11]. As the physico-chemical characteristics vary from one specie to another, it is necessary to lower the cost by adopting suitable pretreatment technologies based on the specific feedstocks [12]. The hydrolysis can be catalyzed by acid or enzyme, but enzymatic hydrolysis is more promising than acid hydrolysis due to its higher potential in technology improvement for the fermentation carried out by yeast or bacteria. The factors that influence the enzymatic hydrolysis of cellulose mainly include substrates, cellulase activity, and reaction conditions (temperature, pH) [5]. Optimizing the loading and the mixtures of enzymes could improve the hydrolysis rate of cellulose. The process of fermentation also has a large impact on enzymatic hydrolysis and ethanol yield. Involved in the fermentation are four processes: Separate hydrolysis and fermentation (SHF), simultaneous saccharification and fermentation (SSF), simultaneous saccharification and co-fermentation (SSCF), and consolidated bioprocessing (CBP). Compared with SHF, the other three processes have the advantages of reducing the loading and the inhibition of sugars and increasing the ethanol yield [5]. In contrast to SSF, SSCF and CBP have another advantage of diminishing the inhibition of xylose and xylan by converting the xylose to ethanol. However, there is no effective method to decrease the inhibition to enzymes and microorganisms for fermentation caused by ethanol. Maybe separation technologies could solve this issue, but further study is needed.

The aim of this work is to review the various unit operations, such as the pretreatment technologies, enzymatic hydrolysis, strains for fermentation, separation techniques and recent advances for ethanol production from lignocellulose.

2. PRETREATMENT

Pretreatment is a critical step in the hydrolysis by enzymes and the subsequent fermentation and it is the most important processing challenge in the biofuel production. Also it is one of the most costly steps in cellulosic ethanol production, accounting for 33% of total processing costs in the base-case NREL design [13]. The purpose of pretreatment is to remove lignin, reduce the crystallinity of cellulose, and increase the accessibility and the specific surface area, so that the cellulose can be easily accessed and hydrolyzed by enzymes or microorganisms. An efficient pretreatment process should meet the following criteria: (1) improving the digestibility of the carbohydrates to form monomeric sugars by enzymatic hydrolysis; (2) resulting in the high recovery of all carbohydrates, and minimizing the degradation of cellulose and hemicellulose; (3) avoiding the formation of inhibitors to subsequent enzymes for hydrolysis and/or fermenting microorganisms; (4) reducing energy demands and minimizing cost. The pretreatment techniques that could change the physicochemical structure of the lignocellulosic biomass and improve hydrolysis rates can be roughly divided into four categories: physical, chemical, physicochemical and biological pretreatments. The pretreatment technologies listed in Table 2 and Table 3 have the promise for cost-effective pretreatment of lignocellulosic biomass for biological conversion to fuels and chemicals.

2.1. Structure of Lignocellulosic Biomass

Lignocellulose is the principal component of plant cell walls. Plant biomass mainly consists of cellulose, hemicellulose, and lignin, with a small amount of protein, pectin, extractives (mainly soluble nonstructural materials such as nonstructural sugars, nitrogenous material, chlorophyll, and waxes), and ash [14]. The components of lignocellulosic materials vary from one species to another and the components of common lignocellulosic biomass are listed in Table 1.

2.1.1. Cellulose

Cellulose is a linear homopolymer composed of repeated cellobiose units linked by β-1,4-glycosidic bonds($C_{6n}H_{10n+2}O_{5n+1}$ (n = degree of polymerization of glucose)) (Figure1a) with an average molecular weight of ~106 Da [19]. The cell walls of plant biomass are composed primarily of cellulose in the form of both amorphous and crystalline. Due to intramolecular (O3-H→O5' and O6-H→O2') and intrastrand (O6-H→O3') hydrogen bonds (Figure1b), cellulose forms long chains, being highly crystallized, insoluble in most common solvents containing water and very resistant to hydrolysis.

Table1. Cellulose, Hemicellulose and Lignin Contents in Common Lignocellulosic Biomass[a]

Lignocellulosic material		Cellulose (%)	Hemicellulose (%)	Lignin (%)
Hard wood	hardwood stems	40-55	24-40	18-25
	hybrid poplar	44.70	18.55	26.44
	eucalyptus	49.5	13.07	27.71
	black locust	41.61	17.66	26.70
Softwood	softwood stems	45-50	25-35	25-35
	pine	44.55	21.90	18.13
Agricultural residual	nut shells	25-30	25-30	30-40
	corn cobs	45	35	15
	rice straw[b]	35	25	12
	cotton seed hairs	80-95	5-20	0
	wheat straw	30	50	15
	sorghum[c]	53.1	17.9	27.9
	leaves	15-20	80-85	0
Grass	grasses	25-40	35-50	10-30
	sugarcane bagasse[b]	39.5-40	20-24	21-25
	switchgrass	45	31.4	12
	miscanthus[c]	48.3	21.3	28.8
	Sida[c]	52.2	26.8	19.1
	coastal bermudagrass	25	35.7	6.4
Waste	paper	85-99	0	0-15
	newspaper	40-55	25-40	18 -30
	waste papers from chemical pulps	60-70	10-20	5-10
	solid cattle manure	1.6-4.7	1.4-3.3	2.7 -5.7
	primary wastewater solids	8-15		
	swine waste	6.0	28	
	sorted refuse	60	20	20

[a] Adapted from ref [14, 15]; [b] Adapted from ref [16, 17]; [c] Adapted from ref [18].

2.1.2. Hemicellulose

Unlike cellulose, hemicelluloses are amorphous and heterogeneous polymers of pentoses (D-xylose, L-arabinose), hexoses (D-mannose, D-glucose, D-galactose), and sugar acids(D-glucuronic, D-galacturonic and methylgalacturonic acids) [21] (Figure2). These sugars are linked together by β-1, 4- and sometimes by β-1, 3-glycosidic bonds. Compared to cellulose, the structure of xylan is more complex, but it does not form tight ordering crystalline structures as cellulose does and is more accessible to enzymatic hydrolysis. A variety of hemicelluloses usually account for 25–35% of the mass of dry wood, 28% in softwoods, and 35% in hardwoods [22]. The main difference between cellulose and hemicellulose is that hemicellulose has branches with short lateral chains consisting of different sugars. Because of the diversity of its sugars, hemicellulose requires a wide range of enzymes to be completely hydrolyzed into free monomers.

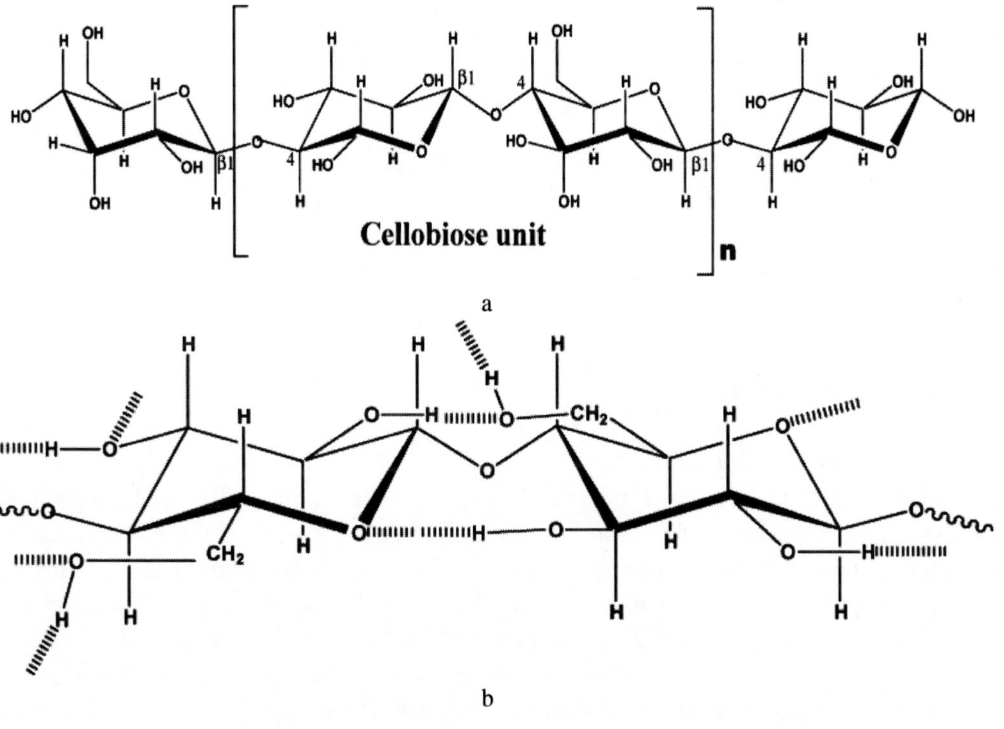

Figure 1. a. Chemical structure of cellulose, b. Intra- and interchain hydrogen-bonded bridging [20].

2.1.3. Lignin

Lignin is the most abundant renewable aromatic material on earth and is second only to cellulose in abundance [24]. Lignin is composed of three phenyl monomers: coniferyl alcohol (guaiacyl propanol), coumaryl alcohol (p-hydroxyphenyl propanol), and sinapyl alcohol (syringyl alcohol). These phenolic monomers are linked together by alkyl -aryl, alkyl-alkyl, and aryl-aryl ether bonds, respectively (Figure 3). In general, herbaceous plants such as grasses have the lowest contents of lignin, whereas softwoods have the highest lignin contents [25]. Most gymnosperm lignins contain primarily guaiacyl units. Angiosperm lignins contain approximately equal amounts of guaiacyl and syringyl units. Both types of lignin generally contain only small amounts of p-hydroxyphenyl units [24]. In the cell walls, lignin is intimately interspersed with hemicelluloses, forming a matrix that surrounds the orderly cellulose microfibrils. Lignin acts as a shield to reinforce the overall structure of lignocellulose and protects the holocellulose from being hydrolyzed by enzymes or microorganisms.

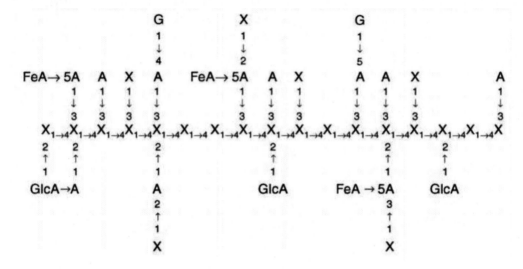

Figure 2. Scheme of the basic structure of hemicellulose. A, arabinose; FeA, ferulic acid; G, galactose; Glc, glucuronic acid; X, xylose [23].

2.2. Physical Pretreatment

2.2.1. Mechanical Comminution

The very first step in ethanol production is to comminute almost all solid lignocellulosic biomass by a combination of milling, grinding or chipping to reduce the particle size, the crystallinity and the degree of polymerization [5]. Mechanical pretreatments, including ball milling, two-roll milling and hammer milling, are primarily used for agricultural residues such as wheat and corn stover [27]. Hammer mills reduce the particle size of biomass by shear and impact action [28]. The cell structure could be broken down and the biomass becomes more accessible to further treatment. The size of the biomass is usually 10-30 mm after chipping and 0.2-2mm after milling or grinding. The final particle size and biomass characteristics, especially the moisture, material properties, feeding rate of the material and machine variables, have great effects on the power needed for mechanical comminution of agricultural materials [28], since power consumption increases rapidly with a decrease in particle size, as shown in Table 2. Size reduction may produce better results [6, 29, 30] but very fine particle size may impose negative effects such as blocking pipeline by generating clumps and increasing the energy consumption in the subsequent process. The energy demand for mechanical comminution to reduce the particle size of hardwood to 1.6 mm is 130 kW·h/ton, which is much higher than 14 kW·h/ton for corn stover [31]. So developing a new pretreatment process to minimize or avoid the energy consumption for size reduction is essential, because the mechanical pretreatment techniques are time-consuming, energy intensive and expensive. Mild torrefaction [32], a new pretreatment process, improves the grindability of fibrous materials through reducing the energy consumption for milling. Further research of this method is needed.

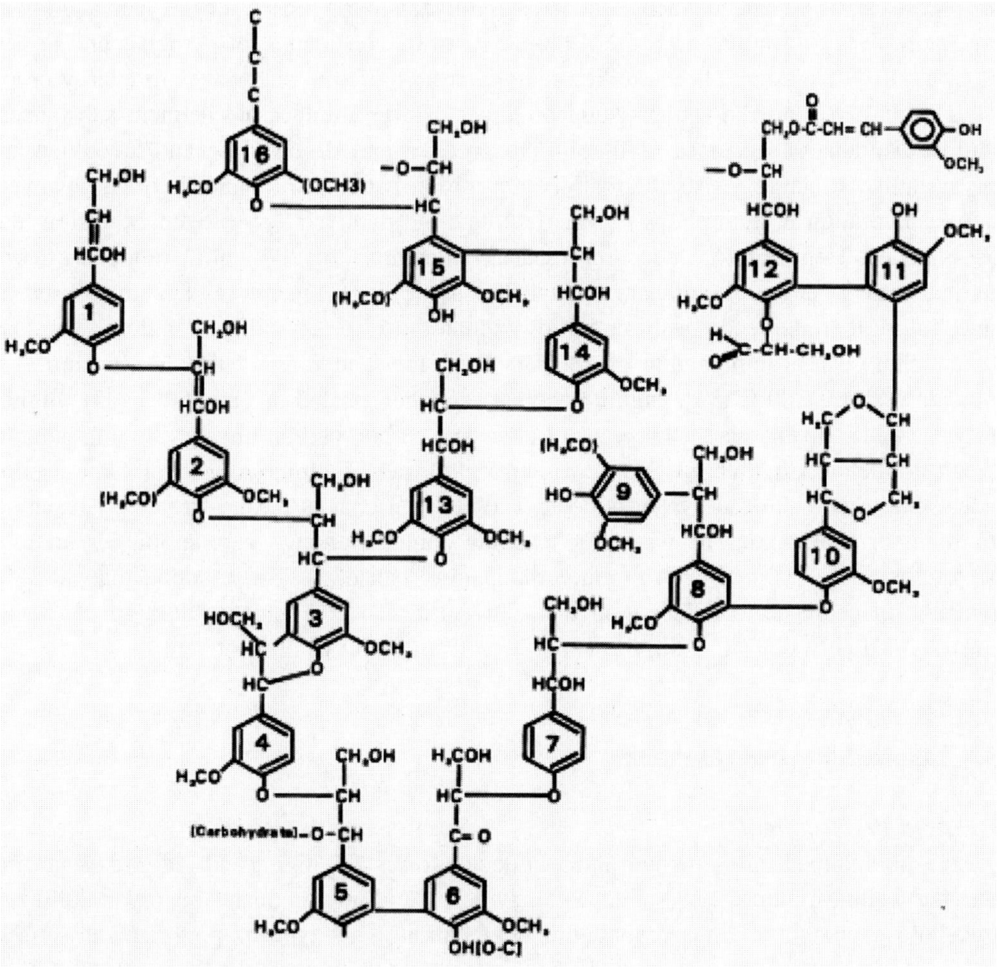

Figure 3. Schematic structural formula for lignin [26].

Table 2. Specific energy consumptions for wheat straw grinding [28]

Material	Moisture content (%wb)	Hammer mill screen opening(mm)	Average specific energy consumption (kw·h·t^{-1})
Wheat straw	8.3	0.8	51.55
	8.3	1.6	37.01
	8.3	3.2	11.36
	12.1	0.8	45.32
	12.1	1.6	43.56
	12.1	3.2	24.66

2.2.2. Pyrolysis

Pyrolysis is the thermal decomposition of materials to gaseous products and residual char at a temperature higher than 300°C [33] in the absence of oxygen or in the presence of oxygen significantly less than required for complete combustion. But Shafizadeh et al. reported that this process can be enhanced when cellulose is pyrolyzed in the presence of oxygen at a lower temperature [33]. Virtually any form of biomass can be considered for pyrolysis and nearly 100 types of biomass have been studied ranging from agricultural residues to energy crops [34]. The pyrolytic breakdown of wood produces a large number of chemical substances. Cellulose degradation occurs at 240-350°C to produce anhydrocellulose and levoglucosan. The crystalline structure resists thermal decomposition better than hemicelluloses; however, waters of hydration in amorphous regions facilitate the pyrolysis for a steam explosion-like process action. Hemicellulose decomposes at a temperature of 200-260°C, producing more volatiles, less tars, and less chars than cellulose [35]. Temperatures for the onset of hemicellulose thermal decomposition are lower than those for crystalline cellulose. The loss of hemicellulose at a temperature of 130-194°C is slow and most of this loss occurs at a temperature above 180°C [34]. Lignin decomposes when heated at 280-500°C [35]. Lignin pyrolysis yields phenols via the cleavage of ether and carbon-carbon linkages. Lignin is more difficult to dehydrate than cellulose or hemicellulose. Lignin pyrolysis produces more residual char than cellulose pyrolysis does. The residual char is further treated by leaching with water or mild acid. The water for leaching contains enough carbon sources, mainly glucose, for microbial growth to produce bioethanol. Fan et al. [36] reported that 80-85% conversion of cellulose to reducing sugars with more than 50% glucose through mild acid hydrolysis (1N H_2SO_4, 97°C, 2.5h) of the products from pyrolysis pretreatment.

2.3. Chemical Pretreatment

2.3.1. Acid Pretreatment

Acid pretreatment aims to hydrolyze lignocellulosics for high yields of sugars and to make the cellulose more accessible to enzymes. Acid pretreatment mainly solubilizes hemicellulose with little impact on lignin degradation. This type of pretreatment can be carried out with concentrated or diluted acid. Concentrated acids have been successfully used to treat agricultural residues and wood for fermentable sugar [37] and temporarily commercialized in World War II. However, concentrated acid is toxic, corrosive and hazardous and requires reactors which are made of costly materials resistant to corrosion. Besides, after hydrolysis, the concentrated acid must be recovered and recycled to make the process economically feasible [38]. Therefore, this method is not suitable for commercial scale because of the high costs for operation and maintenance [39].

Diluted acid pretreatment is considered as a more favorable method for industrial applications and has been successfully used to pretreat a wide range of feedstocks, including softwood, hardwood, herbaceous crops, agricultural residues, wastepaper, and municipal solid waste [40]. Of all acid-based pretreatment methods, like hydrochloric acid, phosphoric acid and nitric acid, diluted H_2SO_4 at a concentration below 4%(w/w) is the most studied acid due to high efficiency and low cost [41]. The dilute sulfuric acid pretreatment results in a high reaction rate and significantly improves the cellulose hydrolysis [42]. Saha et al. [43]

demonstrated that the saccharification yield of wheat straw pretreated with 0.75% v/v of H_2SO_4 at 121°C for 1 h is as high as 74%. There are primarily two types of dilute acid pretreatment processes: low solids loading (5–10 wt. %), high temperature (T >160°C), continuous-flow processes and high solids loading (10–40 wt.%), lower temperature (T < 160 °C), batch processes [5]. Three important contributions made by dilute sulfuric acid pretreatment (0.2–2.0 wt. % sulfuric acid, 121–220°C) of lignocellulose [41] are: (1) hydrolysis of the hemicellulose components to monomeric sugars; (2) exposure of cellulose for enzymatic digestion by removal of hemicellulose and part of the lignin; and (3) solubilization of heavy metals which may be contaminating the feedstock. Additionally, some significant disadvantages of dilute sulfuric acid are [40]: (1) corrosion that requires expensive materials for construction; (2) the neutralization of acidic hydrolysates before fermentation; (3) the reverse solubility of gypsum when neutralized with inexpensive lime; (4) formation of degradation products and inhibitors to enzymatic hydrolysis and fermentation; (5) disposal of neutralization salts; and (6) reduction of biomass particle size.

2.3.2. Alkaline Pretreatment

Alkaline pretreatment is one of the effective chemical pretreatment processes. Some bases, including sodium hydroxide, potassium hydroxide, calcium hydroxide (lime), aqueous ammonia, ammonia hydroxide and sodium hydroxide in combination with hydrogen peroxide or others, have been employed for the pretreatment of lignocellulosic materials, and the effect of alkaline pretreatment depends on the lignin content of the materials. [36, 40, 44, 45] Alkali pretreatment can be carried out at a lower temperature and pressure under ambient conditions, but requires a longer retention time than other pretreatments [11]. The action mechanism is related to the saponification of intermolecular ester bonds crosslinking lignin, hemicellulose and other components. Alkali pretreatments increase cellulose digestibility and they are more efficient for delignification and degradation of minor cellulose and hemicellulose than acid or hydrothermal processes [46]. Alkali treatment also dissolves silica, hydrolyzes uronic and acetic esters, and swells cellulose, and the reduction of crystallinity is due to swelling [47]. Sun et al. [48] reported that the content of lignin in wheat-straw was reduced by 80% using 1.5% NaOH for 144h at 20°C. The effectiveness of the method varies and depends on the substrate and treatment conditions. Generally, alkaline pretreatment is more efficient for hardwood, herbaceous crops and agricultural residues with a low lignin content than for softwood with a high lignin content [6]. The digestibility of NaOH-treated hardwood increased from 14% to 55% with the decrease of lignin content from 24-55% to 20% [49]. Zhu et al. [17] found that the fermentation efficiency of $NH_4OH-H_2O_2$-treated sugarcane bagasse can be improved by 89.08% with reducing the content of lignin from 21.04% to 10.69% under the condition of 53% $NH_4OH-H_2O_2$ for 20h at 63°C. But the effect of NaOH pretreatment on softwoods with lignin content greater than 26% is less powerful. Lime also has been used for lignocellulose pretreatment and widely investigated due to a low regent cost, less safety requirements and easy recovery from hydrolysate by reaction with CO_2 [11, 25]. Playne [50] reported the enzyme digestibility of sugarcane bagasse treated with lime at ambient conditions for 192 h is improved from 20% to 72%. A significant drawback of alkaline pretreatment is the conversion of alkali into irrecoverable salts and/or the incorporation of salts into the biomass during the pretreatment process and thus it is a challenge to dispose of the large amount of salts in the alkaline pretreatment [11, 51].

2.3.3. Wet Oxidation

Wet oxidation(WO), a method of treating lignocellulose with water and air or oxygen at a temperature above 120°C [52], has been effective in pretreating a range of biomass resources [53, 54], and is commonly used in the pretreatment of low-lignin materials, such as agricultural residues. A large fraction of lignin in the materials can be solubilized and oxidized, and the effect can be increased if a small amount of alkali is used as catalyst [55]. Oxygen pressure and the type of raw material used have great effects on the fractionation of the materials, the formation of sugars and by-products, and the cellulose enzymatic convertibility, and with the increase of oxygen pressure, the by-products content increases and the lignin content decreases [53]. And this method is more efficient for pretreating sugarcane bagasse, because after pretreatment and enzymatic hydrolysis, 56.5% of the cellulose in the raw bagasse can be converted to glucose, but it has little effect on the enzymatic hydrolysis of peanut shells, cassava stalks and rice hulls. Formic acid and acetic acid are the major degradation products, accounting for 66-80%(w/w) of the by-products, and acetic acid is formed by not only the initial hydrolysis of acetyl groups in hemicellulose, but also the end product from WO [54].

2.3.4. Sulfite Pretreatment to Overcome Recalcitrance of Lignocelluloses (SPORL)

The SPORL process is a new method for pretreatment of lignocellulose [56, 57]. In this process, wood chips first react with a solution of sodium bisulfite (or calcium or magnesium or other bisulfite) at 160-190°C and pH 2-5 (for about 10-30 min in batch operations). Then the pretreated substrates are fiberized through mechanical milling to generate fibrous substrate for subsequent saccharification and fermentation [58]. Concentrated hemicellulosic sugar stream can be gained and energy consumption of chemical pretreatment is low because of the low ratio of pretreatment liquor to solid biomass (2-3). The action mode is believed to be the combined effects of removal of hemicelluloses, depolymerization of cellulose, partial delignification, partial sulfonation of lignin which increases the hydrophilicity of pretreated substrates and reduces non-productive adsorption of enzymes on lignin, as well as the increase in the surface area produced by mechanic milling [59, 60]. In terms of total sugar production, process energy efficiency, lignin co-product potentials, and commercial scalability, the SPORL process appears to have inherent advantages for woody biomass conversion, especially softwood [58]. The enzymatic conversion of cellulose to glucose is over 90% for softwood and hardwood within 24 to 48h [60]. Tian et al. [57] compared the yields of sugar and ethanol from Aspen pretreated with SPORL and dilute acid, and found that SPORL is more effective, resulting in a higher enzymatic digestibility, sugar and ethanol production, less consumption of energy than its counterpart DA pretreatment, and 91% of theoretical value of enzymatic hydrolysis glucose yield for SPORL (2% [v/v] sulfuric acid, 3% bisulfite at 170°C for 30min). An important drawback of SPORL is the formation of fermentation inhibitors, such as acetic acid, hydroxymethylfurfural (HMF), and furfural, but the amount of these inhibitors is lower than that from dilute acid pretreatment [56].

2.3.5. Organosolv Pretreatment

The organosolv method is based on the principle that lignin can be solubilized in certain organic solvents. With the addition of catalyst, feedstocks are treated at a low temperature (below 180°C) or without catalyst (auto-catalysis) at a higher temperature (185-210°C) [61].

The dissolved lignin can be recovered and used for various purposes, either as a feedstock for chemicals or as fuel [4]. The organosolv pretreatment method involves organic solvents, such as methanol, ethanol, acetone, glycols and phenols, and the effect of the pretreatment can be greatly improved with the addition of mineral acid, such as HCl or H_2SO_4 [62], but a high yield of xylose dissolves in the aqueous solution. Ethanol organosolv pretreatment can effectively remove the recalcitrance of woody biomass for enzymatic cellulose saccharification [62-64], because of delignification and solubilization of hemicellulose with a resulting increase of accessible surface area [61]. Typical organosolv pretreatment conditions for woody biomass are a temperature about 160-190°C, a pretreatment time of 30-60min, and an ethanol concentration of 40-60%. Approximately 74% of the Klason lignin in untreated wood is separated by ethanol organosolv pretreatment at center point conditions (180°C, 60 min, 1.25% H_2SO_4, and 60% ethanol) [63]. Several advantages of the pretreatment are: (1) size-reduction is not necessary even when pretreatment is directly applied to commercial wood chips; (2) it produces a readily digestible cellulose substrate from almost all kinds of feedstock including softwood and hardwood species; and (3) it also produces very high purity and quality lignin with the potential of high-value utilizations [58]. However, complete solvent (ethanol) recovery is a critical issue to make the process cost-effective. Furthermore, removal of solvents from the system is necessary because the solvents might be inhibitory to the growth of microorganisms, enzymatic hydrolysis, and fermentation. Low boiling point alcohol (mainly methanol and ethanol) pretreatment seems to be the most promising organosolv process because of the lower chemical costs and easy recovery of solvents, but it needs a high pressure to remain for pretreatment [61].

Organic acids, mainly formic acid and acetic acid, are also used for the pretreatment of biomass [65, 66]. There are two functions served by an organic acid solvent: one is to dissociate partial hydrogen ion to accelerate delignification and hydrolysis of hemicellulose and the other is to dissolve the lignin fragments [67]. Under optimal conditions (25% acid concentration, 1:15 solid-liquid ratio, 230W microwave intensity, and 5min irradiating time), the content of lignin can be reduced by 46.1% and 51.54%, and the sugar yields can be increased from 35.28% to 71.41% and 80.08% when acetic acid and propionic acid are used as solvents, respectively [67]. However, it appears to have inherent drawbacks that the formylation and acetylation of cellulose inevitably associate with formic acid and acetic acid pretreatment, and organic acid causes serious corrosions [61].

2.3.6. Ionic Liquids Pretreatment

The organic salts of a low melting point are known as ionic liquids (ILs), which have attracted increasing attention in lignocellulose pretreatment. Room temperature ILs are considered as green solvents, due to the advantages of low vapor pressure, wide liquidus range, excellent dissolution power and convenience in recylcling [68, 69]. ILs have tunable physico-chemical properties such as polarity, viscosity and melting point, depending on the selection of various anions and cations [70]. ILs contain the following cations: different imidazolium, pyridinium and pyrrolidinium based cations, choline, tetrabutylammonium, tetrabutylphosphonium ions, and cations of several polycyclic amidine based bases. The anions of choice are halogens, formates, acetates, amides, imides, thiocyanates, phosphates, sulfates, sulfonates, and dichloroaluminates. Carbohydrates and lignin can be simultaneously dissolved in ILs and the mechanism is believed to be the disruption of the inter- and intramolecular hydrogen bonding of cellulose and the formation of new hydrogen bonds between the carbohydrate hydroxyl protons and the anions of the ILs [71]. Dissolved cellulose can be recovered from the ionic liquid solutions by addition of antisolvents like water. When an antisolvent is added, the ions of the ionic liquid (IL) are extracted into the aqueous phase, then the IL molecules are shielded by water molecules, which form hydrodynamic shells around the ionic liquid molecules and thus direct interactions between cellulose and ionic liquid molecules are inhibited; therefore, cellulose is precipitated from ILs [72]. The reconstructed cellulose exhibited typically lower degrees of crystallinity and lignin content than native cellulose. The crystallinity index(CrI) determined from infrared spectra was investigated for the regenerated celluloses obtained via dissolving sugarcane bagasse in [AMIM]Cl [73]. The CrI was decreased from 38.9 to 20.8 with the lignin content being reduced by 41% after IL pretreatment at 100°C for 1h. The regenerated SCB from the combined pretreatment exhibited significantly enhanced enzymatic digestibility with an efficiency of 91.4% after 12 h of hydrolysis, which was 64% higher than the efficiency observed for the regenerated SCB after the individual $NH_4OH–H_2O_2$ pretreatment. In addition, an ethanol yield of 0.42 g/g was achieved with a corresponding fermentation efficiency of 94.5%. The effects of ILs on the capacity of dissolving cellulose mainly depend on the components of ILs and physical properties. The size of anion and cation has significant impacts on the dissolution of cellulose. With an increase in the alkyl chain strength in cation and the size of anion, the solvent power for cellulose decreases. It is reported that not only the properties of anions and cations, but also their combination, makes an ionic liquid a good solvent for cellulose. Additionally, hydrogen bond basicity and viscosity have great effects as well. High hydrogen bond basicity facilitates the dissolution of cellulose in the ILs. The high viscosity of an ionic liquid requires a high dissolution temperature. When an ionic liquid exhibits a low viscosity, cellulose can be dissolved at a low temperature [70]. The dissolution power of ionic liquids for cellulosics also depends on the species type, dissolution time, temperature and solid-to-ionic liquid ratio. Kilpeläinen et al. [74] investigated the dissolution of wood in ionic liquids, and found that the water content and the particle size of the wood chip had great effects on the dissolution efficiency. The dissolution rate is dependent on the wood particle sizes as follows: ball-milled wood powder > sawdust≥ TMP fibers >> wood chips. Depolymerization of cellulose can also occur in ionic liquids. A rapid depolymerization of cellulose took place at 80°C in [AMIM]Cl corresponding to a degree of polymerization about 650 after 30 min [75]. The corrosivity and toxicity of ionic liquids are important with regard to their practical applications in industrial conditions, and further research is needed to improve the economics of ILs pretreatment, including IL recovery methods and low cost synthesis of ILs, before they can be applied on an industrial scale [12]. In addition, techniques for hemicellulose and lignin recovery from solutions after extraction of cellulose need to be developed [76]. Development of ILs pretreatment offers a great potential for future lignocellulose biorefinery.

2.4. Physico-chemical Pretreatment

2.4.1. Stream Explosion (Autohydrolysis)

Steam explosion, one of the hydrothermal pretreatment techniques, is regarded as the most commonly employed and cost-effective process [44, 77]. Two steps are included in this method: pretreatment of biomass with high-pressure saturated steam (160-260 °C, 0.69-4.83 MPa) [25] for a short time, ranging from seconds to several minutes, and swift depressurization, causing the steam to expand rapidly within the feedstock and open its structure. The process causes hemicellulose degradation and lignin transformation due to high temperature, which results in the release of natural acid such as acetic acid, and water also acts as an acid at a high temperature. The removal of hemicelluloses is believed to expose the cellulose fibres and increase accessibility of cellulose to enzyme and microorganisms [78]. Lignin is removed only to a limited extent during the pretreatment because it is redistributed on the fiber surfaces as a result of melting and depolymerization/repolymerization reactions [79]. About 80% of theoretical ethanol converted from wheat straw pretreated with steam explosion at 190°C for 10min can be achieved, and this method can lead to the degradation of cellulose fiber and concentration of Klason lignin under harsh conditions(>180°C) [80]. But, the delignification can be greatly improved by combining steam explosion with ethanol extraction [81]. The high recovery of xylose (45-65%) has made steam-explosion pretreatment economically attractive [15, 82]. Addition of H_2SO_4 (or SO_2) or CO_2 in steam explosion can effectively improve enzymatic hydrolysis, decrease the production of inhibitory compounds, and lead to more efficient removal of hemicellulose [83]. The hemicellulose recovery yield of steam explosion with addition of sulfuric acid is higher (65-85%) than that of steam explosion without acid (57-61%) [80]. The variables that greatly affect steam explosion pretreatment are residence time, temperature, chip size and moisture content [84] and the combined effect of both temperature (T) and time(t), which is described by the severity factor (Ro) [Ro= t*e [T-100/14.75]] with the Ro for the optimal conditions for maximum sugar yield ranging from 3.0 to 4.5 [85]. Some effects of steam explosion treatment on lignocellulosics are reported in literature [86]: (1) steam explosion treatment increases crystallinity of cellulose by promoting crystallization of the amorphous portions, (2) hemicellulose is easily hydrolyzed by steam explosion treatment, and (3) steam explosion promotes delignification. Steam explosion can efficiently increase enzymatic hydrolysis of hardwoods and agricultural residues, but it is less effective for softwoods [87, 88]. Limitations of steam explosion include destruction of partial xylan fraction, incomplete disruption of the lignin-carbohydrate matrix, and generation of inhibitors to enzymes and microorganisms [89].

2.4.2. Liquid Hot Water Method

Liquid hot water (LHW), another hydrothermal pretreatment method, is applied for pretreatment of lignocellulosic materials, and does not need rapid decompression, catalysts or chemicals. LHW subjects biomass to hot water in liquid state at a high pressure during a fixed period, inducing elevated recovery rates for pentoses and generating small amounts of inhibitors. The treatment generally occurs at a temperature range of 170-230°C and a pressure above 5 MPa for 20 min [47]. The objective of the liquid hot water is to solubilize mainly the hemicellulose and alter the structure of the lignocellulose so as to make the cellulose more accessible and avoid the formation of inhibitors [12, 90]. Lignin is partially depolymerized and solubilized during hot water pretreatment, but it is impossible to achieve complete delignification, due to the recondensation of soluble components originating from lignin. During the pretreatment, approximately 40-60% of the total biomass is dissolved, with 4-22% of the cellulose, 35-60% of the lignin, and all of the hemicellulose being removed [25].

Gonzalez et al. [91] compared the enzymatic hydrolysis of olive tree pretreated with liquid hot water and steam explosion, and the results showed that the yield of glucose and total sugars obtained from that treated with LHW was lower than that treated with steam explosion, but LHW had a higher pentosan recovery and less inhibitors when compared with steam explosion. Three types of liquid hot water reactor configurations are being used: cocurrent, counter-current, and flow-through [11]. In cocurrent pretreatment, water and lignocellulose move in the same direction, and the slurry of biomass and water is heated to a desired temperature and held under pretreatment conditions for a desired residence time before being cooled. In counter-current pretreatment, water and lignocellulose move in opposite directions through the pretreatment reactor. In a flow-through reactor, hot water is made to pass over a stationary bed of lignocellulose. Liquid hot water pretreatments are attractive for the potential cost-savings: no catalyst demand and low-cost reactor construction due to the low-corrosion potential, as well as no demand for size reduction of the biomass [92]. The disadvantages of the liquid hot water technique are: demand of high pressure to maintain water in liquid state at an elevated temperature [93], demand of large amount of water and formation of inhibitors. The remaining xylan and acetyl group in the solid residue after pretreatment has a marked effect on substrate degradability [25].

2.4.3. Ammonia Fiber Explosion (AFEX)

Ammonia fiber explosion (AFEX) pretreatment includes liquid ammonia and steam explosion [22]. The difference from steam pretreatment is the use of ammonia instead of water. In the AFEX process, an alkaline thermal pretreatment, the lignocellulosic materials are mixed with liquid anhydrous ammonia at a high temperature and pressure for a period of time and followed by a rapid pressure release. Typically, the dosage of liquid ammonia is 1-2 kg ammonia/kg dry biomass, the temperature is 100°C, and the residence time is 30 min [8].The mechanism of AFEX is considered as follows: The aqueous ammonia reacts primarily with lignin, but not with cellulose, resulting in the cleavage of lignin-carbohydrate linkages, depolymerization of lignin, swelling and physical disruption of biomass fiber. And deacetylation of hemicellulose is also observed. AFEX is reported to be an effective pretreatment process for herbaceous and agricultural residues, but works only moderately well on hardwood; however, it is not an attractive method for softwood due to the high lignin content [11]. The main factors influencing the AFEX process are ammonia loading, temperature, high pressure, moisture content of biomass, and residence time [94]. After the AFEX pretreatment at optimal conditions(loading 1kg ammonia/kg dry matter with 80% moisture content at 100°C for 5min), the enzymatic conversion of glucan in switchgrass can be increased from 16% to 93% , and the ethanol yield of pretreated sample is 2.5 times more than that of the untreated sample [90]. About 80.6% of glucan and 89.6% of xylan conversions of AFEX-treated rice straw (including monomeric and oligomeric sugars) can be achieved with less than 3% sugar loss during the pretreatment [95]. Advantages of the AFEX method are that no inhibitors of the downstream processes are produced and small particle size is not required [5, 11]. Furthermore, as AFEX produces only pretreated solid materials and all of the ammonia quickly evaporates, there is no need to adjust pH and wash the treated materials for subsequent processes. This pretreatment has the drawbacks of being less efficient for biomass containing higher lignin contents, such as softwood and newspaper, and degrading only a very small fraction of the solid material, particularly hemicellulose [94]. Additionally, the AFEX process requires an efficient recovery of ammonia to be economical because of the high cost of ammonia, and evaporation is a possible way to recover the ammonia after the pretreatment.

Another process is ammonia recycle percolation (ARP), in which aqueous ammonia (5-15 wt%) passes through a reactor packed with biomass at a temperature between 140-210°C for up to 90min at a percolation rate about 5ml/min [12, 90]. Like AFEX, ARP is efficient for the delignification of hardwood and agricultural residues, but is somewhat less powerful for softwood. The ARP process is well suitable for wheat straw pretreatment [96]. The wheat straw treated with the ammonia solution (15% v/v) showed a high delignification of 70% at 170°C for 30min, and the effect of temperature on the delignification is more significant than the reaction time. Besides, more than 90% of enzymatic digestibility and 90.66% of the theoretical ethanol yield (35°C, 24h) can be achieved.

2.4.4. CO₂ Explosion

Supercritical CO_2 explosion works in a way similar to the steam and ammonia explosion techniques in the pretreatment of lignocellulosic materials. A supercritical fluid refers to a fluid at a temperature and pressure above its critical point, where distinct liquid and gas phases do not exist. The supercritical CO_2 has a lower temperature than steam explosion and possibly costs less than ammonia explosion, and the low temperature protects monosaccharides from being decomposed by the acid. During the explosive release of the carbon dioxide pressure, the structure of the feedstock is disrupted with increasing the accessible surface area of the substrate to hydrolysis [97]. Moreover, the increase of pressure facilitates faster penetration of CO_2 molecules into the crystalline structures, producing more glucose after the explosion. It is hypothesized that CO_2 would form carbonic acid when dissolving in water and increase the hydrolysis rate [25]. Dale and Moreira [98] reported that 75% of the theoretical glucose is released by enzymatic hydrolysis of alfalfa treated with this method(4 kg CO_2/kg fiber at a pressure of 5.62 MPa). Park et al. [99] treated the cellulose with supercritical CO_2 at 16 MPa for 90 min at 50°C, and obtained a nearly 100% enzymatic conversion of cellulose to glucose. Zheng et al. [100, 101] compared CO_2 explosion with steam and ammonia explosion for the pretreatment of recycled paper mix and sugarcane bagasse, and found that CO_2 explosion is more cost-effective than ammonia explosion. It does not cause the formation of inhibitors as in steam explosion. The main drawback to this pretreatment technology is its high costs [51].

2.5. Biological Pretreatment

Biological pretreatment is a safe and environmentally-friendly method for lignin removal from lignocellulose. And compared to other pretreatment techniques, it boasts low chemicals consumption and energy saving with less mechanical support [5]. Moreover, it is performed in mild environmental conditions [15]. Biological pretreatment processes are carried out by using microorganisms such as brown-, white- and soft-rot fungi to remove lignin and hemicellulose through liberating cellulose [102]. Among the microorganisms used for degrading the cellulosics, the white-rot fungus seems to be the most effective microorganism than other studied microbial groups [47], because the white-rot basidiomycetes can degrade lignin more rapidly and extensively [24]. Brown rots attack cellulose while white and soft rots attack both cellulose and lignin [101].

And lignin undergoes a series of oxidative changes during the mineralization by white rot fungi, including aromatic ring cleavage and the release of low-molecular weight fragments(mainly <1kd) [24]. Lignin degradation by white-rot fungi occurs through the action of lignin-degrading enzymes [103], which mainly include Manganese peroxidase, H_2O_2-producing enzymes and laccase [24]. Fungi have two types of extracellular enzymatic system: one is the hydrolytic system, which degrades polysaccharide, and the other is oxidative ligninolytic system, which degrades lignin and opens phenyl rings [104]. In addition, growing evidence indicates that ligninase (lignin peroxidase) is the key lignin-degrading enzyme of white-rot fungi; however, further studies are needed to clarify the mechanism and the relative importance of aromatic ring cleavage by ligninase [24]. The most promising fungi for straw pretreatment are the two *Pleurotus spp.*: *Pleurotus sp.* 535 and *Pleurotus ostreatus*, while *Pycnoporus cinnabarinus* 115 and *Phlebia radiata* 79 may have the potential for softwood treatment in that they can degrade guaiacyl-type synthetic lignin [105].

Additionally, Lee et al. regarded *S. hirsutum* as an effective potential fungus for hardwood pretreatment [103]. They investigated the effects of biological pretreatment on the Japanese red pine Pinus densiflora with three white-rot fungi: *Ceriporia lacerata*, *Stereum hirsutum*, and *Polyporus brumalis*. Of the three white-rot fungi tested, *S. hirsutum* selectively degraded the lignin fraction of the feedstock, rather than the holocellulose (hemicellulose and cellulose).

The loss of total weight was 10.7%, and the highest lignin loss among the tested samples was 14.5% after 8 weeks of pretreatment with *S. hirsutum*. Besides, the holocellulose loss was 7.8%, lower than that using *C. lacerata* and *P. brumalis*. Extracellular enzymes from *S. hirsutum* showed higher activity of ligninase and lower activity of cellulase than those from other fungi, and the enzymatic sugar yield of the pine treated with *S. hirsutum* was increased to 21.0%, when compared to unpretreated samples. Biological treatments in combination with chemical treatments have been studied as well [105, 106]. Annele [105] compared the fungal treatment with combined fungal and alkali treatment on the enzymatic hydrolysis of wheat straw.

The results showed that the combined fungal and alkali pretreatment is not more efficient than either treatment alone, and the merit of fungal treatment is that it can produce a higher proportion of glucose in sugar mixtures than alkali treatment alone. Furthermore, they also found that oxygen atmosphere can enhance the effect of fungal pretreatment on the enzymatic hydrolysis of biomass by reducing the time for a positive pretreatment effect by about 1 week when compared with the treatments of *P. sordida* 37 and *Pycnoporus cinnabarinus* 115. Balan et al. [107] studied the effect of fungal conditioning of rice straw followed by AFEX pretreatment and enzymatic hydrolysis. They reported that the glucan and xylan conversions of rice straw treated with white-rot fungi and *Pleurotus ostreatus*, followed by AFEX are significantly higher and the AFEX conditions are less severe than those of rice straw treated with AFEX alone.

For the protection of cellulose and hemicellulose, cellulase-less mutant has been developed for the selective degradation of lignin, but in most cases of biological pretreatment, the rate of hydrolysis is very low, impeding its implementation [15].

3. ENZYMATIC HYDROLYSIS

Enzymatic hydrolysis is the second step for bioethanol production, in which complex carbohydrates are converted to simple monomers. The process is compatible with many pretreatment options and it is usually carried out by cellulase enzymes that are highly substrate specific [109].

Compared to acid hydrolysis, enzymatic hydrolysis requires less energy and mild environment conditions(40-50°C, pH 4-5) [82, 110], in addition to low toxicity and low corrosion. Cellulase enzymes can be produced from bacteria or fungi [111]. Cellulase enzymes can be produced from bacteria such as *Clostridium*, *Cellulomonas*, *Bacillus*, *Thermomonospora*, *Ruminococcus*, *Bacteriodes*, *Erwinia*, *Acetovibrio*, *Microbispora* and *Streptomyces*. Fungi such as *Sclerotium rolfsii*, *P. chrysosporium* and species of *Trichoderma*, *Aspergillus*, *Schizophyllum* and *Penicilium* can also be used to produce cellulases as well. Because anaerobes grow very low and need anaerobic growth conditions, most research for commercial cellulase production has focused on fungi [84]. Among the various cellulolytic microbial strains, *Trichoderma* is one of the most extensively studied strains producing cellulase and hemicellulase, and it is able to produce three groups of enzymes, including cellobiohydrolases(E.C.3.2.1.91; CBHs, exoglucanases), endoglucanases (E.C.3.2.1.4, EGs) and β-glucosidases(E.C.3.2.1.21). *T. reesei* can secrete at least two cellobiohydrolases(CBH I and CBH II), five endoglucanases(EG I , EG II , EGIII, EGIV, and EGV) and three endoxylanases [112] (CBH I and CBH II hydrolyse the reducing and non-reducing ends of cellulose chains, respectively), and the enzyme complex of *T. reesei* is deficient in β-glucosidases activity. However, *Aspergillus* is a very efficient β-glucosidase producing strain [113].

Cellulases are a complex system of several enzymes that act synergistically to hydrolyze cellulose. At least three major groups of cellulases are involved in the hydrolysis process: (1) endoglucanase (EG, endo-1,4-D-glucanohydrolase, or EC 3.2.1.4.) which attacks regions of low crystallinity in the cellulose fiber, creating free chain-ends; (2) exoglucanase or cellobiohydrolase(CBH, 1,4- β-D -glucan cellobiohydrolase, or EC 3.2.1.91.) which degrades the molecule further by removing cellobiose units from the free chain-ends; (3) β-glucosidase (EC 3.2.1.21) which hydrolyzes cellobiose to produce glucose [114]. In addition to the three major groups of cellulase enzymes, there are also a number of ancillary enzymes that attack hemicellulose, such as glucuronidase, acetylesterase, xylanase, β-xylosidase, galactomannanase and gluco-mannanase [84]. Hemicellulolytic enzymes are more complex than cellulases as they are a mixture of at least eight enzymes such as endo-1,4-β-$_D$-xylanases, exo-1,4-β-$_D$-xylocuronidases, α-$_L$-arabinofuranosidases, endo-1,4-β-$_D$ mannanases, β-mannosidases, acetyl xylan esterases, α-glucoronidases and α-galactosidases [115]. Endo-β-1-4-xylanases primarily cleave the internal β-1-4 bonds between xylose units, exoxylanase releases xylobiose units and β-xylosidase releases xylose from xylobiose and short chain xylooligosachharides [116].

During the enzymatic hydrolysis, cellulose is degraded by the cellulases, and hemicellulose is hydrolyzed by xylanases to reducing sugars, mainly glucose and xylose, that can be fermented to ethanol by the microbe, such as yeasts or bacteria.

Table 2. Summary of various processes used for the pretreatment of lignocellulosic biomass[a]

Pretreatment and process	advantages	limitations and disadvantages
Mechanical comminution	Reducing cellulose crystallinity and particle size	Higher power consumption than inherent biomass energy
Pyrolysis pulsed electrical field	Producing gas and liquid products in ambient conditions; disrupting plant cells; simple equipment	High temperature; loss of hemicellulose; demand for more research in ash production process
Steam explosion	Causing hemicellulose degradation and lignin transformation; cost-effective	Destruction of a portion of the xylan fraction; incomplete disruption of the lignin-carbohydrate matrix; generation of compounds inhibitory to microorganisms
AFEX	Increasing accessible surface area, removing lignin and hemicellulose to an extent; not producing inhibitors for downstream processes	Not efficient for biomass with high lignin content
CO_2 explosion	Increasing accessible surface area; cost-effective; not causing formation of inhibitory compounds	Not modifying lignin or hemicellulose
SPORL [b]	Energy efficiency; hydrolysis and sulfonation of lignin; depolymerization of cellulose ; reducing nonproductive adsorption of enzymes on lignin	Formation of fermentation inhibitors ; hydrolysis of hemicellulose
Ozonolysis pretreatment	Reducing lignin content; not producing toxic residues	Demand for large amount of ozone; expensive
Wet oxidation [c]	Removing lignin	Formation of by-products; degradation of hemicellulose
Acid pretreatment	Hydrolyzing hemicellulose to xylose and other sugars; altering lignin structure	High cost; equipment corrosion; formation of toxic substances; recovery of acid
Alkaline pretreatment	Removing hemicelluloses, acetyl and lignin; increasing accessible surface area	Demand for long residence times; formation of irrecoverable salts into biomass
Organosolv	Hydrolyzing lignin and hemicellulose	Demand for the drainage of solvents from the reactor, evaporation, condensation, and recycling; high cost
Biological pretreatment	Degrading lignin and hemicellulose; low energy demand	Slow bioconversion

[a] Adapted from Ref [25]; [b] Adapted from Ref [58]; [c] Adapted from Ref [53].

Table 3. Effect of different technologies on the structure of lignocellulosic biomass

Process	Increasing accessible surface area	Decrystalizing cellulose	Removing hemicel-ulose	Removing lignin	Altering lignin structure	Generating toxic compounds
Milling	H	H	-	-	-	-
Steam explosion	H	-	H	M	H	H
LHW	H	ND	H	L	M	L
CO₂ explosion	H	-	H	-	-	-
AFEX	H	H	M	H	H	L
ARP	H	H	L	H	H	M
Oxidative	H	ND	-	M	H	L
Dilute acid	H	-	H	L	H	H
SPORL	H		M	L	H	M
Alkali	H	-	L	H	H	L
Lime	H	ND	M	M	H	M
Organic solvents	H		M	H	H	M

[a] Adapted from Ref [11, 12, 108].

H: high effect; M: moderate effect; L: low effect; ND: not determined.

3.1. Improving Enzymatic Hydrolysis

3.1.1. Substrate

Substrate concentration is one of the main factors that affect the initial rate of enzymatic hydrolysis of cellulose. High substrate concentration can cause substrate inhibition, which substantially lowers the rate of the hydrolysis, and the degree of substrate inhibition depends on the ratio of total substrate to total enzyme [5]. Huang and Penner [117] reported that the substrate inhibition that occurred at the ratio of the microcrystalline substrate Avicel pH 101 to the cellulase from *Trichoderma reesei*(grams of cellulose/FPU of enzyme) is more than 5, and the optimum ratio is 1.25g Avicel pH 105 per FPU of the cellulose from *T. reesei* [77, 118]. The accessibility of lignocellulosic substrates to cellulases depends on the structural features of the substrate including cellulose crystallinity, degree of cellulose polymerization, specific surface area, and content of lignin [5].

Lignin is believed to be a strong inhibitor of cellulases, given that the reduction of lignin content leads to an improvement in the hydrolysis efficiency [44, 77, 119]. Lignin interferes with enzymatic hydrolysis by blocking the access of cellulases to cellulose and nonproductive absorption of cellulases to hydrophobic lignin. Berlin et al. [120] explored the inhibitory nature of lignin for enzymatic hydrolysis and found that lignin interferes with enzymatic hydrolysis by binding to the enzyme and forming a triple complex of enzyme-substrate inhibitor. For lignocellulosic substrates, nonproductive binding and inactivation of enzymes by the lignin component appear to be important factors limiting catalytic efficiency. Moreover, although cellulose-binding domains (CBDs) are hydrophobic and probably participate in lignin binding, the results show that cellulases lacking CBDs also have a high affinity for lignin, suggesting the presence of lignin-binding sites on the catalytic domain. On a concentration basis, lignin inhibition has been found comparable to inhibition by glucose, but not competitive in nature. McMillan [44] reported that the removal of lignin can dramatically enhance the hydrolysis rate. Another method is addition of surfactants during hydrolysis, which is capable of modifying the cellulose surface property by adsorbing lignin onto surfactant and minimizing the irreversible binding of cellulase on lignin, and lowering enzyme loading [121, 122]. Engineering of enzymes with lower lignin affinity could provide a useful strategy for improvement of cellulase activity on lignocellulosic substrates [120]. However, Rollin et al. [123] reported that improving the surface area accessible to cellulase is a more important factor for achieving a high sugar yield, compared with delignification.

3.1.2. End-product Inhibition of Cellulase Activity

Many studies have shown that cellobiose, glucose and ethanol are major inhibitors of cellulases [124]. Particularly, the activity of CBH I is strongly inhibited by its product, cellobiose. The results from FTIR and circular dichroism showed that cellobiose could combine with the tryptophan residue located at the active site of cellobiohydrolase and produce steric hindrance, thereby preventing cellulose chains from diffusing further into the active site of the enzyme. In addition, the change in the conformation of cellobiohydrolase caused by cellobiose binding also reduces the activity of cellobiohydrolase during hydrolysis [125]. For improving the enzyme activity, several methods have been developed to reduce the inhibition, including the addition of high concentrations of enzymes, the supplementation of β-glucosidases or/and xylanases during hydrolysis, and the removal of sugars during hydrolysis by ultrafiltration or simultaneous saccharification and fermentation (SSF) [5]. The addition of β-glucosidases into the *T. reesei* cellulases system can achieve better saccharification than the system without β-glucosidases, because β-glucosidase can degrade the cellobiose to glucose to release the inhibition [126]. Adequate levels of β-glucosidase and xylanase activities appear to be important for enhancing the efficiency of lignocellulose hydrolysis by the enzymatic blends [127]. Immobilization β-glucosidase on an inert carrier offers a prospect for thermostability improvement and significant cost savings by facilitating enzyme recycling through multiple cycles of batch-wise hydrolysis [128].

Podkaminer et al. [129] investigated the effect of ethanol on enzyme inactivation, showing that enzyme activity is moderately stabilized at an ethanol concentration between 0 and 40 g/L, but an ethanol concentration above 40 g/L will accelerate enzyme inactivation, leading to 75% loss of enzymatic activity in 80 g/L ethanol after 4 days. At 37°C, ethanol does not show a strong effect on the rate of enzyme inactivation. Currently, there is still no effective approach to reduce the inhibition by ethanol.

4. FERMENTATION

Cellulose, the major constituent of the plant cell wall, consists of an unbranched polymer of glucose. And pentose is the predominant component in the hemicellulose hydrolysates. Theoretically, functional microorganisms should have the ability to utilize the lignocellulosic hydrolysates, mainly composed of hexose and pentose [5, 11, 128, 129]. To realize the efficient bioconversion of hydrolysates, many fermentation strategies have been established and they can be classified into four main processes: Separate hydrolysis and fermentation (SHF), Simultaneous saccharification and fermentation (SSF), Simultaneous saccharification and fermentation (SSCF), and Direct microbial conversion (DMC). Generally, microorganisms that can secrete cellulase and/or hemicellulase, and utilize hexose and pentose to yield ethanol can be used for lignocellulose-ethanol production. Rates of bacterial-mediated cellulose degradation vary widely, relying on the operation process, organisms, and the type of the lignocellulosic material. And the choice of a most suitable process will be determined upon the kinetic properties of microorganisms and the type of lignocellulosic hydrolysate in addition to process economics [130].

4.1. Separate Hydrolysis and Fermentation (SHF) with Microorganisms

In this process, enzymatic hydrolysis is performed separately from fermentation. The hydrolysis reactors, hexose reactions and pentose fermentation are separated for the convenience of removing the bioethanol by distilling. Absolutely, SHF can carry out each step under optimal conditions. However, the inhibition of cellulase and β-glucosidase enzymes by glucose released would be more obvious, which implies lower solids loadings and higher enzyme loadings to achieve acceptable yields [41]. For the reasons above, in the SHF process, we should use microorganisms with the ability of delignification, lignocellulose hydrolysis, and yielding ethanol with hexose and pentose in good coordination and cooperation by just importing some lignocellulose degradation enzymes instead of cellulolytic bacteria directly.

4.1.1. Direct Microbial Conversion of Cellulose to Ethanol

Degradation of lignocellulose in the environment involves a fairly diverse collection of bacteria, although within environmental niches only a few species are likely to dominate. In the past decade, most studies on the cellulase producers have been focused on aerobic fungus, *Trichoderma reesei* [26]. Besides, *Cellulomonas, Clostridium, Bacillus, Thermomonospora, Ruminococcus, Baceriodes, Erwinia, Acetovibrio, Microbispora*, and *Streptomyces* have been used in cellulolytic biomass hydrolyzation [131]. However, many problems exist in developing the bioprocess for cellulase production from *T. reesei*: ① scale-up of enzyme production due to oxygen transfer in mycelial broth; ② poor mixing due to shear sensitivity of fungus; and ③low cell-bound enzyme activity. So, recently, exploring anaerobic microorganisms in cellulase system has drawn many researchers' attention for its more stringent limitations than aerobes in the overall metabolic activities.

In the past decades, many cellulolytic anaerobes have been isolated and characterized for potential technology development for fuel or chemical production by direct fermentation of biomass components [132, 133]. We can draw a conclusion that anaerobic organisms have priorities over aerobes in many aspects. Firstly, they possess high affinity for cellulose to minimize the wash away of enzymes from substrates. Besides, the hydrolysis pattern often emerges as a cell-enzyme-substrate complex, which can enhance the cell binding towards cellulose and minimize the diffusion of hydrolysis products. Thirdly, it can alleviate the activity of cellulase system by assembling cellulolytic components into a stable complex. Due to these advantages, anaerobic strains were widely studied in cellulose biomass degradation. Among all the anaerobes, thermophilic anaerobes were more popular for their higher growth and metabolic rates. South et al. [134] compared the conversion efficiencies of pretreated hardwood flour to ethanol between *T. reesei*-cellulase /*Saccharomyces cerevisiae* and *C. thermocellum*. The result demonstrated that at a similar range of residence time, the substrate conversion in the latter system (77%) was significantly higher than that in the former one in the simultaneous saccharification and fermentation system (31%).

4.1.2. Microorganisms in Co-fermentation of Lignocellulosic Hydrolyzates

Functional microorganisms hydrolyzed the lignocellulosic biomass into fermentable reducing sugars, mainly composed of hexose-utilized strains and pentose-utilized strains. For the efficiency of bioconversion, co-fermentation using mixed cultures of different compatible species should be mentioned. It has been reported that co-culture of the *R. albus/ M. smithi* provides evidently support for the general rule that mixed cultures can outperform pure cultures [135]. As yeast often demonstrates high ethanol efficiency and tolerance, so screening yeast that can assimilate both hexoses and pentoses become a potential orientation. However, hexose-utilizing microorganisms grow faster than pentose-utilizing microorganisms and thus the conversion of hexoses to ethanol is more significantly elevated. This problem can be solved by employing a respiratory deficient mutant of the hexose-utilizing strain along with rapid pentose-fermenting strains. On the one hand, the former can elevate the fermentation and growth activities of the pentose-utilizing strain. On the other hand, the presence of hexose-utilizing stain can partly reduce the catabolic repression on the pentose consumption [136]. However, the strains that can utilize xylose are not large in number. Scientists at the Solar Energy Research Institute have demonstrated that the production cost of ethanol can be reduced from $1.65 to $1.23 per gallon if all the xylose is utilized for ethanol production.

It has been confirmed that various yeast cultures, i.e. *Candida tropicalis*, *Pichia stipitis*, *Pachysolen tannophilus*, *Schizosaccharomyces pombe*, *Saccharomyces cerevisiae* and co-culture of *P. tannophilus* and *S. cerevisiae*, can degrade Untreated spent sulphite liquor in the presence of xylose isomerases and 4.6 mM azide [137]. Besides, *Escherichia coli*, *Zymononas mobilis*, and *Klebsiella planticola* were also applied in xylose-ethanol production. Another variant of co-fermentation is utilizing a single microorganism capable of assimilating both hexoses and pentoses in an optimal way to achieve a high conversion efficiency and ethanol yield. This can be realized in two ways: modifying the genes of *S. cerevisiae* and *Z. mobilisto*, which allows the assimilation of pentoses, and introducing the gene encoding the metabolic pathways for the production of ethanol to microorganisms that are capable of fermenting both hexoses and pentoses in their native form [138]. Genetic engineering approach is still a hot topic in the lignocellulose bioconversion process and has a bright prospect.

4.2. Simultaneous Saccharification and Fermentation and its Suitable Strains

Compared with SHF, Simultaneous saccharification and fermentation (SSF), which involves saccharification and fermentation process in one vessel, is a more popular option for production of bioethanol from lignocellulosic materials [139]. SSF has the following advantages [5]: (1) higher hydrolysis rate by converting sugars that inhibit the cellulase activity; (2) lower enzyme demand; (3) higher product yields; (4) less risk of contamination and lower requirements for sterile conditions due to instant glucose removal and ethanol production; (5) shorter process time; and (6) less reactor volume due to the use of a single reactor. However, the xylose is much more inhibitory for conversion process than ethanol does. Besides, in SSF, the optimal conditions for hydrolysis and fermentation are different, which means a difficult control and optimization of process parameters [140]. In an SSF process for wheat straw, hydrolyzate may gain a final bioethanol concentration close to 40g/l with a yield being higher than 70% based on total hexose and pentoses in Olofsson K's study [141]. Due to the high cost of the commercial cellulases, the integration of cellulase production process by T. reesei with ethanol fermentation has been proposed. However, most cellulases work in an optimal way at 40-50°C and pH of 4-5 whereas the fermentation of hexoses with *S. cerevisiae* is optimal at 30°C, and the fermentation of pentoses at pH of 5-7 [142]. For this reason, thermotolerant yeasts, such as *Kluyveromyces marxianus*, that can be cultured at 42-43°C in batch SSF process are widely used [143, 144]. On the other hand, pentose utilized yeast, such as Candida acidothermophilum, C. brassicae, S. uvarumand, and Hansenula polymorpha, can also be used to improve the fermentation efficiency. Another proposed approach is the utilization of mixed cultures. To this end, cellulase- and hemicellulose-producing fungus Fusarium oxysporum along with S. cerevisiae was employed to gain co-fermentation about ethanol [145]. Laplace et al. [136] have demonstrated that co-cultures of Saccharomyces cerevisiae CBS 1200 and Candida shehatae ATCC 24 860 can utilize both glucose and xylose, the yields of which can reach 100% and 27% on glucose and xylose, respectively. In the elective mixed cultures grown on cellulosic material at neutral pH, *C. thermocellum* is the dominant species, followed by several prevalent thermophilic *Bacillus* species, and most other thermophiles could not degrade both pentoses and hexoses with high ethanol yields [146, 147]. Logically, the mixed culture approach presents a high complexity during its implementation at an industrial level, but simple operation process and high efficiency show its bright future.

4.3. Simultaneous Saccharification and Co-fermentation with Microorganisms

Simultaneous saccharification and co-fermentation (SSCF) is another frequently-used process. Most of the process is similar to SSF, except that hexose and pentose fermentation are operated in the same vessel. This means microorganisms used in co-fermentation of mixed cultures are also suitable for this process. For example, the co-culture of *P. stipitis* and *Brettanomyces clausennii* has been utilized for the SSCF of aspen at 38°C and pH of 4.8 yielding 369 L EtOH per ton of aspen during 48 h batch process [142]. Besides, recombinant strains also reveal a popular future. For instance, the recombinant *Z. mobilis* assimilating xylose is widely used, due to its high ethanol tolerance and yield. It was reported that the utilization of recombinant *Z. mobilis* can result in a 92% glucose conversion to ethanol and an 85% xylose conversion to ethanol [148]. As in the case of SSF, introducing microbial strains capable of growing at an elevated temperature may improve the efficiency of this process.

4.4. Consolidated Bioprocessing with Microorganisms

Consolidated bioprocessing (CBP), also known as Direct microbial conversion (DCM), combines cellulase production, cellulose hydrolysis and glucose fermentation into a single step. Application of CBP entails no operating cost or capital investment for dedicated enzyme production, reduced diversion of substrate for enzyme production, and compatible enzyme and fermentation systems [55]. In another word, this process requires strains which can secrete cellulase or hemicellulase directly. Lynd et al. (2005) reported the results of the comparative simulation of SSCF and CBP processes assuming aggressive performance parameters, stating that production cost of ethanol for SSCF is 4.99 US cents/L, whereas CBP just 1.11 US cents/L [149].

However, as a relatively new technology, teething pains cannot be avoided. For example, during CBP, cellulase is generated inside the reaction vessel, making this process more complex than simply adding enzymes. Besides, the optimal enzymatic conditions of the enzyme secreted are often in conflict with the growth or metabolic conditions of the strains. Most studies about the CBP approach have concluded that cellulose hydrolysis by microbes is the main obstacle to the enzymatic hydrolysis operation. Therefore, this paradigm justifies the search of microorganisms that actively secrete cellulases as well as grow well in anaerobic conditions. For this reason, the thermophilic bacterium *C.thermocellum* which could effectively degrade cellulose and hemicellulose is widely used. It is reported that the CBP using *C. thermocellum* shows a 31% higher substrate conversion than the system using *T. reesei* and *S. cerevisiae*. Besides, some filamentous fungi as *Monilia sp.*, *Neurospora crassa* and *Paecilomyces sp.* are also able to transform cellulose into ethanol [150]. However, a high ratio of acetic acid to ethanol and even toxic carboxylic acids are generally present in their bioenergy-producing process [151].

To date, to the best of our knowledge, there is no microorganism known that can exhibit the whole combination of features required for CBP. In order to improve the ethanol production through the increase of yield or tolerance, the best strategies to solve this obstacle are genetically engineered microorganisms with a high native cellulolytic activity and a high ethanol yield and tolerance in a CBP configuration. Most studies on the improvement of strains' ethanol tolerance are focused on *C. thermocellum* for its high cellulose hydrolysis efficiency and thermophilic anaerobic culture conditions. Lynd et al. [149] reported that the acclimatization bacteria of *C. thermocellum* can endure the ethanol concentration exceeding 60 g/L. For the method of genetic engineering, the engineering of *E. coli*, *K. oxytoca* and *Z. mobilisand* by the yeast *S. cerevisiae* has been studied and the production of cellulases by bacterial hosts can result in a high yield of ethanol. Undoubtedly, ongoing research on genetic and metabolic engineering will make possible the development of effective and stable strains of microorganisms for converting cellulosic biomass into ethanol.

5. SEPARATION TECHNOLOGY

With the commercial promotion of biomass hydrolysis and fermentation technologies, advancements in production recovery technologies will be required. As is known to all, high concentration of ethanol implies a strong inhibition to both the growth and metabolism of the strains. Separation is a key step where major costs are generated in the industry. Therefore, an optimum separation strategy is the key to decrease the production cost of bioethanol. For the cases in which fermentation products are more volatile than water, recovery by distillation is the usual choice. Generally, separation strategies can be classified into two processes: reaction-separation and separation-separation. Different ethanol production processes would have a different optimal separation method, and by making this process more commercial and cost-effective, we can speed up the day when bioethanol becomes an alternative to fossil oil on a large scale.

5.1. Reaction–separation Integration

Reaction-separation integration is a particularly attractive alternative to the intensification of ethanol fermentation process. Most of the integrated schemes are oriented to the integration of fermentation steps and separation operations, although several steps have already been formulated. In this process, integration of enzymatic hydrolysis and ethanol separation are considered as an integrated pattern. Enzyme immobilization technology and hollow-fiber membrane reactors are generally used in this process to improve the economy. In this pattern, the enzymes are confined inside the reactor allowing the separation of the substrate and the product, which makes the reutilization of the enzyme possible while preserving their activities. Due to its reduction in the inhibition effect of formed sugars on the cellulases and the increase in productivity in continuous operation, this pattern has been widely used in the saccharification of lignocellulosic biomass, the substrate conversion ratio of which can be increased by 53% against a 35% conversion in the traditional batch operations [152]. When ethanol is removed from the culture broth, its inhibition effect on strains' growth and metabolism is diminished or neutralized. Ethanol removal pattern can be achieved by vacuum, gas stripping, membranes, and liquid extraction. Ethanol removal by vacuum often operates with a vacuum chamber combined with the vessel, which allows the partial product removal and the increase of overall process productivity. However, this system requires the addition of pure oxygen and the bleed stream of broth. Another ethanol removal approach is realized by gas stripping. Its advantages lie in increasing the sugar concentration in the stream feeding fermentor. Gong et al. [153] have demonstrated that the simultaneous variant of the fermentation-stripping process using an air-lift reactor with a side arm can improve liquid circulation and mass transfer, and the ethanol concentration can exceed 130 g/L within 24 h in fed-batch regime. Ethanol removal can also be realized through membranes, i.e., the removed ethanol is distilled and the obtained bottoms are recycled to the culture broth resulting in a drastic reduction of generated wastewater. Membrane distillation (MD) is a relatively new process that can be considered as the most energy-saving, low-cost alternative to conventional separation processes [154]. The coupling of *C. thermohydrosulfuricum* that directly converts uncooked starch into ethanol with pervaporation has also been tested obtaining ethanol concentrations in the permeate of 27–32% w/w with silicalite zeolite membranes [155]. Besides, another reasonable approach for increasing the productivity of alcoholic fermentation is removing inhibitory product through an extractive biocompatible agent (solvent) that favors the migration of ethanol to solvent phase, known as extractive fermentation. The selection of the solvent is a crucial factor for extractive fermentation technology. Kang et al. [156] utilized a hollow-fiber membrane reactor for carrying out extractive fermentation with yeast using oleyl alcohol and dibutyl phthalate obtaining a productivity of 31.6 g/(L·h). The distribution coefficient of ethanol between the solvent and aqueous phases should be high to improve both the productivity and the product concentration. Besides, the solvent should be no toxic for fermenting cells. Bruce and Daugulis [157] identified the solvent mixture of oleyl alcohol with 5% (v/v) 4-heptanone as a promising extractant due to its reduced inhibitory effect and increased distribution coefficient with a solvent screen program for evaluating fermentation configurations. Extractive fermentation can also be realized by using aqueous two-phase fermentation where two phases are formed in the bioreactor as a result of the addition of two or more incompatible polymers [158]. In this pattern, ethanol can be accumulated in the upper layer, while the cells in the lower phase, which is convenient for ethanol distilling as well as reducing the inhibition effect. However, this process is rather complex and expensive due to the high cost of polymers and therefore it has not been further developed for ethanol production.

5.2. Separation-Separation Integration

Separation-separation integration technology involved in ethanol production often corresponds to the integration of the conjugated types, such as equipment of closing the flow sheet by fluxes or refluxes. Pinto et al. [159] have employed Aspen Plus for the simulation and optimization of the saline extractive distillation for several substances (NaCl, KCl, KI and $CaCl_2$), applying the NRTL model developed for electrolytic systems to calculate activity coefficients with a considerably low consumption of salts. However, the recovery of salts was not simulated, which means the evaporation and recrystallization of salts were deduced, but energy demand for salts recovery could significantly increase the energetic expenditures. So, the utilization of commercial simulators is the key to a given process configuration for predicting its behavior. It was reported that the dehydration with molecular sieves presented lower operation costs of all the analyzed flowsheets, despite its higher capital cost for the complexity of automation and control system inherent to the pressure swing adsorption technology [160]. A typical pervaporation process often boasts a low operation cost but a high yield of dehydrated ethanol. Compared with azeotropic distillation using benzene, the pervaporation using multiple membrane modules can gain the same ethanol production rate and quality (99.8 wt.%), but the operation cost is approximately 1/3-1/4 that of azeotropic distillation [161]. For the distillation section, compared with azeotropic distillation, configurations involving vapor recompression and feed preconcentration are competitive alternatives. In short, the major obstacles to the industrial application of a separation approach for bioethanol production are still the cost and feasibility.

6. CHALLENGES AND PROSPECTS IN COMMERCIAL PROMOTION OF LIGNOCELLULOSE-ETHANOL

As discussed above, lignocellulosic biomass is the most promising feedstock in renewable energy production due to its great availability and low cost, but a large-scale commercial production of bioethanol has still not been implemented for its high production cost. Therefore, several breakthroughs must be achieved.

For the lignocellulose pretreatment, several problems should be solved: (1) high energy consumption for biomass pretreatment. Although the cost of energy consumption for woody biomass pretreatment can be reduced to the level used for agricultural biomass, improvement in the yield of hemicellulose sugars still should be put on the top agenda [58]; (2) process scalability. In spite of maturity of the technologies of bioethanol transportation and storage, the whole process scalability is still one of the key concerns for commercial production; (3) feedstock versatility. As is discussed in the second part, feedstocks vary from each other in their favorite pretreatment method. The choice of a suitable feedstock is as important as the choice of a pretreatment method in reducing the cost of ethanol production; (4) recovery of pretreatment chemicals and wastewater treatment. For a real clean bioethanol production, related technologies should be developed for the treatment of wastewater.

For the enzyme hydrolysis, exploring an enzyme with a high efficiency and thermostability still remains a great challenge for ethanol production. Concentrating cells by immobilization, adsorption or recycling usually increases the reactor productivity for fermentation of soluble substrates, so does the removal of inhibitory fermentation products. However, it is the rate of cellulose hydrolysis that limits the overall productivity of lignocellulose-derived ethanol fermentations. What's more, most studies available in the literature were performed on a rather small scale and in most cases, the enzymatic hydrolysis process was performed at a low substrate concentration but a high enzyme dosage. Finding robust strains for enzyme secretion, better understanding the mechanism of inhibition and getting rid of the inhibition are the key to realize large-scale bioethanol production.

Besides, exploring a suitable fermentation strategy with robust strains remains a main obstacle to its wide application at an industrial scale. Ethanol is a significant inhibitor to the growth of yeast as well as bacteria, while xylose-fermenting strains often show low ethanol tolerance. Removing ethanol during the course of fermentation is a good solution in this regard. The volatility of ethanol makes in situ ethanol removal possible via gas stripping or vacuum fermentation. Furthermore, the choice of a high efficiency technology for ethanol fermentation is the key to the process. CBP strategy for ethanol production shows its evident advantages both in efficiency and cost. However, establishing suitable CBP strains needs a great deal of work not only on strains domestication but also on strains construction.

In summary, we still have a long way to go in developing a mature bioethanol production process from the choice of lignocellulose feedstock to production separation for large-scale industrial bioethanol production.

CONCLUSION

As discussed in the previous sections, renewable bioethanol production offers the potential for a distributed ethanol supply network model based on on-site ethanol production. In the process, pretreatment, saccharification of feedstocks and conversion of fermentable sugars are the rate-limiting steps, and also the key to lignocellulose-bioethanol. As a freshly developed but promising technology, the ethanol production from lignocellulosic feedstocks on an industrial scale confronts with many challenges. More research efforts need to be made to develop a robust and well-studied bioethanol production technology to solve our human energy crisis and environmental problems and really realize sustainable development on earth.

ACKNOWLEDGMENTS

The authors gratefully acknowledge the financial support of the National Natural Science Foundation of China (No.51278200 and No. 51078147), Guangdong Natural Science Foundation (No. S2012010010380), the Fundamental Research Funds for the Central Universities, SCUT (No.2012ZM0081), the Guangdong Provincial Science and Technology Program (No. 2012B091100163, No. 2011B090400033, and No. 2010A010500005) and the Opening Project of Center of Guangdong Higher Education for Engineering and Technological Development of Speciality Condiments (No.GCZX-B1103).

REFERENCES

[1]   Lynd LR, Wang MQ. A Product-Nonspecific Framework for Evaluating the Potential of Biomass-Based Products to Displace Fossil Fuels. *Journal of Industrial Ecology* 2008;7:17.

[2]   da Costa Sousa L, Chundawat SPS, Balan V, Dale BE. 'Cradle-to-grave'assessment of existing lignocellulose pretreatment technologies. *Current opinion in biotechnology* 2009;20:339.

[3]   Wyman CE, Dale BE, Elander RT, Holtzapple M, Ladisch MR, Lee Y. Coordinated development of leading biomass pretreatment technologies. *Bioresource Technology* 2005;96:1959.

[4]   Galbe M, Zacchi G. Pretreatment: The key to efficient utilization of lignocellulosic materials. *Biomass and Bioenergy* 2012;46:.

[5]   Sun Y, Cheng J. Hydrolysis of lignocellulosic materials for ethanol production: a review. *Bioresource Technology* 2002;83:1.

[6]   Bjerre AB, Olesen AB, Fernqvist T, Plöger A, Schmidt AS. Pretreatment of wheat straw using combined wet oxidation and alkaline hydrolysis resulting in convertible cellulose and hemicellulose. *Biotechnology and Bioengineering* 2000;49:568.

[7]   Yeh AI, Huang YC, Chen SH. Effect of particle size on the rate of enzymatic hydrolysis of cellulose. *Carbohydrate Polymers* 2010;79:192.

[8]   Balat M. Production of bioethanol from lignocellulosic materials via the biochemical pathway: A review. *Energy Conversion and Management* 2011;52:858.

[9]   Dehkhoda A. Concentrating lignocellulosic hydrolysate by evaporation and its fermentation by repeated fedbatch using flocculating *Saccharomyces cerevisiae*. 2008.

[10] Galbe M, Zacchi G. A review of the production of ethanol from softwood. *Applied microbiology and biotechnology* 2002;59:618.

[11] Mosier N, Wyman C, Dale B, Elander R, Lee Y, Holtzapple M, et al. Features of promising technologies for pretreatment of lignocellulosic biomass. *Bioresource Technology* 2005;96:673.

[12] Alvira P, Tomás-Pejó E, Ballesteros M, Negro M. Pretreatment technologies for an efficient bioethanol production process based on enzymatic hydrolysis: A review. *Bioresource Technology* 2010;101:4851.

[13] Lynd LR. Overview and evaluation of fuel ethanol from cellulosic biomass: technology, economics, the environment, and policy. *Annual review of energy and the environment* 1996;21:403.

[14] Jørgensen H, Kristensen JB, Felby C. Enzymatic conversion of lignocellulose into fermentable sugars: challenges and opportunities. *Biofuels, Bioproducts and Biorefining* 2007;1:119.

[15] Hamelinck CN, Hooijdonk G, Faaij APC. Ethanol from lignocellulosic biomass: techno-economic performance in short-, middle-and long-term. *Biomass and Bioenergy* 2005;28:384.

[16] Saha BC. Hemicellulose bioconversion. *Journal of industrial microbiology & biotechnology* 2003;30:279.

[17] Zhu ZS, Zhu MJ, Xu WX, Liang L. Production of bioethanol from sugarcane bagasse using NH_4OH-H_2O_2 pretreatment and simultaneous saccharification and co-fermentation. *Biotechnology and Bioprocess Engineering* 2012;17:316.

[18] Michalska K, Miazek K, Krzystek L, Ledakowicz S. Influence of pretreatment with Fenton's reagent on biogas production and methane yield from lignocellulosic biomass. *Bioresource Technology* 2012;119:72.

[19] Dadkhah-Nikoo A, Bushnell D. Analysis of wood combustion based on the first and second laws of thermodynamics. *Journal of energy resources technology* 1987;109:129.

[20] Mohan D, Pittman Jr CU, Steele PH. Pyrolysis of wood/biomass for bio-oil: a critical review. *Energy & Fuels* 2006;20:848.

[21] Limayem A, Ricke SC. Lignocellulosic biomass for bioethanol production: Current perspectives, potential issues and future prospects. *Progress in Energy and Combustion Science* 2012;38:449.

[22] Balat M, Balat H, Öz C. Progress in bioethanol processing. *Progress in Energy and Combustion Science* 2008;34:551.

[23] Gray KA, Zhao L, Emptage M. Bioethanol. *Current opinion in chemical biology* 2006;10:141.

[24] Kirk TK, Farrell RL. Enzymatic" combustion": the microbial degradation of lignin. *Annual Reviews in Microbiology* 1987;41:465.

[25] Kumar P, Barrett DM, Delwiche MJ, Stroeve P. Methods for pretreatment of lignocellulosic biomass for efficient hydrolysis and biofuel production. *Industrial & Engineering Chemistry Research* 2009;48:3713.

[26] Lee J. Biological conversion of lignocellulosic biomass to ethanol. *Journal of Biotechnology* 1997;56:1.

[27] Fan L, Gharpuray MM, Lee YH, Aiba S, Fiechter A, Klein J, et al. Cellulose hydrolysis: Springer-verlag Berlin, Germany; 1987.

[28] Mani S, Tabil LG, Sokhansanj S. Grinding performance and physical properties of wheat and barley straws, corn stover and switchgrass. *Biomass and Bioenergy* 2004;27:339.

[29] Pandey A. Handbook of plant-based biofuels: CRC; 2008.

[30] Jin S, Chen H. Superfine grinding of steam-exploded rice straw and its enzymatic hydrolysis. *Biochemical engineering journal* 2006;30:225.

[31] Ruffell J. Pretreatment and hydrolysis of recovered fibre for ethanol production. University of British Columbia; 2008.

[32] Chiaramonti D, Rizzo AM, Prussi M, Tedeschi S, Zimbardi F, Braccio G, et al. 2nd generation lignocellulosic bioethanol: is torrefaction a possible approach to biomass pretreatment? *Biomass Conversion and Biorefinery* 2011;1:9.

[33] Shafizadeh F, Bradbury A. Thermal degradation of cellulose in air and nitrogen at low temperatures. *Journal of Applied Polymer Science* 2003;23:1431.

[34] Runkel ROH. Zur Kenntnis des thermoplastischen Verhaltens von Holz. *European Journal of Wood and Wood Products* 1951;9:41.

[35] Soltes E, Elder T. Pyrolysis in Organic Chemicals from Biomass. CRC Press, Boca Raton, FL; 1981;320.

[36] Fan L, Gharpuray M, Lee Y. Cellulose hydrolysis. *Biotechnology monographs*. Volume 3. 1987.

[37] Vedernikov N, Karlivans V, Roze I, Rolle A. Mechanochemical destruction of plant raw materials-polysaccharides in presence of small amounts of concentrated sulfuric-acid. *Sibirskii Khimicheskii Zhurnal* 1991;5:67.

[38] von Sivers M, Zacchi G. A techno-economical comparison of three processes for the production of ethanol from pine. *Bioresource Technology* 1995;51:43.

[39] Wyman C. Handbook on bioethanol: production and utilization: CRC; 1996.

[40] Zheng Y, Pan Z, Zhang R. Overview of biomass pretreatment for cellulosic ethanol production. *International journal of agricultural and biological engineering* 2009;2:51.

[41] Silverstein RA. A comparison of chemical pretreatment methods for converting cotton stalks to ethanol. 2005.

[42] Karimi K, Emtiazi G, Taherzadeh MJ. Ethanol production from dilute-acid pretreated rice straw by simultaneous saccharification and fermentation with *Mucor indicus*, *Rhizopus oryzae*, and *Saccharomyces cerevisiae*. *Enzyme and Microbial Technology* 2006;40:138.

[43] Saha BC, Iten LB, Cotta MA, Wu YV. Dilute acid pretreatment, enzymatic saccharification and fermentation of wheat straw to ethanol. *Process Biochemistry* 2005;40:3693.

[44] McMillan JD. Pretreatment of lignocellulosic biomass. ACS symposium series: ACS Publications; 1994, p. 292.

[45] Himmel ME, Adney WS, Baker JO, Elander R, McMillan JD, Nieves RA, et al. Advanced bioethanol production technologies: a perspective. ACS Symposium Series: ACS Publications; 1997, p. 2.

[46] Carvalheiro F, Duarte LC, Gírio FM. Hemicellulose biorefineries: a review on biomass pretreatments. 2008;67:849.

[47] Sarkar N, Ghosh SK, Bannerjee S, Aikat K. Bioethanol production from agricultural wastes: An overview. *Renewable Energy* 2012;37:19.

[48] Sun R, Lawther JM, Banks W. Influence of alkaline pre-treatments on the cell wall components of wheat straw. *Industrial crops and products* 1995;4:127.

[49] Millett MA, Baker AJ, Satter LD. Physical and chemical pretreatments for enhancing cellulose saccharification. Biotechnol Bioeng Symp;(United States): Dept. of Agriculture, Madison, WI; 1976;6:125.

[50] Playne M. Increased digestibility of bagasses by pretreatment with alkalis and steam explosion. *Biotechnology and Bioengineering* 1984;26:426.

[51] Chiaramonti D, Prussi M, Ferrero S, Oriani L, Ottonello P, Torre P, et al. Review of pretreatment processes for lignocellulosic ethanol production, and development of an innovative method. *Biomass and Bioenergy* 2012;46:25.

[52] McGinnis GD, Wilson WW, Mullen CE. Biomass pretreatment with water and high-pressure oxygen. The wet-oxidation process. *Industrial & Engineering Chemistry Product Research and Development* 1983;22:352.

[53] Martín C, Thomsen AB. Wet oxidation pretreatment of lignocellulosic residues of sugarcane, rice, cassava and peanuts for ethanol production. *Journal of Chemical Technology and Biotechnology* 2007;82:174.

[54] Klinke HB, Ahring BK, Schmidt AS, Thomsen AB. Characterization of degradation products from alkaline wet oxidation of wheat straw. *Bioresource Technology* 2002;82:15.

[55] Schmidt AS, Thomsen AB. Optimization of wet oxidation pretreatment of wheat straw. *Bioresource Technology* 1998;64:139.

[56] Wang G, Pan X, Zhu J, Gleisner R, Rockwood D. Sulfite pretreatment to overcome recalcitrance of lignocellulose (SPORL) for robust enzymatic saccharification of hardwoods. *Biotechnology progress* 2009;25:1086.

[57] Tian S, Zhu W, Gleisner R, Pan X, Zhu J. Comparisons of SPORL and dilute acid pretreatments for sugar and ethanol productions from aspen. *Biotechnology progress* 2011;27:419.

[58] Zhu J, Pan X. Woody biomass pretreatment for cellulosic ethanol production: technology and energy consumption evaluation. *Bioresource Technology* 2010;101:4992.

[59] Shuai L, Yang Q, Zhu J, Lu F, Weimer P, Ralph J, et al. Comparative study of SPORL and dilute-acid pretreatments of spruce for cellulosic ethanol production. *Bioresource Technology* 2010;101:3106.

[60] Zhu J, Pan X, Wang G, Gleisner R. Sulfite pretreatment (SPORL) for robust enzymatic saccharification of spruce and red pine. *Bioresource Technology* 2009;100:2411.

[61] Zhao X, Cheng K, Liu D. Organosolv pretreatment of lignocellulosic biomass for enzymatic hydrolysis. *Applied microbiology and biotechnology* 2009;82:815.

[62] Pan X, Arato C, Gilkes N, Gregg D, Mabee W, Pye K, et al. Biorefining of softwoods using ethanol organosolv pulping: Preliminary evaluation of process streams for manufacture of fuel-grade ethanol and co-products. *Biotechnology and Bioengineering* 2005;90:473.

[63] Pan X, Gilkes N, Kadla J, Pye K, Saka S, Gregg D, et al. Bioconversion of hybrid poplar to ethanol and co-products using an organosolv fractionation process: Optimization of process yields. *Biotechnology and Bioengineering* 2006;94:851.

[64] Pan X, Xie D, Yu RW, Saddler JN. The bioconversion of mountain pine beetle-killed lodgepole pine to fuel ethanol using the organosolv process. *Biotechnology and Bioengineering* 2008;101:39.

[65] Kootstra AMJ, Beeftink HH, Scott EL, Sanders JPM. Comparison of dilute mineral and organic acid pretreatment for enzymatic hydrolysis of wheat straw. *Biochemical engineering journal* 2009;46:126.

[66] Gong G, Liu D, Huang Y. Microwave-assisted organic acid pretreatment for enzymatic hydrolysis of rice straw. *Biosystems engineering* 2010;107:67.

[67] McDonough T. The chemistry of organosolv delignification. *Tappi journal* 1993;76:186.

[68] Rogers RD, Seddon KR. Ionic liquids--solvents of the future? *Science* 2003;302:792.

[69] Phillips DM, Drummy LF, Conrady DG, Fox DM, Naik RR, Stone MO, et al. Dissolution and regeneration of Bombyx mori silk fibroin using ionic liquids. *Journal of the American Chemical Society* 2004;126:14350.

[70] Mäki-Arvela P, Anugwom I, Virtanen P, Sjöholm R, Mikkola JP. Dissolution of lignocellulosic materials and its constituents using ionic liquids—A review. *Industrial crops and products* 2010;32:175.

[71] Fort DA, Remsing RC, Swatloski RP, Moyna P, Moyna G, Rogers RD. Can ionic liquids dissolve wood? Processing and analysis of lignocellulosic materials with 1-n-butyl-3-methylimidazolium chloride. *Green Chemistry* 2007;9:63.

[72] Zavrel M, Bross D, Funke M, Büchs J, Spiess AC. High-throughput screening for ionic liquids dissolving (ligno-) cellulose. *Bioresource Technology* 2009;100:2580.

[73] Zhu Z, Zhu M, Wu Z. Pretreatment of sugarcane bagasse with NH4OH-H2O2 and ionic liquid for efficient hydrolysis and bioethanol production. *Bioresource Technology* 2012.

[74] Kilpeläinen I, Xie H, King A, Granstrom M, Heikkinen S, Argyropoulos DS. Dissolution of wood in ionic liquids. *Journal of agricultural and food chemistry* 2007;55:9142.

[75] Zhang H, Wu J, Zhang J, He J. 1-Allyl-3-methylimidazolium chloride room temperature ionic liquid: A new and powerful nonderivatizing solvent for cellulose. *Macromolecules* 2005;38:8272.

[76] Hayes DJ. An examination of biorefining processes, catalysts and challenges. *Catalysis Today* 2009;145:138.

[77] Himmel ME, Baker JO, Overend RP. Enzymatic conversion of biomass for fuels production: American Chemical Society Washington, DC. ACS Symposium Series 566; 1994;371.

[78] Kabel MA, Bos G, Zeevalking J, Voragen AGJ, Schols HA. Effect of pretreatment severity on xylan solubility and enzymatic breakdown of the remaining cellulose from wheat straw. *Bioresource Technology* 2007;98:2034.

[79] Li J, Henriksson G, Gellerstedt G. Lignin depolymerization/repolymerization and its critical role for delignification of aspen wood by steam explosion. *Bioresource Technology* 2007;98:3061.

[80] Ballesteros I, Negro MJ, Oliva JM, Cabañas A, Manzanares P, Ballesteros M. Ethanol production from steam-explosion pretreated wheat straw. Twenty-Seventh Symposium on Biotechnology for Fuels and Chemicals: Springer; 2006, p. 496.

[81] Hongzhang C, Liying L. Unpolluted fractionation of wheat straw by steam explosion and ethanol extraction. *Bioresource Technology* 2007;98:666.

[82] das Neves MA, Kimura T, Shimizu N, Nakajima M. State of the art and future trends of bioethanol production. *Dynamic Biochemistry, Process Biotechnology and Molecular Biology* 2007;1:1.

[83] Morjanoff P, Gray P. Optimization of steam explosion as a method for increasing susceptibility of sugarcane bagasse to enzymatic saccharification. *Biotechnology and Bioengineering* 1987;29:733.

[84] Duff SJB, Murray WD. Bioconversion of forest products industry waste cellulosics to fuel ethanol: a review. *Bioresource Technology* 1996;55:1.

[85] Alfani F, Gallifuoco A, Saporosi A, Spera A, Cantarella M. Comparison of SHF and SSF processes for the bioconversion of steam-exploded wheat straw. *Journal of industrial microbiology & biotechnology* 2000;25:184.

[86] Jeoh T. Steam explosion pretreatment of cotton gin waste for fuel ethanol production. Virginia Polytechnic Institute and State University; 1998.

[87] Clark T, Mackie K. Steam explosion of the softwood Pinus radiata with sulphur dioxide addition. I. Process optimisation. *Journal of wood chemistry and technology* 1987;7:373.

[88] Ballesteros I, Oliva J, Navarro A, Gonzalez A, Carrasco J, Ballesteros M. Effect of chip size on steam explosion pretreatment of softwood. *Applied Biochemistry and Biotechnology* 2000;84:97.

[89] Mackie K, Brownell H, West K, Saddler J. Effect of sulphur dioxide and sulphuric acid on steam explosion of aspenwood. *Journal of wood chemistry and technology* 1985;5:405.

[90] Alizadeh H, Teymouri F, Gilbert TI, Dale BE. Pretreatment of switchgrass by ammonia fiber explosion (AFEX). *Applied Biochemistry and Biotechnology* 2005;124:1133.

[91] Cara C, Moya M, Ballesteros I, Negro MJ, González A, Ruiz E. Influence of solid loading on enzymatic hydrolysis of steam exploded or liquid hot water pretreated olive tree biomass. *Process Biochemistry* 2007;42:1003.

[92] Weil J, Sarikaya A, Rau SL, Goetz J, Ladisch CM, Brewer M, et al. Pretreatment of yellow poplar sawdust by pressure cooking in water. *Applied Biochemistry and Biotechnology* 1997;68:21.

[93] Tomas-Pejo E, Oliva J, Ballesteros M. Realistic approach for full-scale bioethanol production from lignocellulose: a review. *Journal of Scientific and Industrial Research* 2008;67:874.

[94] Talebnia F, Karakashev D, Angelidaki I. Production of bioethanol from wheat straw: An overview on pretreatment, hydrolysis and fermentation. *Bioresource Technology* 2010;101:4744.

[95] Zhong C, Lau MW, Balan V, Dale BE, Yuan YJ. Optimization of enzymatic hydrolysis and ethanol fermentation from AFEX-treated rice straw. *Applied microbiology and biotechnology* 2009;84:667.

[96] Han M, Moon SK, Kim Y, Chung B, Choi GW. Bioethanol production from ammonia percolated wheat straw. *Biotechnology and Bioprocess Engineering* 2009;14:606.

[97] Zheng Y, Lin HM, Wen J, Cao N, Yu X, Tsao GT. Supercritical carbon dioxide explosion as a pretreatment for cellulose hydrolysis. *Biotechnology letters* 1995;17:845.

[98] Dale BE, Moreira MJ. Freeze-explosion technique for increasing cellulose hydrolysis. Biotechnol Bioeng Symp;(United States): Colorado State Univ., Fort Collins; 1982;12:31.

[99] Park CY, Ryu YW, Kim C. Kinetics and rate of enzymatic hydrolysis of cellulose in supercritical carbon dioxide. *Korean Journal of Chemical Engineering* 2001;18:475.

[100] Zheng Y, Lin HM, Tsao GT. Pretreatment for cellulose hydrolysis by carbon dioxide explosion. *Biotechnology progress* 2008;14:890.

[101] Prasad S, Singh A, Joshi H. Ethanol as an alternative fuel from agricultural, industrial and urban residues. *Resources, Conservation and Recycling* 2007;50:1.

[102] Schurz J, Ghose T. Bioconversion of Cellulosic Substances into Energy Chemicals and Microbial Protein Symposium Proceedings. IIT, New Delhi 1978:37.

[103] Lee J, Gwak K, Park J, Park M, Choi D, Kwon M, et al. Biological pretreatment of softwood Pinus densiflora by three white rot fungi. *Journal of Microbiology-Seoul-* 2007;45:485.

[104] Sánchez C. Lignocellulosic residues: biodegradation and bioconversion by fungi. *Biotechnology advances* 2009;27:185.

[105] Hatakka AI. Pretreatment of wheat straw by white-rot fungi for enzymic saccharification of cellulose. *Applied microbiology and biotechnology* 1983;18:350.

[106] Itoh H, Wada M, Honda Y, Kuwahara M, Watanabe T. Bioorganosolve pretreatments for simultaneous saccharification and fermentation of beech wood by ethanolysis and white rot fungi. *Journal of Biotechnology* 2003;103:273.

[107] Balan V, da Costa Sousa L, Chundawat SPS, Vismeh R, Jones AD, Dale BE. Mushroom spent straw: a potential substrate for an ethanol-based biorefinery. *Journal of industrial microbiology & biotechnology* 2008;35:293.

[108] Lynd LR, Weimer PJ, Van Zyl WH, Pretorius IS. Microbial cellulose utilization: fundamentals and biotechnology. *Microbiology and molecular biology reviews* 2002;66:506.

[109] Béguin P, Aubert JP. The biological degradation of cellulose. *FEMS microbiology reviews* 1994;13:25.

[110] Ferreira S, Duarte AP, Ribeiro MHL, Queiroz JA, Domingues FC. Response surface optimization of enzymatic hydrolysis of Cistus ladanifer and Cytisus striatus for bioethanol production. *Biochemical engineering journal* 2009;45:192.

[111] Mani S, Tabil LG, Opoku A. Ethanol from agricultural crop residues-An overview. The society for engineering, food and biological systems MBSK 2002:02.

[112] Xu J, Takakuwa N, Nogawa M, Okada H, Morikawa Y. A third xylanase from *Trichoderma reesei* PC-3-7. *Applied microbiology and biotechnology* 1998;49:718.

[113] Taherzadeh MJ, Karimi K. Enzymatic-based hydrolysis processes for ethanol from lignocellulosic materials: A review. *Bioresources* 2007;2(4):707.

[114] Coughlan M, Ljungdahl L. Comparative biochemistry of fungal and bacterial cellulolytic enzyme systems. FEMS symposium-Federation of European Microbiological Societies; 1988, p. 11.

[115] Jørgensen H, Kutter JP, Olsson L. Separation and quantification of cellulases and hemicellulases by capillary electrophoresis. *Analytical biochemistry* 2003;317:85.

[116] Saha BC, Bothast RJ. Enzymology of xylan degradation. ACS Symposium Series: ACS Publications; 1999, p. 167.

[117] Huang X, Penner MH. Apparent substrate inhibition of the *Trichoderma reesei* cellulase system. *Journal of agricultural and food chemistry* 1991;39:2096.

[118] Penner MH, Liaw ET. Kinetic consequences of high ratios of substrate to enzyme saccharification systems based on Trichoderma cellulase. ACS symposium series: ACS Publications; 1994, p. 363.

[119] Lu Y, Yang B, Gregg D, Saddler JN, Mansfield SD. Cellulase adsorption and an evaluation of enzyme recycle during hydrolysis of steam-exploded softwood residues. *Applied Biochemistry and Biotechnology* 2002;98:641.

[120] Berlin A, Gilkes N, Kurabi A, Bura R, Tu M, Kilburn D, et al. Weak lignin-binding enzymes. *Applied Biochemistry and Biotechnology* 2005;121:163.

[121] Eriksson T, Börjesson J, Tjerneld F. Mechanism of surfactant effect in enzymatic hydrolysis of lignocellulose. *Enzyme and Microbial Technology* 2002;31:353.

[122] Tu M, Zhang X, Paice M, McFarlane P, Saddler JN. Effect of surfactants on separate hydrolysis fermentation and simultaneous saccharification fermentation of pretreated lodgepole pine. *Biotechnology progress* 2009;25:1122.

[123] Rollin JA, Zhu Z, Sathitsuksanoh N, Zhang YHP. Increasing cellulose accessibility is more important than removing lignin: A comparison of cellulose solvent-based lignocellulose fractionation and soaking in aqueous ammonia. *Biotechnology and Bioengineering* 2011;108:22.

[124] Saddler J. Factors limiting the efficiency of cellulase enzymes. *Molecular Microbiology* 1986;3:84.

[125] Zhao Y, Wu B, Yan B, Gao P. Mechanism of cellobiose inhibition in cellulose hydrolysis by cellobiohydrolase. *Science in China Series C: Life Sciences* 2004;47:18.

[126] Xin Z, Yinbo Q, Peiji G. Acceleration of ethanol production from paper mill waste fiber by supplementation with β-glucosidase. *Enzyme and Microbial Technology* 1993;15:62.

[127] Maeda RN, Serpa VI, Rocha VAL, Mesquita RAA, Anna LMMS, De Castro AM, et al. Enzymatic hydrolysis of pretreated sugar cane bagasse using Penicillium funiculosum and*Trichoderma harzianum* cellulases. *Process Biochemistry* 2011;46:1196.

[128] Tu M, Zhang X, Kurabi A, Gilkes N, Mabee W, Saddler J. Immobilization of β-glucosidase on Eupergit C for lignocellulose hydrolysis. *Biotechnology letters* 2006;28:151.

[129] Podkaminer KK, Shao X, Hogsett DA, Lynd LR. Enzyme inactivation by ethanol and development of a kinetic model for thermophilic simultaneous saccharification and fermentation at 50°C with Thermoanaerobacterium saccharolyticum ALK2. *Biotechnology and Bioengineering* 2011;108:1268.

[130] Chandel AK, Chan E, Rudravaram R, Narasu ML, Rao LV, Ravindra P. Economics and environmental impact of bioethanol production technologies: an appraisal. *Biotechnology and Molecular Biology Review* 2007;2:14.

[131] Lo YC, Saratale GD, Chen WM, Bai MD, Chang JS. Isolation of cellulose-hydrolytic bacteria and applications of the cellulolytic enzymes for cellulosic biohydrogen production. *Enzyme and Microbial Technology* 2009;44:417.

[132] Zeikus J. Thermophilic bacteria: ecology, physiology and technology. *Enzyme and Microbial Technology* 1979;1:243.

[133] Zeikus J. Chemical and fuel production by anaerobic bacteria. *Annual Reviews in Microbiology* 1980;34:423.

[134] South C, Hogsett D, Lynd L. Continuous fermentation of cellulosic biomass to ethanol. *Applied Biochemistry and Biotechnology* 1993;39:587.

[135] Weimer P, French A, Calamari Jr T. Differential fermentation of cellulose allomorphs by ruminal cellulolytic bacteria. *Applied and environmental microbiology* 1991;57:3101.

[136] Laplace JM, Delgenes JP, Moletta R, Navarro JM. Cofermentation of glucose and xylose to ethanol by a respiratory-deficient mutant of Saccharomyces cerevisiae co-cultivated with a xylose-fermenting yeast. *Journal of fermentation and bioengineering* 1993;75:207.

[137] Lindén T. The Fermentation of Lignocellulose Hydrolysates with Xylose Isomerases and Yeasts. 1992.

[138] Cardona CA, Sánchez ÓJ. Fuel ethanol production: Process design trends and integration opportunities. *Bioresource Technology* 2007;98:2415.

[139] Bertilsson M, Olofsson K, Lidén G. Prefermentation improves xylose utilization in simultaneous saccharification and co-fermentation of pretreated spruce. *Biotechnol Biofuels* 2009;2:8.

[140] Claassen P, Van Lier J, Lopez Contreras A, Van Niel E, Sijtsma L, Stams A, et al. Utilisation of biomass for the supply of energy carriers. *Applied microbiology and biotechnology* 1999;52:741.

[141] Olofsson K, Bertilsson M, Lidén G. A short review on SSF-an interesting process option for ethanol production from lignocellulosic feedstocks. *Biotechnol Biofuels* 2008;1:1.

[142] Olsson L, Hahn-Hägerdal B. Fermentation of lignocellulosic hydrolysates for ethanol production. *Enzyme and Microbial Technology* 1996;18:312.

[143] Ballesteros M, Oliva J, Negro M, Manzanares P, Ballesteros I. Ethanol from lignocellulosic materials by a simultaneous saccharification and fermentation process (SFS) with *Kluyveromyces marxianus* CECT 10875. *Process Biochemistry* 2004;39:1843.

[144] Hari Krishna S, Janardhan Reddy T, Chowdary G. Simultaneous saccharification and fermentation of lignocellulosic wastes to ethanol using a thermotolerant yeast. *Bioresource Technology* 2001;77:193.

[145] Mamma D, Christakopoulos P, Koullas D, Kekos D, Macris B, Koukios E. An alternative approach to the bioconversion of sweet sorghum carbohydrates to ethanol. *Biomass and Bioenergy* 1995;8:99.

[146] Donnison A, Brockelsby C, Morgan H, Daniel R. The degradation of lignocellulosics by extremely thermophilic microorganisms. *Biotechnology and Bioengineering* 1989;33:1495.

[147] Hudson JA, Morgan HW, Daniel RM. The cellulase activity of an extreme thermophile. *Applied microbiology and biotechnology* 1991;35:270.

[148] Wooley R, Ruth M, Sheehan J, Ibsen K, Majdeski H, Galvez A. Lignocellulosic biomass to ethanol process design and economics utilizing co-current dilute acid prehydrolysis and enzymatic hydrolysis current and futuristic scenarios. DTIC Document; 1999.

[149] Lynd LR, Zyl WH, McBRIDE JE, Laser M. Consolidated bioprocessing of cellulosic biomass: an update. *Current opinion in biotechnology* 2005;16:577.

[150] Szczodrak J, Fiedurek J. Technology for conversion of lignocellulosic biomass to ethanol. *Biomass and Bioenergy* 1996;10:367.

[151] Raman B, Pan C, Hurst GB, Rodriguez M, McKeown CK, Lankford PK, et al. Impact of pretreated switchgrass and biomass carbohydrates on *Clostridium thermocellum* ATCC 27405 cellulosome composition: a quantitative proteomic analysis. *PLoS One* 2009;4:e5271.

[152] Gan Q, Allen S, Taylor G. Design and operation of an integrated membrane reactor for enzymatic cellulose hydrolysis. *Biochemical engineering journal* 2002;12:223.

[153] Gong C, Cao N, Du J, Tsao G. Ethanol production from renewable resources. *Recent Progress in Bioconversion of Lignocellulosics* 1999:207.

[154] Lawson KW, Lloyd DR. Membrane distillation. *Journal of membrane Science* 1997;124:1.

[155] Mori Y, Inaba T. Ethanol production from starch in a pervaporation membrane bioreactor using *Clostridium thermohydrosulfuricum*. *Biotechnology and Bioengineering* 2004;36:849.

[156] Kang W, Shukla R, Sirkar K. Ethanol production in a microporous hollow-fiber-based extractive fermentor with immobilized yeast. *Biotechnology and Bioengineering* 2004;36:826.

[157] Bruce L, Daugulis A. Extractive fermentation by *Zymomonas mobilis* and the use of solvent mixtures. *Biotechnology letters* 1992;14:71.

[158] Banik R, Santhiagu A, Kanari B, Sabarinath C, Upadhyay S. Technological aspects of extractive fermentation using aqueous two-phase systems. *World Journal of Microbiology and Biotechnology* 2003;19:337.

[159] Pinto R, Wolf-Maciel M, Lintomen L. Saline extractive distillation process for ethanol purification. *Computers & Chemical Engineering* 2000;24:1689.

[160] Montoya M, Quintero J, Sánchez O, Cardona C. Efecto del esquema de separación del producto en la producción biotecnológica de alcohol carburante. II Simposio sobre biofábricas Avances de la biotecnología en Colombia Medellín, Colombia; 2005.

[161] Tsuyumoto M, Teramoto A, Meares P. Dehydration of ethanol on a pilot-plant scale, using a new type of hollow-fiber membrane. *Journal of membrane Science* 1997;133:83.

In: Biomass Processing, Conversion and Biorefinery   ISBN: 978-1-62618-346-9
Editors: Bo Zhang and Yong Wang   © 2013 Nova Science Publishers, Inc.

Chapter 21

BIOPROCESSING: THE USE OF THERMOPHILIC AND ANAEROBIC BACTERIA

*Jing-Rong Cheng and Ming-Jun Zhu**

School of Bioscience and Bioengineering, South China University of Technology,
Guangzhou Higher Education Mega Center, Panyu, Guangzhou, P.R. China

ABSTRACT

Lignocellulosic waste is the most cost-effective raw material for producing high-value products, especially biofuel, by fermentation, which can be expected to ameliorate many problems associated with the consumption of petroleum–based fuels and gas pollution. The main obstacle to the utilization of lignocellulosic feedstocks for fuel production is the cost, and consolidated bioprocessing (a direct fermentation of lignocellulosic feedstocks by microorganisms), a new bioenergy-generating technology, can contribute to it. The present review is aimed at summarizing both the bottlenecks and innovative strategies employed in biodegrading lignocellulosic feedstocks with kinds of thermophilic and anaerobic bacteria. Moreover, co-culture of microorganisms, a promising technology to degrade lignocellulosic biomass into clean and affordable renewable energies has been discussed. The microbial consortium provides a valuable platform for further study on interactions between multi-microbial species, and has a potential biotechnological application in lignocellulosic biomass degradation and bioconversion for the promising biofuel.

Keywords: thermophilic and anaerobic bacteria; lignocellulosic waste; biofuel; co-culture

1. INTRODUCTION

Energy has become a key consideration in discussions of sustainable development. High dependence on fossil fuels, a prototype of unsustainable energy, has led to serious energy crisis and environmental problems. Therefore, renewable energy sources will probably

---
* Author to whom correspondence should be addressed; E-Mail: mjzhu@scut.edu.cn; Tel.: +86-020-39380623.

become one of the most attractive substitutes in the drawing near future [1]. Most of the researchers focus on renewable energy technologies (RET) for sustainable development and they believe that daily energy needs for long lasting life on the planet earth should be met through the waste-to-energy routes (WTER), which does not cause negative impacts on the society and environment [1].

Lignocellulosic materials, such as agricultural residues, forestry wastes, paper wastes and wood chips, are both abundant and sustainable. In spite of significant progress in advancing conversion of lignocelluloses to bioenergy, one of the major barrier to achieving widespread commercialization is lacking cost-competitive processes for biofuel production from mixed sugar hydrolysates [2-7]. The key points for an economical lignocellulosic bioenergy process include: (1) efficient pretreatment, (2) low-cost and available enzymatic hydrolysis, and (3) efficient fermentation.

There is no doubt that choosing appropriate fermentation strategies with robust strains is a key step for this approach. Several studies report that some anaerobic bacteria have the ability of degrading cellulose, and producing ethanol, hydrogen and organic acids as products [8-12], while few studies have demonstrated bioenergy production from microorganism directly [12], perhaps for its low efficiency. Recently, it is reported that mesophilic anaerobic bacteria cannot utilize cellulose effectively, but it can just hydrolyze cellulose and generate H_2 with the addition of exogenous cellulase [13]. It is encouraging to find that thermophilic anaerobic bacteria can effectively utilize cellulose [9], which shows great potential for bioenergy production from biomass without the addition of exogenous cellulase.

This present review summarizes kinds of thermophilic anaerobic bacteria which exhibit robust ability of directly converting lignocellulose to bioenergy.

2. HYDROLYTIC ENZYMES CLUSTER AND CELLULOSOME

Based on the complex structure of lignocellulose, most researches indicate that the hydrolysis of cellulose is the major rate-limiting step for the lignocellulosic biomass degradation. When it comes to high-efficient degradation of lignocellulosic materials, enzymatic hydrolysis methods should be presented. Cellulases and hemicellulases play important roles in the enzymatic conversion of cellulose and hemicellulose into soluble fermentable sugars. Many microorganisms can secrete cellulose and/or hemicellulose-degrading enzymes. Usually, endoglucanase, exoglucanase and ancillary enzymes are individually secreted and act synergistically to degrade lignocellulose in many aerobic fungi and bacteria, such as *Trichoderma reesi* [14]. Besides, several anaerobic microorganisms have evolved distinct enzyme systems, which are called cellulosome. Currently, most researches find that bioenergy production can be achieved by adding some cellulases, but the cost is really high and thus it is impossible to be applied in large-scale bioenergy production. However, it could be employed as the starting point of using microorganisms to produce enzymes and fermentation (such as co-culture), in which the production cost can be remarkably reduced. Cellulosome is a large, extracellular enzyme complex which can degrade lignocellulose efficiently and is a best candidate for the lignocelluloses' hydrolysis.

2.1. Enzymatic Hydrolysis of Cellulose and Hemicellulose

Due to the fact that the native forms of lignocellulose are resistant to biochemical attack, focus is now fixed on lignocellulose pretreatments which can increase the susceptibility of a material to enzymatic degradation. Though chemical processes are more mature technically, significant technical progress for enzymatic processes reveals priorities over costs in terms of investments and operations. Enzymatic hydrolysis has some important advantages such as high yield of fermentable sugars acquired under mild process conditions and no corrosion problems [15, 16]. Complete enzymatic degradation of lignocellulosic material is achieved through the coordinated activities of several polypeptides (Figure 1). These polypeptides may exist in the form of a single multifunctional aggregate, such as the case for cellulosome of Clostridium thermocellum, or the polypeptides may exist separately in solution. With efficient pretreatments, most lignin and part of the hemicellulose have been removed, leaving the cellulose and hemicellulose in the biomass. Subsequently, xylan, the main part of the hemicellulose, is catalyzed into xylose through xylanase, while cellulose is catalyzed into glucose by different enzymes, such as cellulase, carboxymethyl cellulase (CMCA), β-glucosidase, and cellobiohydrolase. In this reaction, endoglucanase (1, 4-β-D-glucan glucanhydrolase, EC 3.2.1.4, cellulase, CMCase) creates C-1 hydroxyl groups (reducing chain ends) by internal cleavage of β-1, 4-glycosidic bonds. It can act on CMC and amorphous celluloses but not on crystalline cellulose. Exocellobiohydrolases (1, 4-β-D-glucan cellobiohydrolase, EC3.2.1.9.1, exoglucanase, cellobiohydrolase, cellobiosidase, exocellulase) cleave glucose, cellobiose and cellotriose from the non-reducing end of the glycosidic chain, creating only minor changes in the celluloses degree of polymerization. Exocellobiohydrolases cleave amorphous as well as crystalline cellulose like Avicel. Besides, β-glucosidase (EC3.2.1.21, cellobiase) cleaves the dimers and trimers produced by the exocellobiohydrolases into monomer sugars [17]. The three major enzymes make up the primary enzymes for cellulose degradation, but they cannot degrade hemicellulose completely. The primary pentosan component of hemicellulose is a chain of 1, 4-linked β-D-xylopyranosyl residues with side chains of arabinose, glucuronic acid, or methyglucuronic acid. Endoxylanases (1, 4-β-D-xylan xylanhydrolase, EC 3.2.1.8, xylanase) can degrade the hemicellulose into xyloolignosaccharides, which are further cleaved by β-xylosidase (1, 4-β-D xylan xylohydrolase, EC 3.2.1.37, xylobiase).

2.2. Nanomachines for Efficient Lignocellulose Degradation in Thermophilic and Anaerobic Bacteria

Unlike fungi and some other bacteria, the cellulases and hemicellulases in the thermophilic anaerobic bacteria usually present in a form of multienzymes complexes called cellulosome. Cellulosome, an extracellular supramolecular machine, can efficiently degrade crystalline cellulosic substrates and associated plant cell wall polysaccharides [18]. Anaerobic cellulolytic bacteria (e.g. Clostridium and Ruminococcus genera) and fungi (e.g. Chytridomycetes) biosynthesize complex cellulase systems called cellulosome [19-22]. Cellulosome was first discovered in C. thermocellum, which has a macromolecular structure similar to ribosome. Due to its ability of orderly and effective degradation of cellulosic materials, it has been studied widely. Due to the biodegradation of lignocellulose highly depends on efficient multiple enzymes complex, cellulosome, generally appearing in anaerobic bacteria, can modify the metabolic activities accurately and offset the lack of capacity in anaerobic fermentation [23]. Therefore, the use of cellulosome in the degradation of lignocellulosic materials to produce biofuel can not only reduce the cost of enzymes, but also improve the degradation efficiency.

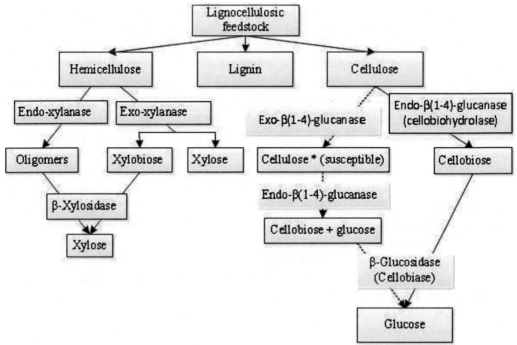

— The main enzymatic pathway;
--- The secondary enzymatic pathway;
* Small cellulose segment or amorphous cellulose.

Figure 1. Mode of action of lignocellulose degradation enzymes.

2.2.1. Multisubunit, Multimodular Cellulosome Structure and its Application in Lignocellulose Degradation

In *C. thermocellum*, the cellulosome complex contains many different types of glycosyl hydrolases which contribute to the efficient degradation of lignocellulosic materials [24, 25]: (1) Endoglucanase: it can cut off the link between amorphous cellulose, such as CMC and TNP-CMC; (2) Exoglucanase; (3) Cellobiose phosphorylase: it can be used to transform the cellobiose into glucose and 1-phophate glucose; (4) Cellodextrin phosphorylase and two β-glucosidase: β-glucosidase A is periplasmic and β-glucosidase B is cytoplasmic; (5) Hemicellulase, such as xylanase and mannase; (6) Chitinase. All of them are bound to, as well as regulated by, a major polypeptide called scaffoldin or cellulosome-integrating protein, CipA. The multiple roles of scaffoldin, namely the cellulose-binding and cell-anchoring functions, as well as organization of the enzyme subunits in the cellulosome complex, were recognized in the early stages of cellulosome research [26]. A schematic view of the cellulosome and its interaction with cellulose and the cell surface has been demonstrated by Sholam with the following Figure 2 [18]. The scaffoldin, which promotes the activity of a cellulosomal enzyme subunit, contains many functional modules dictating its various activities: (i) improving complex affinity for the substrate and catalytic efficiency via CBM (carbohydrate-binding module); (ii) recruiting catalytic proteins by multiple cohesin domains that interact with GHs (glucan hydrolases) dockerin domains; and (iii) binding the cellulosome to the cell wall through covalent (sortase-mediated) or non-covalent (through surface layer homology domains) interactions by anchoring protein. The cellulosomal enzymes are also modular in nature. The cohesin–dockerin interaction therefore governs the assembly of the complex, while the interaction of the complex with cellulose is mediated by the scaffoldin-borne CBD (carbohydrate binding domain). Though the CBD and cohesin domains have a similar type of fold, their functional components are completely different. As is reported that scaffoldins from different cellulosome-producing species are inherently cross-reactive [27], and these properties might be used to identify new scaffoldins [28-31].

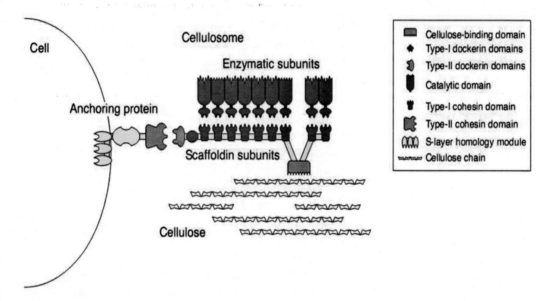

Figure 2. Structure of cellulosome and the adherent module on cellulose.

The cellulolytic systems in the thermophilic anaerobic bacterium can be classified into cellulosomes. The cellulosomal enzymes are bound noncovalently to cellulosome-integrating protein, which carries a CBD. The CBM (carbohydrate binding module) has been considered as the limiting factor in the hydrolysis of cellulose [18]. The more inclusive term CBM has evolved to reflect the diverse ligand specificity of these modules. CBMs are divided into families based on amino acid sequence similarity with different ligand specificity. Generally, CBMs have three roles with respect to the function of their cognate catalytic modules: (i) a proximity effect, (ii) a targeting function and (iii) a disruptive function [32]. Hemicellulases are multi-domain proteins and contain structurally discrete catalytic and non-catalytic modules. Non-catalytic modules generally consist of CBD, interdomain linkers, and dockerin modules. CBD plays a pivotal role in facilitating the targeting of the enzyme to the polysaccharide, and dockerin modules can mediate the binding of the catalytic domain via cohesion-dockerin interactions, either to the microbial cell surface or to cellulosome [32].

It is just for the high efficiency that the cellulosome secreted strains were widely studied. Cellulosome in the *C. thermocellum* shows a high catalysis rate, in which the hydrogen yield can reach 1.6 mol H_2 per mol glucose using delignified wood fibers [33]. In the study of cellulosome from *C. thermocellum*, NG [24] and Huang [25] pointed out that maximal production of cellulase was at its exponential phase of growth, while maximal endo- and exoglucanase activities were presented at early stationary phase. It has been reported that based on the well study of cellulosome of *C. thermocellum* JYT01, the mono-culture can degrade 137 g L^{-1} cellulose and completely convert it into sugars, with ethanol concentration of 491 mM, which has almost reached its maximal ethanol tolerance [34].

3. FUNCTIONAL MICROORGANISMS

The chemical composition of the lignocellulosic feedstock plays an important role in determining efficiency of bioenergy production during the conversion processes. However, lignocellulose is a complex substrate, which is composed of a mixture of carbohydrate polymers (cellulose and hemicellulose) and lignin; these components build up about 90% of dry matter in lignocelluloses, with the rest consisting of extractive and ash [35]. So functional microorganisms hydrolyzed the lignocellulosic biomass to bioenergy are mainly composed of hexose and pentose utilizing strains [36-39].

Since some anaerobic bacteria have shown the ability of degrading cellulose to ethanol, hydrogen and organic acids, the novel approach to using anaerobic bacteria for the bioconversion of lignocellulose to bioenergy attracts the scientific community's attention. This novel approach of bioenergy production is very simple, which can be clarified into five steps: (1) the synthesis of cellulase and hemicellulase; (2) the enzymolysis of cellulose and hemicellulose; (3) the intaking of monosaccharide and oligosaccharides; (4) hexose fermentation; (5) pentose fermentation. Using anaerobic bacteria can realize the complex steps in just one bioreactor, which could simplify the process as well as reduce the operation cost. In the past, several cellulolytic anaerobes have been isolated and characterized for potential technology development for biofuel production using biomass directly [40, 41]. Using thermophilic anaerobic bacteria shows great advantages in the process: (1) high stability of the strains; (2) high thermal stability of the secreted enzymes; (3) contamination-free for the high temperature beyond 55°C; (4) simplified process of separation and recovery of volatile substrate, such as ethanol for high operation temperature; (5) the high level metabolic rate of cellulase and hemicellulase; (6) reduction of the cost of cooling process.

3.1. Pentose Fermenting Microorganisms

Since xylose is the main pentose fraction in hemicellulose, screening pentose fermentative microorganisms is essential to develop commercial bioenergy production process from lignocellulosic biomass. Most strains for bioethanol and biohydrogen production could not utilize xylose directly, so present processes often face the difficulty of low products yields [42]. It has been reported that making full use of the xylose fraction of lignocellulosic raw material would increase ethanol yield of 25% in the ethanol production

[43, 44]. Researches also have demonstrated that the production cost of ethanol can be reduced from $1.65 to $1.23 per gallon if all the xylose were utilized for ethanol production [43]. Therefore, efficient utilization of xylose is a main obstacle of lignocellulose-bioenergy production. Compared to hexose-fermenting microorganisms, only a few pentose-fermenting microorganisms have been identified. Table 1 lists some thermophilic anaerobic microorganisms which have been reported for bioenergy production from xylose or hemicellulose biomass. Among those known pentose–fermenting microorganisms, the thermophilic *Caldicellulosiruptor saccharolyticus* and *T. thermosaccharolyticum* W16 can utilize xylose efficiently with a yield of 2.24 mol H_2 per mol xylose [45] and 2.19 mol H_2 per mol xylose [46, 47] respectively. In the future research, isolation of the pentose fermenting microorganisms still has a long way to go [19].

Table 1. Several hemicellulose or xylose-utilizing thermophilic anaerobic strains for bioenergy production

Microorganism	Substrate	Conditions	Bioenergy	performance	Reference
Pure cultures					
T. thermosaccharo-lyticum W16	Xylose	Batch, 60 °C	Hydrogen	2.19 mol H_2 mol^{-1} xylose	[50]
Clostridium sp. strain no. 2	Hemicellulose	Continuous, 60 °C	Hydrogen	2.06 mol H_2 mol^{-1} xylose	[51]
Caldicellulosiruptor saccharolyticus	Paper sludge hydrolysate	Batch,70 °C, pH 6.4	Hydrogen	6.0 mmol H_2 L^{-1}h^{-1}	[45]
Caldicellulosiruptor saccharolyticus	Xylose	Batch, 70 °C, pH 7.2	Hydrogen	2.24 molH_2 mol^{-1}xylose	[45]
Thermoancerobacterium aoteariense △ldh	Xylose	Fed-batch, 55 °C	Ethanol	0.25 g g^{-1} xylose	[52]
Thermoanaerobacterium saccharolyticum	50 g/l cassava starch	Fed-batch, 55 °C	Ethanol	5.83 g L^1	[53]
Mix cultures					
Microflora	Xylose	Batch, 75 °C	Hydrogen	0.54 mol H_2 mol^{-1} xylose	[54]
Sewage sludge	Xylose	Continuous, 50 °C	Hydrogen	1.4 mol H_2 mol^{-1} xylose	[55]
Compost	Xylose	Fed-batch, 55 °C	Hydrogen	1.7 mol H_2 mol^{-1} xylose	[56]

3.2. Cellulose and Hexose Fermenting Microorganisms

Hexose, the predominant component of the cellulose hydrolysates, can be utilized by most of the microorganisms. However, there are only few strains which can degrade cellulose directly to produce bioenergy. Table 2 summarizes some reported thermophilic anaerobic microorganisms that can produce bioenergy from cellulose directly or its hydrolysates. The highest H_2 yield is approximately 83% of its theoretical value, which is found in the thermophilic anaerobic bacteria *Caldicellulosiruptor saccharolyticus* [48]. Wang [49] has reported that *Clostridium acetobutylicum* X9 can degrade acetic stream-exploded corn stalks to hydrogen with specific hydrogen production rate of 3.4 mmol g^{-1} steam-exploded corn stalks. Ren [42, 46, 47] has reported that *T. thermosaccharolyticum* W16 could simultaneously ferment the mixture of glucose and xylose with a hydrogen yield up to 2.37 mol H_2 mol^{-1} substrate. However, up to now, strains isolated to hydrolyze native cellulosic biomass is limited due to time-consuming and complicated techniques.

Table 2. Several cellulose or hexose-utilizing theromphilic anaerobic strains for bioenergy production

Microorganism	Substrate	Conditions	*Bioenergy*	Yield	Reference
Thermotoga elfii	Glucose	Batch, 65 °C	Hydrogen	3.3 mol H_2 mol^{-1}glucose	[48]
Caldicellulosiruptor saccharolyticus	Sucrose	Batch, 70 °C	Hydrogen	5.9 mol H_2 mol^{-1} sucrose	[57]
Thermoanaerobacterium	Cellulose	Batch, pH 6.5 and 55 °C	Hydrogen	7.56 mg H_2 g^{-1} cellulose	[58]
C. thermohydrosulfuricum strain 39E	Glucose	Batch, pH 6.5 and 65 °C	Ethanol	1.9mol ethanol mol^{-1} glucose	[59]
C. thermocellum 24705	Delignified wood fibers	Batch, 60 °C	Hydrogen	1.6 mol H_2 mol^{-1}glucose	[33]
C. thermocellum SS19	Filter paper	Batch, 60 °C	Bioethanol	0.41g ethanol g^{-1} substrate	[60]
C. thermocellum SS21	Alkali extracted paddy straw	Batch, 60 °C	bioethanol	0.37g ethanol g^{-1} substrate	[60]
Sludge compost	Waste water	Continuous, 60 °C	hydrogen	8.3mmol H_2 L^{-1}h^{-1}	[61]

4. APPLICATION OF THERMOPHILIC ANAEROBIC BACTERIA IN DIFFERENT HYDROLYSIS AND FERMENTATION STRATEGIES

Lignocellulosic biomass is potentially sustainable and the most attractive substrate for 'biorefinery strategies' to produce high-value products (e.g. fuels, bioplastics or enzymes) through fermentation processes [62-64]. Commonly, the pretreated biomass can be processed by variety of fermentation strategies such as separate hydrolysis and fermentation (SHF), simultaneous saccharification and fermentation (SSF), simultaneous saccharification and co-fermentation (SSCF) and consolidated bioprocessing (CBP). Figure 3 shows the general fermentation strategies and consolidated bioprocessing and Table 3 lists the advantages and disadvantages of several fermentation processes. Due to its low cost and commercial feasibility, most researchers prefer the SSF route, in which enzymes and fermentative organisms are added to the same vessel to produce bioenergy from the mixture of sugars released from biomass catalyzed by the added enzymes [65]. Furthermore, the main advantage of SSF for ethanol production is increasing hydrolysis rate of lignocellulosic biomass for removing sugar end product inhibition by simultaneous fermentation. Due to its direct conversion of lignocellulose into bioenergy, CBP has been proposed as a single and cheap process. CBP can combine all biochemical steps involved in cellulosic bioenergy production into a single bioconversion process, showing a great potential in minimizing the processing steps, enhancing efficiency [19] and reducing cost for capital, substrate and other raw materials, and utilities associated with cellulase production [32].

4.1. Advantages and Challenges of CBP

The Department of Energy (DOE) of United States claims that CBP is the ultimate low-cost configuration for hydrolysis and fermentation of lignocellulosic material because of more simplified feedstock processing, lower energy inputs, and higher conversion efficiency than less integrated processes (e.g., simultaneous saccharification and co-fermentation) [66]. CBP utilizes recombinant microorganisms to produce enzymes, hydrolyze lignocellulose, and coferment hexose and pentose in a single step [19]. A number of advantages can be shown on this process. Firstly, it can decrease the contamination possibility during transmission and reduce a large amount of capital cost. Secondly, it inherits the capability of fermenting both pentose and hexose from SSCF. However, as a new technology, teething pains are unavoidable. For example, cellulase is generated inside the reaction vessel for CBP, making this process more complex than SSCF. Enzyme-microbe synergy has been found in some thermophilic anaerobic organisms and complex cellulase systems. CBP requires efficient hydrolysis enzymes and obtains the hydrolysis products under anaerobic and high temperature conditions.

In spite of intensive attempts, biomass-to-bioenergy CBP is not widely used on an industrial scale because of technical limitations in identifying appropriate cellulolytic and bioenergy producing bacteria. For instance, the thermophilic bacterium *C. thermocellum* could effectively degrade cellulose and hemicellulose; however, a high ratio of acetic acid to ethanol and even toxic carboxylic acids appeared in the bioenergy-producing process [67]. As yet, no organisms or compatible combinations of microorganisms have reported to be available for both high enzyme production efficiency and high production tolerance, although various organisms have already combined multiple functions. Proper microbes with specific traits, the mixture of cellulosic substrates, and optimum fermentation conditions can help to deal with the troubles present in CBP. Right now, the improvement of native microorganisms has already become a hot topic of the technology. This orientation can be divided into three parts: (1) improving several product-related properties such as yield and tolerance for the native cellulolytic microorganisms; (2) building engineering cellulolytic microorganisms with high product yield and tolerance for biomass-bioenergy production; (3) co-culture strategy. Different kinds of organisms are used to ferment hexose and pentose in one reactor, and can gain a high yield of objective production. Among the three strategies, the co-culture technology is the most economic-friendly and popular method without genetic modifications.

4.2. Application of the Co-Culture System in CBP

As a feasible strategy to realize efficient CBP, co-culture systems are widely studied. After a series of enzymatic hydrolysis, the main fractions of biomass, cellulose and hemicellulose, are degraded into kinds of sugars. Glucose and xylose are the two dominant sugars in lignocellulosic hydrolysates. In order to gain a high yield of bioenergy, the sugars need to be fermented efficiently. However, current approaches are inefficient because no native microorganism can convert all sugars into ethanol or hydrogen at a high yield [68]. Currently, the biodegradation of cellulosic biomass through microbial co-cultures or complex communities have been proposed as a highly efficient approach, which avoids the problems of biotechnological application, feedback regulation as well as metabolite repression resulted from isolated single strains [69-71]. Poor yields in lignocellulose-derived ethanol and hydrogen fermentations are primarily due to incomplete hydrolysis of cellulose or organisms that could not use a variety of pentose and hexose found in the hydrolysates. Identification and development of organisms capable of utilizing a broad range of substrates (carbonhydrates), particularly pentose, would have a significant impact on the economics of bioenergy production and have been the focus of significant research activity [72-76]. Studies on microbial consortia and their mixed enzyme systems could provide an important foundation for understanding the complex interactions of natural lignocellulosic degradation and establishing a platform for biotechnological application involving biomass degradation in composting, anaerobic digestion and enzymatic biomass saccharification [77]. Figure 4 presents a simplified process of lignocellulosic degradation using co-culture of cellulose-degrading microorganisms and other microorganisms. Most of the co-culture systems are established on *C. thermocellum*, combining other *T. tengcongensis*, such as *T. saccharolyticum*, *T. ethanolicus*, *C. thermohydrosulfuricum* and *T. brockii*. Depending on synergism of microorganisms and enzymes, cellulose is degraded into cellobiose and glucose by *C. thermocellum*, which can be converted into kinds of bioenergy; at the same time, hemicellulose can be hydrolyzed into pentose, which can be converted into target products by other strains. However, the compatibility among the microorganisms is the biggest obstacle for this technology.

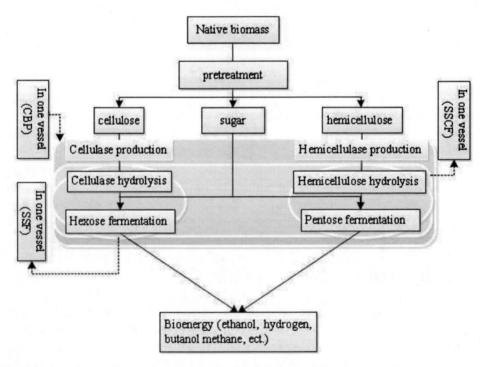

Figure 3. Fermentation strategies and consolidated bioprocessing.

Table 3. Advantages and disadvantages of several fermentation processes

Production process	Advantages	Disadvantages
Direct fermentation	Fewer equipment requirements and low cost	Low yield and produce some by-product such as organic acid
Indirect fermentation	Technology maturity and fewer equipment requirements	Higher cost than direct fermentation, end-product inhibition
SSF (Simultaneous saccharification and fermentation)	Fewer equipment requirements Short producing time and high yield	Xylose inhibition, hydrolysis and fermentation temperature discordance and ethanol inhibition
NSSF (Nonisothermal simultaneous fermentation)	Temperature discordance in both reactors, less cellulose needed	More *Saccharomyces cerevisiae* dead in the reactor for constant temperature changes
Immobilized cells fermentation	High cell concentration, repeated use of cells, high ethanol concentration	High research cost, and few related-studies
Consolidated bioprocessing	Simplified feedstock processing, lower energy inputs, and higher conversion efficiencies	More complex process

The strategies of co-culture provide a new way to improve the cellulose hydrolysis, enhance the use of substrates, and thus increase the output of products. There are several advantages for the bioenergy production by this strategy. Firstly, the effects caused by exogenous factors can be weakened by partial bacteria of the system. In this way, it can simplify the complex, metabolic pathway among the bacteria [78]. Furthermore, through co-culture, the glucose produced by hydrolysis can be consumed by the member of the co-culture, which can reduce the inhibition between the products and enzymes to achieve a high conversion. In brief, the co-culture strategies have combined different kinds of metabolic ways that can reduce negative effects of the harmful products caused by members of the system, and have simplified the fermentation process. Due to the complex structure of lignocellulose, when we use different bacteria, we can make full use of their advantages and enhance the fermentation efficiency. There are two main strategies about co-culture: (i) different kinds of bacteria exist in the fermentation liquid and they can use different carbohydrate substrates; (ii) using different bacterial characteristics to increase the fermentation rate. Kohji Miyazaki [65] used the aerobic and anaerobic bacteria for co-fermentation, in which aerobic microbes can reduce the available oxygen and create anaerobic conditions that promote the growth of anaerobic or microaerophilic strains. So far, the advantages can be reflected in many studies. Lei [79] demonstrated that the thermophilic anaerobic co-culture system, *C. themocellum* and *C. thermolacticum*, showed great potential for cellulose-ethanol production, in which the highest ethanol yield (as a percentage of the theoretical maximum) of 75% could be reached for MCC. Qian [80] reported that a method utilizing the co-culture of *C. thermocellum* and *C. thermosaccharolyticum* could be used to improve hydrogen production via the thermophilic fermentation of cornstalk waste. The hydrogen yield in the co-culture fermentation process reached 68.2 mL g^{-1} cornstalk, which was 94.1% higher than that in the mono-culture. Ng et al. [8] isolated a new strain of *C.*

thermocellum LQR1 from the contaminated culture of strain LQ8. When the isolate was co-cultured with *C. thermohydrosulfuricum* strain 39E isolated by Zeikus et al.[81], a stable co-culture system was established and showed a 3-fold enhancement in ethanol production and a 2-fold decrease in the acetate acid production. A similar co-culture system of *C. thermocellum* and *C. saccharoperbutylacetonicum* showed the higher-level production of ethanol and butanol, and both xylose and cellobiose were fermented by *C. saccharoperbutylacetonicum*.

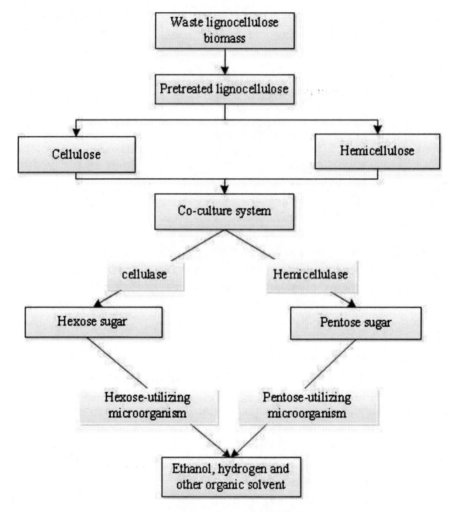

Figure 4. Simplified process of using co-culture for degradation of lignocellulose.

4.3. Obstacles of Co-Culture Strategy in Consolidated Bioprocessing

4.3.1. Construction of Co-Culture Systems

Establishing a stable co-culture system is a tedious process, in which, all the culture conditions, such as temperature, pressure and carbon sources, need to be adjusted to the optimum for each strain. Besides, the growth condition, the metabolic relationship among the strains (such as synergy and substrate competition) and other factors (such as growth promoters and growth inhibitors) should also be taken into consideration. A strain, used in the co-culture, should meet several conditions. All species must be able to be transformed at least 20 times without undergoing significant changes, i.e., neither excessive growth nor growth phenomena will occur after inoculation [82].

Based on insights into how commercial and natural systems function, a potential co-culture approach for bioenergy production means the combination of a hexose-fermenting strain and a pentose-fermenting strain to utilize cellulose and hemicellulose simultaneously. Table 4 shows different kinds of co-culture systems used in biorefinery. Degradation of lignocellulose in the environment involves a number of bacteria, and although within the environmental niches only a few species are likely to dominate. Among the lignocellulose degraders, the relatively popular strains are *C. termitidis*, a cellulolytic bacterium isolated from the termite gut, *Acidothermus cellulolyticus*, isolated from the cellulose-rich soils of acidic hot springs, and *Ruminococcus albus*, the dominant cellulolytic bacteria in the bovine rumen [83, 84]. While only a few cellulolytic bacteria, such as *C. thermocellum*, have been studied for their industrial potential, additional studies of bacterially mediated cellulose degradation have been undertaken [85]. However, *C. thermocellum* can only utilize hexose in high efficiency but not pentose. If a xylose-utilizing bacterium is co-cultured with *C. thermocellum*, a high-efficiency co-culture system can be established to degrade lignocellulose. Li [86] demonstrated that *C. thermocellum* and *T. aotearoense* can be established as a stable co-culture system to degrade lignocellulose. Cassava pulp as the substrate, the glucose generated by the co-culture system can reach 13.65 ± 0.45 g L^{-1}, which was 1.75 and 1.17-fold greater than that produced by monocultures of *C. thermocellum* and *T. aotearoense*, respectively.

4.3.2. Interactions between Different Microorganisms in the Co-Culture System

Co-fermentation is not a simple combination of different bacteria. Compared with mono-culture, interactions between different microorganisms play a critical role in co-culture system. Specifically, a stable co-culture system could be controlled by metabolic interactions (i.e., syntrophic relationships, or competition for substrates) and other interactions (i.e., growth promoters or inhibitors) [23]. Positive interactions in co-cultures may take place through decreasing the available oxygen by aerobic microbes, creating anaerobic conditions which promote the growth of anaerobic or microaerophilic strains. Negative interactions could happen when two microorganisms compete for the same resource, such as space and limited nutrient. Therefore, understanding the interactions between associated strains in a co-culture system is quite important. Currently, very little research has been done, which perhaps offers new avenues for future research work.

5. POTENTIAL BENEFITS AND CHALLENGES TO USING CO-CULTURE FOR LIGNOCELLULOSE ON AN INDUSTRIAL SCALE

As discussed in the previous section, all the co-culture systems used in CBP for degrading lignocellulose to bioenergy have been performed in the laboratory. Its high conversion efficiency and yield attract many researchers, and the investment on this valuable project is increasing in some way. However, the use of co-culture for bioenergy production on an industrial scale still has a long way to go due to some serious challenges.

Table 4. Summary of different thermophilic anaerobic co-culture systems used in biorefinery

Coculture system	Substrate	Conditions	Bioenergy	performance	Reference
Klebsiella pneumonia - C. thermocellum	Steam exploded aspen-wood (1%)	Batch, 55 °C	Butanol Ethanol	1.6 g L^{-1} 0.5 g L^{-1}	[95]
C. thermocellum, Zymononas mobilis – Acetivibrio cellulolyticus	Cellulose (2.5%)	Batch, 50 °C	Ethanol	2.5 g L^{-1}	[96]
Z. anaerobia – C. thermocellum	1% steam exploded wood	Batch, 50 °C	Ethanol	1.6 mg mL^{-1}	[97]
C. thermocellum CT2 – C. thermosaccharolyticum HG8	Alkali treated banana waste (100 g L^{-1})	Batch, 60 °C	Ethanol	22g L^{-1}	[98]
C. thermocellum LQRI *– C. thermohydrosulfuricum* 39E	Various saccharides derived from cellulosic biomass	Continuous, 65 °C	Ethanol	1.8 mol of anhydroglucose unit in MN300 cellulose	[8]
Microflora	Cellulose	Batch, 55 °C	Hydrogen	4.55 mmol H$_2$ g^{-1} cellulose	[58]
Sludge compost	Waste water	Continuous, 60 °C	Hydrogen	8.3 mmol H$_2$ L^{-1}h^{-1}	[61]
C. thermocellum JN4 *– T. thermosaccharolyticum* GD17	Microcrystalline cellobiose	Batch, 60 °C	Hydrogen	1.8 mol H$_2$ mol^{-1} glucose	[12, 99]
Thermosaccharolyticum HG8 *– C.thermocellum* CT2	Alkali treated banana waste	Batch, 60 °C	Ethanol	0.41g g^{-1} substrate	[98]
C.thermocellum – T.saccharolyticum	92g l^{-1} Avicel	Batch, 55 °C	Ethanol	38g L^{-1} ethanol	[100]
C.thermocellum – T.aotearoense	40g l^{-1} cassava pulp	Batch, 55 °C	Hydrogen	4.06 mmol H$_2$	[86]

Firstly, increasing the bioenergy production efficiency by the functional microorganisms is essential. It has been reported that hydrogen production rates are much lower from cellulose (0.11 m^{3}m^{-3}d^{-1}) than from glucose (1.23 m^{3}m^{-3}d^{-1}) [42]. Metabolic modifications at both molecular and physioecological levels are necessary. At the molecular level, future studies should focus on modifying the key enzymes and functional pathways, controlling the organic matter transportation, and even regulating or reconstructing the entire metabolic network in order to increase the bioenergy yield and lignocellulose degradation. Another important challenge to the co-culture process is the apparently low ethanol tolerance of xylose-fermenting strains [87-90]. The third major challenge is finding the optimal process parameters (pH, temperature, nutrition, and oxygen demand) and the acceptable ranges of substrates that can enable optimal activity of each strain in co-culture system. Improvements in technology for immobilizing cells and innovative fermentor designs could help to solve those problems [91, 92].

Secondly, the ethanol tolerance of available strains for biorefinery is generally less than 2% (v/v), which is considered as a main limiting factor for industrial exploitation [93, 94]. Improvement of the bioenergy production, such as bioethanol and hydrogen, may depend upon decreasing the influence of ethanol through selection of ethanol-tolerant strains or ethanol removal systems coupled with the fermentation [87]. Development of an economically viable lignocellulose-based ethanol fermentation process requires the microorganism with a characteristic of high ethanol tolerance. In this regard, the ethanol tolerance of thermophilic anaerobic bacteria is considered as an important characteristic in the industrial bioethanol production [94]. Using continuous mode of operation to improve the strain ethanol tolerance provides an opportunity for long-term adaptation to high ethanol concentrations [101]. At the physioecological level, selection of ethanol-tolerant strains, knocking out branch metabolic related genes, establishing engineering strains, and cell immobilization technology are still the hot topics.

Thirdly, several requirements must be fulfilled to establish a stable co-culture system. One is that the strains of the system must be compatible and able to grow together in a similar environment. Compared with pure culture, interactions between the different microorganisms play a critical role in co-culture systems. Specifically, a stable co-culture system could be controlled by metabolic interactions (i.e., syntrophic relationships, or competition for substrates) and other interactions (i.e., growth promoters or inhibitors such as antibiotics) [66]. Perhaps due to the complexity of the systems, very little research has been done so far, which offers new avenues for future research.

Right now, most of the lignocellulosic hydrolysis process would result in a broad range of inhibitory and toxic compounds, which have a significant negative impact on pure culture fermentation, and even co-culture fermentation. The composition and concentration of inhibitory and toxic compounds often depend on the type of lignocellulosic materials and the pretreatment and hydrolysis processes. In a synthetic medium, most laboratory co-cultures do not encounter these problems, while it could be quite tough problems for industrial production. Currently, the usual solutions to these problems are to remove the inhibitory or toxic compounds or apply detoxification by liming or steam stripping before the fermentation process starts. Unfortunately, this approach adds considerably to the operational costs. Another approach is adaptation, which has been shown to be an efficient method for improvement of tolerance to inhibitors from lignocellulosic hydrolysates [102, 103].

In short, to develop a mature and effective bioenergy production technology, bioconversion performance from lignocellulosic biomass is supposed to be further improved in conversion rates, cost-effectiveness, and system scale-up. Rapid advancements in the development of multi-processes, coupled with mechanistic research as well as additional funding, could make the commercialization of this new bioenergy technology in the future.

CONCLUSION

Based on the above review, it is clear that integrating thermophilic anaerobic strains in CBP strategy in the bioconversion of the second generation feedstocks (such as lignocellulosic materials) to bioenergy shows a very feasible solution to renewable energy production via biotechnology. Thermophilic anaerobic bacteria have a distinct advantage over conventional yeasts for bioenergy production in their ability to use a variety of organic waste biomass feedstocks and to withstand temperature extremes. Application of co-culture strategy entails neither operation cost nor capital investment for dedicated enzyme production and reduced diversion of substrate for enzyme production. Establishment of stable co-culture system is rate-limiting step to realize high efficiency CBP, and the key point of co-culture system is selecting appropriate thermophilic anaerobic strains. Despite the limited information available for co-culture research, the examples available have shown that co-culture fermentation could be utilized for bioenergy production from lignocellulosic biomass. However, more efforts need to be made in developing robust and well-studied co-culture systems that can be applied in industrial production. If a mature lignocellulose-to-energy system is established with a co-culture strategy, we can completely decrease not only our dependence on petroleum, but also gas pollution. Aside from the scientific assessment, a careful and detailed economic programming is needed to put this entire blueprint into practice.

ACKNOWLEDGMENTS

The authors gratefully acknowledge the financial support of the National Natural Science Foundation of China (No.51278200 and No. 51078147), Guangdong Natural Science Foundation (No. S2012010010380), the Fundamental Research Funds for the Central Universities, SCUT (No.2012ZM0081), the Guangdong Provincial Science and Technology Program (No. 2012B091100163, No. 2011B090400033, and No. 2010A010500005) and the Opening Project of Center of Guangdong Higher Education for Engineering and Technological Development of Speciality Condiments (No.GCZX-B1103).

REFERENCES

[1] Kothari R, Tyagi V, Pathak A. Waste-to-energy: A way from renewable energy sources to sustainable development. *Renewable and Sustainable Energy Reviews* 2010;14:3164.

[2] Cardona CA, Sánchez ÓJ. Fuel ethanol production: Process design trends and integration opportunities. *Bioresource Technology* 2007;98:2415.

[3] Gray KA, Zhao L, Emptage M. Bioethanol. *Current opinion in chemical biology* 2006;10:141.

[4] Ingram L, Gomez P, Lai X, Moniruzzaman M, Wood B, Yomano L, et al. Metabolic engineering of bacteria for ethanol production. *Biotechnology and bioengineering* 2000;58:204.

[5] Lin Y, Tanaka S. Ethanol fermentation from biomass resources: current state and prospects. *Applied Microbiology and Biotechnology* 2006;69:627.

[6] Mielenz JR. Ethanol production from biomass: technology and commercialization status. *Current Opinion in microbiology* 2001;4:324.

[7] Stephanopoulos G. Challenges in engineering microbes for biofuels production. *Science* 2007;315:801.

[8]   Ng TK, Ben-Bassat A, Zeikus J. Ethanol production by thermophilic bacteria: fermentation of cellulosic substrates by cocultures of *Clostridium thermocellum* and *Clostridium thermohydrosulfuricum*. *Applied and environmental microbiology* 1981;41:1337.

[9]   Lynd LR, Grethlein HE, Wolkin RH. Fermentation of cellulosic substrates in batch and continuous culture by *Clostridium thermocellum*. *Applied and environmental microbiology* 1989;55:3131.

[10] Sato K, Goto S, Yonemura S, Sekine K, Okuma E, Takagi Y, Hon-Nami K, Saiki T. Effect of Yeast Extract and Vitamin B12 on Ethanol Production from Cellulose by *Clostridium thermocellum* I-1-B. *Applied and environmental microbiology* 1992;58:734.

[11] Desvaux M, Guedon E, Petitdemange H. Cellulose catabolism by *Clostridium cellulolyticum* growing in batch culture on defined medium. *Applied and environmental microbiology* 2000;66:2461.

[12] Liu Y, Yu P, Song X, Qu Y. Hydrogen production from cellulose by co-culture of *Clostridium thermocellum* JN4 and *Thermoanaerobacterium thermosaccharolyticum* GD17. *International journal of hydrogen energy* 2008;33:2927.

[13] Datar R, Huang J, Maness PC, Mohagheghi A, Czernik S, Chornet E. Hydrogen production from the fermentation of corn stover biomass pretreated with a steam-explosion process. *International journal of hydrogen energy* 2007;32:932.

[14] Waeonukul R, Kosugi A, Tachaapaikoon C, Pason P, Ratanakhanokchai K, Prawitwong P, et al. Efficient Saccharification of Ammonia Soaked Rice Straw by Combination of *Clostridium thermocellum* Cellulosome and *Thermoanaerobacter brockii* β-Glucosidase. *Bioresource Technology* 2012;107:352.

[15] Duff SJB, Murray WD. Bioconversion of forest products industry waste cellulosics to fuel ethanol: a review. *Bioresource Technology* 1996;55:1.

[16] Hamelinck CN, Hooijdonk G, Faaij APC. Ethanol from lignocellulosic biomass: techno-economic performance in short-, middle-and long-term. *Biomass and Bioenergy* 2005;28:384.

[17] Wood TM. Properties of cellulolytic enzyme systems. *Biochemical Society Transactions* 1985;13:407.

[18] Shoham Y, Lamed R, Bayer EA. The cellulosome concept as an efficient microbial strategy for the degradation of insoluble polysaccharides. *Trends in microbiology* 1999;7:275.

[19] Lynd LR, Weimer PJ, Van Zyl WH, Pretorius IS. Microbial cellulose utilization: fundamentals and biotechnology. *Microbiology and molecular biology reviews* 2002;66:506.

[20] Bayer EA, Lamed R, White BA, Flint HJ. From cellulosomes to cellulosomics. *The Chemical Record* 2008;8:364.

[21] Doi RH. Cellulases of mesophilic microorganisms. *Annals of the New York Academy of Sciences* 2008;1125:267.

[22] Rincon MT, Dassa B, Flint HJ, Travis AJ, Jindou S, Borovok I, et al. Abundance and diversity of dockerin-containing proteins in the fiber-degrading rumen bacterium, *Ruminococcus flavefaciens* FD-1. *PLoS One* 2010;5:e12476.

[23] Maki M, Leung KT, Qin W. The prospects of cellulase-producing bacteria for the bioconversion of lignocellulosic biomass. *International journal of biological sciences* 2009;5:500.

[24] Ng T, Weimer P, Zeikus J. Cellulolytic and physiological properties of *Clostridium thermocellum*. *Archives of microbiology* 1977;114:1.

[25] Huang HJ, Ramaswamy S, Tschirner U, Ramarao B. A review of separation technologies in current and future biorefineries. *Separation and Purification Technology* 2008;62:1.

[26] Bayer EA, Morag E, Lamed R. The cellulosome—a treasure-trove for biotechnology. *Trends in Biotechnology* 1994;12:379.

[27] Lamed R, Naimark J, Morgenstern E, Bayer E. Specialized cell surface structures in cellulolytic bacteria. *Journal of bacteriology* 1987;169:3792.

[28] Shoseyov O, Takagi M, Goldstein MA, Doi RH. Primary sequence analysis of *Clostridium cellulovorans* cellulose binding protein A. *Proceedings of the National Academy of Sciences* 1992;89:3483.

[29] Gerngross T, Snell K, Peoples O, Sinskey A, Csuhai E, Masamune S, et al. Overexpression and purification of the soluble polyhydroxyalkanoate synthase from *Alcaligenes eutrophus*: evidence for a required posttranslational modification for catalytic activity. *Biochemistry* 1994;33:9311.

[30] Kakiuchi M, Isui A, Suzuki K, Fujino T, Fujino E, Kimura T, et al. Cloning and DNA Sequencing of the Genes *Encoding Clostridium* josui Scaffolding Protein CipA and Cellulase CelD and Identification of Their Gene Products as Major Components of the Cellulosome. *Journal of bacteriology* 1998;180:4303.

[31] Costerton JW, Stewart PS, Greenberg EP. Bacterial biofilms: a common cause of persistent infections. *Science* 1999;284:1318.

[32] Menon V, Rao M. Trends in bioconversion of lignocellulose: Biofuels, platform chemicals & biorefinery concept. *Progress in Energy and Combustion Science* 2012;38:522.

[33] Levin DB, Islam R, Cicek N, Sparling R. Hydrogen production by *Clostridium thermocellum* 27405 from cellulosic biomass substrates. *International journal of hydrogen energy* 2006;31:1496.

[34] Xu C, Qin Y, Li Y, Ji Y, Huang J, Song H, et al. Factors influencing cellulosome activity in Consolidated Bioprocessing of cellulosic ethanol. *Bioresource Technology* 2010;101:9560.

[35] Dehkhoda A. Concentrating lignocellulosic hydrolysate by evaporation and its fermentation by repeated fedbatch using flocculating *Saccharomyces cerevisiae*. 2008;3.

[36] Chen CY, Lu WB, Liu CH, Chang JS. Improved phototrophic H$_2$ production with *Rhodopseudomonas palustris* WP3-5 using acetate and butyrate as dual carbon substrates. *Bioresource Technology* 2008;99:3609.

[37] Chen M, Zhao J, Xia L. Enzymatic hydrolysis of maize straw polysaccharides for the production of reducing sugars. *Carbohydrate Polymers* 2008;71:411.

[38] Mosier N, Wyman C, Dale B, Elander R, Lee Y, Holtzapple M, et al. Features of promising technologies for pretreatment of lignocellulosic biomass. *Bioresource Technology* 2005;96:673.

[39] Sun Y, Cheng J. Hydrolysis of lignocellulosic materials for ethanol production: a review. *Bioresource Technology* 2002;83:1.

[40] Zeikus J. Thermophilic bacteria: ecology, physiology and technology. *Enzyme and Microbial Technology* 1979;1:243.

[41] Zeikus J. Chemical and fuel production by anaerobic bacteria. *Annual Reviews in Microbiology* 1980;34:423.

[42] Ren N, Wang A, Cao G, Xu J, Gao L. Bioconversion of lignocellulosic biomass to hydrogen: potential and challenges. *Biotechnology advances* 2009;27:1051.

[43] Lee J. Biological conversion of lignocellulosic biomass to ethanol. *Journal of Biotechnology* 1997;56:1.

[44] Jeffries T, Kurtzman C. Taxonomy, genetics and strain selection of xylose-fermenting yeasts. Enz *Microb Technol* 1994;16:922.

[45] Kádár Z, de Vrije T, van Noorden GE, Budde MAW, Szengyel Z, Réczey K, et al. Yields from glucose, xylose, and paper sludge hydrolysate during hydrogen production by the extreme thermophile *Caldicellulosiruptor saccharolyticus*. *Applied Biochemistry and Biotechnology* 2004;114:497.

[46] Ren N, Wang B, Huang JC. Ethanol-type fermentation from carbohydrate in high rate acidogenic reactor. *Biotechnology and bioengineering* 1997;54:428.

[47] Ren Z, Ward TE, Regan JM. Electricity production from cellulose in a microbial fuel cell using a defined binary culture. *Environmental science & technology* 2007;41:4781.

[48] Van Niel E, Budde M, De Haas G, Van Der Wal F, Claassen P, Stams A. Distinctive properties of high hydrogen producing extreme thermophiles, *Caldicellulosiruptor saccharolyticus* and *Thermotoga elfii*. *International journal of hydrogen energy* 2002;27:1391.

[49] Wang A, Ren N, Shi Y, Lee DJ. Bioaugmented hydrogen production from microcrystalline cellulose using co-culture—*Clostridium acetobutylicum* X9 and *Ethanoigenens harbinense* B49. *International journal of hydrogen energy* 2008;33:912.

[50] Ren N, Cao G, Wang A, Lee DJ, Guo W, Zhu Y. Dark fermentation of xylose and glucose mix using isolated *Thermoanaerobacterium thermosaccharolyticum* W16. *International journal of hydrogen energy* 2008;33:6124.

[51] Taguchi F, Yamada K, Hasegawa K, Taki-Saito T, Hara K. Continuous hydrogen production by *Clostridium* sp. strain no.2 from cellulose hydrolysate in an aqueous two-phase system. *Journal of fermentation and bioengineering* 1996;82:80.

[52] Ke Z, Wang H, Cai Y, Hang Z, Mei Z, Wang J. Optimization of culture medium for ethanol production of *Thermoanaerobacterium aotearoense* engineering strain by response surface methodology. *China Brewi* 2011:59.

[53] Chen S, Gong X, Zhu M. The Growth and Fermentation Characteristics of *Thermoanaerobacterium saccharolyticum*. *Microbiology China*2010;37:1105.

[54] Yokoyama H, Moriya N, Ohmori H, Waki M, Ogino A, Tanaka Y. Community analysis of hydrogen-producing extreme thermophilic anaerobic microflora enriched from cow manure with five substrates. *Applied Microbiology and Biotechnology* 2007;77:213.

[55] Lin CY, Wu CC, Hung CH. Temperature effects on fermentative hydrogen production from xylose using mixed anaerobic cultures. *International journal of hydrogen energy* 2008;33:43.

[56] Calli B, Schoenmaekers K, Vanbroekhoven K, Diels L. Dark fermentative H_2 production from xylose and lactose—Effects of on-line pH control. *International journal of hydrogen energy* 2008;33:522.

[57] Van Niel CW, Feudtner C, Garrison MM, Christakis DA. Lactobacillus therapy for acute infectious diarrhea in children: a meta-analysis. *Pediatrics* 2002;109:678.

[58] Liu H, Zhang T, Fang HHP. Thermophilic H_2 production from a cellulose-containing wastewater. *Biotechnology letters* 2003;25:365.

[59] Zinder S, Anguish T, Cardwell S. Effects of temperature on methanogenesis in a thermophilic (58°C) anaerobic digestor. *Applied and environmental microbiology* 1984;47:808.

[60] Sudha Rani K, Swamy M, Seenayya G. Production of ethanol from various pure and natural cellulosic biomass by *Clostridium thermocellum* strains SS21 and SS22. *Process Biochemistry* 1998;33:435.

[61] Ueno Y, Otsuka S, Morimoto M. Hydrogen production from industrial wastewater by anaerobic microflora in chemostat culture. *Journal of fermentation and bioengineering* 1996;82:194.

[62] Octave S, Thomas D. Biorefinery: Toward an industrial metabolism. *Biochimie* 2009;91:659.

[63] FitzPatrick M, Champagne P, Cunningham MF, Whitney RA. A biorefinery processing perspective: Treatment of lignocellulosic materials for the production of value-added products. *Bioresource Technology* 2010;101:8915.

[64] Tan T, Shang F, Zhang X. Current development of biorefinery in China. *Biotechnology advances* 2010;28:543.

[65] Cheng CL, Lo YC, Lee KS, Lee DJ, Lin CY, Chang JS. Biohydrogen production from lignocellulosic feedstock. *Bioresource Technology* 2011;102:8514.

[66] DOE U. Breaking the biological barriers to cellulosic ethanol: a joint research agenda. 2006.

[67] Raman B, Pan C, Hurst GB, Rodriguez M, McKeown CK, Lankford PK, et al. Impact of pretreated switchgrass and biomass carbohydrates on *Clostridium thermocellum* ATCC 27405 cellulosome composition: a quantitative proteomic analysis. *PLoS One* 2009;4:e5271.

[68] Chen Y. Development and application of co-culture for ethanol production by co-fermentation of glucose and xylose: a systematic review. *Journal of industrial microbiology & biotechnology* 2011;38:581.

[69] Soundar S, Chandra T. Cellulose degradation by a mixed bacterial culture. *Journal of industrial microbiology & biotechnology* 1987;2:257.

[70] Mori Y. Characterization of a symbiotic coculture of *Clostridium thermohydrosulfuricum* YM3 and *Clostridium thermocellum* YM4. *Applied and environmental microbiology* 1990;56:37.

[71] Haruta S, Cui Z, Huang Z, Li M, Ishii M, Igarashi Y. Construction of a stable microbial community with high cellulose-degradation ability. *Applied Microbiology and Biotechnology* 2002;59:529.

[72] Häggström L, Förberg C. Significance of an extracellular polymer for the energy metabolism in *Clostridium acetobutylicum*: a hypothesis. *Applied Microbiology and Biotechnology* 1986;23:234.

[73] Hinman ND, Wright JD, Hogland W, Wyman CE. Xylose fermentation. *Applied Biochemistry and Biotechnology* 1989;20:391.

[74] Keim CR, Venkatasubramanian K. Economics of current biotechnological methods of producing ethanol. *Trends in Biotechnology* 1989;7:22.

[75] Lovitt RW, Kim BH, Shen G, Zeikus J, Phillips JA. Solvent production by microorganisms. *Critical reviews in biotechnology* 1988;7:107.

[76] Schneider H, Jeffries TW. Conversion of pentoses to ethanol by yeasts and fungi. *Critical reviews in biotechnology* 1989;9:1.

[77] Wongwilaiwalin S, Rattanachomsri U, Laothanachareon T, Eurwilaichitr L, Igarashi Y, Champreda V. Analysis of a thermophilic lignocellulose degrading microbial consortium and multi-species lignocellulolytic enzyme system. *Enzyme and Microbial Technology* 2010;47:283.

[78] Teixeira LC, Linden JC, Schroeder HA. Simultaneous saccharification and cofermentation of peracetic acid-pretreated biomass. *Applied Biochemistry and Biotechnology* 2000;84:111.

[79] Xu L, Tschirner U. Improved ethanol production from various carbohydrates through anaerobic thermophilic co-culture. *Bioresource Technology* 2011;102:10065.

[80] Li Q, Liu CZ. Co-culture of *Clostridium thermocellum* and *Clostridium thermosaccharolyticum* for enhancing hydrogen production via thermophilic fermentation of cornstalk waste. *International journal of hydrogen energy* 2012;35:8945.

[81] Zeikus J, Ben-Bassat A, Hegge P. Microbiology of methanogenesis in thermal, volcanic environments. *Journal of bacteriology* 1980;143:432.

[82] Kato S, Haruta S, Cui ZJ, Ishii M, Igarashi Y. Network relationships of bacteria in a stable mixed culture. *Microbial ecology* 2008;56:403.

[83] Baker JO, Adney WS, Nleves RA, Thomas SR, Wilson DB, Himmel ME. A new thermostable endoglucanase, *Acidothermus cellulolyticus* E1. *Applied Biochemistry and Biotechnology* 1994;45:245.

[84] Ford T, Sacco E, Black J, Kelley T, Goodacre R, Berkeley R, et al. Characterization of exopolymers of aquatic bacteria by pyrolysis-mass spectrometry. *Applied and environmental microbiology* 1991;57:1595.

[85] Weimer PJ. Cellulose degradation by ruminal microorganisms. *Critical reviews in biotechnology* 1992;12:189.

[86] Li P, Zhu M. A consolidated bio-processing of ethanol from cassava pulp accompanied by hydrogen production. *Bioresource Technology* 2011;102:10471

[87] Delgenes J, Escare M, Laplace J, Moletta R, Navarro J. Biological production of industrial chemicals, ie xylitol and ethanol, from lignocelluloses by controlled mixed culture systems. Industrial Crops and Products 1998;7:101.

[88] Laplace J, Delgenes J, Moletta R, Navarro J. Ethanol production from glucose and xylose by separated and co-culture processes using high cell density systems. *Process Biochemistry* 1993;28:519.

[89] Laplace J, Delgenes J, Moletta R, Navarro J. Combined alcoholic fermentation of D-xylose and D-glucose by four selected microbial strains: process considerations in relation to ethanol tolerance. *Biotechnology letters* 1991;13:445.

[90] Lebeau T, Jouenne T, Junter GA. Continuous alcoholic fermentation of glucose/xylose mixtures by co-immobilized *Saccharomyces cerevisiae* and *Candida shehatae*. *Applied Microbiology and Biotechnology* 1998;50:309.

[91] Fu N, Peiris P, Markham J, Bavor J. A novel co-culture process with *Zymomonas mobilis* and *Pichia stipitis* for efficient ethanol production on glucose/xylose mixtures. *Enzyme and Microbial Technology* 2009;45:210.

[92] Taniguchi M, Itaya T, Tohma T, Fujii M. Ethanol production from a mixture of glucose and xylose by a novel co-culture system with two fermentors and two microfiltration modules. *Journal of fermentation and bioengineering* 1997;84:59.

[93] Lynd L. Production of ethanol from lignocellulosic materials using thermophilic bacteria: critical evaluation of potential and review. *Lignocellulosic materials* 1989;38:1.

[94] Georgieva TI, Mikkelsen MJ, Ahring BK. High ethanol tolerance of the thermophilic anaerobic ethanol producer *Thermoanaerobacter* BG1L1. *Central European Journal of Biology* 2007;2:364.

[95] Yu EKC, Chan MKH, Saddler JN. Butanediol production from lignocellulosic substrates by *Klebsiella pneumoniae* grown in sequential co-culture with *Clostridium thermocellum*. Applied *Microbiology and Biotechnology* 1985;22:399.

[96] Saddler J, Chan MKH, Louis-Seize G. A one step process for the conversion of cellulose to ethanol using anaerobic microorganisms in mono-and co-culture. *Biotechnology letters* 1981;3:321.

[97] Saddler J, Chan MKH. Optimization of *Clostridium thermocellum* growth on cellulose and pretreated wood substrates. *Applied Microbiology and Biotechnology* 1982;16:99.

[98] Harish K, Srijana M, Madhusudhan R, Gopal R. Coculture fermentation of banana agro-waste to ethanol by cellulolytic thermophilic *Clostridium thermocellum* CT2. *African Journal of Biotechnology* 2012;9:1926.

[99] Liu W, Wang A, Ren N, Zhao X, Liu L, Yu Z, et al. Electrochemically Assisted Biohydrogen Production from Acetate. *Energy & Fuels* 2007;22:159.

[100] Saddler J, Chan MKH. Conversion of pretreated lignocellulosic substrates to ethanol by *Clostridium thermocellum* in mono- and co-culture with *Clostridium thermosaccharolyticum* and Clostridium thermohydrosulphuricum. *Canadian journal of microbiology* 1984;30:212.

[101] Lynd LR, Ahn HJ, Anderson G, Hill P, Sean Kersey D, Klapatch T. Thermophilic ethanol production investigation of ethanol yield and tolerance in continuous culture. *Applied Biochemistry and Biotechnology* 1991;28:549.

[102] Bothast RJ, Nichols NN, Dien BS. Fermentations with new recombinant organisms. *Biotechnology progress* 1999;15:867.

[103] Taniguchi M, Tohma T, Itaya T, Fujii M. Ethanol production from a mixture of glucose and xylose by co-culture of *Pichia stipitis* and a respiratory-deficient mutant of *Saccharomyces cerevisiae*. *Journal of fermentation and bioengineering* 1997;83:364.

SECTION 3. INTEGRATED BIOREFINERY PROCESSES

In: Biomass Processing, Conversion and Biorefinery   ISBN: 978-1-62618-346-9
Editors: Bo Zhang and Yong Wang   © 2013 Nova Science Publishers, Inc.

Chapter 22

ECONOMIC ANALYSIS OF WASTE TO POWER: A CASE STUDY OF GREENSBORO CITY

Ransford R. Baidoo, Abolghasem Shahbazi,
Matthew Todd and Harith Rojanala*

Department of Natural Resources and Environmental Design,
North Carolina A and T State University, NC, US

ABSTRACT

Waste is a rejected or unwanted leftover of materials after an economic activity takes place. These include municipal solid waste (MSW), and construction and demolition (C&D) waste. Managing these wastes involve collection, transportation, re-use, recycling, land filling and Waste-to-Energy (WTE). The three WTE technologies considered for this study are combustion, thermal gasification and plasma gasification. This study sought to determine the electrical energy that could be obtained from each technology, to do the benefit and cost analysis, and to show whether it would be economical to convert waste to power. It was shown that combustion process produces carbon dioxide (CO_2), which has negative impact to the environment. Thermal gasification produces less CO_2 and comparatively produces less electrical power output. Plasma gasification however gives less CO_2 but gives more power output. Results also show that, with the application of WTE, the payback period for combustion is about 9 years, conventional gasification is 8 years and for plasma gasification is about 10 years.

1. INTRODUCTION

1.1. City of Greensboro

Greensboro is the largest city of the "triad", the rest being Winston-Salem and High Point, and occupies an area of about 131.75 square miles at an elevation of 897 feet above sea

---

* Corresponding author: Ransford R. Baidoo rkbaidoo@yahoo.co.uk.

level [1]. The City of Greensboro now consumes about 700 MW of power [2], mainly from coal, natural gas and hydroelectric [3]. The City of Greensboro has a population of about 170,000, and generates about 280,000 tons of waste per year. Initially, most of the MSW generated was sent to the "White Street landfill", near the city for disposal. With the current impasse between the city authorities and the residents at that area, the waste is now conveyed to the "Transfer Station" at a tipping fee of $41 per ton, from where it is sent to Mt. Gilead, in Mountgomry county, about 73 miles, south of Greensboro for disposal in a landfill. Figure 1 shows C&D waste being spread at the White Street landfill site.

Figure 1. Bulldozer Spreading C&D waste at the White Street landfill site.

1.2. Waste Generation and Management

There is no doubt that the quantity of waste generated by a community depends on the population of that community. Thus the waste generated by the City of Greensboro has increased astromomically since the incorporation of the City in 1829, which poses a big challenge to the City Authorities. These include municipal solid waste (MSW), construction and demolition (C&D) waste, and wastewater. Managing these waste streams generally involve collection, transportation, re-use, recycling, land filling and Waste-to-Energy (WTE). Land filling which is a means of burying the waste in the ground, has been identified to have other negative effects on the environment. Some of these environmental effects are contamination of ground water, attracting rodents and birds, emitting foul odor, as well as producing methane gas. This gas is said to be about twenty times more potent than carbon dioxide which causes global warming [4].

Often these waste streams possess energy (BTU or Joule) content. It is therefore possible and desirable to transform the energy to a form that will allow for the performance of work toward the ultimate state of disposal called Waste-to-Energy.

The three main methods of WTE are physical, biological and thermal. Under physical, the waste is converted to Refuse-Derived-Fuel (RDF) for use in thermal processes. Under biological, the waste is digested or fermented and converted to biofuels. Under thermal, the waste is incinerated to boil water to run a steam turbine or it could also be converted to synthesis gas (syngas) for use in a gas turbine or engine to generate electrical power. Using the chemical equation of biomass or waste $C_6H_{10}O_4 + 2H_2O_4$ [5], David Lide's CRC handbook of thermophysical and thermochemical data, Table 1, [6] was used to determine the heating values of these technologies and their results were used to analyze the potential power to be derived from each of the technologies.

The aim of this research is to determine how much electrical energy could be obtained through each technology and to do the benefit and cost analysis, to show that in total it would be beneficial and economical to convert the waste to power, rather than burying it in a landfill.

2. Wastes to Energy Technologies

2.1. Wastes-to-Energy Technology Tree

Figure 2 depicts the waste to energy technology tree in which we have physical, thermal and biological processes of conversion [7]. In this research, only the waste to energy extraction through the thermal process has been examined, namely combustion, thermal or conventional gasification and plasma gasification.

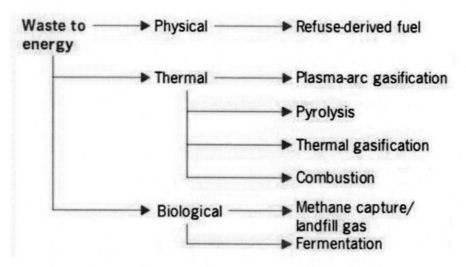

Figure 2. Waste-to-Energy Technology Tree.

2.2. Combustion

Combustion is the process of burning biomass or waste in an oxygen rich environment. This process emits more carbon dioxide, which hurt the environment by promoting global warming and health hazards. They operate at temperatures between 600 and 800 °C [8]. Equations (1) to (3) show the chemical equations used to determine the heating value of the combustion process. The $6CO_2$ compound on the product side of equation (2) shows that the process emits carbon dioxide.

Table. 1. Relevant Thermochemical Data

No.	Symbol	$\Delta_f H°$	No.	Symbol	$\Delta_f H°$
1	O	+249.18	10	H_2 (g)	0.00
2	O_2	0.00	11	H_2O(l)	-285.83
3	CO	-110.5	12	H_2O(g)	-241.8
4	CO_2 (g)	-393.5	13	CH_4	-74.6
5	H	+217.18	14	C	0.00
6	H^+	0.00	15	C_2	+830.47
7	H_2 (l)	0.00	16	e^+	-802.4
8	$C_6H_{10}O_4$	-994.3	17	CH_2O	-108.6
9	C^+	716.68	18	O^+	249.18

$$\text{Biomass Equation}: C_6H_9O_4 + 2H_2O \rightarrow \frac{13}{2}H_2 + 6CO \tag{1}$$

$$\text{Modified for Combustion}: C_6H_{10}O_4 + 6.5O_2 \rightarrow 6CO_2 + 5H_2C \tag{2}$$

$$\Delta_f H° = 6CO_2 + 5H_2O - \left(C_6H_{10}O_4 + 6.5O_2\right) \tag{3}$$

$$\Delta_f H° = -393.5x6 - 241.8x5 - \left(-994.3 - 6x0\right)$$

$$\Delta_f H° = -2361 - 1209 + 994.3 = -3570 + 994.3$$

$$\Delta_f H° = = -3570 + 994.3 = 2575 \text{kJ/mol} = 17.64 MJ/kg$$

Figure 3 depicts how the combustion of waste or biomass takes place in a combustion chamber. The heat energy generated is used to heat water to convert it to steam. The steam is then expanded through a steam turbine to generate electricity, which is then connected to the transmission line. The exhaust heat from the turbine is then fed back to the combustion chamber through the heat recovery steam generator so as to improve the efficiency of the power plant.

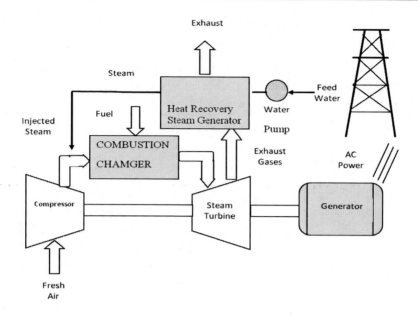

Figure 3. Steam Injected Turbine.

2.3. Thermal Gasification

Gasification is an oxygen starved process whereby carbonaceous materials such as woody products, vegetation and solid waste are converted to synthesis gas (syngas). This is a mixture of hydrogen and carbon mono oxide (H_2+CO), whose energy content is similar to natural gas (CH_4). Syngas has the advantage of being more useful than the original solid fuel because it could be converted to other forms of energy such as ethanol or biodiesel [9]. Theoretically, gasification processes produce little or no carbon dioxide and operate between 800 and 1200 °C [9]. Equations (4) to (5) show the chemical equations used to determine the heating value of the thermal gasification process.

Modified for Thermal Gasification:

$$C_6H_{10}O_4 + 2H_2O \rightarrow 7H_2 + 6CO \tag{4}$$

$$\Delta_f H° = 7H_2 + 6CO - (C_6H_{10}O_4 + 2H_2O)$$

$$\Delta_f H° = 7 \times 0 - 6 \times 110.5 - (-994.3 - 2 \times 285.83)$$

$$\Delta_f H° = -663 - (-994.3 - 571.66) = -663 - (-1565.96)$$

$$\Delta_f H° = = -663 + 1565.96 = 902.96 \, kJ/mol$$

$$\text{Heating Value} = 6.2\,\text{MJ/kg} \tag{5}$$

As compared to combustion, the heating value is less because even though the breaking of molecular bonds releases heat energy, some energy is utilized for the formation of new compounds; hence the net energy is reduced.

2.4. Plasma Gasification

Like thermal gasification, plasma gasification operates in an oxygen starved environment and produces syngas. Unlike thermal gasification, plasma gasification requires electrical power for its operation. It operates at temperatures between 1200 and 2000 °C [10]. Equations (6) and (7) show the chemical equations used to determine the heating value of the plasma gasification process, while equations (8) and (9) give the ratio of "plasmafication" to combustion and thermal gasification, respectively.

Modified for Plasmafication:

$$C_6H_{10}O_4 + 2H_2O \rightarrow 7H_2 + 6CO \tag{6}$$

$$\Delta_f H^\circ = 7H_2 + 6CO - \left(C_6H_{10}O_4 + 2H_2O\right)$$

$$= 6C^{\pm} + 6O^{\pm} + 14H + 14e^{-} - \left(C_6H_{10}O_4 + 2H_2O\right)$$

$$= 6 \times 716.68 + 6 \times 249.18 + 14 \times 0 - 14 \times 802.4 - \left(C_6H_{10}O_4 + 2H_2O\right)$$

$$= 4300.08 + 1495.08 - 11233.6 - \left(-994.3 - 2 \times 285.83\right)$$

$$= 5795.16 - 11233.6 - \left(-1565.96\right) = -11233.6 + 7361.12$$

$$= -3872.48\,\text{kJ/mol} \quad (26.52\,\text{MJ/kg}) \tag{7}$$

$$\text{Plasma/Combustion} = 3872.48 / 2575 = 1.5 \tag{8}$$

$$\text{Plasma/Gasifier} = 3872.48 / 902.96 = 4.3 \tag{9}$$

The heating value in the plasma process was found to be higher because, in that process, all the compounds are broken down to their constituent elements and as such no new compounds were formed and therefore all the heat energy released made available for use.

2.5. Gasification Process of Waste-to-Energy

Figure 4 shows the steps from tipping of waste and its flow through the gasifier to the turbines where electrical power is generated [11].

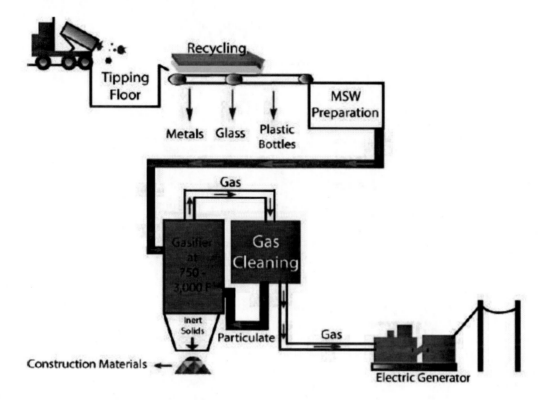

Figure 4. Gasification of Waste to Power.

Figure 5 depicts the schematic of the fluidized-bed gasifier, which is one of the four main types of gasifiers in use today [12]. In this gasifier, air stream is blown under the flow of the biomass material, causing it to rotate in the gasifier in the form of a fluid. Thus heat in the gasifier evenly disintegrates the material and converts it to synthesis gas. This is cleaned through cyclones and filters before reaching the gas turbine.

In Figure 6, controlled compressed air is passed through a fluidized-bed gasifier which forms the gasification chamber. The Synthesis gas so produced is then expanded through a gas turbine. This causes the alternator to generate electricity. The exhaust from the gas turbine is then passed through a heat recovery steam generator and the steam so generated expanded through a steam turbine to also generate electricity. The two sources of electricity are then synchronized onto infinite or common bus-bars before being connected to the transmission line.

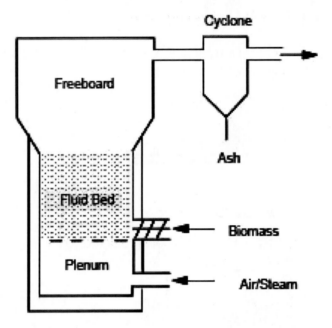

Figure 5. Fluidized Bed Gasifier.

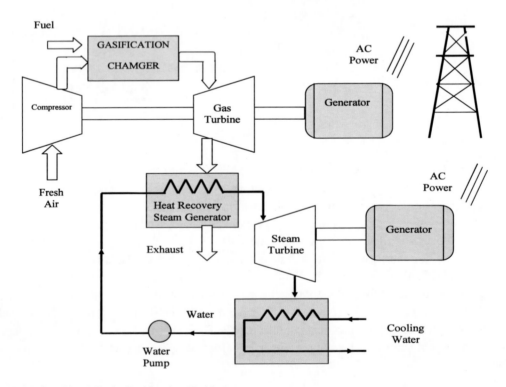

Figure 6. Combined Cycle Gasification Turbine.

3. CONVERSION OF GREENSBORO WASTE TO ELECTRICAL POWER

3.1. Methodology

In order to estimate the electrical power that could be extracted from the waste generated at the City of Greensboro, the heating values of the waste stream was determined as in equations (3), (5) and (7) for combustion, thermal gasification and plasma gasification respectively.

The other parameters required to do the analysis are the efficiency (η) of the steam turbine 33% [13] and that of the combined cycle gas turbine (CCGT) 49% [14], as well as the annual quantity of waste generated W_t. Equation (10) below was written in excel to calculate the electrical power P_G, that could be generated from the quantity of waste.

$$P_G = \frac{W_t(tons) * 907(kg/ton) * H_{HV}(kJ/kg) * \eta}{330 * 24 * 3600(sec\,s.)} \, kW$$

(10)

It was assumed that the plant operated for 330 days in a year, at 24 hours a day and 3600 seconds an hour. Since the weight of the waste is in US tons, the total weight was multiplied by 0.907185 to convert it to tones (SI). The power was multiplied by 8760 (hours in a year) to give the annual energy generated from the waste as shown in equation (11).

$$E_G = P_G * 8760(hrs)\ kWl$$

(11)

The annual revenue was obtained by multiplying the energy generated, by the avoidable cost of electricity in the state, (0.04$/kWh) [15], as shown in equation (12) below,

$$Annl\_Revn = E_G(kWh) * X(\$/kWh)\ US\$$$

(12)

where X is the total cost of electricity in $/kWh.

In order to determine the power capacity (PC), the annual quantity of waste generated, shown against year 2010 in Table 2 column 2, was divided by the number of days in a year (366). The result was also divided by the plant availability, which is about 90%.

$$PC = \frac{279,568}{366 * 0.9} = 849MW$$

(13)

Since power plants are manufactured according to plant capacity, it is desirable to choose one near the theoretical PC. From research the near available PC is 750 MW, so this PC was chosen for this research.

Data from NC Recycling and Solid Waste Management Plan [16] shown in Table 2 columns 1 and 2 were used to calculate the power, energy and revenue from Greensboro MSW (using equation 10-12), as shown in Table 2 columns 3 to 5 respectively. It was assumed that about 25% of the total MSW was recycled and the rest 75% used for the recovery process [17].

Table 2. Power, Energy and Revenue from MSW

Municipal Solid Waste (MSW) - Recycling (25%)					MSW + C&D
YEAR	Combustion	Power (kW)	Energy	Revenue (US$)	Totl Engy (kWh)
2008	276106.80	38346.92	335919022.28	13436760.89	379411462.30
2009	276362.17	38382.39	336229712.35	13449188.49	368988821.30
2010	279568.87	38827.75	340131070.55	13605242.82	361439945.55
2011	261809.30	36361.22	318524295.96	12740971.84	338466942.17
YEAR	Gasification	Power (kW)	(kWh)	Revenue (US$)	
2008	276106.80	19963.32	174878651.27	6995146.051	197520713.04
2009	276362.17	19981.78	175040395.79	7001615.831	192094710.70
2010	279568.87	20213.63	177071433.67	7082857.347	188164783.78
2011	261809.30	18929.57	165822997.74	6632919.91	176205092.36
YEAR	Plasmafication	Power (kW)	(kWh)	Revenue (US$)	
2008	276106.80	85602.72	749879837.61	29995193.5	846969021.90
2009	276362.17	85681.89	750573398.27	30022935.93	823702318.26
2010	279568.87	86676.08	759282490.82	30371299.63	806850787.54
2011	261809.30	81170.00	711049185.92	28441967.44	755567618.39

Table 3. Power, Energy and Revenue from C&D

Construction and Demolition (C&D) - Recycling (72%)					MSW + C&D
YEAR	Combustion	Power (kW)	Energy	Revenue (US$)	TOTAL (US$)
2008	95754.57	4964.89	43492440.02	1739697.60	15176458.49
2009	72123.67	3739.62	32759108.96	1310364.36	14759552.85
2010	46914.41	2432.52	21308875.00	852355.00	14457597.82
2011	43906.47	2276.56	19942646.22	797705.85	13538677.69
YEAR	Gasification	Power (kW)	(kWh)	Revenue (US$)	
2008	95754.57	2584.71	22642061.77	905682.47	7907298.30
2009	72123.67	1946.84	17054314.91	682172.60	7765029.94
2010	46914.41	1266.36	11093350.10	443734.00	7076653.91
2011	43906.47	1185.17	10382094.62	415283.78	28857251.22
YEAR	Plasmafication	Power (kW)	(kWh)	Revenue (US$)	
2008	95754.57	11083.24	97089184.29	3883567.37	33878760.88
2009	72123.67	8348.05	73128919.99	2925156.80	32948092.73
2010	46914.41	5430.17	47568296.72	1902731.87	32274031.50
2011	43906.47	5082.01	44518432.46	1780737.30	30222704.74

Table 3 columns 1 and 2 also show the NC Recycling and Solid Waste Management Plan data for C&D waste [16]. This data was also used to calculate the power, energy and revenue generated in Greensboro (using equation 10-12), as shown in columns 3 to 5 of Table 2.

It was assumed that about 28% of the total C&D waste contained wood or combustibles [18] and was thus added to the recovery process. Table 2 column 6 shows the total revenue expected from both MSW and C&D, which is the sum of Table 2 column 5 and Table 3 column 5. Table 3 column 6 also shows the total energy generated from MSW and C&D which is the sum of values in Table 2 column 4 and Table 3 column 4.

3.2. Payback Period

In order to calculate the payback period (PBP), there was the need to determine the cash inflow and cash outflow of each technology. For the cash inflow aspect, a renewable energy certificate (REC) of $6/MWh as well as $2.5M per each installation is paid as renewable energy incentive to the renewable energy producer [15]. This is in addition to the avoided cost of (0.04$/kWh). Thus values of the total energy generated for each technology, for year 2010, shown in Table 3 column 6 was transferred to Table 4 columns 2 to 4 for the analysis. The renewable energy certificate and the renewable energy credit were added to give the total cash inflow as shown in Table 4, under cash inflows.

Table 4. Cash flows for MSW and C&D waste

ANNL QUANTITY OF WASTE (US Tons	270000		
CAPACITY OF PLANT (750 TPD)	750		
CASH INFLOWS	**Combustion**	**Gasification**	**Plasmafication**
Sale of annual electrical energy (US $)	14457597.82	7076653.91	32274031.50
Renewable energy certificate ($6/MWh	2168639.67	1128988.703	4841104.725
Renewable energy taxe credit (Paid once)	2500000	2500000	2500000
Total	**19126237.50**	**10705642.62**	**39615136.23**
CASH OUTFLOWS	**Combustion**	**Gasification**	**Plasmafication**
Cost of Incinerator/ Gasifier US($)	90000000	30000000	250000000
Project Devp'mt Cost (30%, 20%) of cost	27000000	9000000	50000000
Siting Permitting and Interconnection	13000000	13000000	13000000
Sub Total Cost (Fixed cost)	**130000000**	**52000000**	**313000000**
Insurance (9%) of fixed cost	11700000	4680000	28170000
O&M Cost ($45/Ton)	33750	33750	33750
Ash dispo 30%, 15%, 10% resp at $41/Ton	332100	1660500	1107000
Emission control equipment (US $)	1000000	500000	500000
Sub Total Cost (Variable cost)	**13065850.00**	**6874250.00**	**29810750.00**

For the outflow cost, incineration equipment was researched and found to be $90,000,000.00 [19]. Thermal gasification cost was found to be $30,000,000.00 [20], and plasma gasification cost found to be $250,000,000.00 [21]. Other cost researched are project development cost (30%) of equipment cost; citing, permitting and interconnection, about $13M, operation and maintenance cost about ($45/Ton) of daily installed capacity. The rest are interest rate and insurance about (9%), ash disposal about 30% for combustion 15% for thermal gasification, and 10% for plasma gasification of annual waste generated.

Emission control equipment cost for combustion was $1M and $500,000.00 each for thermal and plasma gasifications [22]. These are summarized as shown in Table 4 under outflows.

Table 5 was constructed out of Table 4. Here the revenue generated in year 1 (column3) under inflows, was deducted from the total cost of each technology in column 2 under outflows to give the net cash flow. Similarly revenue in year 2 was deducted from that of year 1. This process was continued until year 15. The cumulative cash flow in Table 5 was found by summing all the cash inflows and outflows for all preceding years and the current year [23]. Note that the renewable energy tax credit of $2.5M was not part of the inflow after the first year.

Table 5. Calculation of Payback Period

YEARS	0	1	2	3
INFLOW		($M)	($M)	
COMBUST		19.13	16.63	16.63
GASIFY		10.71	8.21	8.21
PLASMA		39.62	37.12	37.12
OUTFLOW ($M)		CUMULATIVE CASH FLOW ($M)		
COMBUST	(143.07)	(123.94)	(107.31)	(90.69)
GASIFY	(58.87)	(48.17)	(39.96)	(31.76)
PLASMA	(342.81)	(303.20)	(266.08)	(228.97)

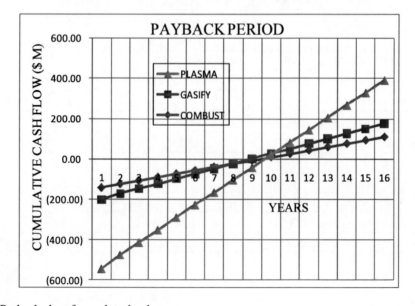

Figure 7. Payback chart for each technology.

Figure 7 shows the chart for the payback period which is the result obtained from the net cash flow of each technology.

Table 6 however displays the annual income and operating cost whose difference gives the annual profit.

Table 6. Annual profits for each technology

ANNUAL	COMBUST	GASIFY	PLASMA
INCOME	16.63	8.21	37.12
OPERATING COST ($M	13.06585	6.87425	29.81075
PROFIT ($M)	3.56	1.33	7.30

4. RESULTS

The results show that, for the same capacity of power plant, combustion can generate a revenue of about $16.6M with a payback period of about 9 years, that of gasification can generate a revenue of about $8.2M with a payback period of about 8 years and that of "plasmafication" to generate a revenue of about $37.1M with a payback period of about 10 years.

CONCLUSION

The environmental challenges facing cities and communities due to waste generation could be mitigated by the application of waste-to-Energy technologies such as combustion, thermal gasification or plasma gasification. The results show that Greensboro could make an income of at least US$16.6M a year from using combustion WTE technology, with a PBP of about 9 years and an annual profit of about $3.6M. Gasification WTE technology could also realize an income of at least US 8.2 M, at a PBP of 8 years and an annual profit of about US $1.3M. However from equation (2) combustion emits carbon dioxide, while from equation (4) gasification does not.

The city authorities would therefore have to decide whether to make more profit at the expense of the environment or otherwise. The results also show that income from "plasmafication" is at least US$ 37.1 M, a year, at a PBP of 10 years, at a profit of about US$7.3M. It is evident from equation (6) that "Plasmafication" does not emit carbon dioxide. This is therefore the best choice of WTE technology due to its high revenue generation as well as its environmental sustainability.

"Plasmafication" technology however has an initial high cost and at the same time the technology is yet to be commercialized. Slag from both gasification technologies could also be sold to increase the revenue.

ACKNOWLEDGMENTS

We wish to acknowledge funding for this research by the US-Army Research Office, contract # W911NF-08-1-0384.

REFERENCES

[1]   *Greensboro, NC Facts. (2010). Greensboro Convention and Visitors Bureau.*
[2]   *State Electric Profiles. (2011). Energy Information Administration.*
[3]   Edwards. W. Davis. (2010). *Electricity In Energy Trading & Investing.* McGraw- Hill (New York). Pp. 111-125.
[4]   US-EPA. Methane-Climate change. (*http://www.epa.gov/methane/*).
[5]   Nicolas J. Themelis et al. (2002). Energy Recovery from New York City Wastes. ISWA *Journal of Waste Management and Research.* pp. 1-7.
[6]   Lide R. David, Kehiaian V. Henry. *(1994). CRC handbook of Thermophysical and Thermochemical data.*
[7]   Eco Green Sites. Biofuels and Waste-to-Energy. (*www.ecogreensites.com/tech.html*).
[8]   Aitkin County Plasma Gasification Study (2008). *Plasma Gasification vrs Incineration.*
[9]   James D. McMillan. (2004). *Biotechnological Routes to Biomass Conversion.* National Bioenergy Center.
[10] Mountouris A. et al. *Waste Plasma Gasification: Equilibrium model development and energy analysis.* Sch. of Chemical Eng. National University of Athens.
[11] Advanced Energy Strategies Inc. (2004). *Investigations into municipal solid waste Gasification for power generation.* Alameda. Power & Telecom Utilities Board. Pp. 6-11.
[12] Franklin H. Holcomb et al. (2008*). Proceedings of the 1st Army Installation Waste to Energy Workshop.* US Army Corps of Engineers. Pp. 39-40.
[13] Masters M. Gilbert. (2004). The Electric Power Industry. In *Renewable and efficient electric power Systems.* (John Wiley & sons Inc). pp. 127-132.
[14] Masters M. Gilbert. (2004). The Electric Power Industry. In *Renewable and efficient electric power Systems.* (John Wiley &sons Inc). pp. 132-137.
[15] US-DOE. Database *of State Incentive for Renewables and Efficiency (DSIRE) Review.* (2010).
[16] *NC Recycling and Solid Waste Management Plan.* (1992). Vol. 1. pp. 6-17.
[17] Blue Ridge Environmental Defense League. (2009). *Gasification impacts on the environment.* P.13.
[18] Recycling Business Assistance Centre. (2010). *C&D wood in North Carolina.*
[19] Environment people law. (2011). *Waste incineration Plant.*
[20] World Waste Technology. (2006). *Thermal gasification.* P. 4-52
[21] Michael, Behar. (2007). Prophet of garbage. *Popular science.*
[22] Oneida-Herkiner solid waste Authority. (2007).*Waste to Energy Analysis Report.* PP. 7-12.
[23] Schmidt, J. Marty. (2012). Payback period. *Encyclopedia of Business Terms and Methods.*

In: Biomass Processing, Conversion and Biorefinery   ISBN: 978-1-62618-346-9
Editors: Bo Zhang and Yong Wang   © 2013 Nova Science Publishers, Inc.

Chapter 23

PROCESS DESIGN FOR BIOLOGICAL CONVERSION OF CATTAILS TO ETHANOL

Bo Zhang[1,2,*]

[1] School of Chemical Engineering and Pharmacy, Wuhan Institute of Technology, Hubei, China
[2] Biological Engineering Program, Department of Natural Resources and Environmental Design, North Carolina A & T State University, Greensboro, NC, US

ABSTRACT

The effects of different pretreatment technologies, including sulfuric acid, hot-water, NaOH and MgCl2 pretreatments, on the fermentation of xylose and glucose released from cattails to ethanol by *Saccharomyces cerevisiae* ATCC 24858 and *Escherichia coli* KO11 were investigated. Glucose from cattail cellulose was able to be fermented to ethanol using *S. cerevisiae*, resulting in 85-91% of the theoretical ethanol yield. Glucose and xylose released from cattail cellulose and hemicellulose were able to be fermented to ethanol using *E. coli* KO11, resulting in approximately 85% of the theoretical ethanol yield. Among four pretreatment methods, the dilute acid pretreatment was found to be superior, and approximately 85% of original sugars in the cattails were converted to ethanol.

The effects of different biomass species (cattail harvested in different seasons and corn stover) and their chemical composition on the overall process efficiency and economic performance considering feedstock availability and feedstock costs to manufacture ethanol from lignocellulose was studied. Cattails harvested in fall have the highest cellulose and xylan content, leading to the lowest Minimum Ethanol Selling Price (MESP) of $2.28, and the highest ethanol production of 57.2 MM gal/yr from processing 773,000 dry U.S. ton feedstock.

Carbon dioxide emission from processing cattails is in the range of 2450-2650 Kmol/hr, which is higher than the CO_2 emission from corn stove (~2110 Kmol/hr). Because of high lignin content in cattails, it's very important to explore emerging technologies such as pyrolysis and gasification to produce multi-products and obtain higher values from cattail lignin. The potential environmental impacts including CO,

---

[*] Phone 336-334-7787. Fax 336-334-7270. E-mail: bzhang@ ncat.edu.

SO_2, NOx emission from the biorefinery process and waste water treatment were also discussed.

1. INTRODUCTION

For a number of reasons, there has recently been increasing interest in converting biomass to liquid fuels. Some of those reasons include limited availability and increasing demand for fossil fuels, especially in developing countries; increasing price; the need for national energy independence and safety; and the need for reduction in greenhouse gas (GHG) emissions [1]. To this end, the federal government has been calling for research into ethanol production from a number of cellulosic sources [2]. Ethanol is considered the most potential next generation transportation fuel, and significant quantities of ethanol are currently being produced from corn and sugar cane via a fermentation process. Utilizing lignocellulosic biomass as a feedstock is seen as the next step towards significantly expanding the ethanol production capacity. However, technological barriers (such as pretreatment, saccharification of cellulose and hemicellulose matrix, and simultaneous fermentation of hexoses and pentoses), economic feasibility, and environmental impacts need to be addressed to efficiently convert lignocellulosic biomass into bioethanol [3].

In this chapter, cattail was used as the feedstock. The use of aquatic plant cattails to produce biofuel will add value to land and reduce emissions of greenhouse gases by replacing petroleum products. The effects of different pretreatment technologies, including dilute sulfuric acid, hot-water, NaOH and $MgCl_2$ pretreatments, on the fermentation of xylose and glucose to ethanol by the baker's yeast *Saccharomyces cerevisiae* ATCC 24858 and *Escherichia coli* KO11 were investigated. The effects of different biomass species (cattail harvested during different seasons and corn stover) and their chemical composition on the overall process efficiency and economic performance considering feedstock availability and feedstock costs to manufacture ethanol from lignocellulose was studied. The potential environmental impacts including CO_x, SO_2, NO_x emission from the production processes and waste water treatment were also discussed.

2. METHODOLOGY

2.1. Feedstock and Biomass Analytical Procedures

The aerial portions of cattails, *Typha latifolia*, were chopped with pruning shears, dried at room temperature, and ground in a Wiley mill to 1 mm mesh size.

Compositional analysis of biomass was carried out using the laboratory analytical procedures (LAPs) developed by the National Renewable Energy Laboratory. The moisture content of the biomass was determined by the LAP #001 method, and the ash content of the biomass was determined by the LAP #005 method. Structural analyses of the samples were carried out according to the LAP #002 method.

For the techno-economical analysis, cattails harvested during different seasons and corn stover are considered. The composition of cattails and corn stover is listed in Table 2.

2.2. Pretreatment Procedures

The pretreatment conditions were as follows: 4% NaOH at room temperature for 24 hours, 180°C hot-water for 15 minutes, MgCl$_2$ pretreatment with 0.4 M concentration at 180°C for 15 minutes, and dilute sulfuric acid pretreatment with 0.5% concentration at 180°C for 5 minutes. The detailed descriptions of NaOH, hot water, H$_2$SO$_4$ and MgCl$_2$ pretreatment processes are presented elsewhere [4-7].

All experiments and analysis were performed in triplicate. Dry matter recoveries and compositional analyses of solids and liquids after the pretreatment step were used to develop a component balance for the pretreatment processes. The remaining soluble mass in the hydrolysate liquid was determined by difference.

2.3. Bacterial Strains and Fermentation

Saccharomyces cerevisiae ATCC 24858 was the yeast organism used to ferment the enzymatically released glucose. Stock cultures were maintained on YM medium. *Escherichia coli* KO11 was used to ferment the enzymatically released glucose and xylose. Chloramphenicol acyl transferase (cat) and the *Z. mobilis* genes for ethanol production (pdc, adhB) are integrated into the chromosome of this strain. Stock cultures were maintained on modified Luria-Bertani (LB) medium containing (per liter): 5 g NaCl, 5 g yeast extract, 10 g tryptone, 20 g xylose, 15 g agar, and 600 mg chloramphenicol [8].

For ethanol production, 4 mL of *S. cerevisiae* or *E. coli* seed culture were used to inoculate 40 mL YM or LB medium in a 250-mL Erlenmeyer flask. The cultures were incubated in a shaker at 30°C or 37°C and 200 rpm, and grown aerobically overnight. The cells were harvested at room temperature by centrifugation at 2600 RCF for 15 minutes. The supernatant was discarded, and the cells were transferred to 250-mL screw-capped Erlenmeyer flasks containing 100 mL of hydrolysate.

The initial cell mass concentration prior to fermentation in each experiment was 4 to 6 g dry weight/L. The cultures were placed in a shaker and incubated at 30°C or 37°C. Fermentation samples were filtered through 0.2 μm nylon membranes and analyzed by HPLC to determine the presence of ethanol and sugars.

2.4. Method for Process Design and Economic Analysis

The process for techno-economic analysis reported here is based on the benchmark case study developed by NREL [9]. This conceptual process design reports ethanol production economics as determined by 2012 conversion targets (90% conversion at 20 wt % total solids) and "nth-plant" project costs and financing. It mainly consists of nine areas: feedstock storage and handling, pretreatment and hydrolyzate conditioning, on-site enzyme production, separated saccharification and fermentation, product separation and purification, wastewater treatment, product storage, lignin combustion for production of electricity and steam, and all other utilities (Figure 1).

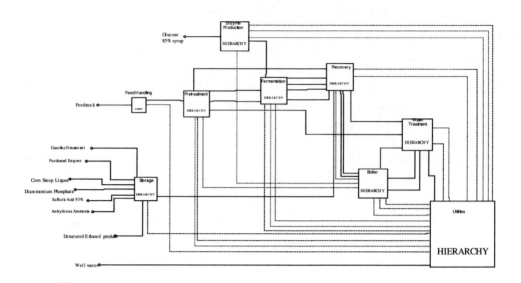

Figure 1. Flow diagram of the overall process for a lignocellulose to ethanol biorefinery.

3. RESULTS AND DISCUSSION

3.1. Fermentation of Pretreated Cattails with *Saccharomyces cerevisiae*

Cattails were pretreated with 4% NaOH for 24 h, with hot water at 180°C for 15 minutes, with 0.4 M MgCl$_2$ at 180°C for 15 minutes, or with 0.5% H$_2$SO$_4$ at 180°C for 5 minutes. All pretreated cattails were washed with deionized water. The pretreated biomass first was enzymatically hydrolyzed for 2 days using a cellulase loading of 15 FPU/g glucan, followed by a 2-day Simultaneous Saccharificiaton and Fermentation (SSF) using *S. cerevisiae* (ATCC 24858).

When diluted pretreated cattails (1 to 2 g glucan/100 mL volume) were used, glucose to ethanol yields were approximately 90%, 91%, 88.7%, and 87% of the theoretical yield for H$_2$SO$_4$, NaOH, hot water, and MgCl$_2$ pretreatments.

3.2. Fermentation of Pretreated Cattails and Cattail Hydrolysates with *E. coli* KO11

A Separate Hydrolysis and Fermentation (SHF) process was used to ferment pretreated cattails to ethanol. First the pretreated cattails were hydrolyzed for two days, and then the resulting hydrolysate was fermented at 37°C for 48 h by *E. coli* KO11. The initial cell mass concentration prior to fermentation in each experiment was 4 to 6 g dry weight/L. The cultures were placed in a shaker and incubated at 37°C for 48 h. Cellulase loading was 15 FPU/glucan. During the SHF processes, the amount of xylose present in the hydrolysate was fermented within 6 hours by *E. coli*. The glucose also was rapidly converted to ethanol. The ethanol reached the highest concentration within 6 hours (Figure 2). The sugars to ethanol yields were approximately 82.6%, 85.9% and 87.5% of the theoretical yield for hot-water, NaOH and MgCl$_2$ pretreatments, respectively.

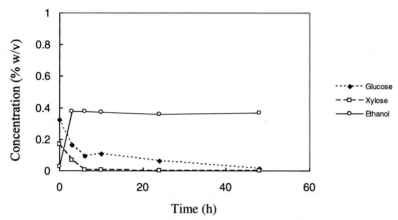

Figure 2. Glucose, xylose and ethanol profiles during the Separate Hydrolysis and Fermentation (SHF) fermentation process of hot water pretreated cattail.

The severe pretreatment could produce inhibitory compounds for enzyme hydrolysis and fermentation processes. To detoxify the cattail hydrolysates from a sulfuric acid pretreatment process, the extracts were over-neutralized by adding lime (Ca(OH)$_2$) until the pH value was 10. Then the pH is adjusted back to 7 by the addition of HCl solution. Before the enzyme hydrolysis step, the pH of detoxified hydrolysates was further adjusted to 5.0 by the addition of sodium citrate buffer. The extracts after detoxification were hydrolyzed using cellulase and hemicellulase, resulting in a slightly increased sugar yield. *E. coli* was pre-cultured in 40 mL LB medium containing 2% xylose, then *E. coli* cells were transferred into the hydrolysates.

Figure 3 shows the glucose, xylose, and ethanol profiles during the fermentation of cattail hydrolysates from the sulfuric acid pretreatment process. The glucose was rapidly converted to ethanol within 10 h of fermentation. The xylose was fermented much more slowly, reaching a minimum fermentable xylose concentration after 24 hours of fermentation. The ethanol yields from this fermentation were approximately 87% of the theoretical yield.

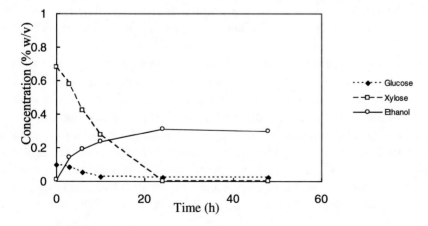

Figure 3. Glucose, xylose, and ethanol profiles of SHF fermentation of cattail hydrolysates from the sulfuric acid pretreatment process.

3.3. Comparison of Four Biomass Pretreatment Methods

For each pretreatment technology, the cellulose to ethanol yields and hemicellulose to ethanol yields are summarized in Table 1. The selection of the fermentation strains needs to match the pretreatment technologies and the feedstock used as well as the process. For example, if a dilute acid pretreatment is used, most of the hemicellulose is degraded, the pretreated biomass can be fermented using baker's yeast *S. cerevisiae*, but pentose sugars rich hydrolysates must be fermented by a microorganism that uses both hexose and pentose sugars. In our calculations, all streams containing both hexose and pentose sugars were assumed to be fermented using *E. coli* KO11. When using a dilute sulfuric acid pretreatment, about 85% of original sugars in the cattails were converted to ethanol. While using other pretreatment technologies, approximately 60% of original sugars were converted to ethanol.

Table 1. Comparison of Four Biomass Pretreatment Methods

Yield (%)	Cellulose to ethanol		Hemicellulose to ethanol		Total sugars to ethanol
	Pretreatment	Fermentation	Pretreatment	Fermentation	
NaOH	77.5	85.9	43.6	85.9	59.2
Hot-water	77.6	85.8	70	82.6	64.2
MgCl$_2$	61.7	87.5	90	87.5	60.2
H$_2$SO$_4$	97.1	90	90	87	85.1

3.4. Process Design Basis

The process for techno-economic analysis reported here is based on NREL's 2011 generic process model [9]. The biorefinery processes 2,205 dry ton/day at 76% theoretical ethanol yield. The rigorous mass and energy balance of the biorefinery process was calculated with Aspen Plus software (version 7.3). It mainly consists of nine areas: feedstock storage and handling, pretreatment and hydrolyzate conditioning, on-site enzyme production, separate saccharification and fermentation, product separation and purification, wastewater treatment, product storage, lignin combustion for production of electricity and steam, and all other utilities (Figure 1).

Cattails harvested in different seasons and corn stover are considered as the feedstock. Cattails are assumed to be planted on the constructed wetlands, which are similar to rice fields. During the harvest season, the wetlands are drained, so a regular stove harvest system could be used to harvest the cattails. The feedstock would be air-dried, milled, and stored in a satellite depot location. The milled feedstock is delivered to the biorefinery occurring six days a week by truck or possibly by rail. Only minimum storage and feed handling are required. The cost of cattails and corn stover assumed is $58.50/dry ton, which is the number used in the NREL's 2011 report. The composition of cattails and corn stover is listed in Table 2. It is assumed, as per the common practice, that cattails and corn stover contain 20% moisture per weight unit.

Table 3. Biomass composition

Component (dry wt %)	Corn Stover	Cattail harvested on			
		12/20/2010 Winter	3/15/2011 Spring	6/28/2011 Summer	10/1/2011 Fall
Glucan	35.05	29.9	31.5	29.5	31.8
Xylan	19.53	13.2	21.1	18.7	21.3
Lignin	15.76	35.6	34.7	34.2	32.2
Ash	4.93	5.1	4.1	7.7	6.8
Acetate	1.81	1.81	1.81	1.81	1.81
Protein	3.1	n.d.	n.d.	n.d.	n.d.
Extractives	14.65	n.d.	n.d.	n.d.	n.d.
Arabinan	2.38	3.4	3.1	3	2.8
Galactan	1.43	-	-	-	-
Mannan	0.6	-	-	-	-
Sucrose	0.77	n.d.	n.d.	n.d.	n.d.

n.d.: not determined

-: not detectable

The feedstock type and composition have significant impacts on the overall process design, for example, the pretreatment reactor. Our previous studies confirmed the technical feasibility of converting cattails to cellulosic ethanol. The results of the dilute acid pretreatment and fermentation are similar to those of corn stover. In the current techno-economical analysis, major process operating conditions for the lignocellulose to ethanol biorefinery are the same as the conditions reported by NREL and given in Table 4 [9]. The plant is assumed to process 2,205 dry U.S. ton feedstock/day, and operate 8,410 operating hours per year (96% uptime). The annual feedstock requirement is 773,000 dry U.S. ton.

3.5. Effects of Biomass Species and Chemical Composition on Process and Economic Analysis

Cattails are assumed to be processed using the same faculties as that described in the NREL's report. In this biorefinery process, the total equipment cost is first computed. Then direct and indirect overhead cost factors (e.g., installation expenses and project contingency) are applied to determine a total capital investment (TCI). The TCI, along with the plant operating costs, is used in a discounted cash flow rate of return (DCFROR) analysis to determine the minimum ethanol selling price (MESP, i.e. plantgate price) for a given discount rate. The minimum ethanol selling price (MESP, $/gallon) is required to obtain a net present value (NPV) of zero for a 10% internal rate of return after taxes.

Table 4. Operating conditions

Feedstock		Enzymatic Hydrolysis	
Moisture Content	20%	Enzyme Loading (mg/g cell)	19.9
Pretreatment		Total Solids (wt%)	20.0%
Acid Concentration (wt%)	0.82	Insoluble Solids (wt%)	10.6%
Acid Loading (mg/g dry biomass)	22.1	Temperature (°C)	32
Total Solids (wt%)	30.0%	Pressure (atm)	1.0
Insoluble Solids in (wt%)	22.4%	Residence Time (days)	3.5
Insoluble Solids out (wt%)	16.6%	**Conversions:**	**Overall**
Temperature (°C)	158	Cellulose to Oligomer	5.0%
Pressure (atm)	5.7	Cellulose to Glucose	90.0%
Conversions	**Overall**	**Fermentation**	
Cellulose to Oligomer	0.3%	Total Solids (wt%)	19.9%
Cellulose to Glucose	9.9%	Insoluble Solids (wt%)	5.2%
Cellulose to HMF	0.3%	Temperature (°C)	32
Xylan to Xylose	90.0%	Pressure (atm)	1.0
Xylan to Furfural	5.0%	Residence Time (days)	1.5
Mannan to Oligomer	2.4%	Conversions:	
Mannan to Mannose	90.0%	Glucose to Ethanol	95.0%
Mannan to HMF	5.0%	Glucose to Zymo (cell mass)	2.0%
Galactan to Oligomer	2.4%	Xylose to Ethanol	85.0%
Galactan to Galactose	90.0%	Xylose to Zymo	2.0%
Galactan to HMF	5.0%	Xylose to Xylitol	5.0%
Arabinan to Arabinose	90.0%	Arabinose to Ethanol	85.0%
Arabinan to Furfural	5.0%	Arabinose to Zymo	1.9%
Acetate to Acetic Acid	100.0%	Arabinose to Glycerol	0.3%
Furfural to Tar	100.0%	Arabinose to Succinic Acid	1.5%
HMF to Tar	100.0%	Corn Steep Liquor loading wt%	0.25%
Lignin to Soluble Lignin	5.0%	Contamination Loss	3.0%
Conditioning		Ethanol Out of Fermenters (wt%)	5.3%
Ammonia Loading (g/L hydrolyzate)	4.7		
Sugar Losses:			
Xylose	1.0%		
Arabinose	1.0%		
Glucose	0.0%		
Galactose	1.0%		
Mannose	1.0%		
Cellobiose	1.0%		

The feedstock can vary in composition and moisture content due to species variety, region, weather, soil type, fertilization practices, harvesting and storage practices, time in storage, and so on. From a sugar content perspective, feedstock composition clearly affects the ethanol yield. Cattails harvested during spring and fall have the highest cellulose and xylan content, resulting in the highest ethanol production and the lowest MESP. When harvesting in summer, cattails are green, and contain more nutrients. Cattails wither between late fall and winter in the field. Cattails harvested during the winter have the lowest cellulose and xylan content, and this may be due to microbial degradation. From the ethanol production point of view, cattails harvested during spring and fall will be the good choice. However, a spring harvest may interrupt the growth of cattails. The fall harvest is preferred.

Table 5. Ethanol Production Process Engineering Analysis and Minimum Ethanol Selling Price (MESP)

	Corn Stover	Cattail harvested on			
		12/20/2010 Winter	3/15/2011 Spring	6/28/2011 Summer	10/1/2011 Fall
Ethanol Yield (gal/dry US ton Feedstock)	78.3	61.5	73.8	67.8	74.0
Ethanol Production (MM gal/yr)	60.5	47.5	57.0	52.4	57.2
MESP ($/gal)	2.16	2.75	2.29	2.49	2.28

3.6. Environmental Impacts

The major waste streams from the lignocellulose to ethanol biorefinery for cattails and corn stover include COx/NOx/SOx/Waste water. Of the carbon inputs to the process, 29% of carbon atoms in corn stove and 18% of carbon atoms in cattails leave as ethanol. The combustor stack and the scrubber vent are two major exits for the rest carbon atoms. Most byproducts of this process, such as lignin, are burned in the combustor to form CO_2. The off gas leaving the scrubber vent is also in the form of CO_2 that is produced during fermentation processes. Cattails contain ~35% lignin, while corn stover has a lignin content of 15.7%. Higher lignin content in cattails results in a CO_2 emission in the range of 2450-2650 Kmol/hr, which is higher than the CO_2 emission from processing corn stover (2110 Kmol/hr). Because lignin is an inexpensive feedstock, it is commonly used by combustion to provide heat and/or power. In general, however, using lignin for recovery of heat and power is not so economical because it can be used as a source of aromatic compounds. Therefore, it is very important to explore emerging technologies such as pyrolysis and gasification to produce multi-products and obtain higher values. For example, liquefaction of lignin extracted from aspen wood resulted in a 90% yield of liquid, representing a significant added value [2].

All sulfur containing streams are fed to the combustor, and the sulfur is converted to sulfur dioxide (SO_2). Since SO_2 concentration in the flue is >1,800 ppmw, the flue gas desulfurization (FGD) is required. Lime (calcium hydroxide) is sprayed into the flue gas as a 20 wt % slurry at 20% stoichiometric excess. FGD process converts 92% of the SO_2 into calcium sulfate, which falls out the bottom of the spray dryer. In addition, 1% of the SO_2 is converted to sulfuric acid. The water in the slurry is vaporized and exits with the flue gas to the baghouse, in which particulate ash is removed. Both ash and calcium sulfate are landfilled. The scrubbed gas is exhausted through a stack.

In the designed process, carbon monoxide (CO) is assumed to be generated at a rate of 0.31 kg/MWh (0.2 lb/MMBtu), and Nitrogen oxide (NOx) is assumed to be generated at 0.31 kg/MWh (0.2 lb/MMBtu). The mechanism of NOx formation is complicated. One study showed that nitrogen source is mainly from air [10], however other factors such as the feedstock composition, combustion temperatures, combustor design, and flue gas cleanup devices like FGD, may also be involved.

A number of wastewater streams including condensed pretreatment flash vapor, boiler blowdown, cooling tower blowdown, and the pressed stillage water streams, are generated by the bioethanol process. In the current design, the wastewater streams are processed by anaerobic digestion and aerobic digestion to digest organic matter in the stream. Anaerobic digestion produces biogases that are rich in methane. Aerobic activated-sludge lagoons produce a relatively clean water stream that can be reused in the process as well as a sludge mainly composing of cell mass. Both biogases and the sludge are fed to the combustor, and burned.

CONCLUSION

Glucose from cattail cellulose was able to be fermented to ethanol using *S. cerevisiae*, resulting in 85-91% of the theoretical ethanol yield. Glucose and xylose released from cattail cellulose and hemicellulose were able to be fermented to ethanol using *E. coli* KO11, resulting in approximately 85% of the theoretical ethanol yield. Among four pretreatment methods, the dilute acid pretreatment was found to be superior, and approximately 85% of original sugars in the cattails were converted to ethanol.

The effects of different biomass species (cattail harvested in different seasons and corn stover) and their chemical composition on the overall process efficiency and economic performance considering feedstock availability and feedstock costs to manufacture ethanol from lignocellulose was studied. Cattails harvested in fall have the highest cellulose and xylan content, leading to the highest ethanol production of 57.2 MM gal/yr and the lowest Minimum Ethanol Selling Price (MESP) of $2.28.

Carbon dioxide emission from processing cattails is in the range of 2450-2650 Kmol/hr, which is higher than the CO_2 emission from corn stove (~2110 Kmol/hr). Because of high lignin content in cattails, it is very important to explore emerging technologies such as pyrolysis and gasification to produce multi-products and obtain higher values from cattail lignin. The potential environmental impacts including CO, SO_2, NOx emission from the biorefinery process and waste water treatment were also discussed.

ACKNOWLEDGMENTS

This research was supported by a grant from the USDA-NIFA Evans-Allen program.

REFERENCES

[1]   Huang H, Ramaswamy S, Tschirner U, Ramarao B. A review of separation technologies in current and future biorefineries. *Separation and Purification Technology* 2008;62:1.

[2]   Zhang B, von Keitz M, Valentas K. Maximizing the liquid fuel yield in a biorefining process. *Biotechnology and Bioengineering* 2008;101:903.

[3]   Zhang B, Shahbazi A. Recent Developments in Pretreatment Technologies for Production of Lignocellulosic Biofuels. *J Pet Environ Biotechnol* 2011;2:108.

[4]   Zhang B, Shahbazi A, Wang L, Diallo O, Whitmore A. Alkali Pretreatment and Enzymatic Hydrolysis of Cattails from Constructed Wetlands. *Am J Eng Applied Sci* 2010;3:328.

[5]   Zhang B, Shahbazi A, Wang L, Diallo O, Whitmore A. Hot-water pretreatment of cattails for extraction of cellulose. *Journal of Industrial Microbiology & Biotechnology* 2011;38:819.

[6]   Zhang B, Wang L, Shahbazi A, Diallo O, Whitmore A. Dilute-sulfuric acid pretreatment of cattails for cellulose conversion. *Bioresource Technology* 2011;102:9308.

[7]   Zhang B, Shahbazi A, Wang L, Whitmore A, Riddick BA. Fermentation of glucose and xylose in cattail processed by different pretreatment technologies. BioResouces 2012;7:2848.

[8]   Moniruzzaman M, York SW, Ingram LO. Effects of process errors on the production of ethanol by Escherichia coli KO11. *Journal of Industrial Microbiology & Biotechnology* 1998;20:281.

[9]   Humbird D, Davis R, Tao L, Kinchin C, Hsu D, Aden A, et al. Process Design and Economics for Biochemical Conversion of Lignocellulosic Biomass to Ethanol: Dilute-Acid Pretreatment and Enzymatic Hydrolysis of Corn Stover. Technical Report: NREL/TP-5100-47764. Available at: www.nrel.gov/docs/fy11osti/47764.pdf. Accessed 05 June 2012.; 2011.

[10]   Huang H-J, Ramaswamy S, Al-Dajani W, Tschirner U, Cairncross RA. Effect of biomass species and plant size on cellulosic ethanol: A comparative process and economic analysis. *Biomass and Bioenergy* 2009;33:234.

In: Biomass Processing, Conversion and Biorefinery
Editors: Bo Zhang and Yong Wang

ISBN: 978-1-62618-346-9
© 2013 Nova Science Publishers, Inc.

Chapter 24

GREEN BIOREFINING OF GREEN BIOMASS

Shuangning Xiu* and Abolghasem Shahbazi

Biological Engineering Program, Department of Natural
Resources and Environmental Design, North Carolina
A and T State University, Greensboro, NC, US

ABSTRACT

A green refinery is an integrated concept to utilize green biomass as an abundant and versatile raw material for the manufacturing of industrial products. It represents an innovative approach to alternative application of surplus grassland biomass. An overview of the main aspects, activity and processing technologies are discussed in this chapter. Recent developments on the green biorefinery in both Europe and North America are also discussed. A focus for future R&D work in that field is recommended.

1. INTRODUCTION

Non-renewable fossil fuels, coal, oil and natural gas are being consumed worldwide at a very fast pace [1]. The United States consumes more than 25% of international oil production but possesses just 1.6% of its oil reserves [2]. If not replaced by alternative energy sources, fossil fuels will eventually be depleted while CO_2 in the atmosphere will rise to levels with dangerous consequences. Fossil fuels are the major source of environmental pollutants, greenhouse gases and ocean acidification. Sustainable development requires the use of renewable resources as alternative feedstocks for chemical, fuel and material production.

Biomass presents an attractive source for the production of fuels and chemicals due to its versatility, renewable nature and low environmental impacts. Considerable attention has been given to lignocellulosic biomass such as agri-residues and energy crops for biofuel production. DOE and USDA projected that the US biomass resources could provide

* Phone 336-334-7787. Fax 336-334-7270. E-mail: xiu@ ncat.edu.

approximately 1.3 billion dry tons of feedstock for biofuels, which would meet about 40% of the annual US fuel demand for transportation [3].

The overall goal is to replace 20% of the United States transportation fuel imports (35 billion barrels per year) by 2017 [4]. Obviously, a significant need exists for exploring its efficient utilization.

There has been an increasing interest in the use of perennial grasses as energy crops in the US and Europe since the mid-1980s. Over time, various conversion technologies have been used to produce biofuels from perennial grass. Treatment can be subdivided into two categories: thermochemical treatment (e.g., pyrolysis, gasification, hydrothermal liquefaction and combustion) and biological treatment (e.g., fermentation, anaerobic digestion). Water present in green biomass poses negative effect on the thermochemical treatment, as it requires high heat of vaporization. In general, drying the "nature-wet" raw material is needed for the thermochemical treatment, which limits the options for green biomass as feedstock and overall process economy.

Research also found that biomass from semi-natural grasslands is difficult to exploit in conventional bioenergy-converting systems, as the chemical composition is detrimental for both conventional anaerobic digestion as a results of high fibre concentrations [5-6] and for cellulosic bioethanol fermentation due to its resistance to enzymatic attack. Cellulosic biomass must be pretreated before it can be enzymatically hydrolyzed [7].

Despite advances in technology, the ability to produce biofuel as a single source of revenue remains infeasible [8] and exhibits a potential for further cost reductions. One possibility for ameliorating this problem is to adopt a biorefinery approach which diversifies the output streams of a biofuel facility by generating diverse co-products along side biofuel for increased revenue generation [7].

Analogous to the petroleum refinery that processes a barrel of crude oil into many petroleum based products including gasoline, diesel fuels and petroleum based chemicals, the biorefinery converts biomass into multiple biofuels and bioproducts. The ability to process raw materials into a varied product stream greatly enhances the versatility of the biorefinery to meet the economic demands of the free market and compete globally. This pioneering concept was first examined for several feedstocks, including alfalfa and Bermuda grass for biofuel and protein recovery applications [9-11].

The technology concept of green biorefinery represents an innovative approach to alternative applications of green biomass. This article comprehensively reviews the state of the art, the feedstocks, and the processing technologies that are used to produce multiproduct in the green biorefinery.

Besides, research development status in both Europe and North America are also discussed. It is also recommended future RandD work in the green biorefinery field.

2. GREEN BIOREFINERY

2.1. Concept of Green Biorefinery

Green biorefineries are also multiproduct systems, which utilize green biomass as an abundant and versatile raw material for the manufacture of industrial products. The concept is currently in an advanced stage of development in several European countries, especially Germany, Denmark, Switzerland, the Netherlands and Austria.

The basic idea of this concept is to utilize the whole green biomass like grass, alfalfa and various other sources and to generate a variety of products that are either valuable products themselves or form the basis for further production lines. Besides biobased materials, fuels and energy may be supplies by this technology. The feasibility of a biorefinery based on grass has been successfully demonstrated by Grass and Hansen [12].

In green biorefinery, careful wet or green fractionation technology is used as the first step (primary refinery) to isolate the green biomass substances in their natural form. Thus, green biomass are separated into a fiber-rich press cake and a nutrient-rich press juice (Figure1). Both fractions have an economic value. Figure1 illustrates the array of potential products of a green biorefinery that can be generated by the downstream processing of press juice and press cake (fiber fraction).

2.2. Feedstocks for Green Biorefinery

Currently, the feedstocks used for green biorefinery are mainly green grass, for example, grass from cultivation of permanent grass lands, closure fields, nature preserves, or green crops such as lucerne, clover, and immature cereals from extensive land cultivation [13]. The interesting valuable components of fresh biomass are proteins, soluble sugars, and the fiber fraction (cellulose, hemicelluloses, and lignin part).

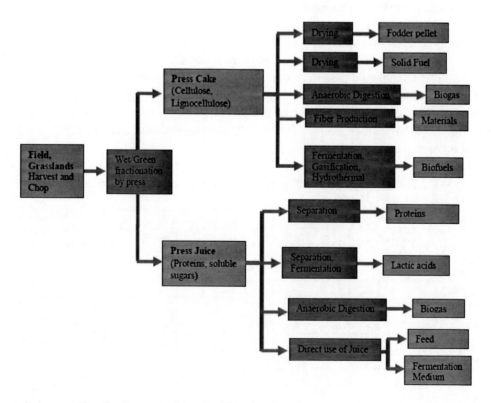

Figure 1. A green biorefinery system for green biomass utilization.

Table 1 shows the chemical composition of representative types of green biomass.In general, biomass is typically composed of 75-90% of sugar polymers, with the other 10-25% of biomass principally being lignin [16]. For herbaceous grass plant, typical contents in the dry mater are in the range of 6-15% for protein, 20-55% water soluble extracts containing 5-16% soluble sugars and for the raw fiber part (cellulose 20-30%, hemicelluloses 15-25%, lignin 3-10%) [18].

In the past, most of attention has focused on the cellulose, hemicelluloses, and lignin present in biomass materials. However, there are also considerable amounts of protein available in these biomass materials as Table 1 indicated. For the biomass species listed, an aquatic plant has higher protein content than other biomass resources. *Spirulina* algae have very high protein content around 60%. Since a kilogram of protein is generally much more valuable than an equal weight of carbohydrate, aquatic plant can be used as a good candidate for green biorefinery in respect of high value protein recovery.

Microalgae are viewed as next generation biofuel feedstocks because of their superior photosynthetic efficiencies and higher carbon capturing capabilities compared to terrestrial plants. Microalgae are usually composed of lipids, proteins, nucleic acids and no-cellulosic carbohydrates.

Conversional algae-to-biodiesel technology requires drying of the algal biomass followed by solvent and or/mechanical extraction. The energy consumed in drying process is more than 75% of the total energy consumption [19]. In the green refinery, the wet fraction of algae can be employed to reduce the need for energy-intensive drying operations and to utilize nutrient-rich green juice for value-added co-product developments (e.g., protein) that would otherwise be destroyed in downstream processing.

Besides algae, energy crops such as switchgrass and grant miscanthus can be good candidates as feedstocks for green biorefinery due to the high yields of biomass per hectare and year, low fertilization and pesticide requirements, broad adaptability, and greater ability to sequester carbon in the oil than most other grasses. Switchgrass (*Panicum virgatum L.*) is a perennial native grass adapted to the prairies of North America. The DOE identified switchgrass as a primary species for development as an energy crop. It has the ability to grow well on marginal croplands without heavy fertilization or intensive managements and has a potential for high fuel yield [20].

Table 1. Composition of representative biomass resources

Biomass type	Aquatic plant		Herbaceous		Woody	
Name	Spirulina algae	Duckweed	Bermuda grass	Switch grass	Poplar	Pine
Component (dry wt.%)						
Celluloses	<1	11.9	31.7	37	41.3	40.4
Hemicellulose	1	13.8	40.2	29	32.9	24.9
Lignin	<1	3.2	4.1	19	25.5	34.5
Crude protein	64	35.1	12.3	3	1.7	0.7
Crude lipid	5	5	11.9	--	--	--
Ash	11	16.5	5	6	0.8	0.5
Total	100	92.4	93.3	94	102.9	101.0
Reference	[14]	[15]	[16]	[17]	[16]	[16]

Grant Miscanthus (*Miscanthus x giganteus*) is a tall perennial grass that reproduces by underground rhizomes and has been evaluated in Europe during the past 10 years as a bioenergy crop. It is characterized as having broad adaptability, high water and fertilizer use efficiency, excellent pest resistance, and tremendous biomass production. Tonnages of between 6-20 tonnes/acre are anticipated from miscanthus. Miscanthus stems may be used as fuel for the production of heat and electric power and as a feedstock for the production of cellulosic ethanol. In general, the advantages of using energy crop as feedstock is its high biomass profit per hectare and a good coupling with national and state's priority to use energy crop as second generation biofuel feedstock.

3. CURRENT PROCESSING TECHNOLOGIES FOR GREEN BIOREFINERY

3.1. Primary Separation

The initial fractionation of the green biomass remains an essential operation for green biorefinery process. Green crop fractionation is now studied in about 80 countries [21]. Water contents in tropical grasses can be as high as 80-85% or similarly ~90% for aquatic species. Those "nature-wet" raw materials require a fast primary processing or the use of preservation methods, like silage or drying. But due to the energy consumption of thermal dryer, the biomass is seldom transformed after drying, so that mechanical fractionation of the web material is usually the first unit operation in green biorefinery plant. Machinery, screw express has been primarily used to press the green juice out of the green biomass. For vegetative biomass like alfalfa, clover and grass, screw presses remove approximately 55-60% of the inherent liquid [22]. Nevertheless other means of preprocessing has also been applied, such as thermal mechanical dewatering method, simultaneous application of a pulsed electric field, and superimposition of ultrasounds [23]. Wachendorf et al. [24] investigated the hydrothermal conditioning at different temperatures (5, 60 and 80 °C) and mechanical dehydration on mass flows of plant compounds into the press fluid for five grassland pastures typical of mountain areas of Germany. The underlying objectives are to obtain a juice in larger quantity and of better quality.

3.2. Processing of Press Juice to Value-Add Products

The green juice contains proteins, free amino acids, organic acids, dyes, enzymes, hormones, other organic substances and minerals [13]. The green juice is a raw material for high quality fodder proteins, cosmetic proteins, human nutrition or platform chemicals like lactic acid and lysine or can be used as substrate for bio-gas production [22]. The main focus is directed to products such as lactic acid and corresponding derivatives, amino acids, ethanol, and proteins.

The best known products which can be obtained from grassland biomass are proteins. Proteins are essential element of the nutrition for humans and animals. They also are important raw materials for adhesives and for the pharmaceutical and cosmetics industries. Protein fractions from fresh alfalfa could be isolated already in 1980s' [25, 9-10].

High-quality fodder proteins and proteins for cosmetic industry can be produced through the fractionation of the green juice proteins in different separation and drying process. The fodder proteins could be a complete substitute for soy proteins. They even have a nutritionally physiologic advantage due to their special amino acid pattern. The general approach to leaf protein production is to harvest fresh green plant material, grind the plant tissue, squeeze out a protein rich juice and then heat the juice to precipitate and recover the protein [26].

Recently, L-lactic acid received increased attention as a feedstock monomer for a biodegradable polymer, polylactide (PLA). Lei et al. have demonstrated that, after separation of proteins from alfalfa press juice, the supernatant can be used as fermentation media for the production of the ammonium lactate, L-lysine-L-Lactate, which acts as intermediates for the production of lactic acid sequence products like lactide [27]. Danner et al. evaluated the possibility of recovery of lactic acid from various grass silages. The crude liquid press extract was pre-treated with ultrafiltration membranes followed by purification with mono-polar electrodialysis [28]. Thang and Novalin [29] separated lactic acid from grass silage juice by chromatography using neutral polymeric resin.

Alternatively, press juice also can be directly used as fermentation media for organic acids production. Andersen and Kiel [30] carried out the fermentation experiments of brown juice with different lactic acid bacteria. Their result shows that juices from grass, clover and alfalfa can easily be converted to a stable universal fermentation media by adding more carbohydrates or for production of other organic acids or amino acids in the second stage fermentation. Their research presents a novel way of simultaneous preservation and utilization of plant juice for fermentation purposes.

3.3. Processing of Press Cake to Value-Add Products

The press cake can be used for the production of green feed pellets/fodder pellet, as a raw material for the production of chemicals, such as levulinic acid, and for the conversion to syngas and hydrocarbons (synthetic biofuels). Richter et al. [31] evaluated the properties of the press cakes as solid fuel, which was derived from five species-rich, semi-natural grasslands via thermal mechanical dewatering method.

Results show that compared with the grassland silages used as parent materials, the press cakes showed significantly lower concentrations of elements detrimental for combustion. The ash softening temperature of the press cakes increased significantly, which is comparable to beech wood. Overall, the solid fuel from press cakes is of superior quality compared with conventional hay. However, the increase in the higher heating value of these press cakes was not statistically significant.

The main focus is directed to fodder pellets and biogas production. The residues of substantial conversion are suitable for the production of biogas, combined with the generation of heat and electricity. In Europe, the green crop drying industry produces fodder pellets by drying crops such as perennial ryegrass, Italian ryegrass, clover grass and alfalfa [32-33]. However, this drying is energetically very expensive. From an energetic point of view, it is far better to use the press cake as a silage feed and/or bioenergy.

Since the press cake is rich in fiber, researchers have also commented on the potential use press cake for fibre applications (e.g., insulation materials, fibre boards, horticultural substrates, and pulpandpaper) [18].

However, little or no attention has been devoted to removing and upgrading the carbohydrate fraction of press cake for fibre applications. Therefore, there is a need to develop and demonstrate technology for utilizing green biomass for various fiber applications.

Biomass conversion to fuels and chemicals is receiving a great deal of attention. The need for a billion tonnes of biomass to produce enough biofuel to replace 30% of US petroleum consumption has been reported by the US Department of Energy (DOE) [20]. The integrated green biorefinery system will certainly enhance the economics of biofuel production by generating value-added co-product from biomass. The research contributed to this effort has been primarily focused on protein recovery in bioethanol production [9, 34-35]. Dale has summarized various ways in which protein might be coproduced with fuels and other chemicals from biomass, using alfalfa as the example crop [9].

Besides fuel ethanol production, other biofuels and chemicals can be produced from press cake as well. The thermochemical conversion process is another promising method of converting press cake for energy purposes. The syngas produced can be converted to fuel alcohols via the Fischer-Tropsch process [36].

4. Research Development Status in Green Biorefinery

4.1. Research Status of Europe

The most notable successes in research and development in the field of green biorefinery system research have been in several European countries. The pilot scale processing trials in Austria (Utzenaich), Denmark (Esbjerg), Switzerland (Obre), and Germany (Brandenburg) have produced lactic acid, amino acids, leaf juice protein concentrate, fibre products and biogas by setting up integrated refineries [18]. In 2001 the first Green Biorefinery started operation in Switzerland with a design load of 5000 tonnes dry matter of grass per year and a combined output of fibres (0.4 tonnes per ton input), protein (160 t/t) and bioenergy (500kWh/t) [37]. A green biorefinery demonstration plant in Havelland (Austria) is under construction, which will have an annual capacity of 20, 000 tones alfalfa and grass biomass and can be diversified in modules for the production of platform chemicals and synthesis gas. The demonstration facility will directly link to the existing green crop drying plant [38].

Since 1999, Dr. Kromus's research group in Australia has started to develop the Austrian Green biorefinery and the results lead to the installation of a basic pilot plant in 2004 [39]. A key element in the Austrian concept is the utilization of fermented grass (Grass Silage) instead of fresh biomass.

However, the majority of the biorefineries in Europe are based on wet fractionation process to isolate the substrate fractions in their natural form (i.e. pressed fibre and nutrient-rich juice from fresh grass or silage) and to develop platform chemicals from the green/brown juice.

The press cake has been primarily used for production of fodder pellet and biogas. Little or no attention has yet been focused on the utilization of the press fiber for fuel purpose, especially as raw materials for biofuel production.

4.2. Research Status of North America

The research development in green biorefinery in North American is not as significant as in Europe. The research has been primarily focus on protein recovery in ethanol production. Various groups in North American have been examining various grasses for their potential as feedstocks for biofuel production. Haque et al. [40] in Oklahoma evaluated productivity of four grasses (Bermuda, faccid, love and switch) and reported that switchgrass is the most viable species for biofuel production. Woodward [20] evaluated the potential for alfalfa, switchgrass, and miscanthus as biofuel corps in Washington State. Research group in Nebraska also found that switchgrass-derived ethanol produced 540% more energy than was required to manufacture the fuel. One ace (0.4 ha) of grassland could, on average, deliver 320 gallons of bioethanol [41]. Another research group in Hawaii has developed an innovative biorefinery approach to green processing of tropical banagrass into biofuel and biobased products [42]. Kammes et al. [43] has extracted leaf protein form orchard grass and switchgrass, which is indented to use as a source of protein for livestock.

Table 2 summarized various green biorefinery developments through the world. As shown in table 2, the most promising products which have been identified from green juice are 1) proteins: as feed or food (low price), hydrolyzed as amino acids for cosmetics or pharmaceutical industry, and 2) lactic acid: for neutralization/buffering, as solvent (ethyl lactate), bioplastics (PLA=polylactide).

For example, fresh green biomass with medium to high protein content-up to 20% crude protein in dry matter-(clover/Grass, Ryegrass, Permanent Pasture, Alfalfa and Cocksfood) have been investigated by Kromus [39] in Austria. Kammes et al. [35] extracted the grass leaf protein from orchard-grass and switchgrass, as coproduct of cellulosic ethanol production. They found that the leaf protein from orchard-grass has potential as a protein supplement for livestock.

Table 2. Green biorefinery of different green biomass in literature

| Grass Type | Main products | | Country | Reference |
	Press Juice	Press Cake		
Alfalfa, mixed grass	Protein, fermentation media	Animal feed, biogas	Germany	[38] [31]
Grass silage (red clover and ryegrass)	Protein Lactic acid	fibres	Austria	[28]
Alfalfa	protein	ethanol	US	[9]
Tropical banagrass	Edible fungus for aquaculture feed supplement	ethanol	US	[34]
Orchard grass and switchgrass	Protein as animal feed	ethanol	US	[35]
Grass, clover and alfalfa	Fermentation media	Fodder pellets for animal feed	Denmark	[30]
silage	Protein as animal feed	Fibre, biogas	Switzerland	[37]
Switch grass	protein	n/a	US	[42]

To date, the economic viability of green biomass fractionation press, for instance the ProXan-Process [43] used to produce a marketed leaf protein concentrate, depends, to a large extent, on the utilization of the solid fraction as high quality ruminant feed [44]. The production of only one product such as, for example, a protein product would not create sufficient revenue to cover the cost of the raw material and the related processing costs.

In the area of the press cake utilization, the research interest is directed to fodder pellets and biogas production in Europe and ethanol production in US, respectively (Table 2). The alternative use of grass press cake as a potential raw material for second-generation biofuels has not been worked on in detail so far but it is indeed worthwhile exploring. Co-generation of biofuel from the green biorefinery would certainly "upgrade" the green biorefinery approach in the minds of strategists and decision-makers. Besides biofuel, the press cake also can be as feedstock for generation of bioproducts, such as lactic acid (which is a base chemical for many applications, from food additive to solvents to plastics). Therefore, there is a need to develop and evaluate the state-of-art biomass conversion technologies for generating biofuels and bioproducts from grass press cake.

CONCLUSION

A green biorefinery offers a solution to generate a variety of products and energy without directly interfering in the food supply chain by utilizing surplus resources. Some RandD work has been done during the last few years and technologies have been pushed forward to demonstration plants in several European countries. However, there is no real impact from an industrial perspective so far. Some problems affecting its industrial utilization and recommendations are described as follows:

1 Feedstock selection and evaluation. Currently, the feedstocks used for green biorefinery are mainly green grass and/or green crops. Better management of species and varieties including the use of second generation biofuel feedstocks (e.g., microalgae) could reduce the associated carbon footprint and improve the production efficiency of the green biorefinery.

2 The development of technologies for the production of biofuels and bioproducts from press cake. In the area of the press cake utilization, the research interest is directed to fodder pellets and biogas production in Europe and ethanol production in US, respectively. Little attention has been devoted to the alternative use of press cake as a raw material for second-generation biofuels and bioproducts production. Therefore, there is a growing demand for developing and evaluating the state-of-art biomass conversion technologies for generating biofuels and bioproducts from grass press cake. Thus, the overall economic efficiency of a green biorefinery will improve.

3 Development of industrial green biorefinery technologies. Over the years, different green biorefinery schemes have been proposed and investigated. However, none of them has commercialized yet. Research should focus on the further development of alternative products and pave the way for the implementation of a green biorefinery.

4 Evaluation of the overall sustainability of process technologies. Evaluate these aspects of sustainability considering costs and emissions involved with input

production, feedstock production, feedstock harvest, transport, storage, pre processing, processing, product transport to consumption and consumption.

REFERENCES

[1]   Shafiee, S., Topal, E. When will fossil fuel reserves be diminished? *Energy Policy* 2009, 37 (1), 181-189.

[2]   Department of Energy. US Crude Oil, Natural Gas, and Natural Gas Liquids Reserves, 2007 Annual Report, DOE/EIA-0216.

[3]   Perlack, R. D., L. L. Wright, A. F. Turhollow, R. L. Graham, B. J. Stokes, and D. C. Erbach. Biomass as feedstock for a bioenergy and bioproducts industry: The technical feasibility of a billion-ton annual supply. Sponsored by USDOE and USDA,2005.

[4]   Biomass Research and Development Board (BRDB). National Biofuels Action Plan 2008: United States Department of Energy (USDOE). Available at: <http://www1. eere.energy.gov/biomass/pdfs/nbap.pdf>.

[5]   Prochnow, A., Heiermann, M., Drenckhan, A., and Schelle, H. Seasonal patten of biomethanisation of grass from landscape management. In: *Agricultural Engineering International; The CIGR E-journal, vol. VII, Manscript ee 05 011.* Available at: http:// cigrjournal.org/index.php/Ejournal.

[6]   Richter, F., R. Gra, T. Fricke, W. Zerr, M. Wachendorf. Utilization of seminatural grassland through integrated generation of solid fuel and biogas from biomass. Part II. Effects of hydrothermal conditioning and mechanical dehydration on anaerobic digestion of press fluids. *Grass and Forage Science* 2009, 64, 354–363.

[7]   Takara, D., Khanal, S. K. Green processing of tropical banagrass into biofuel and biobased products: An innovative biorefinery approach. *Bioresource Technology* 2011, 102, 1587-1592.

[8]   Biofuels in the US Transportation Sector. US Energy Information Administration 2007. Available at: www.eia.doe.gov/oiaf/analysispaper/biomass.html

[9]   Dale, B. E. Biomass refining: protein and ethanol from alfalfa. *Ind. Eng. Chem. Prod. Res. Dev.* 1983, 22, 466–472.

[10]   Dale, B. E., Matsuoka, M. Protein recovery from leafy crop residues during biomass refining. *Biotechnol. Bioeng.* 1981, 23, 1417–1420.

[11]   Buentello, J. A. and Gatlin, D. M. Evaluation of coastal Bermuda gress protein isolate as a substitute for fishmeal in practical diets for channel catfish Ictalurus punctatus. *Journal of the world aquaculture society* 1997, 28(1), 52-61.

[12]   Grass, S., Hansen, G. Production of ethanol or biogas, protein concentrated and technical fibers from grass/clover. In: *Proceedings of the Second International Symposium.* The Green Biorefinery, Feldbach, Austria, 13-14 October, 1999.

[13]   Kamm, B., Kamm, M. International biorefinery systems. *Pure Appl. Chem.* 2007, 79 (11), 1983-1997.

[14]   Vardon, D. R., Sharma, B. K., Blazina, G. V., Rajagopalan, K., and Strathmann, T. J. Thermochemical conversion of raw and defaltted algal biomass via hydrothermal liquefaction and slow pyrolysis. *Bioresource Technology* 2012, 109:178-187.

[15] Xiu, S., Shahbazi, A., Croonenberghs, J., Wang, L. Oil production from duckweed by thermochemical liquefaction. *Energy Sources, Part A: Recovery, Utilization, and Environmental Effects* 2010, 32(14):1293-1300.

[16] Huber, G. W. and J. A. Dumesic. An overview of aqueous-phase catalytic processes for production of hydrogen and alkanes in a biorefinery. *Catalysis Today* 2006, 111(1-2), 119-132.

[17] US Department of Energy. Biomass feedstock composition and property database. Department of Energy, Biomass Program, 2006. Available at : http://www.eere.energy. gov/biomass/progs/search1.cgi.

[18] Mandl, M. Status of green biorefinery in Europe. *Biofuels Bioprod. Biorefin.* 2010, 4, 268-274.

[19] Lardon, L., A. Helias, B. Sialve, J. P. Stayer and O. Bernard. Life-Cycle Assessment of Biodiesel Production from Microalgae. *Environ. Sci. Technol.* 2009, 43, 6475-6481.

[20] Woodward, W. T. W. The potential for Alfalfa, Switchgrass and Miscanthus as Biofuel Crops in Washington. In: *2008 Proceedings Washington State Hay Grower Association Annual Conference and Trade Show*, January 16-17, 2008. CBCAG802.

[21] Singh, N. In: *Green vegetation Fractionation Technology*, Science Publishing 1996, Lehbanon, NH.

[22] Kamm B., P. Schönicke, M. Kamm. Biorefining of green biomass – technical an energetic considerations, *Clean* 2009, 37, 27–30.

[23] Arlabosse, P., Blanc, M., Kerfai, S., and Fernandez. Production of green juice with an intensive thermo-mechanical fractionation process. Part I: Effects of processing conditions on the dewatering kinetics. *Chemical Engineering Journal* 2011, 168, 586-592.

[24] Wachendorf, M., Richter, F., Fricke, T., Graβ, R., and Neff, R. Utilization of semi-natural grassland through integrated generation of solid fuel and biogas from biomass. I. Effects of hydrothermal conditioning and mechanical dehydration on mass flows of organic and mineral plant compounds, and nutrient balances. *Grass and Forage Science*, 2009, 64, 132-143.

[25] Pirie, N. W. Leaf protein after forty year. *BioEssays*, 1986, 5,174–175.

[26] Pirie, N. W. Leaf protein: Agronomy, Quality, Preparation and Use. *International Biological Program Handbook* No.20, 1971, Blackwell, Oxford.

[27] Leiß, S., J. Venus, B. Kamm. Fermentative Production of *L*-Lysine-*L*-lactate with Fractionated Press Juices from the Green Biorefinery. *Chemical Engineering and Technology* 2010, 33(12), 2102-2105.

[28] Danner, H., Madzingaidzo, L., M. Holzer, L., Mayrhuber, and Braun. Extraction and purification of lactic acid from silages. *Bioresource Technology* 2000, 75, 181-187.

[29] Thang, V. H. and Novalin, S. Green biorefinery: Separation of lactic acid from grass silage juice by chromatography using neutral polymeric resin. *Bioresource Technology* 2008, 99, 4368-4379.

[30] Anderson, M., Kiel, P. Integrated utilization of green biomass in the green biorefinery. *Ind. Crops Prod.* 2000, 11, 129-137.

[31] Richter, F., R. Gra, T. Fricke, W. Zerr, M. Wachendorf. Utilization of seminatural grassland through integrated generation of solid fuel and biogas from biomass. Part II. Effects of hydrothermal conditioning and mechanical dehydration on anaerobic digestion of press fluids. *Grass and Forage Science* 2009, 64, 354–363.

[32] Walker, H. G., G. O. Kohler, W. N. Garrett. Comparative feeding value of alfalfa press cake residues after mechanical extraction of protein. *Journal of Animal Science* 1982, 55, 498–504.

[33] Thomsen, M. H. Complex media from processing of agricultural crops from microbial fermentation. *Appl. Microbiol. Biotechnol.*, 2005, 68, 598-606.

[34] Takara, D., Khanal, S. K. Green processing of tropical banagrass into biofuel and biobased products: An innovative biorefinery approach. *Bioresource Technology* 2011, 102, 1587-1592.

[35] Kammes, K. L., Bals, B. D., Dale, B. E., M. S. Allen. Grass leaf protein, a coproduct of cellulosic ethanol production, as a source of protein for livestock. *Animal Feed Science and Technology* 2011, 164:79-88.

[36] Boateng, A. A., Jung, H. G., Adler, P. R. Pyrolysis of energy crops including alfalfa stems, reed canarygrass, and eastern gamagrass. *Fuel* 2006, 85(17-18):2450-2457.

[37] Baier, U., Grass, S. Bioraffination of Grass, Anaerobic Digestion 2001 – 9th World Congress for Anaerobic Conversion for Sustainability, Antwerpen 2001.

[38] Kamm, B., Hille, C., Schonicke, P., Teltow, E. V. Green biorefinery demonstration plant in Havelland (Germany). *Biofuels, Bioprod. Bioref.* 2010, 4, 253-262.

[39] Kromus, S., B. Wachter, W. Koschuh, M. Mandl, C. Krotscheck, M. Narodoslawsky. The green biorefinery Austria – development of an integrated system for green biomass utilization, *Chemical and Biochemical Engineering Quarterly* 2004, 18, 7–12.

[40] Haque, M., Epplin, F. M., Taliaferro, C. M. Nitrogen and harvest frequency effect on yield and cost for four perennial grasses. *Agric. J.* 2009, 101, 1463-1469.

[41] Schemer, M. R., Vogel, K. P., Michell, R. B., Perrin, R. K. Net energy of cellulosic ethanol from switchgrass. *Proc. Natl. Acad. Sci.,* 2008, 105, 464-469.

[42] Bals, B. Teachworth, L., Dale, B., Balan,V. Extraction of proteins from switchgarss using aqueous ammonia within an intergrated biorefinery. *Applied Biochemistry and Biotechnology* 2007,143, 187-198.

[43] Edwards, R. H., De Fremery, D., Miller, R., Kohler, G. O. Pilot plant production of alfalfa leaf protein concentrate. *Chem. Eng. Prog. Symp. Ser.* 1978, 24(172), 158-166.

[44] Sinclair, S. Protein extraction from pasture. *Literature review Part A: The plant fractionation bio-porcess and adaptability to farming systems*, 2009. Available at: www.maf.govt.nz/sff/about-projects/search/C08-001/literature review.pdf.

INDEX

#

1,3 propanediol, 302
1,4-dioxane, 160, 161, 162
3-hydroxypropionate, 309, 315, 317, 324

A

Abengoa Bioenergy, 336, 344
abrasion resistance, 131
acetic acid, 24, 144, 145, 155, 194, 195, 198, 202, 204, 206, 208, 210, 211, 218, 219, 223, 229, 247, 296, 306, 340, 356, 357, 359, 370, 393
acetone, 15, 17, 24, 141, 145, 155, 160, 162, 199, 223, 294, 295, 296, 299, 333, 357
acetyl-CoA carboxylase, 307, 308, 311
acid value, 147, 258, 268
acid-catalyzed transesterification, 256
Acinetobacter calcoaceticus, 308
activated sludge, 281
acyl-CoA chain elongation, 313
adsorption, 24, 83, 229, 304, 356, 364, 372, 373, 380
AGCO, 106, 109
ageing, 141, 149, 200
agricultural residue, 6, 8, 13, 17, 28, 34, 38, 39, 63, 65, 66, 67, 71, 83, 96, 125, 135, 171, 337, 348, 352, 354, 355, 356, 359, 360, 361, 386
air-lift reactor, 371
Alaskan frontier, 5
Alaskan Natural Wildlife Reserve, 66
Alberta, 65
alcohol, 16, 23, 28, 29, 33, 34, 168, 177, 197, 198, 200, 208, 217, 229, 232, 237, 240, 241, 248, 251, 255, 256, 259, 260, 264, 284, 290, 291, 292, 296, 309, 312, 313, 314, 322, 323, 326, 351, 357, 371, 383
aldehydes, 143, 155, 198, 202, 205, 225, 227, 229, 308, 326

alfalfa, 8, 38, 87, 96, 361, 436, 437, 439, 440, 441, 442, 444, 446
algae, 5, 6, 8, 15, 17, 24, 26, 31, 63, 153, 154, 160, 162, 264, 277, 281, 285, 287, 320, 438
alkali, 34, 35, 81, 95, 154, 155, 157, 158, 160, 162, 178, 209, 228, 242, 256, 258, 299, 329, 331, 332, 340, 342, 355, 356, 362
alkali metal oxide, 209
alkali-catalyzed transesterification, 256
alkaline, 17, 24, 28, 35, 79, 81, 82, 83, 91, 147, 155, 178, 258, 297, 299, 331, 332, 339, 340, 342, 355, 360, 374, 376, 377
alkaline methanolysis, 258
alkane, 233, 241, 308
alkenes, 308, 309, 316
alternator, 415
alumina, 173, 230, 294, 298
anaerobic, ix, 7, 290, 295, 310, 311, 327, 363, 367, 370, 381, 385, 386, 388, 389, 390, 391, 392, 393, 394, 395, 397, 398, 399, 403, 404, 405, 406, 432, 436, 444, 445
anaerobic bacteria, ix, 381, 385, 386, 388, 390, 395, 399, 403
ancillary enzymes, 363, 386
anhydrocellulose, 143, 354
animal biomass, 7, 9, 13, 23, 26, 29
animal fat, 154, 251, 254, 264, 282, 312
animal manure, 9, 67, 153, 154, 160, 162
anion-exchange resin, 209
anti-aging cream, 278
aquatic biomass, 16
aqueous ammonia, 331, 334, 343, 360, 361, 446
aqueous-phase HDO, 240
aqueous-phase hydrotreating, 221, 223, 240
arabinose, 16, 69, 144, 154, 170, 351, 352, 387
arches, 85
aromatics, 17, 195, 196, 207, 223, 228, 229, 237, 240, 290, 291, 295
Arthrospira, 278

artificial drying, 99, 107, 108
ash, 13, 18, 22, 23, 24, 25, 26, 27, 29, 31, 32, 33, 34, 35, 83, 126, 132, 142, 145, 150, 155, 173, 174, 175, 177, 184, 202, 253, 349, 364, 390, 420, 424, 431, 440
Aspen Plus, 372, 428
ASTM, 251, 253, 254, 263, 266, 271, 272, 273
ASTM D6751, 253, 263, 271, 272, 273
atomized liquid, 174
attrition, 112, 116, 119
autohydrolysis, 75, 79, 334, 342
auto-ignition temperature, 64

B

Bacillariophyta, 277
bagasse, 6, 8, 21, 38, 63, 333, 335, 336, 339, 342, 343, 345, 355, 356, 381
ball mill, 74, 93, 111, 112, 120, 352
ball milling, 74, 120
barley, 6, 8, 15, 49, 58, 65, 67, 86, 88, 89, 96, 97, 106, 123, 375
barley straw, 6, 15, 88, 97, 123, 375
benzene, 17, 155, 195, 237, 238, 239, 291, 372
Bermuda grass, 82, 436
big blue stem, 38
bimetallic, 215, 229, 238, 239, 245, 247
bio-based chemicals, ix, 301, 304, 305, 307, 320
biobutanol, 5, 7, 30, 296
biochemical, ix, 3, 5, 6, 23, 25, 26, 27, 28, 33, 40, 80, 153, 154, 162, 241, 281, 288, 374, 387, 392
bioconversion, 14, 23, 27, 41, 168, 301, 303, 304, 320, 321, 334, 343, 364, 366, 368, 375, 377, 379, 380, 382, 385, 390, 392, 399, 402
biodegradability, 27, 72, 74
biodegradable de-icers, 149
biodiesel, 5, 7, 8, 40, 133, 213, 217, 222, 251, 252, 253, 254, 255, 260, 263, 264, 265, 266, 267, 268, 269, 270, 271, 272, 273, 274, 275, 277, 280, 282, 283, 284, 285, 286, 287, 290, 305, 308, 309, 310, 312, 413, 438
bioethanol, 5, 7, 8, 16, 23, 30, 33, 64, 66, 67, 68, 79, 94, 96, 241, 277, 287, 288, 322, 323, 329, 330, 348, 354, 363, 367, 368, 370, 372, 373, 374, 375, 376, 378, 379, 380, 381, 390, 392, 399, 424, 432, 436, 441, 442
biohydrogen, 5, 8, 381, 390
biological, x, 9, 17, 27, 28, 33, 39, 67, 69, 71, 72, 84, 90, 99, 100, 107, 169, 289, 290, 320, 327, 329, 330, 331, 348, 349, 362, 376, 380, 402, 404, 411, 436
biological drying, 107
biological pretreatment, 84, 90, 362

biomass classification, 6, 7
biomass composition, 13
biomass efficiency, 3
biomass form, 37, 53
biomass fractionator, 340
biomass species, 3, 423, 424, 432, 433, 438
biomass-derived fuels/products, ix
blending, 49, 52, 58, 181, 187, 189, 264, 289
BlueFire Ethanol Incorporated, 337
Botryococcus, 283
Boudouard reaction, 185, 210
Brazil, 7, 20, 38, 45, 252, 348
bridge, 38, 97, 143, 170
British, 8, 65, 127, 172, 375
Brønsted acid, 195, 207, 239, 240, 246, 292
bulk density, 18, 27, 38, 44, 53, 54, 56, 72, 74, 86, 87, 120, 122, 128, 132, 134
bundled materials, 100
butanol, 9, 15, 27, 40, 160, 198, 199, 208, 231, 235, 254, 284, 290, 291, 294, 295, 296, 297, 304, 312, 322, 326, 333, 396
by-product, 39, 52, 67, 80, 90, 125, 141, 251, 256, 268, 310, 312, 356, 364, 395

C

Caldicellulosiruptor saccharolyticus, 391, 392, 403
calorific value, 25, 26, 27, 35, 86, 181, 182, 271, 284
Canadian, v, 31, 32, 58, 61, 65, 66, 91, 109, 406
Canadian prairies, 65, 91
CANMET surfactant, 197, 204
CAPCONN model, 46
carbohydrate binding domain, 389
carbonization, 139, 141, 173
carbonization zone, 173
carboxylic acid, 155, 194, 197, 208, 227, 232, 233, 245, 307, 333, 370, 393
case study, 57, 58, 59, 425
catalytic cracking, ix, 193, 194, 195, 196, 199, 207, 208, 216, 217
catalytic reforming, 209, 212, 219, 242
catalytic steam gasification, 181, 183, 184, 189, 190
cation-exchange resin, 208
CatLiq process, 156
cattail, 331, 335, 423, 424, 427, 432, 433
Ceiba Pentandra, vi, 263, 265, 266, 267, 268, 269, 270, 271, 272, 273
cellulase, 14, 28, 94, 329, 331, 333, 334, 335, 339, 341, 342, 348, 362, 363, 365, 366, 367, 368, 369, 370, 380, 381, 382, 386, 387, 390, 393, 402, 426, 427
cellulosome, 382, 386, 387, 388, 389, 390, 401, 402, 404

centrate, 281
cereal, 8, 48, 49, 50, 55, 65, 75, 91, 105
cetane number, 197, 266, 271
chemical treatment, 9, 24, 281
chip, 12, 21, 27, 115, 358, 359, 379
chipper, 102, 103, 111, 113, 114, 129
Chlamydomonas reinhardtii, 282, 287
Chloramphenicol, 425
Chlorella, 278, 279, 281, 283, 286, 287
Chlorella vulgaris, 287
chlorophyll, 6, 17, 284, 349
Chlorophyta, 277
chopper, 105
chromatography, 196, 244, 440, 445
Citrobacters, 306
cleavage, 75, 144, 146, 156, 157, 162, 231, 232, 239,
    241, 354, 360, 362, 387
Clostridium, 306, 314, 315, 320, 326, 327, 382, 383,
    387, 391, 401, 402, 403, 404, 405, 406
Clostridium acetobutylicum X9, 391, 403
Clostridium thermocellum, 382, 401, 402, 404, 405,
    406
cloud filter plugging point, 272
cloud point, 272, 275
CO2 concentration, 4, 187, 189, 280, 283, 308
CO2 explosion, 28, 80, 84, 361
coastal Bermuda grass, 333, 334, 342
cobalt, 214, 225, 246, 281
coconut, 8, 89, 97, 254
co-culture, 368, 369, 385, 386, 393, 395, 396, 397,
    398, 399, 400, 401, 403, 404, 405, 406
co-current, 167, 173, 382
co-fermentation, 304, 322, 349, 368, 369, 382, 392,
    404
co-gasification, 182, 191
cohesin–dockerin interaction, 389
coke, 194, 195, 196, 205, 207, 208, 210, 211, 226,
    228, 229
cold biomass pelletizing, 131, 135
Columbia, 8, 65, 127, 375
combine harvester, 105
combustion chamber, 133, 174, 271, 412
combustion reaction, 169
combustion zone, 173, 177
comminuted biomass materials, 100
comminution, 56, 84, 111, 113, 121, 122, 352, 364
composite biomass, 7, 9, 12
composite residue logs, 100
compression, 48, 64, 74, 85, 87, 88, 112, 116, 125,
    254
compressive resistance, 131, 132
concentrated acid, 82, 337, 354

condensation, 76, 140, 141, 148, 157, 161, 164, 195,
    233, 234, 240, 247, 296, 299, 309, 316, 323, 364
configuration, 37, 39, 44, 46, 54, 55, 173, 175, 199,
    337, 370, 372, 393
coniferyl alcohol, 144, 236, 351
constructed wetland, 428
contamination, 13, 23, 24, 115, 205, 228, 230, 279,
    281, 368, 393, 410
continuous transesterification, 251
conventional hydrodeoxygenation, 205, 221
conventional multi-pass forage harvest system, 104
conventional pyrolysis, 140
cooking oil, 251, 254, 261, 273, 274, 282
corn stover, 6, 38, 56, 58, 76, 79, 81, 82, 83, 86, 87,
    88, 94, 95, 96, 97, 104, 105, 107, 109, 123, 151,
    172, 248, 332, 333, 334, 335, 336, 339, 341, 342,
    343, 344, 345, 352, 375, 401, 423, 424, 428, 429,
    431, 432
cost effectiveness, 99, 101, 107
cotton, 8, 15, 38, 266, 284, 350, 376, 379
counter-current, 167, 173, 360
covalent, 16, 70, 389
cracking, 113, 146, 147, 161, 185, 195, 196, 207,
    209, 216, 218, 225, 248, 291, 296
creosote oil, 155, 160
CRL, 100
crude oil, 66, 196, 207, 222, 225, 254, 258, 285, 289,
    301, 436
crystallization, 79, 304, 359
culture condition, 278, 282, 311, 370, 396
Cyanobacteria, 277
Cyanophyceae, 277
cyanotoxin, 280
cyclization, 157

D

deamination, 157
decarboxylation, 157, 207, 232, 235, 309
decompose, 146, 252
decomposition, 6, 24, 32, 79, 89, 139, 140, 144, 146,
    154, 157, 158, 159, 162, 185, 187, 210, 339, 345,
    354
dehydration, 52, 157, 158, 196, 198, 207, 216, 217,
    231, 232, 233, 234, 235, 237, 239, 240, 246, 247,
    291, 293, 296, 307, 372, 439, 444, 445
dehydrocyclization, 291
dehydrogenation, 75, 157, 232
densification, 37, 38, 39, 41, 43, 48, 50, 53, 61, 62,
    67, 72, 85, 86, 87, 88, 89, 125, 132, 135
densified bales, 44
density, 27, 29, 30, 44, 50, 52, 85, 86, 87, 88, 89, 97,
    111, 112, 120, 122, 123, 125, 126, 127, 132, 141,

155, 156, 159, 160, 161, 198, 225, 232, 238, 266, 271, 279, 284, 286, 289, 292, 308, 312, 315, 405

deoxygenation, 157, 196, 214, 217, 221, 223, 224, 225, 228, 230, 237, 238, 240, 250

depolymerization, 28, 29, 75, 76, 79, 154, 155, 157, 162, 202, 236, 248, 356, 358, 359, 360, 364, 378

derivative, 314

diatoms, 8, 277

diesel, ix, 148, 167, 171, 197, 199, 200, 204, 209, 213, 225, 231, 235, 240, 241, 243, 251, 252, 253, 254, 260, 264, 271, 274, 277, 283, 284, 287, 289, 291, 293, 296, 309, 312, 330, 333, 436

diesel engine, 148, 252

differential HHV, 26

dilute-acid pretreatment, 80, 151, 344, 377

direct fermentation, 367, 385, 395

disc mill, 111, 119

disc mower, 105

disk chipper, 113, 114

distillation, 68, 141, 200, 204, 217, 235, 253, 370, 371, 372, 382, 383

distribution, 13, 15, 39, 44, 45, 46, 47, 49, 50, 51, 52, 53, 55, 58, 59, 65, 81, 91, 111, 122, 123, 131, 147, 161, 163, 207, 211, 224, 291, 292, 312, 371

double involuted disk chunker, 115, 116

downdraft, 173

downdraft gasifier, 173

drum chipper, 114, 115

durability, 87, 88, 89, 96, 130, 131, 132, 134

durene, 291

E

Eberbach, 118

economic analysis, ix, 425, 428

electrical, 134, 177, 183, 364, 409, 411, 414, 415, 417

electricity, 3, 4, 5, 7, 40, 67, 127, 134, 148, 167, 168, 171, 176, 177, 179, 229, 282, 412, 415, 417, 425, 428, 440

elemental analysis, 22, 227

emollient, 278

emulsification, 193, 194, 204, 213

EN 14214, 266, 271, 272, 273

Enamora Gasification Plant, 177

endoglucanase, 14, 363, 386, 387, 405

energy cane, 336

energy consumption, 4, 5, 81, 83, 85, 88, 89, 111, 112, 122, 127, 131, 199, 201, 211, 212, 222, 258, 352, 353, 356, 372, 377, 438, 439

energy content, 27, 39, 74, 241, 285, 413

energy crops, 6, 8, 29, 38, 48, 63, 92, 354, 435, 436, 438, 446

energy demand, 3, 62, 72, 116, 119, 264, 352, 372

Energy Independence and Security Act, 290

engine, 49, 64, 148, 151, 176, 177, 199, 204, 212, 215, 241, 252, 253, 254, 267, 272, 273, 287, 288, 411

entrained bed gasifier, 167

environmental benefit, 62, 251, 260

environmental impact, 32, 37, 45, 48, 49, 50, 53, 58, 172, 338, 341, 348, 381, 423, 424, 432, 435

enzymatic catalyst, 259

enzymatic hydrolysis, 16, 28, 33, 71, 74, 79, 80, 81, 82, 93, 95, 332, 333, 336, 338, 342, 343, 344, 345, 348, 355, 356, 359, 360, 361, 362, 363, 365, 367, 370, 371, 373, 374, 377, 379, 380, 382, 386

enzyme kinetics, 332

equilibrium reaction, 259

Escherichia coli, 322, 323, 325, 326, 327, 423, 424, 425, 433

esterification, 161, 193, 194, 197, 198, 199, 208, 209, 217, 218, 263, 268, 273, 275

ethanol dehydration, 293

ethylene glycol, 155, 160, 219

ethylene oligomerization, 293, 294, 295, 298

ethylene selectivity, 292, 293

EU-ALTENER, 134

Eucalyptus, 11, 19, 38

eukaryotic, 277

Europe, 102, 127, 252, 435, 436, 439, 441, 443

European Committee for Standardization, CEN, 133

exoglucanase, 14, 363, 386, 387, 390

expansion, 4, 9, 62, 75, 76, 77, 79, 135, 334, 340, 342, 343, 348

extraction, 17, 18, 26, 35, 66, 81, 99, 101, 107, 144, 159, 163, 166, 235, 242, 263, 274, 283, 343, 358, 359, 371, 378, 411, 433, 438, 446

extractives, 9, 13, 15, 17, 25, 26, 27, 32, 34, 68, 71, 75, 131, 349

extrinsic moisture, 22, 29

extrusion, 52, 72, 75, 89, 131

F

FAME, 253, 259, 264, 267, 282

fast pyrolysis, 32, 56, 135, 140, 142, 146, 147, 149, 150, 151, 155, 156, 163, 193, 197, 200, 201, 202, 212, 213, 214, 215, 216, 217, 218, 223, 224, 225, 240, 243, 244, 245, 249

fatty acid, 154, 157, 251, 252, 254, 256, 258, 259, 263, 264, 265, 267, 275, 278, 282, 283, 285, 307, 308, 309, 310, 312, 313, 316, 319, 323

fatty acid biosynthesis pathway, 313

feedstock quality, 37, 39, 53

feedstock supply facility, 37

feedstock versatility, 373
feller buncher, 101
felling, 99, 101, 104, 107
fermentation strategy development, 304
fermentation-derived, ix, 289, 290, 292, 297
fermentation-stripping process, 371
fine solid, 167, 174
fire ignition temperature, 54
Fischer-Tropsch, 148, 167, 171, 175, 177, 209, 441
fixed bed gasifier, 172
fixed carbon, 18, 22, 24, 25, 29
fixed carbon content, 25
flax, 6, 8, 12, 65, 106
flax straw, 6, 65
fluidization, 167, 173
fluidized bed gasifier, 167, 172, 173, 175, 177, 182,
    183, 184, 190
fluxome, 304
Food and Agriculture Organization, 31, 65, 91, 96
food processing waste, 153, 154, 160, 162
forage, 67, 99, 105, 108, 333, 342
forager, 105
forest residue, ix, 3, 5, 7, 23, 63, 67, 347
formic acid, 145, 164, 194, 202, 204, 205, 206, 234,
    235, 357
forwarder, 103
fossil fuel, ix, 3, 4, 29, 38, 61, 172, 203, 222, 273,
    283, 294, 307, 329, 330, 444
fresh food, 46, 47, 48, 55
Friedel–Crafts alkylation reaction, 339
fuel slurry, 174
fungi, 17, 71, 84, 306, 361, 362, 363, 370, 386, 387
furfural, 15, 76, 78, 79, 157, 203, 206, 223, 229, 232,
    233, 234, 245, 246, 247, 340, 356

G

galactoglucomannan, 144
galactose, 16, 69, 144, 154, 170, 351, 352
galacturonic acid, 144
garbage reverse distribution, 51, 52
gas phase HDO, 229, 238
gasification zone, 173, 184
Gaudin-Schuhmann equation, 122
GC, 147, 229, 267
GC-MS, 267
gene synthesis, 304
genome, 304, 322, 326
geothermal, 5, 63, 168, 348
GHG, 4, 29, 62, 282, 424
global warming, 3, 28, 61, 66, 182, 410, 412
glucan hydrolases, 389
glucopyranose, 14, 143, 169

glycerol, 15, 154, 155, 157, 160, 161, 163, 220, 246,
    247, 251, 252, 256, 258, 259, 306, 310, 311, 317,
    319, 324, 325, 327
glycosyl hydrolase, 388
GPS, 49
grass chute, 105
gravity, 64, 118, 141, 167, 172, 202, 203
green biomass, x, 348, 435, 436, 437, 438, 439, 441,
    442, 443, 445, 446
green biorefining, x
green fractionation, 437
green juice, 438, 439, 440, 442, 445
green refinery, 435
greenhouse gas, 3, 67, 182, 222, 241, 285, 302, 306,
    424
greenhouse gas emission, 3, 67, 222, 241, 285, 302,
    306
Greensboro, vii, 409, 410, 417, 419, 421, 422
gross calorific value, 25, 198
ground water, 410
guaiacol, 206, 214, 215, 227, 229, 237, 238, 239,
    242, 245, 246, 249
Güssing Plant, 176

H

hair care, 278
hammer mill, 39, 111, 112, 115, 116, 117, 129, 352
Harboore Plant, 177
harvest, 6, 7, 8, 10, 13, 22, 29, 31, 38, 39, 41, 43, 45,
    48, 49, 51, 53, 54, 55, 57, 58, 65, 99, 100, 101,
    103, 104, 105, 106, 107, 108, 109, 114, 334, 428,
    431, 440, 444, 446
harvester, 103, 105, 106, 109
hay, 8, 44, 54, 85, 105, 440
HCL CleanTech, 337
heating rate, 140, 141, 143, 144, 145, 146, 151, 158,
    160, 162
heating value, 22, 23, 25, 26, 27, 29, 35, 86, 125,
    132, 140, 141, 148, 155, 158, 161, 171, 194, 197,
    198, 200, 201, 202, 203, 205, 208, 209, 221, 223,
    225, 271, 272, 411, 412, 413, 414, 417, 440
hemicellulase, 16, 363, 366, 370, 390, 427
Hesston, 106
heterogeneous catalyst, 155, 157, 158, 160, 162, 205,
    206, 208, 258
heteropolyacid, 208
heterotrophic, 278, 279, 286
hexanol, 254, 312, 313, 326
high calorific value, 8, 130
high energy radiation, 75
high-efficiency modern, 6
higher heating value, HHV, 25

Hillco, 107, 109
history, 150, 169, 281, 337
hollow-fiber membrane reactor, 371
holocellulose, 27, 28, 29, 347, 348, 352, 362
homogenous catalysts, 160
hot-water, 329, 331, 335, 423, 424, 425, 427
hybrid poplar, 113, 332, 335, 343, 377
hydrodesulphurization, 225
hydroelectric, 410
hydrogenation, 161, 171, 196, 205, 206, 209, 225, 227, 228, 229, 231, 232, 233, 234, 235, 236, 237, 238, 239, 240, 242, 244, 245, 246, 247, 249, 291, 293
hydrotreating, 156, 205, 206, 207, 223, 225, 227, 230, 240, 243, 244
hydroxyacetaldehyde, 149, 204
hydroxyfatty acids, 309, 324
HZSM-5, 194, 195, 199, 200, 207, 208, 216, 217, 239, 240, 242, 243, 250, 291, 292, 293, 295, 296, 298

I

impact, 29, 31, 44, 65, 112, 116, 121, 131, 132, 133, 146, 177, 292, 301, 347, 349, 352, 354, 394, 399, 409, 443
impact mill, 112, 121
impact resistance, 131, 132
in silico modeling, 304
in situ product recovery, 295, 304
Inbicon, 335
INDEBIF, 134
Indian grass, 38
industrialization, 4, 198, 282
infrastructure, 37, 39, 40, 44, 53, 54, 83, 86, 231, 297, 321
inorganic acid salt, 208, 209
inorganic component, 13, 17, 23
inorganic salt, 329, 331, 339, 345
instability, 141, 198, 202, 208, 225, 230, 310
installation expenses, 429
intrinsic moisture, 22
involuted disk chunker, 115, 116
iodine value, 259, 266
Iogen, 336
ion-exchange, 208, 217, 299, 304
ionic liquid, 28, 32, 206, 208, 209, 329, 337, 338, 339, 344, 345, 357, 378
isobutanol, 296, 302, 306, 312, 322
isoeugenol, 194
isomerization, 157, 158, 232, 246, 293, 296
isoprene, 306, 307, 321, 322
isotope, 278

J

Jathropa, 254, 260
Jatropha curcas, 273, 275
jet mill, 111, 112, 121

K

Kapok, 265, 274
Karl-Fischer, 267
Kekabu, 265, 274
kernel processor, 105
kerosene, 252
kinematic viscosity, 271, 272
kinetics, 34, 77, 228, 246, 251, 322, 338, 342, 345
Klebsiella, 306, 310, 325, 327, 368, 398, 406
knife mill, 111, 113, 117, 118, 122, 123

L

lactic acid, 15, 302, 306, 311, 315, 322, 439, 440, 441, 442, 443, 445
Lactobacilli, 305, 306
land-based biomass, 7
landfill, 52, 182, 410, 411
leaf seasoning, 107
leaf trash, 38
levoglucosan, 24, 143, 144, 145, 149, 195, 204, 223, 354
levulinic acid, 234, 235, 248
Lewis acid, 195, 232, 339, 341
Lewis adduct, 339
Lewis base, 339
LHV equation, 25
life cycle analysis, 50, 51
light intensity, 280
lignin-derived compounds, 221, 223, 237, 238, 239, 241
lignin-derived oligomer, 202, 204
lignocellulosic feedstock, ix, 15, 34, 37, 38, 39, 40, 43, 44, 50, 53, 54, 55, 60, 171, 172, 373, 382, 385, 390, 404
Lignol, 340
lime, 52, 83, 331, 332, 333, 337, 342, 355, 427
linseed, 106
lipid, 254, 279, 280, 282, 283, 286, 287, 323
LIPS, 156
liquefaction, ix, 153, 154, 155, 156, 157, 158, 159, 160, 161, 162, 163, 164, 165, 166, 222, 230, 241, 431, 436, 444, 445
liquid biofuel, 37, 38, 39, 41, 44, 53

loading, 39, 41, 43, 45, 46, 48, 49, 50, 54, 55, 61, 74, 79, 99, 101, 104, 107, 232, 239, 240, 328, 329, 331, 333, 334, 335, 336, 347, 348, 355, 360, 366, 379, 426
lower heating value, LHV, 25
Luria-Bertani, 425

M

Malaysia, 181, 182, 252, 263, 265
mannase, 388
mannose, 16, 69, 144, 154, 170, 351
Maritimes, 65
market, 40, 49, 85, 102, 126, 127, 134, 135, 142, 252, 265, 278, 301, 302, 303, 304, 306, 309, 312, 321, 436
Markov chain, 47
Mascoma, 336
McLeod, 49, 58
mechanical comminution, 74, 352
mechanical fractionation, 439, 445
mechanical processing, 39, 54, 338
mechanism, 17, 111, 158, 216, 245, 251, 256, 290, 291, 292, 294, 296, 298, 316, 332, 338, 339, 355, 357, 360, 362, 373, 431
medium, 31, 32, 38, 44, 76, 120, 131, 141, 153, 154, 156, 162, 167, 239, 258, 281, 304, 308, 314, 327, 399, 401, 403, 425, 427, 442
membrane, 274, 304, 308, 371, 372, 382, 383
membrane separation, 304
mesoporous catalyst, 207
MESP, 423, 429, 430, 431, 432
metabolic engineering, 301, 302, 304, 307, 312, 314, 320, 321, 370
metabolic relationship, 396
metabolome, 304
metal catalyst, 219, 239, 244, 245, 250, 324
metal-acid/base bifunctional catalyst, 209
metallic oxide, 205, 208, 209, 211
methanation reaction, 206
methane, 7, 9, 40, 171, 173, 174, 176, 177, 183, 185, 186, 187, 188, 375, 410
methanol, 24, 40, 141, 144, 145, 149, 160, 161, 196, 203, 205, 206, 209, 217, 218, 220, 239, 252, 255, 256, 258, 259, 263, 268, 270, 271, 273, 274, 275, 284, 285, 290, 291, 357
microalgae, ix, 8, 29, 159, 161, 164, 165, 254, 277, 278, 279, 280, 281, 282, 283, 284, 285, 286, 287, 288, 308, 323, 443
microalgal biotechnology, 260, 278, 283, 286
microalgal oil, 279, 282, 284, 285
microbial conversion, ix, 366, 369
microwave-chemical pretreatment, 81

mineral content, 145
Miscanthus, 19, 35, 38, 48, 56, 254, 439, 445
Miscanthus sinensis, 35, 38
Miscanthus x giganteus, 38, 439
miscibility, 141
MixAlco, 333
mixotrophic, 278, 279, 280
Modified Dulong's equation, 25
moisture, 7, 18, 22, 23, 24, 25, 35, 38, 76, 77, 85, 86, 87, 88, 89, 107, 116, 117, 118, 122, 129, 130, 131, 132, 135, 150, 177, 227, 258, 333, 352, 359, 360, 424, 428, 430
molybdenum, 225, 228, 246
mono-functional compounds, 232, 233
Moringa oleifera, 273
moving bed gasifier, 167, 172, 173
MTG-based process, 293
municipal organic waste, 3, 29

N

NaOH, 15, 83, 157, 159, 209, 259, 331, 332, 333, 340, 355, 423, 424, 425, 426, 427, 428
naphthalene, 244, 249, 291
natural gas, 4, 29, 91, 127, 162, 171, 177, 241, 264, 410, 413, 435
NC Recycling and Solid Waste Management Plan, 417, 419
net calorific value, 25
nickel-promoted, 225
noble metal catalyst, 205, 212, 215, 218, 229, 238, 245
non-covalent, 16, 389
nonisothermal catalyst bed, 227
NREL, 34, 151, 331, 349, 425, 428, 429, 433
NSSF, 395
nutrient, 8, 31, 279, 280, 286, 287, 314, 397, 437, 438, 441, 445

O

oak, 21, 38
oat, 65, 86, 89, 96
octane, 64, 209, 228, 232, 233, 240, 246, 290
octanol, 161, 204, 254, 313
oil crop, 50, 55, 254, 282
oil production, 5, 156, 159, 212, 213, 222, 282, 283, 287, 435
oil-seed, 38
olefin, 227, 293, 295, 323
oligomerization, 144, 248, 291, 293, 294, 295, 296, 297, 298

omics, 304
Ontario, 3, 65
organic extraction, 204
organosolv, 28, 356, 377, 378
overhead cost, 429
oxidation, 24, 26, 27, 81, 82, 95, 144, 167, 168, 208, 253, 271, 294, 307, 308, 310, 313, 320, 324, 326, 328, 340, 356, 364, 377
oxidative stability, 267, 271, 272
oxygenates, ix, 162, 194, 207, 216, 223, 240, 248, 289, 290, 297
oxygenic photosynthesis, 6, 277

P

packaging, 46, 58, 128, 305
palm, 8, 146, 151, 159, 161, 165, 181, 182, 183, 189, 190, 254, 272, 273, 274, 275
Panicum virgatum, 38, 438
paraffins, 291, 295
particle size, 27, 29, 32, 67, 74, 88, 89, 94, 111, 112, 115, 116, 117, 119, 120, 121, 131, 132, 146, 151, 174, 183, 230, 341, 352, 358, 360, 374
pathway engineering, 304
payback period, 409, 419, 421
p-coumaryl alcohol, 236
pelletization, ix, 125, 127
pelletizing process, 130, 131
pentose fermenting, 391
perennial grasses, 8, 17, 436
pervaporation, 371, 372, 383
pH, 10, 32, 141, 147, 148, 198, 202, 203, 208, 209, 280, 343, 348, 356, 360, 363, 365, 369, 391, 392, 398, 404, 427
pH shift, 280
PHA, 314, 315, 319, 320, 327, 328
PHB, 314, 315, 320, 327, 328
phenol, 155, 160, 162, 163, 166, 204, 215, 229, 237, 238, 239, 240, 249, 250
phenylpropane, 144, 170, 236
photoautotrophic, 278, 286
photobioreactor, 286
photoheterotrophic, 278, 280, 286
photosynthesis, 6, 14, 29, 63, 133, 279
Phylaris arundinacea, 38
physical, 3, 9, 23, 28, 38, 39, 56, 60, 67, 71, 72, 74, 75, 76, 79, 83, 87, 88, 89, 90, 92, 97, 99, 100, 107, 112, 123, 144, 147, 155, 157, 202, 230, 280, 284, 302, 314, 315, 320, 331, 339, 348, 349, 358, 360, 375, 411
physical pretreatment, 74, 75, 112
physico-chemical, 5, 72, 75, 79, 87, 90, 347, 348, 349, 357

pilot, 107, 148, 154, 155, 171, 181, 183, 184, 189, 212, 336, 337, 340, 383, 441
pin mill, 111, 112
pine, 7, 16, 19, 27, 33, 74, 147, 159, 227, 228, 243, 244, 344, 350, 362, 376, 377, 381
Pinus L., 38
pipeline supply, 44
plantgate price, 429
plasma, 409, 411, 414, 417, 419, 420, 421
plasma gasification, 409, 411, 414, 417, 419, 420, 421
plastic deformation, 85, 96
polycondensation, 205
polyethylene glycol, 336, 344
polyhydroxyalkanoate, 328, 402
polyhydroxybutyrate, 314
polymerization, 16, 27, 69, 74, 78, 83, 84, 120, 157, 170, 205, 217, 226, 227, 243, 306, 307, 315, 350, 352, 358, 365, 387
polyols, 160, 162, 220, 229, 232, 247
poplar, 8, 19, 31, 38, 63, 79, 142, 225, 333, 334, 342, 343, 350, 379
Populus, 38, 78, 94
pour point, 272, 275
power capacity, 417
power output, 409
prairie cord grass, 38
precipitation, 15, 22, 304, 338
pressurized solvent, 156
process design, x, 177, 382, 425
process scalability, 373
processing technologies, ix, 435, 436
product inventory, 50
project contingency, 429
prokaryotic, 277
proteome, 304
ProXan-Process, 443
PureVision, 340
pyrolysis oil/biofuel, ix
pyrolysis vapor, 140, 148, 193, 194, 195, 196, 207

Q

Quebec, 65

R

radical, 143, 144, 161
rail transportation, 44
rapeseed, 146, 151, 254, 258, 259, 264, 272, 275
reactive solvent solvolysis, 154
rearrangement, 85, 154, 157, 162, 256

recalcitrance, 27, 28, 32, 33, 67, 75, 303, 321, 344, 345, 357, 377
reciprocating knives, 105
reed canary grass, 8, 38, 166
regenerant, 278
regulation, 39, 44, 307, 394
Renergi LTK Dryer, 131
renewable energy, 3, 5, 32, 55, 60, 63, 66, 153, 167, 169, 178, 193, 201, 252, 282, 289, 372, 385, 399, 400, 419, 420
Renewable Fuels Association, 66
residence time, 76, 77, 78, 82, 84, 87, 140, 141, 145, 146, 147, 158, 159, 202, 333, 340, 359, 360, 364, 367
residue, 8, 21, 23, 76, 104, 141, 154, 158, 159, 160, 161, 162, 168, 203, 253, 360, 366
rice straw, 75, 81, 83, 95, 142, 350, 360, 362, 375, 376, 377, 379
road map, 40
Route LogiX, 49
rye, 6, 10, 82, 95, 106
rye grass, 6, 10

S

saccharide, 144, 339
saccharification, 32, 56, 68, 82, 95, 124, 304, 320, 322, 330, 336, 338, 344, 345, 349, 355, 356, 357, 366, 368, 369, 371, 373, 375, 376, 377, 378, 380, 381, 382, 392, 393, 394, 395, 405, 424, 425, 428
sale, 50, 133
saline extractive distillation, 372
Salix spp., 38
saponification value, 266
Saskatchewan, 3, 30, 61, 65, 91, 92, 96
Saskatchewan Ministry of Agriculture, 65, 91
saturated cyclics, 291
sawdust, 7, 8, 9, 12, 21, 27, 63, 127, 128, 129, 131, 133, 135, 159, 163, 195, 227, 243, 358, 379
saw-like blade, 105
scaffoldin, 388
scaffoldin-borne, 389
SCF Technologies, 156
screening, 29, 304, 305, 368, 378, 390
screw express, 439
semi-biomass, 7, 9, 10, 13, 23, 26, 29
semi-whole crop harvesting and transport model, 49
separation, 83, 94, 111, 112, 122, 141, 171, 190, 204, 217, 234, 256, 258, 259, 268, 278, 304, 306, 308, 343, 347, 349, 370, 371, 372, 373, 390, 402, 425, 428, 432, 440
severity factor, 77, 78, 359
shear, 76, 77, 112, 113, 116, 352, 367

Shell Research Laboratory, 156
Siberia, 5
silage, 9, 105, 439, 440, 441, 442, 445
silage harvester, 105
silica sand, 173
sinapyl alcohol, 16, 236, 351
single pass harvesting system, 48, 106
size reduction, ix, 37, 39, 41, 43, 50, 74, 89, 111, 112, 115, 117, 120, 122, 124, 341, 352, 360
skidder, 101, 102, 103
skin care, 278
slagging mode, 173
slow pyrolysis, 140, 144
small-scale timber harvesting system, 99, 104, 108
SO2-catalyzed, 336, 341, 344
sodium hydroxide, 147, 165, 258, 259, 331, 332, 355
sohxlet, 268
solar, 6, 8, 60, 63, 168, 348
solid acid, 198, 208, 231, 240, 246, 274, 294, 298, 299
solid base, 198, 208
solid fuel, 18, 22, 413, 440, 444, 445
solvent, 15, 17, 26, 80, 153, 154, 159, 160, 161, 162, 166, 196, 235, 247, 295, 304, 306, 327, 338, 340, 357, 358, 371, 378, 381, 383, 438, 442
solvent extraction, 80, 304
solvent-based pre-treatment, 340
solvolysis, 151, 153, 155, 160, 162, 163
sortase-mediated, 389
southern pine, 38
soybean, 96, 254, 264, 282
spiral-head wood chunker, 115, 116
SPORL, 356, 364, 365, 377
stain development, 304
standard procedure, 17, 55
standardization, 37, 39, 55
starch, 27, 66, 87, 89, 131, 170, 222, 254, 290, 302, 303, 371, 383, 391
steam explosion, 28, 75, 76, 77, 78, 79, 80, 91, 147, 336, 344, 359, 360, 361, 378, 379
steam turbine, 411, 412, 415, 417
steam/feedstock ratio, 181, 188
subcritical, 154, 156, 157, 160, 161, 162, 164, 166
substrate, 72, 80, 230, 295, 304, 308, 309, 310, 312, 313, 315, 325, 340, 355, 356, 357, 360, 361, 363, 365, 367, 370, 371, 373, 380, 389, 390, 392, 396, 397, 398, 400, 439, 441
succinic acid, 302, 305, 322
sugar cane, 38, 45, 55, 290, 329, 330, 343, 381, 424
sugarcane bagasse, 6, 34, 75, 78, 94, 151, 333, 350, 355, 356, 358, 361, 375, 378
sulfided Mo-based catalyst, 230
sulfur content, 227

sulfuric acid, 17, 82, 234, 235, 268, 336, 337, 344, 354, 359, 425, 427, 428, 433
sun protection, 278
sunflower stalk, 38, 163
SunOpta BioProcess, 336
supercritical, 28, 32, 94, 95, 157, 161, 163, 164, 165, 208, 209, 215, 217, 218, 259, 273, 275, 361, 379
supercritical condition, 208
supply logistics, ix, 37, 39, 40, 45, 50, 51, 55
support, x, 6, 30, 65, 199, 205, 214, 230, 232, 237, 238, 239, 249, 321, 361, 368, 374, 400
surfactant, 197, 199, 204, 254, 307, 366, 380
sweet sorghum, 17, 38, 333, 339, 342, 345, 382
switchgrass, 8, 32, 38, 63, 87, 88, 94, 97, 104, 123, 172, 179, 303, 321, 333, 334, 335, 336, 338, 342, 343, 344, 345, 350, 360, 375, 379, 382, 404, 438, 442, 446
syngas, 7, 8, 23, 148, 168, 171, 172, 177, 181, 182, 186, 187, 188, 189, 190, 330, 411, 413, 414, 440, 441
synthetic regulatory circuits, 304
syringol, 237
system integration, 53, 55
systems-level, 304

T

Taguchi, 263, 268, 269, 275, 403
tar, 140, 155, 157, 158, 173, 174, 176, 177, 183, 184, 185, 189, 226
TDP process, 156
technoeconomic assessment, 156
techno-economical analysis, 424, 429
ternary phase, 254
Terrabon, 333
terrestrial biomass, 8
tetralin, 155, 160, 161, 162, 163, 196
theoretical ethanol yield, 361, 423, 428, 432
thermal cracking, 187, 264
thermal gasification, 409, 413, 414, 417, 420, 421
Thermoanaerobacterium, 381, 391, 392, 401, 403
thermochemical, ix, x, 3, 5, 6, 23, 24, 27, 34, 40, 133, 153, 156, 162, 163, 165, 167, 168, 191, 221, 248, 289, 290, 295, 330, 411, 436, 441, 445
thermodynamic equilibrium, 194
thermophilic, ix, 367, 369, 370, 381, 382, 385, 386, 387, 389, 390, 391, 393, 395, 398, 399, 401, 403, 404, 405, 406
thermophilic anaerobic bacteria, 386, 387, 390, 391, 399
thermosaccharolyticum W16, 391, 403
third-generation, 277, 278, 282
timothy grass, 6

tobacco, 121, 274
torrefaction, 39, 40, 131, 135, 353, 376
transalkylation, 217, 238, 249, 296
transcriptome, 304
transesterification, 251, 252, 255, 256, 257, 258, 259, 260, 264, 266, 267, 268, 273, 274, 284, 285
transition metal, 195, 207, 229, 245, 247, 293, 339
transition metal catalyst, 229, 247, 293
transpiration drying, 99, 107, 108
trucking, 53, 99, 101, 107
two-pass harvesting, 48, 104
two-stage hydrotreating, 227
Typha latifolia, 424
tyre, 181, 182, 185, 187, 189

U

unconsolidated logging residues, 100
unicellular, 277
updraft, 173, 175, 177
updraft gasifier, 173, 177
upgrading, 141, 142, 149, 155, 156, 162, 164, 193, 194, 195, 196, 197, 198, 199, 200, 201, 202, 203, 206, 209, 212, 213, 215, 216, 217, 218, 221, 223, 225, 236, 241, 242, 243, 244, 247, 248, 441
urea, 332
UV fluoresence, 196

V

vegetable oil, 8, 154, 164, 251, 252, 254, 256, 258, 259, 264, 268, 273, 274, 284, 312
Verenium, 336
Vermeer, 107, 109
Viking Gasifier, 177
viscosity, 64, 141, 147, 148, 155, 161, 197, 198, 202, 203, 204, 213, 225, 244, 253, 254, 258, 264, 272, 284, 357
volatile, 18, 22, 24, 143, 145, 146, 147, 161, 185, 204, 231, 281, 306, 312, 370, 390
volatile organic acid, 204

W

waste delivery system, 55
waste reverse logistic delivery, 53
waste to power, ix, 409, 411
waste tyre, 181, 182, 183, 184, 187, 188, 189, 190
waste water treatment, 424, 432
waste-to-energy, 177, 386
wastewater concentration, 280
water resistance, 131, 132

water-based biomass, 7
water-gas shift reaction, 158
wet oxidation, 82, 374, 377
wheat, 6, 8, 17, 38, 49, 58, 65, 67, 75, 76, 82, 83, 84, 86, 88, 89, 91, 94, 95, 96, 97, 106, 117, 123, 128, 142, 190, 332, 337, 342, 348, 350, 352, 353, 355, 359, 361, 362, 369, 374, 375, 376, 377, 378, 379, 380
wheat straw, 6, 17, 38, 75, 76, 82, 83, 84, 86, 89, 91, 94, 95, 96, 117, 123, 142, 332, 337, 342, 350, 353, 355, 359, 361, 362, 369, 374, 376, 377, 378, 379, 380
white-rot fungi, 84, 362, 380
Wiley mill, 118, 424
willow, 38, 63, 76
wind, 5, 60, 63, 65, 86, 107, 168, 348
wood chunker, 111, 115

WTE, 409, 410, 411, 421

X

xylan, 16, 69, 76, 79, 82, 84, 144, 246, 329, 331, 334, 335, 343, 349, 351, 359, 360, 362, 363, 364, 378, 380, 387, 423, 430, 432
xylanase, 16, 333, 339, 363, 366, 380, 387, 388

Z

zeolite, 158, 194, 195, 199, 200, 207, 216, 217, 220, 225, 238, 240, 249, 250, 291, 297, 298, 299, 371
zeolite cracking, 225
Ziegler, 293